Challenging Problems in Plant Health

Edited by
Thor Kommedahl and Paul H. Williams

Published by
The American Phytopathological Society
St. Paul, Minnesota

End sheets: Headquarters of the
American Phytopathological Society, St. Paul, Minnesota

Library of Congress Catalog Card Number: 83-70966
International Standard Book Number: 0-89054-057-8

Printed in the United States of America

The American Phytopathological Society
3340 Pilot Knob Road
St. Paul, Minnesota 55121, USA

Contributors

John H. Andrews	Department of Plant Pathology, University of Wisconsin, Madison, WI 53706
Robert Aycock	Department of Plant Pathology, North Carolina State University, Raleigh, NC 27650
Kenneth F. Baker	U.S. Department of Agriculture, Ornamental Plants Research Laboratory, Corvallis, OR 97330
George H. Berg	Food and Agriculture Organizations, United Nations, Rome, Italy
Eileen Brennan	Cook College, Rutgers-The State University, New Brunswick, NJ 08903
James L. Brewbaker	Department of Horticulture, University of Hawaii, Honolulu, HA 96822
J. A. Browning	Department of Plant Science, Texas A&M University, College Station, TX 77843
John A. Bruhn	Biochemicals Department, E.I. DuPont de Nemours & Co., Inc., Wilmington, DE 19898
Ivan W. Buddenhagen, II	Department of Agronomy and Range Science, University of California, Davis, CA 95616
Barbara Christ	Department of Botany, University of British Columbia, Vancouver, Canada V6T 2B1
Stella M. Coakley	National Center for Atmospheric Research, Boulder, CO 80307
J. M. Daly	Department of Biochemistry and Nutrition, University of Nebraska, Lincoln, NE 68583
C. J. Delp	Biochemistry Department, E.I. DuPont de Nemours & Co., Inc., Wilmington, DE 19898
Stephen Diachun	220 Tahoma Road, Lexington, KY 40503
Gerald E. Edwards	Department of Botany, Washington State University, Pullman, WA 99164
L. F. Elliott	USDA-ARS, Department of Agronomy and Soils, Washington State University, Pullman, WA 99164
Emanuel Epstein	Department of Land, Air, and Water Resources, University of California, Davis, CA 95616

Howard Ferris	Department of Nematology, University of California, Riverside, CA 92521
Richard A. Fleming	Department of Environmental Engineering, Cornell University, Ithaca, NY 14853
J. G. Foster	Department of Botany, Washington State University, Pullman, WA 99164
Deborah R. Fravel	Department of Plant Pathology, North Carolina State University, Raleigh, NC 27650
W. E. Fry	Department of Plant Pathology, Cornell University, Ithaca, NY 14853
John F. Fulkerson	U.S. Department of Agriculture, Room 6444 South, Washington, D.C. 20250
R. G. Grogan	Department of Plant Pathology, University of California, Davis, CA 95616
James G. Horsfall	Connecticut Agricultural Experiment Station, New Haven, CT 06504
R. D. Jackson	U.S. Soil and Water Conservation Laboratory, Phoenix, AZ 85040
A. L. Jones	Department of Botany and Plant Pathology, Michigan State University, East Lansing, MI 48824
Thor Kommedahl	Department of Plant Pathology, University of Minnesota, St. Paul, MN 55108
S. B. Ku	Department of Botany, Washington State University, Pullman, WA 99164
André Läuchli	Department of Land, Air, and Water Resources, University of California, Davis, CA 95616
Ida Leone	Department of Plant Pathology, Cook College, Rutgers-The State University, New Brunswick, NJ 08930
Sharon R. Long	Department of Biological Sciences, Stanford University, Stanford, CA 94305
E. S. Luttrell	Department of Plant Pathology, University of Georgia, Athens, GA 30602
David R. MacKenzie	Department of Plant Pathology, Pennsylvania State University, University Park, PA 16802
C. E. Main	Department of Plant Pathology, North Carolina State University, Raleigh, NC 27650

D. H. Marx	Institute for Mycorrhizal Research, Southeastern Forest Experiment Station, U.S. Department of Agriculture, Athens, GA 30602
Michael V. McKenry	San Joaquin Valley Agricultural Research and Extension Center, Parlier, CA 93648
William Merrill, Jr.	Pennsylvania State University, University Park, PA 16802
Steven C. Nelson	American Phytopathological Society, St. Paul, MN 55121
John S. Niederhauser	2474 Camino Valle Verde, Tucson, AZ 85715
Dale M. Norris	Department of Entomology, University of Wisconsin, Madison, WI 53706
R. J. Oshima	California Department of Food and Agriculture, Sacramento, CA 95814
William C. Paddock	Tropical Agricultural Development, Palm Beach, FL 33480
Robert I. Papendick	USDA-ARS, Department of Agronomy and Soils, Washington State University, Pullman, WA 99164
Clayton Person	Department of Botany, University of British Columbia, Vancouver, Canada V6T 2B1
D. I. Rouse	Department of Plant Pathology, University of Wisconsin, Madison, WI 53706
Joseph M. Russo	Department of Horticulture, Pennsylvania State University, University Park, PA 16802
Roy M. Sachs	Department of Environmental Horticulture, University of California, Davis, CA 95616
N. C. Schenck	Department of Plant Pathology, University of Florida, Gainesville, FL 32611
Robert C. Seem	Department of Plant Pathology, New York Agricultural Experiment Station, Geneva, NY 14456
Luis Sequeira	Department of Plant Pathology, University of Wisconsin, Madison, WI 53706
James F. Shepard	Department of Plant Pathology, Kansas State University, Manhattan, KS 66506
Fred W. Slife	Department of Agronomy, University of Illinois, Urbana, IL 61801

Mark A. Smith Animal and Plant Health Inspection Service, U.S. Department of Agriculture, Washington, DC 20523

Paul S. Teng Department of Plant Pathology, University of Minnesota, St. Paul, MN 55108

H. D. Thurston Department of Plant Pathology, Cornell University, Ithaca, NY 14853

B. G. Tweedy CIBA-GEIGY Corporation, Greensboro, NC 27409

Anne K. Vidaver Department of Plant Pathology, University of Nebraska, Lincoln, NE 68583

Paul E. Waggoner Connecticut Agricultural Experiment Station, New Haven, CT 06504

H. E. Wheeler Department of Plant Pathology, University of Kentucky, Lexington, KY 40546

G. B. White Department of Agricultural Economics, Cornell University, Ithaca, NY 14853

Paul H. Williams Department of Plant Pathology, University of Wisconsin, Madison, WI 53706

Miles Wimer American Phytopathological Society, St. Paul, MN 55121

M. S. Wolfe Plant Breeding Institute, Cambridge CB2 2LQ, England

Milton Zaitlin Department of Plant Pathology, Cornell University, Ithaca, NY 14853

George A. Zentmyer Department of Plant Pathology, University of California, Riverside, CA 92521

Foreword

The year 1983 marks the 75th anniversary of the American Phytopathological Society, its Diamond Jubilee. In 1979, in preparation for this anniversary, the Council of the American Phytopathological Society invited both the Society of Nematologists and the Mycological Society of America to meet with them in 1983 in Ames, Iowa, in a joint meeting to share in this Diamond Jubilee celebration.

The Diamond Jubilee Program Organizing Committee appointed in 1980 was composed of D. F. Bateman, W. E. Fry, R. G. Grogan, T. Kommedahl, S. H. Smith, M. E. Stanghellini, S. A. Tolin, B. G. Tweedy, H. E. Waterworth, and P. H. Williams representing the American Phytopathological Society, O. J. Dickerson and J. R. Bloom representing the Society of Nematologists, and A. J. Jaworski representing the Mycological Society of America. The committee decided that as we were celebrating this 75th anniversary we should look strongly to our future while recognizing our past. With this concept in mind, it set about organizing a Diamond Jubilee volume to be entitled "Challenging Problems in Plant Health." Over 200 topics were suggested by the members of these societies for inclusion in this volume, which was to be distributed at the Diamond Jubilee meetings in 1983.

The committee met in November of 1980 at the Nittany Lion Inn, University Park, PA, to consider all of these suggestions and to develop an outline for the volume. Following considerable discussion, the topics to be included and the authors were selected. T. Kommedahl and P. H. Williams, who agreed to be coeditors of the volume, were to contact the potential authors asking them to deliver their manuscripts by the end of 1982. It was also decided at this meeting that a series of symposia following the general outline of the volume would be organized for presentation at the Diamond Jubilee meeting. The committee intended that the initial distribution of the Diamond Jubilee volume at the 1983 meetings, coupled with these symposia, would enable the 75th anniversary to be truly a celebration looking toward an exciting future while recognizing a distinguished past.

The field of phytopathology has come a long way in 75 years and has been closely intertwined with the fields of nematology and mycology. It is indeed fitting that societies representative of these fields should join together in celebrating both their past accomplishments and their future challenges.

The future is ours, with the hope that at our 100th anniversary celebration we will be able to say that the 25 years following our Diamond Jubilee were the most exciting because we turned our challenges into opportunities.

Samuel Smith, Chair
Diamond Jubilee Program Organizing Committee
Department of Plant Pathology
The Pennsylvania State University
University Park, Pennsylvania

Preface

E Pluribus Unum! This familiar motto inscribed on coins of the United States signifies the coming together of many peoples of the world to form one nation. This book reflects that same concept in that many sciences have come together to form the discipline of plant pathology. From the Table of Contents, it is evident that biology, chemistry, and physics, as well as such seemingly unrelated fields as meteorology and economics, bear on plant health in some way. This is reflected in the division of the volume into five sections: history; nature, identification, and management of losses; biological and physical constraints to plant health (genetic, environmental, ecological, and epidemiological); disease control; and socioeconomic factors. The sciences and mathematics are woven into the fabric.

The intent of the committee that planned this book was to select writers who would espouse a point of view rather than exhaustively review a particular topic. We wanted to look backward briefly to ascertain where we are in a particular discipline and then to look forward to the problems that will challenge us in the next decade or two. To a greater or lesser degree this was done. One will not read very far into the book before encountering conclusions or viewpoints that are contradictory or argumentative; however, this should stimulate thought and discussion. Ideas that will be challenged vehemently by some will undoubtedly be applauded by others. Thus, one should read the entire book in sequence.

Throughout the book, despite the seeming specificity of some chapter titles, there is an undercurrent of concern that plant pathology should work in some way to solve the problem of food supply in relation to human population growth—this problem is referred to in 11 of the 48 chapters. Solutions range from the esoteric to the mundane; 14 chapters allude to the possibilities of genetic engineering and 12 chapters to some form of pest management. Twelve chapters refer to the use of or need for computers for these and other applications, illustrating the increasing application of new technologies to solve problems in plant health.

The following paragraphs give a sample of the many concepts and findings about plant health that make up this volume.

At the beginning, we as a Society indulge ourselves somewhat by reliving our 75 years, recalling how our relationships with similar societies and organizations developed and how our national headquarters was established. However, we rapidly move on to the current "age of plants," as Browning puts it, only to encounter the startling thought that healthy plants may be a threat to civilization, as espoused by Paddock.

We are experiencing renewed interest in crop losses and their assessment, due in part to new technologies. We can use satellites to detect pests and computers to deliver the information to interested groups or individuals.

Constraints to plant health are legion, and some of the new ways of alleviating these constraints are genetic, morphological, or physiological. To produce cultivars tolerant to physiological factors in air or soil, alteration of the genotype by genetic engineering has been proposed. Intergeneric and interspecific protoplast fusion can provide a way to transfer resistance genes from donor to recipient species even when the two are sexually incompatible.

Environmental constraints are discussed in six chapters. Efforts are under way to maintain plant health under variable climatic conditions that may someday become predictable. To monitor this environment, a high-altitude, powered platform located over a major agricultural area can provide continuous surveillance and report to farmers in that area on weather, pests, and diseases. Orbiting satellites supplement this information. Farmers are likely to use microprocessors and microcomputers not only to receive information but to predict disease incidence. Mathematical models aid in this prediction. Growth models and systems that center on plant stress can show how environmental factors affect growth and yield and suggest ways to implement integrated pest management.

The selection and breeding of cultivars tolerant to mineral stresses are needed, requiring the cooperation of plant pathologist, plant physiologist, and geneticist. Also cultivars are needed that use water efficiently.

Host specificity is not well understood, and more must be known about recognition phenomena. Chemical events in pathogenesis need to be determined and the gene-for-gene concept, thought sometimes to be universal, needs critical testing. Highly specific gene products have yet to be demonstrated to everyone's satisfaction.

Mycorrhizal fungi are important in plant health. However, too much emphasis has been put on laboratory work and not enough on field work. Mycorrhizal technology has the potential for solving problems in the supply of food, fuel, and fiber.

Control methods suggested vary from the use of conventional chemicals to the use of naturally produced chemicals that inhibit spore germination or block enzymes necessary to penetration of host tissues and that have potential use as fungicides. Problems of pesticide resistance haunt the chemical control industry. Several authors argue that integrated control is necessary in crop management. Techniques in breeding durably resistant crops are described, as are emerging problems in genetic homogeneity that must be addressed.

Socioeconomic factors—economics, governmental regulations, international organizations, and education—affect plant health. Techniques useful in research and extension have application also to the classroom. A new doctorate degree has been suggested in three chapters (Chapters 1, 5, and 48).

These are but a few of the portents of things to come that are given in this book—others will be discovered by the reader. We hope that the book will reveal to the student the many faces of plant pathology and give to the practicing scientist an overview of the current and future state of the art and science of plant health.

Thor Kommedahl
Paul H. Williams

Contents

PART I

Plant Health in Perspective

PART II

Nature of Losses

PART III

Biological and Physical Constraints to Plant Health

Genetic Considerations

Potential for Alleviation of Environmental Constraints

PART V
Socioeconomic Factors Affecting Plant Health Maintenance

PART I

Plant Health in Perspective

The birth of plant pathology was in very truth a blessed event although it has no natal day. For the course of history is so complex and continuous that it cannot always clearly mark the birthdays of its most significant events. Exactly when plant pathology was born is a moot question; why it was born is no question at all. It was born of necessity.... For better or for worse, plant pathology had its genesis in fields and granaries more than in halls of ivy.

—E. C. Stakman, 1958

As we are now exploiting the products of basic research in agriculture done 40 and 50 years ago, so we should also be doing now the basic research that will provide the new techniques for producing food needed to feed the world's population in the year 2000.

—Paul C. Mangelsdorf, 1966

CHAPTER 1

A LOOK AT THE PAST

JAMES G. HORSFALL

For its Golden Jubilee Volume in 1958, the Society asked me to write a chapter, A LOOK TO THE FUTURE. I was young enough and brash enough to undertake the job despite Victor Borge's dictum that "Forecasting is a difficult business, especially forecasting the future." Perhaps it is fitting, now that I am older and less brash, that I should write for the Diamond Jubilee Volume, A LOOK AT THE PAST.

This is not a history of world plant pathology. With few exceptions I shall discuss only the events and trends in plant pathology that have been helped along by members of the Society since it was established in 1908. The general history of our science has been well treated by other historians like Parris (8), Baker (2), and Ainsworth (1). I shall try to make mine a more philosophical discussion of the subject. Since I have been a member of the Society for 58 of its 75 years, my look at the past will inevitably have an autobiographical flavor. For this, I beg your forgiveness.

SOME FIRSTS

Since the history of science is a history of scientists, let us set down a few first scientists. As far as I know the first paper by a professional plant pathologist, not an amateur, was Farlow's paper on the "potato rot" in 1876. The first dozen plant pathologists in the United States were J. C. Arthur, C. E. Bessey, T. J. Burrill, W. R. Dudley, W. G. Farlow, B. T. Galloway, B. D. Halsted, W. A. Kellerman, F. Lamson-Scribner, S. M. Tracy, R. Thaxter, and A. F. Woods.

The first university lecture on plant disease was delivered by the Reverend Chauncy F. Goodrich at Yale in 1860 (Phytopathology 59:1760, 1969). The first university courses on plant disease were organized in 1871 by Bessey at Iowa State College and by Burrill in Illinois. The first organizational unit with "pathology" in its name was the section of Vegetable Pathology in the U.S. Department of Agriculture, established by Lamson-Scribner in 1886. The first university department to be called plant pathology was established by F. D. Chester at Delaware in 1888. (California please note!)

In 1887, Effie A. Southworth (nominated by Erwin F. Smith) became the first lady plant pathologist. Subsequently, she married Volney Spalding, who sent Erwin F. Smith and L. R. Jones on into plant pathology. B. D. Halsted, as Farlow's first student, was granted the first doctorate (D.Sc.) in the field. The second doctorate in plant pathology and Cornell's first was awarded to J. C. Arthur in 1886. B. M. Duggar wrote the first American textbook on plant pathology in 1909.

Burrill rode with John Wesley Powell in the first boat to dare the rolling, boiling white water in the Grand Canyon of the Colorado River, and David G. Fairchild, a very early plant pathologist, and nephew of Halsted, distinguished himself by marrying a daughter of Alexander Graham Bell. He went on to become a famous plant explorer, eventually publishing a book, *The World Is My Garden*.

Cynthia Westcott in 1933 became the first consulting plant pathologist. She helped with the health of plants that make our cities beautiful.

And similarly B. M. Duggar is the only plant pathologist to rate an historical roadside marker. This one marks his birthplace alongside U.S. Route 80 at Gallion, Alabama. It says in part: "Son of a beloved country doctor, he carried a dedicated spirit into the frontiers of science." His honor was not for plant pathology, however, but for the discovery of aureomycin.

THE SOCIETY IS ORGANIZED

Because Americans love organization and bureaucracy, the plant pathologists created their own organization at a meeting of a few of the faithful in Baltimore in December, 1908. It was another splinter group from the botanical main stem. Not everybody present at the meeting was in favor of the splintering. In his biography of A. D. Selby, Stover (10) makes it clear that the splintering created considerable noise and pain. He said,

> Professor Selby was one of the first and most persistent advocates of the American Phytopathological Society and took a very active part in the stormy discussion that attended its birth.

The vote was 32 to 12, not exactly a unanimous decision. Selby was rewarded with the Secretary-Treasurership in the first year and with the Presidency in the third. One hundred and thirty persons, including 12 women, became charter members.

SOME SEMINAL IDEAS THAT HAVE SHAPED OUR HISTORY

However philosophical this "Look at the Past" may be, it must enumerate some seminal ideas that have been born during the 75 years and nurtured more or less by members of our Society. The following are my choices. You will want to make your own. Since priorities are always controversial, I shall not assign names, much as I would like to. This should fend off some "sticks and stones."

Chemistry: Chemical basis of resistance, the nucleic acid nature of viruses, chemotherapy, organic fungicides, the chelation principle in biology, phytoalexins, lectins.

Epidemiology: Insect transmission of pathogens, temperature, moisture, and pH control, transfer of seed production to the arid west, Rachel Carson episode, computer modeling, and forecasting.

Genetics: Heterothallism in fungi, gene-for-gene hypothesis, the "Green Revolution," genetic vulnerability, transduction, recombinant DNA, multilines, fungal genetics, mutation of viruses, sex in the rusts, differential cultivars, resistance to pesticides, genetic engineering.

Resistance and susceptibility: Induced resistance, plasmids in crown gall, recognition phenomena, significance of mycorrhizae, chemical nature of resistance, phytoalexins, tolerance, disease escape.

Etiology: Bacteria, viruses, viroids, nematodes, multiple pathogens, ice nucleation, environmental stress factors, smog, mineral deficiencies, mycoplasmas.

Control: Biological control, suppressive soils, epiphytic antagonists, organic fungicides, meristem and single cell cultures, certified seed, heat therapy, hyperparasites, hypovirulence.

SOME SEMINAL METHODS THAT HAVE SHAPED OUR HISTORY

Lord Kelvin once said in effect that you cannot understand what you cannot measure. I must add that you cannot measure without adequate methods. The classic case of understanding in plant pathology is Stanley's Nobel Laureate work on "crystallizing" tobacco mosaic virus. His whole thesis rested on his ability to measure the titre of TMV. He did this with the local lesion method of Holmes.

Herewith my list of powerful methods developed within the 75 years of the Society: local lesions to measure virus titre, serology to identify viruses, gradient centrifugation to assay viruses, *Neurospora* breeding leads to biochemistry of genetics, growing seeds in an arid climate, climate control chambers, use of differential cultivars, remote sensing, computer hardware and software for epidemiology, pathometry of disease, genetic engineering.

SOME KUDOS FOR OUR LABORS

Plant pathologists working on plant health have been awarded their share of kudos both by their fellow plant pathologists and by others.

Our Society has its own system of awarding kudos. By 1981 we had elected 161 fellows of the society. We had given out 24 Ruth Allen Awards and four Awards of Distinction, five Campbell awards, seven Ciba-Geigy awards, and five Lee M. Hutchins awards.

Election to the National Academy of Science carries prestige among all the sciences. In 1908 we had 130 members in the Society, and one in the Academy = 0.77%; 10 years later, 400 members, four in the Academy = 1.00%; 25 years later, 774 members, seven in the Academy = 0.90%; 50 years later, 1937 members, seven in the Academy = 0.36%; 73 years later, 3168 members, 14 in the Academy = 0.44%.

We were high (I think) to begin with because we were mostly in botany departments and were, therefore, noticeable to the Botany section of the Academy. (I was in a botany department for my first dozen years.) As we began to split out from botany departments, we fell from view of botanists and smelled to them like "applied scientists." As a result, our memberships in the Academy declined.

The neglect of applied sciences was corrected for biologists with the creation in 1964 of the Section of Applied Biology. On that account our membership in the Academy has doubled and our percentage has risen from 0.36 to 0.44.

I am unable to find a good list of those that have been elected to foreign Academies. To top off the honors in science, two of our members or former members have been made Nobel Laureates, W. M. Stanley and N. E. Borlaug, and The Alexander von Humbolt award has been won by T. O. Diener and the Wolf Prize by J. C. Walker and Karl Maramorosch.

KUDOS OUTSIDE PLANT PATHOLOGY

Plant pathologists have been honored for work outside plant pathology. In 1943, L. R. Jones was asked by President F. D. Roosevelt to serve on his Science Advisory Board. E. C. Stakman served later on the Governing Board of the National Science Foundation and as President of the AAAS. Of the first 10 presidents of the Botanical Society, five were members of our Society: R. Thaxter, E. F. Smith, W. G. Farlow, L. R. Jones, and R. A. Harper. Later that Society elected (in order) J. C. Arthur, B. M. Duggar, W. Crocker, A. H. R. Buller, Neil Stevens, and A. F. Blakeslee. None has been elected since 1950.

B. M. Duggar, S. Rich, and J. G. Horsfall have been presidents of the Society of Industrial Microbiology, and J. G. Dickson, G. L. McNew, and G. A. Gries have been presidents of the American Institute of Biological Sciences. J. G. Harrar has been President of the prestigious Rockefeller Foundation.

At least four plant pathologists (that I know of) became university presidents: Frank Poole (Clemson), J. H. Jensen (Oregon State), J. W. Oswald (Pennsylvania State), and R. A. Young (Nebraska). At least four have become vice-presidents: T. J. Burrill (Illinois), J. B. Kendrick, Jr. (California), D. F. Crossan (Delaware), and N. N. Winstead (North Carolina State). I can count at least 16 that have become deans. All four Managing Directors of the Boyce-Thompson Institute have been members: W. Crocker, G. L. McNew, R. H. Wellman, and R. A. Young.

SOME COUNTERPRODUCTIVE ITEMS

Life in American plant pathology has not been all beer and skittles, however. We have had to reverse course along several trails.

The Battle of Kittery Point

The decision to eradicate *Ribes* to save white pines from blister rust became a case of reversed direction. An early harbinger of the decision to reverse was the "Battle of Kittery Point." Roland Thaxter, who stands so tall in early plant pathology, had a summer place at Kittery Point on the coast of Maine. When the all-wise father in Washington sent his scouts to destroy the gooseberries in Thaxter's garden, Thaxter got out his trusty old blunderbuss and blocked the invaders at his garden gate. Thaxter did not need any learning in statistical theory nor a computer to tell him that it was a waste of effort to try to eradicate all *Ribes*. Of course, he lost his *Ribes* and the battle, but he won the war because the *Ribes* eradication program was itself eradicated after millions had been spent on it. Too bad Thaxter died too soon.

The March of Chestnut Blight

Similarly, F. C. Stewart, one of Thaxter's students, became a pariah a few years later in Pennsylvania when he told the eradicators in the Governor's office in Harrisburg that they, too, were wasting taxpayers' money trying to eradicate chestnut blight. *Endothia parasitica* vindicated Stewart by marching right down the spine of the Appalachians.

The Rachel Carson Episode

In 1962, Rachel Carson, being distressed by the widespread dumping of pesticides from airplanes, wrote a landmark book, *Silent Spring*, in which she damned the entomologists (and by implication us) as Stone Age scientists (p. 297). The book changed some of our research priorities. For example, the percentage of plant pathologists in fungicide research fell sharply and that in plant breeding rose just as sharply. In 1963, one year after her book, the Directory of Members showed only five workers in biological control. So many climbed on the Carson bandwagon, however, that by 1968, only five years later the number had risen 15-fold to 75. It has now fallen back to 41, still eight times as many as when Carson's book was printed.

A New Disease Comes Out of the Woodwork

In 1970 the fabulous corn blight seemed to come out of nowhere and this news invaded the *Wall Street Journal* and other popular periodicals. To save detasseling costs, the corn companies had converted most of the corn of the country to the male-sterile type. Unconsciously they made it highly susceptible to a *Helminthosporium*, presumably a mutant. In 1970 the disease killed enough corn in the United States that, if fed to cattle, the lost corn could have produced enough beef for 30 million hamburgers. F. C. Bawden told me once that this disease caused the British dairy farmers to shift from maize to barley as cattle feed. This raised the price of barley, used to make Scotch whiskey, by 30%.

The disease had hit the Phillipines first, but we did not test our hybrids there. We waited until the blight hit us.

Wallace Disperses Plant Pathologists Among Crop Divisions

Perhaps the most dramatic counterproductive move in plant health during the history of the Society was made by Henry Agard Wallace, Secretary of Agriculture under F. D. Roosevelt. In 1933, he proposed to disperse the plant pathologists and entomologists of the USDA among the crop departments. At the Christmas meetings, in Boston that year, of the American Association for the Advancement of Science, the entomologists appointed a group to call on Wallace.

I can still see H. H. Whetzel stomping up and down the halls and lobby of the hotel trying to persuade the plant pathologists to do likewise, but they did not.

Why not? The data were stark and clear. Science and politics clashed in plant pathology, not in entomology. Four of the seven councilors for our Society, as published in PHYTOPATHOLOGY, were employees of the USDA. No such influence was present among the entomologists.

The entomologists went to Washington and persuaded Wallace not to split them up. Science triumphed over politics. No pathologist went to Washington and politics triumphed over science.

Secretary Wallace's move scattered the researchers in plant pathology so far apart that the synergistic flow of ideas was seriously impeded. As a result the entomologists by 1960 were publishing more papers per scientist than the plant pathologists. This situation has since been corrected.

The *Plant Disease Reporter* Dies

One of the saddest cases of reversed direction is the killing of the *Plant Disease Reporter*. Its fate comprises a footnote to history. For six decades it steadily and competently performed the Federal function, for which it was established, of keeping us advised as to what diseases were damaging our crops. Most states delegated a person to report from time to time on the serious diseases for each State. I can remember that in 1946 I followed with some apprehension in the pages of the *Reporter* as *Phytophthora infestans* adapted itself to tomatoes and marched up the Atlantic seaboard toward my tomatoes. Alas, this function is no more.

Over time the journal slowly shifted toward an outlet for research papers. In due course the inevitable happened; the editor converted it to a "refereed" journal. I can recall telling the editor that converting it to a refereed journal would kill it. I remembered full well that the USDA had killed the *Journal of Agricultural Research* because, in the view of the USDA, publication of scientific journals is not a Federal function.

I can remember when I was president of the Society that the USDA complained loudly about paying the newly instituted page charges.

Some have said that the *Plant Disease Reporter* was "born again" in the guise of our own journal, PLANT DISEASE. The latter does continue the function of publishing research in useful plant pathology, but it does not continue the central function of reporting disease in "real time" as the computer experts call it.

Intellectual Status of PHYTOPATHOLOGY

Recently R. R. Nelson (7) wrote a Letter to the Editor saying "the health [of PHYTOPATHOLOGY] is poor. Editorial barriers are erected in front of new ideas or concepts."

This "somber thought" of his has prompted me to test his hypothesis by citability. How does citability of PHYTOPATHOLOGY compare with that of other journals? Garfield (5) has compared it with 61 botanical journals. His measure of quality is "impact factor," the proportion of articles that are cited somewhere. PHYTOPATHOLOGY ranks 21st among the 61 journals. Even so it is well ahead of its fellow journals in *agricultural* botany. What is worse is that the impact factor of PHYTOPATHOLOGY has fallen from 1.2 in 1974–75 to 0.9 in 1978, a drop of about a quarter in four years.

How do plant pathologists cite their own journal? Has this fallen also? We can turn to two sources that cover the whole field of plant pathology, the *Annual Review of Phytopathology* and *Plant Disease*—an Advanced Treatise. For these two we shall calculate the proportion of references that derive from

PHYTOPATHOLOGY to the total number of references. This parameter we will call the "citation ratio."

The citation ratio in the *Annual Review of Phytopathology* dropped 25% in the 16 years between the 1964–65 volumes and the 1980–81 volumes. I took disease physiology as an example from the two Advanced Treatises. The citation ratio here also fell by a fourth between the first edition in 1960 and the second in 1980.

However you look at it, Dr. Nelson, the intellectual standing of our journal does seem to be going downhill. Is this because plant pathologists are producing fewer citable ideas or are publishing them elsewhere? I suspect the latter. If the former, we are in deep trouble.

I understand that the new editorial board is revising the policy of screening new manuscripts. I hope that bright intuitive ideas will be encouraged. I say, "Let us gamble a bit." Some of our competitors do.

SOME PROFOUND CHANGES IN PHILOSOPHY

During its life, the society has certainly seen some profound changes in philosophy as it strove to improve plant health.

Plant pathology operates from three base stations, one at each corner of the disease triangle: the weather, pathogen, and host. I am fascinated by the way we have wandered philosophically around that triangle.

The Weather

Before de Bary, most thinkers in plant pathology operated from the weather base. Bad weather produces disease. Any thinker could see that. Ireland had phenomenally cold, wet weather in 1845 and 1846. Of course the potatoes blighted!!

The Pathogen

Then de Bary appeared, turned the corner, and moved over to the pathogen base. In 1853, he put the capstone on the proof that a plant disease is induced by a living entity called a fungus. This sent us all scurrying down the vitalist path. Now it was *living* things that induce disease. In order to make certain of their prime role, George K. K. Link in the 1930s said that they "incite" disease, a phrase that I don't like. One who incites a riot can die in the ensuing melee, but the riot will continue. Not so in plant disease.

deBary made fungi so popular that doctor's thesis after doctor's thesis was based on a new fungal disease. Many of us pushed a vitalist variant. We would kill the pesky fungi with fungicides, a term (by the way) coined by Halsted about the turn of the century. Traveling this path we shifted from bordeaux mixture to organic fungicides including the ethylenebisdithiocarbamates which became the most widely used organic fungicides in the world today.

Bacteria

de Bary, bright as he was, became so enamoured by his fungi that he could not see for a long, long time that another living organism, a bacterium, could induce disease. One of his students, Alfred Fischer, carried on a bitter polemic with

Erwin F. Smith on the subject. Smith, of course, won. Bacteria do indeed induce disease.

Nematodes

In 1914, N. A. Cobb introduced the study of parasitic nematodes into the United States. He promptly created the term, nematology, and thus became the first nematologist. He went on to become Mr. Nematologist in the Country. Like entomology, mycology, and virology, nematology has heavily emphasized the organism, not its damage to the host. The host is merely pabulum.

Viruses

Right after bacteria were allowed entry into the hall of pathogens, we came across a new and mysterious pathogen. We found no bacteria or fungi in the diseased tissue, but the sap was infectious even after passage through a Berkefeld filter to remove any bacteria.

The vitalist theory had us in thrall. Obviously either the sap or a particle in the sap was alive. Any thinker could see that. The entity could reproduce itself. Only living things can reproduce themselves. Beijerinck gave us a name for it. He called it *contagium vivum fluidum*—the contagious living fluid.

For 81 years the vitalists held the field. During the 1920s, they happily ignored the innovative work of Duggar, Mulvany, and Vinson, who were holding that the virus is not living. Viruses could be manipulated like any other chemical.

But in the 1930s, the vitalists lost the battle to the viruses. Stanley concluded that a virus is a "crystallized" protein. Viruses, then, are not ultramicroscopic bacteria. Soon we learned that a virus is a nucleic acid with protein as the encapsulating coat. I see no possible excuse for cluttering the language with "encapsidate." A capsid is a sucking insect.

It was inevitable that one day an ultrasmall virus would be discovered, and it was. T. O. Diener named it a "viroid." It is naked. It has no protective sheath.

Mycoplasmas

If the vitalistic view inhibited the development of viruses, then viruses inhibited the development of mycoplasmas. My colleague E. M. Stoddard and I were caught in this net.

X-disease of peach is a disease with infectious sap. Ipso facto, X-disease of peach is caused by a virus!! Even the eminent Bawden said so. In the 1940s, Stoddard (9), who held no Union card in virology, tried to improve the health of peaches. He showed that X-disease could be cured with several chemicals. I suggested to Stoddard that he try the sulfanilamide system to see if his "virus" would act like bacteria. He did and it did. Sulfanilamide was effective and could be reversed with *p*-aminobenzoic acid, just as in bacteria.

The experts residing on the virological Mt. Olympus laughed this off by saying, "You can't cure viruses with sulfanilamide." They probably thought that Stoddard was massaging his data. Nearly 20 years later Bawden (4) still called Stoddard's work "an unsubstantiated claim." The syllogism went this way. "X-disease is caused by a virus. You can't cure a virus disease with sulfanilamide. Therefore, you can't cure X-disease with sulfanilamide."

To this day I am astonished that we allowed ourselves to be intimidated by the experts. We should have written our own syllogism. "Bacterial diseases can be cured with sulfanilamide. X-disease can be cured with sulfanilamide. Therefore

X-disease is caused by a bacterium."

After Stoddard died, the evidence came in. Now X-disease is known to be induced by a living organism, a mycoplasma. Stoddard was vindicated.

Back to the Weather

In 1950 we became recidivists. We fell back to pre-de Bary times. Weather again became an inducing agent of bad health in plants. In that year Middleton et al (6) published the finding that smog, in addition to SO_2 and fluorides, is a pathogen. They called attention for the very first time to the price that plants must pay to live among the citizens in smoggy Los Angeles, California. Eventually oranges and zinfandel grapes paid the ultimate price. They died.

Shortly thereafter, Vanderplank (11) pulled the plug from the epidemiology barrel. He became the great expositor of the principles of quantitative epidemiology. He emphasized "r"—the rate of disease development. This term probably came naturally to him as a former chemist who had had to deal with many rate constants.

A few years later came the demonstration that a computer can be used to simulate epidemics and thus to assist farmers in deciding when to spray.

The Health of the Host

And now we are finally emphasizing the host. We have planted our feet on the host base of the triangle. We are speaking of plant health, hence the title of this book. I hope that this emphasis will occasionally send some of us back to farmers' fields, back to the real world where plants grow.

THE SHIFT FROM STATIC TO DYNAMIC

We have made a dramatic and powerful shift in thinking during the 75 years in the life of the Society. We have shifted 180 degrees from static to dynamic thinking. When the Society was born, we took snapshots of our subject. Now we make movies. Then we described things. Now we discuss mechanisms. How and why do diseases do what they do?

Then we described new diseases by the score, nay by the hundreds. We had to. We had to learn the territory. The titles of books reflected the descriptive or static phase: Diseases of Vegetables. Diseases of Citrus. Even the books labeled Plant Pathology divided their chapters by crops or pathogens. Now text books have dynamic titles like "Physiological Plant Pathology."

In 1902, a few years before our Society was born, A. D. Selby concluded that plant pathology comprised a systematic body of facts, but plaintively inquired, "Has it assumed scientific rank?" I think it has not, Dr. Selby.

A half century after Selby, Brierley (in England) could only echo Selby: "Up to the present the study of plant disease has remained merely an aggregate of data.... It has not attained the status of a science."

In the days of Selby and Brierley we dealt with the statics. Now we deal with rates. Before we could deal with rates, however, we had to quantify the details of our subject. Let us take Vanderplank as an example. His whole reputation in epidemiology (and I salute him) rests on "r," the rate of disease development with time. As far as I know, Barratt (3) working on his master's thesis, first used the rate of disease development as a quantitative measure. He applied it to varietal resistance of tomatoes to *Alternaria solani*. Barratt, in turn, could not develop a rate curve without quantifying his disease. This he did with the well known Horsfall-Barratt grading system.

With the arrival of quantification in plant pathology it became a science. Selby and Brierley, please note.

THE FLIGHT FROM THE FIELD

In 1969 I called attention to a striking trend in plant pathology, the flight from the field. The percentage of papers in our journal devoted to field research declined from 54% in 1945 to 18% in 1969. Dr. C. L. Campbell, of North Carolina State University writes me that in 1980 we have recovered somewhat back to 30%. Professor Zadoks tells me that the down trend occurs in Holland as well.

CONSULTANTS BEGIN TO FUNCTION

As university professors flee from the fields, they leave behind a partial vacuum in plant health. Into this is flowing a stream of consultants. The stream is small but growing. In 1933, Cynthia Westcott helped fill a similar vacuum in city gardening. She became the first consultant. She was followed a quarter of a century later by Robert S. Cox, who began in 1957 to provide services to farmers in Florida.

THE OPTIMIST SPEAKS

Some thinkers, viewing the past of plant pathology have been pessimistic about the future. In 1902 Erwin F. Smith wrote early in his career that "... the facts that lie on the surface...have now been pretty well picked up." In 1913 the great W. G. Farlow, as president of the Botanical Society, reminiscing about his part in the changes that had taken place, predicted that at the meeting 40 years hence (that would be 1953)... "it can hardly be said that the president will have to record any such radical and complete transformations as it has been my privilege to present to you this evening."

I hardly hold with the pessimists. I buy the old saying that says in effect, "The ocean holds more good fish than have ever been caught."

THE END

I raise my glass in a toast to you who hold the future in your hands. May the health of our plants continue to improve!

ACKNOWLEDGMENTS

I am deeply grateful to C. L. Campbell, R. G. Grogan, B. G. Tweedy, and P. E. Waggoner for reading the manuscript and offering elegant suggestions. I bled the following people for ideas, for which I thank them: J. F. Adams, S. L. Anagnostakis, R. Baker, D. F. Bateman, L. M. Black, R. J. Cook, E. B. Cowling, J. Elliston, W. Merrill, M. N. Schroth, A. F. Sherf, P. Tsao, C. E. Yarwood, and G. A. Zentmyer.

LITERATURE CITED

1. Ainsworth, G. C. 1981. Introduction to the History of Plant Pathology. Cambridge Univ. Press, Cambridge. 315 pp.
2. Baker, K. F. 1965. A plant pathogen views history. Pages 36–70 in: History of Botany. Clark Mem. Library, Univ. Calif., Los Angeles.
3. Barratt, R. W. 1945. Intraseasonal advance

of disease to evaluate fungicidal and genetical differences. (Abstr.) Phytopathology 35:64.
4. Bawden, F. C. 1964. Plant Viruses and Virus Diseases, 4th ed. Ronald Press, New York. 361 pp.
5. Garfield, E. 1980. Journal Citation Studies, 33. Botany Journals Part 1. Curr. Contents 31:5-13.
6. Middleton, J. T., Kendrick, J. B., Jr., and Schwalm, H. W. 1950. Injury to herbaceous plants by smog or air pollution. Plant Dis. Rep. 34:245-252.
7. Nelson, R. R. 1980. Some thoughts on the current and future health of Phytopathology and the American Phytopathological Society. Phytopathology 70:364-365.
8. Parris, G. K. 1968. A chronology of plant pathology. Privately Printed. Starkville, MS. 167 pp.
9. Stoddard, E. M. 1947. The X-disease, of peach and its chemotherapy. Conn. Agric. Exp. Stn. Bull. 506.
10. Stover, W. B. 1925. Augustine Dawson Selby. Phytopathology 15:1-10.
11. Vanderplank, J. E. 1963. Plant Diseases: Epidemics and Control. Academic Press, New York. 349 pp.

CHAPTER 2

OUR RELATIONSHIPS TO OTHER SCIENTIFIC SOCIETIES AND ORGANIZATIONS IN THE PAST 75 YEARS

ROBERT AYCOCK and DEBORAH R. FRAVEL

The founders of the American Phytopathological Society (APS) envisioned an organization that would be an invaluable aid in promoting the development of plant pathology in America and thus serve a rapidly expanding agriculture. Moreover, it was hoped that its influence would have international significance. Fears of invasion by dangerous exotic pathogens that do not respect national boundaries were entirely justified.

The APS was an offspring of the parent Botanical Society of America in 1908, and some doubted the advisability of breaking away at that time. However, pressing questions regarding the outbreak of destructive epidemics and the spread of plant diseases, whether from farm to farm or over wide areas, were subjects of great concern to plant pathologists. The society provided a vehicle for voicing these concerns and for organizing resources and efforts to deal with them. These tasks were quite difficult, however, owing to the diversity of backgrounds and interests of the early members of the group. Even at that time plant pathology was, to some extent at least, multidisciplinary with respect to both agricultural and biological subject matter. It was only natural that certain groups wished to protect the integrity of their own territories and resisted attempts to disturb the status quo. Mindful of the benefits of interdisciplinary cooperation, however, the first Council meeting, held in Washington, DC, on 26 March 1909, considered, among other things, the questions of affiliation with other societies.

Owing to the extraordinary efforts of the leaders of the APS during the first decade, plant pathology emerged as a strong, independent discipline with its own body of knowledge and principles, its own journal, and its own centers of excellence in research, graduate training, and extension. As the stature of the profession grew, the scope of its involvement in allied sciences expanded and the character of the science became even more multidisciplinary. The need for highly specialized resource persons in the physical, biological, and agricultural sciences and related technologies increased. However, the phenomenon of disease and the dynamics of disease in plant populations were focal points that maintained bonds of unity and loyalty. Plant pathologists in a general sense continued to enjoy close ties with other scientists based upon their mutual or overlapping interests and professional respect.

PLANT PATHOLOGY AND BIOLOGY

Although the spirit of give and take generally prevailed and relatively few instances of selfish isolationism impeded progress, leaders of the new Society were well aware of the divisive potential of situations where interests either overlap or come into conflict. L. R. Jones, writing in the first article published in PHYTOPATHOLOGY (2), argued the case for a separate society for plant pathologists, while emphasizing at the same time the relationships of plant pathology to all of biology. He commented that the work of the phytopathologist is mainly with fungus and bacterial parasites, yet the bacteriologists and mycologists might justifiably view as unscientific attempts to subdivide their fields by separating the organisms parasitic upon higher plants into a distinct group. Dr. Jones also called attention to the fact that the embarrassments of the phytopathologist would not stop there. Agronomists and horticulturists each have their own fields and these must be invaded if the work of the pathologist is to be pursued.

> Moreover, these overlappings were not to be considered temporary. The advances of scientific knowledge would not more clearly segregate these related branches of knowledge along natural lines of change. Indeed the *reverse* would be true. Segregation into a separate society or profession was justified *only* on the basis that *greater efficiency in service* would be secured. If plant pathologists are permitted to segregate problems from their more natural associations, on the grounds of epidemiology, fundamental relationships are not in reality affected—and such separation *increases* the responsibility of those who make the move (italics ours).

Jones concluded that advances in plant pathology would be in proportion to the clear recognition that the principal problems to be pursued are biological, rather than economic. Variety and complexity would increase with the progress of the science, and the plant pathologist would have to realize the complex interrelationships of scientific phenomena;

> in his work he is to act *not as an individual*, but as one of a large company, coordinating what he is doing with what *has* been done, cooperating in the present with associates and colleagues, and patiently and painstakingly preparing for future work to be done by those who are to follow (italics ours).

The interrelationships of plant pathology with other services were reemphasized many times. E. M. Freeman, in a paper read before the Society in 1935 at its summer meeting at St. Paul (1), called attention to the fact that a century of scientific achievement had not altered plant pathology's relationship to its sister sciences, although the enormous expansion and increasing complexity of plant science had steadily constricted the scope of operation of the individual. An Anton de Bary—"superman" in many fields—

> is no longer humanly possible, and it needs no divine power of prophecy to predict that he will never again appear. . .the superman must be surrounded by intelligent, generous, genuine, and spontaneous cooperation.

Freeman, in his argument for broad understanding and appreciation of the sister sciences, called attention to the trap that many plant pathologists fall into when they succumb to the alluring, but unjustified analogies that are drawn between the practices of medicine and plant disease control. The bewildering diversity of plant species from mushrooms to great trees, the multiplicity of pathogens that affect one plant, and the inextricable connections of control practices to methods of culture, soil conditions, climate, etc., are in sharp contrast to the body of knowledge on the practice of medicine, which concerns *one* host species, in which evaluations of treatment are generally aided by intelligent response and communication.

Dr. Freeman also addressed the problem of cooperation in institutions between and among departments with their attendant problems of territorial rights and responsibilities. "No administrational scheme can possibly eliminate the overlapping of sciences and departments," he noted. "It is the chief business of the administration to see that these overlapping borders are well oiled to prevent excessive friction." While recognizing the need for cooperation, he maintained that plant pathology needs to be recognized as a "professional guild" not subordinated to single crop plants. Such a subordination demotes professional scientists to skilled labor functions that delimit the scope, activity, and service of plant pathologists and discourages their cooperation with other scientists.

INTERNATIONAL INTERESTS

In addition to cooperation with other biological specialties and their organizations, there was early interest in international cooperation. At the fourth annual meeting in Cleveland,[1] an important resolution adopted by the Society read in part:

> Resolved, that the American Phytopathological Society, recognizing the fact that plant diseases do not recognize national limits or geographical boundaries. . . respectfully recommends that administrators of research institutions recognize the importance of establishing closer international relations and take such steps as may be practicable to secure this end, including not only more frequent visits of American investigators to foreign countries. . .but also securing. . .the engagement of the best foreign experts in plant pathology (in this country).

In reviewing the first decade of the APS at the annual meeting in December, 1918, C. L. Shear (upon whose personal suggestion a committee of plant pathologists from the U.S. Department of Agriculture [USDA] met to undertake the organization of an APS in 1908) envisioned in the years ahead a great international plant pathological society that would provide closer union of interests and organizations (5). (Previous plans for participation in an International Conference on Phytopathology proposed for Rome in 1914 were considered by a committee of L. R. Jones, W. A. Orton, and C. L. Shear, but no record of APS's participation is recorded, and whether or not the conference was ever held is open to question.)

[1] Annual Reports of the APS, from which much of the information in this article was taken, are to be found in PHYTOPATHOLOGY, vols. 1–60 for 1911–1970; PHYTOPATHOLOGY NEWS, vols. 4–7 for 1970–1973 and 13–14 for 1979–1980; PROCEEDINGS OF THE AMERICAN PHYTOPATHOLOGICAL SOCIETY, vols. 1–4 for 1974–1979; and PLANT DISEASE, vol. 65 for 1981.

In the afterglow of the Armistice of November, 1918, Dr. Shear called upon American plant pathologists to accept the "great responsibility and improve the wonderful opportunity now offered for world service in advancing science and promoting the brotherhood of mankind." He was emotionally moved by the conclusion of the world's greatest catastrophe and looked to the new foundations upon which the structure of science, ethics, and politics were to be built. Without a spirit of brotherly love and service, future structures would be but another Tower of Babel whose downfall would be greater than that of the present.

RELATIONSHIPS WITH AMERICAN SCIENTIFIC SOCIETIES

Affiliation and cooperation with other societies was always a matter of concern. In fact, the original organizing group broke away from their botanical colleagues with some difficulty. The vote to form a separate society was far from unanimous (32 to 12). Neither was the departure of plant pathologists from their botanical colleagues taken lightly by the botanists. The Botanical Society of America requested in 1912 that the newly formed APS consider the matter of closer affiliation, and a committee consisting of the president and secretary-treasurer was appointed to consider the matter (*Phytopathology*, Vol. 2, p. 44). A report of an APS Committee on Affiliation in Relation to Other Societies in 1915, while appreciating the courteous offer of the Botanical Society, recommended that a closer affiliation not be pursued because of financial problems and other unspecified reasons (*Phytopathology*, Vol. 5, p. 128). A brief but excellent summary of our associations with other organizations is given by McCallan in a book (4) published upon the occasion of our 50th anniversary. He calls attention particularly to the long-standing contacts with the Botanical Society of America, the Mycological Society of America, and the Potato Society of America.

APS met for many years routinely with the American Association for the Advancement of Science (AAAS). Indeed, the Society was organized when that group met in Baltimore from 28 December 1908 to 2 January 1909. The last meeting with AAAS took place in New York in 1949. As indicated by McCallan, meetings during the years of World War II had demonstrated the advantages of meetings held at times other than Christmas and jointly with other organizations at times convenient to them.

The APS became affiliated with the American Institute of Biological Sciences (AIBS) in 1949. It was thought that this affiliation would help our society to speak with a stronger voice and provide an opportunity for participation with more scientists whose interests were much closer to ours than was possible with the huge and diverse AAAS. Relationships with AIBS have varied. It is not surprising that the marriage of basic and applied or agriculturally oriented interests has not always been a happy one. This has been particularly true since major concerns over world hunger and food production often seem opposed to the maintenance of an unpolluted environment. Nonetheless, real efforts have been made by AIBS in recent years to include and consider the interests of all biologists. Significant changes in the governance of the society allowing more appropriate representation from the constituent societies has been one positive step in enhancing the quality of these professional relationships. APS has continued to meet with AIBS on an intermittent basis since 1952.

The Society has always been interested in working with other organizations on specific problems and in more effectively providing for communication among members with common interests. In the early years these projects included such things as the development of Pan American literature lists (published in a number of volumes of PHYTOPATHOLOGY), culture collections, lists of research projects, and sponsorship of *Biological Abstracts*, originally through membership in the Union of American Biological Societies. Often, special committees were named to handle these projects.

The number of societies, committees, councils, commissions, etc., with which our society has cooperated directly through meetings, correspondence, and reports is too great to present in detail here. Some with whom we have had relationships over a period of years include, in addition to those already mentioned, the American Grassland Council, the American Standards Institute, the American Type Culture Collection, the Biological Stain Commission, the Food and Container Institute, the International Shade Tree Conference, the National Research Council, the Tropical Research Foundation, and the Union of American Biological Societies. In more recent times, our Society played a key role in organizing the International Society for Plant Pathology and cohosted its second meeting with the Society of Nematologists at the University of Minnesota in September 1973.

SOME SPECIAL EFFORTS BETWEEN THE WARS

The Advisory Board of the APS was a direct successor of the War Emergency Board of World War I, which had been formed at the Pittsburgh meetings in late December of 1917. The War Board had been appointed to promote cooperation in war work, particularly in the production of foodstuffs. It worked closely with similar committees appointed by the Botanical Society of America and Section G of AAAS.

Apparently the cooperation stimulated after only one year's experience with the War Emergency Board was a new experience for American plant pathologists. "The scientific workers' training and experience tend to develop their individuality and to render it difficult for them to work in groups after the manner of businessmen," was the way G. R. Lyman (3) evaluated the situation at that time. "While few plant pathologists oppose cooperation in principle," he continued, "many are fearful of some of its effects when put into general practice." He even quoted Mr. Elihu Root, a noted statesman of the period, who had recently charged that scientists had organized everything except themselves.

Some of the specific functions of the new Advisory Board were to: 1) represent plant pathologists before the National Research Council; 2) confer with workers in related fields, i.e., entomology, genetics, horticulture, and agronomy, in order to promote joint efforts on common problems; and, 3) promote international relationships. The Board, whose first chairman was G. R. Lyman, was very effective for several years in presenting the case for plant pathology, both at home and abroad. Not surprisingly, one of the principal problems was the securing of funds to support the various cooperative projects planned. A big push was made for the establishment of a Phytopathological Institute, a project that never succeeded, and activities were discontinued after the establishment of the Boyce Thompson Institute. Other early activities included cooperative tests with fungicides and their methods of application and the planning and arrangement of summer tours. As McCallan pointed out (4), eventually many of the duties of the

Advisory Board were assumed by the APS Council and the numerous proliferating committees associated with the growth and maturing of the Society.

Plant pathologists, like most everyone else, learned to appreciate the benefits of cooperation. "The value of cooperation between different branches of plant science was brought out in joint sessions. . ." is a phrase which recurs word for word in several annual reports. Usually such organizations as the Potato Association of America, the American Association of Entomologists, AAAS (Section G), affiliated botanical societies, the American Society for Horticultural Science, the American Society of Plant Physiologists, and others are mentioned.

These joint sessions and the cooperative efforts of our Society through its members and committees often provided the impetus to advance our knowledge in special areas of interest. One regional example is the work of the Tobacco Disease Council and the Plant Nematode Council. The Tobacco Disease Council was organized and became affiliated with the Tobacco Workers Conference in Greensboro, NC, in November 1935. The purpose of the organization was to promote cooperation in energetic scientific investigation leading to effective tobacco disease control. The Plant Nematode Council was organized originally as the Root Knot Nematode Committee in February 1937 by federal and state workers at Nashville, TN. This group promoted research on plant nematode problems and assisted each other in research. These two groups called attention to the very serious nematode problems in the South (they met jointly in 1941 at Tifton, GA) and played a key role in the gradual development of comprehensive nematode disease research programs in the Southeast.

A significant step in coming to grips with practical problems that required cooperation was participation in the formation of a Crop Protection Institute in June 1921 at Rochester, NY. This institute was an outgrowth of the work of the Advisory Committee and reflected the "spontaneous desire on the part of plant pathologists, economic entomologists, and certain businessmen to secure united attack on certain common problems (*Phytopathology,* Vol. 11, p. 198). The purpose of the Institute was: 1) to promote the general welfare through efficient control of injurious insects and plant diseases, 2) to promote control of insects and plants injurious to humans, animals, and animal products, 3) to support and direct research upon other problems of similar nature, 4) to further cooperation among entomologists, scientific workers, plant pathologists, manufacturers of insecticides, fungicides, and similar materials, and the manufacturers of appliances, and 5) to assist in the dissemination of scientifically correct information regarding the control of injurious insects and plant diseases. Thirty-two individuals were present in the organizing group and represented 1) the National Research Council, 2) the American Association of Economic Entomologists, 3) the APS, and 4) manufacturers of fungicides and insecticides. Administration was placed in the hands of a Board of Trustees of 13 members, representing the above groups and the Association of Official Agricultural Chemists.

The institute received proposals for research and functioned as a review board and intermediary between the scientist and the company that provided funds. Reports of the work of the Institute were provided to the Society more or less on an annual basis, and a summary of 12 years of work was published in volume 24 of PHYTOPATHOLOGY. Some of the research of interest to plant pathologists that was funded included studies on crown gall, copper fungicides, flotation sulfur, and organic chemical byproducts. The Institute appeared to be very active for about 20 years, up until World War II, after which industrial concerns developed other means for screening and evaluating the multitude of new chemicals that were being developed.

CURRENT EFFORTS

The most significant step in recent times toward improving cooperation and communication among crop protection disciplines was taken on 8 August 1975 in the Shamrock Hotel, Houston, TX on the occasion of the APS meeting. In a parlor overlooking the Astrodome, Kenneth T. Knight, President of the Entomological Society of America (ESA) presented a draft of the Articles of Confederation for a proposed Intersociety Consortium for Plant Protection to representatives of the Weed Science Society of America (WSSA), the Society of Nematologists (SON), the Entomological Society of America, and the APS. Those in attendance in addition to Dr. Knight were the presidents of APS (James Tammen), SON (Walter Thames), and WSSA (C. L. Foy), the president-elect of APS (Robert Aycock), and other Society representatives: David Schlegel (APS), Ray F. Smith (ESA), and James Horsfall (APS).

The draft presented by Dr. Knight was revised and adopted and referred back to the constituent societies for ratification. Dr. Tammen acted as the chairman of the Provisional Executive Council. The Articles of Confederation were duly ratified by all societies. At a meeting of the Executive Council on 5 April 1976, goals for the Consortium were more precisely defined and several committees were established: Survey and Detection, Pest Management, Crop Loss Assessment, and Education. The Council has since consulted and cooperated on matters of mutual interest with representatives of the Animal and Plant Health Inspection Service, the Cooperative Research Service, and the Science and Education Administration of the USDA; the Environmental Protection Agency; and other governmental, educational, and research agencies. Each committee was authorized to make direct contact with relevant federal agencies, keeping the Executive Council informed of their activities. The representatives of the societies were to keep their constituents informed through their organizational newsletters, meetings, etc.

After 6 years, the Consortium has been useful in a number of ways. It has provided a very effective and continuing link of communication among the crop protection disciplines that was never available before. In addition to the usual reports, correspondence, etc., the Consortium maintains a list of scientists who have expertise in many subject areas of concern. These are available for the development of symposia, conferences, workshops, etc., that are of mutual interest.

The Consortium has also provided a much higher degree of visibility for the crop protection disciplines in Washington. In a relatively short time, members of the Executive Council and the various committees have established relationships with members of Congress and their aides and with members of the executive branch of government as well.

The Consortium, through its conversations and discussions in Washington, has on occasion been able to function as a liaison or intermediary between and among U.S. governmental agencies whose missions and interpretations of responsibilities sometimes appear to be in conflict. Substantial assistance has been rendered in assisting the USDA-Science and Education Administration to prioritize research and extension needs in the field of crop protection. Two additional committees, Biological Control, and Pesticides, have been established to enhance the Consortium's activities along this line.

Specific documents have been developed for governmental organizations and agencies. A document on integrated pest management was developed under Consortium auspices for the Experiment Station Committee on Policy by a study committee chaired by J. L. Apple. Corn, peach, and nectarine pest management production guides have been prepared for the Environmental Protection Agency by George Bird.

CONCLUSION

From the early days of our profession, we have been confronted on the one hand with the need to specialize and on the other the need to maintain our essential unity with biology. Our only justification for specialization and segregation, according to L. R. Jones (2) was for greater efficiency in service.

These concerns have been reaffirmed and spoken of to me by latter day saints of our "professional guild" as well. To quote J. C. Walker (7),

> Already we see cults developing who refer to themselves as plant virologists, plant disease physiologists. . .plant nematologists and I presume just around the corner, plant disease molecular biologists. This is all to the good because to make basic progress we must not only specialize but reach out to gain the advantage of mingling in other fields. . .I am not so much concerned that plant pathology will disappear like the exploding atom. There will always be plant disease problems and crop losses from disease. What I am concerned about is that these specialty groups will lose plant pathology. There is real danger of being cut off in space without landing gear. We are already showing signs of building a Tower of Babel within our science, wherein plant pathologists

will not understand each other or their techniques and philosophies.

E. C. Stakman (6) considered the relative importance of curiosity and sense of service in the motivation of such pioneers as Fontana, Targioni, Prevost, and Kühn. "They were exceptionally curious," he noted, "but they were exceptionally curious about the phenomena that impinged directly on the interests of the people generally."

American plant pathologists have shown their commitment both to service and to the pursuit of basic knowledge. In general we have recognized the need to relate to other disciplines and to other organizations. We have reached out to "mingle in allied fields." To quote Walker (7) once again, "we were started on a fabulous journey by de Bary in 1863—there are many stimulating challenges still awaiting us.

ACKNOWLEDGMENTS

We are grateful to C. J. Nusbaum, William Neal Reynolds Emeritus Professor of Plant Pathology at North Carolina State University, for many helpful suggestions and to G. W. Bird, Professor and Coordinator of Pest Management Programs at Michigan State University, who supplied much useful information on recent activities of the Intersociety Consortium for Crop Protection.

LITERATURE CITED

1. Freeman, E. M. 1936. Phytopathology—And its future. Phytopathology 26:76-82.
2. Jones, L. R. 1911. The relations of plant pathology to the other branches of botanical

science. Phytopathology 1:39-44.

3. Lyman, G. R. 1919. The Advisory Board of American Plant Pathologists. Phytopathology 9:202-206.
4. McCallan, S. E. A. 1959. The American Phytopathological Society—The first fifty years. Pages 24–31 in: Plant Pathology Problems and Progress 1908–1958. C. S. Holton, G. W. Fisher, R. W. Fulton, Helen Hart, and S. E. A. McCallan, eds. Univ. Wisc. Press, Madison. 588 pp.
5. Shear, C. L. 1919. First decade of the American Phytopathological Society. Phytopathology 9:163-170.
6. Stakman, E. C. 1964. Opportunity and obligation in plant pathology. Annu. Rev. Phytopathol. 2:1-12.
7. Walker, J. C. 1963. The future of plant pathology. Annu. Rev. Phytopathol. 1:1-4.

CHAPTER 3

PROMOTING PLANT HEALTH THROUGH COMMUNICATION

THOR KOMMEDAHL, STEVEN C. NELSON, and MILES WIMER

Scientific societies exist to promote professional activities of their members and to publish journals as outlets for research findings. Increased use of journal publication has been apparent recently because of the general trend away from in-house publications by institutions to increased reliance on refereed journal articles. Agricultural scientists look at professional journals as essential research resources that serve as the primary outlets for research as well as providing criteria for choosing research topics.

Editors are appointed as guardians of the science to protect the integrity of the profession. They and their editorial boards control access to publication, the dissemination of research findings, and discussion of ideas. The American Phytopathological Society (APS) is no exception. The Society has had one journal, PHYTOPATHOLOGY, since 1911 and a second, PLANT DISEASE, since 1980. For most of our history, PHYTOPATHOLOGY has been the main outlet for communicating plant pathology to plant pathologists and to agricultural scientists in general; during this period this journal has established a distinguished record in scientific publication.

Growth of journal activities can be shown by the increase in the size of the editorial board of PHYTOPATHOLOGY. From 1911 to 1982, the number of senior editors (exclusive of the editor-in-chief) varied from 0 to 4. However, the number of associate editors ranged from 12 to 18 in 1911–1969, then increased to 24 in 1970, and varied from 37 to 42 in 1971–1982. The reasons for the increase are the greater number of manuscripts submitted and the use of at least two reviewers (instead of one) for each manuscript.

Our plan for this chapter is to review briefly the events leading to the establishment of Society headquarters (an important step because it facilitated publication), to describe the current situation in publication of APS materials, and then to peer into the near future for events that appear likely or probable for scientific publishing in general.

EVENTS LEADING TO A SOCIETY HEADQUARTERS

Early editors of PHYTOPATHOLOGY depended mainly on department staff for secretarial and technical editing assistance or else did these chores themselves. However, at least by 1944, when Helen Hart became editor-in-chief, a part-time editorial assistant, Eunice Brown, was employed using Society funds. This practice continued for subsequent editors until 1964, when Thor Kommedahl was appointed editor-in-chief. At this time, during the presidency of A. E. Dimond, a bid from Association Services, Inc., St. Paul, MN, was presented to Council. The bid was to handle redactory and secretarial services for $9,000 per year for a 3-year period, which was about $1,000 less than the competitive cost of hiring a secretary and editorial assistant. Association Services, Inc., was a partnership formed by R. J. Tarleton (then executive vice-president for the American Association of Cereal Chemists [AACC]) and Jack Morris (head of an advertising and public relations firm). Council accepted this bid, and this decision set in motion a series of events that led eventually to the establishment of a national headquarters for the Society.

By 1966, the decision was made to provide partial assistance to the Secretary of the Society using staff of Association Services, Inc., when J. P. Fulton held this office. About this time, R. J. Tarleton, because of increasing responsibilities with AACC, severed his connection with Association Services, Inc., and the Society was faced with the decision to continue with Association Services under different management (in part) or to establish a relationship with another organization for the same services. President G. A. Zentmyer appointed a Committee on New and Future Use of Services (CONFUS), which was chaired by A. E. Dimond. After considering several other societies and agencies, Council voted to accept the committee's proposal to form a partnership with the AACC and establish joint headquarters in the Twin City area of Minnesota.

In 1967, Council, in Washington, DC, under President A. Kelman, agreed to cooperate with the AACC on a 1-year trial basis and to establish a headquarters at 1955 University Avenue, St. Paul. Treasurer and Business Manager M. F. Kernkamp transferred financial records from outgoing Treasurer D. H. Marsden to headquarters, and R. J. Tarleton was appointed executive vice-president for APS in addition to serving in this capacity for AACC. The headquarters staff at this time consisted of the executive vice-president, a director of publications, an office manager, two business clerks, one manuscript records clerk, two part-time student employees, and one technical editor.

In 1968, the two societies found the trial period to be mutually advantageous, and they signed a 5-year agreement to extend the lease at 1955 University Avenue. Shortly thereafter a fire at 2:30 A.M. on 22 October destroyed the headquarters building. Only 15 manuscripts were lost, although some were either charred or water-soaked. Also some records were burned. A new location at 1821 University Avenue (the Griggs-Midway Building) was selected, and a new lease was signed to share space with AACC. In 1969, Secretary J. P. Fulton completed the transfer of secretarial files to APS headquarters.

Governing bodies of both societies sought a permanent location for headquarters. D. E. Ellis, as President of APS, appointed a committee to work with a similar committee of AACC. APS members were M. F. Kernkamp (chair), G. A. Brandes, A. Kelman, J. B. Kendrick, Jr., and D. W. Rosberg. M. F. Kernkamp for APS, D. G. McPherson for AACC, and R. J. Tarleton searched for possible sites in the Twin City area. The site selected was a 3-acre plot in Eagan, a suburb of St. Paul. The APS Finance and Investment Committee chaired by D. H. Marsden recommended purchase as a sound, financial investment, and land was acquired at a cost of $15,000 per acre.

Bids were sought for a building, and the one chosen was for $306,000. The

Society was canvassed by mail, and the 1,200 who voted approved acceptance of the bid by a ratio of 9 to 1. The APS cost (half of the $306,000) was amortized by the sale of $50,000 of Society stocks and by funds obtained in a building campaign headed by J. C. Walker. The contract was signed 26 March 1971, and ground-breaking ceremonies were held that month, with participation by APS President T. Kommedahl, AACC President E. J. Bass, and Committee Chairpersons M. F. Kernkamp for APS and D. G. McPherson for AACC. APS and AACC moved into the completed building 8 November 1971. Additional details regarding the operation of headquarters is reported in the 1979 Directory of Members, in annual reports of the Society, and in PHYTOPATHOLOGY NEWS.

CREATION OF PUBLICATION OPPORTUNITIES

With the establishment of a headquarters building, the Society had a place for modern, cost-effective equipment for publication, as well as effective work space for the technical staff needed for redactory and business services. Economies in publication were realized with these additional facilities at headquarters.

The responsibilities of editors of publications began to change as the headquarters' staff assumed more and more responsibility for correspondence, promotion, and editing of manuscripts. The last editor-in-chief to handle contacts with authors and reviewers for all manuscripts submitted was Thor Kommedahl. Following his term of 4 years, manuscripts were assigned by the editor-in-chief to senior editors, and the editor-in-chief assumed a policy-making role in Society publications, as well as settling disputes, resolving complaints regarding manuscripts, and handling letters to the editor. At the same time, experts in editing and graphics hired by the Society had increasing input into the form and style of Society publications, and member editors of publications concentrated mainly on the scientific quality of the manuscripts submitted. These changes have lessened the direct role of member editors in the editorial process in both journal and nonjournal publications. Member editors usually edit manuscripts in their spare time. They are scientists first and maybe editors somewhere down the line; in fact, they may be amateurs when it comes to editing. Obviously we want editors to be scholars who are active in the science itself, but more and more we see the need also for professionals who know the craft of technical editing and graphics production. Currently on the APS headquarters' staff are professional editors and graphics personnel who have such skills and who have been active in all phases of publication by the Society, including production of compendia and PLANT DISEASE. Such expertise results in readers getting a readable, attractive, and informative publication.

The technological developments in changing from hot-metal type and letterpress printing to cold type and offset printing led to economies in publication that helped finance the growing service function of headquarters. The result has been, and will continue to be, that professional writers, editors, and graphics specialists have increasing and considerable influence on the final appearance of the published article.

Professional and scientific communications have been perceived as preeminent needs and responsibilities of the Society. We have assumed more responsibility for the entire communication process than have other scientific societies comparable to our own, and we have utilized the skills and technology needed to accomplish our communication goals. For the purposes of economy and quality control, our staff has become involved and skilled in processes such as phototypesetting, proofing, keylining, design, promotion, and distribution; other societies contract for these activities and accept the services that are available. Moreover, in addition to publication of journals, APS has attempted to meet the needs of the scientific audience by publishing monographs, classics, symposia, and manuals and to meet the needs of grower and special interest groups by publishing books, compendia, slide sets, brochures, and directories. In other words, APS has established a facility and staff that can handle virtually all aspects of communication. Our Society is now in complete control of the process (printing and binding are by contract).

To reach objectives encompassing all phases of publication, a well-equipped facility and a staff of technical editors, graphics specialists, and marketing experts has been developed. This is feasible largely because of the joint partnership with the AACC. By using resources available to both societies, the scope of publication has been increased and costs have been reduced in producing the printed page, which is still the goal over the short run. In the long run, it will be possible to develop electronic publication that will substitute, perhaps in part only, for the printed page.

NEW TECHNOLOGIES AT HEADQUARTERS

New technologies are being applied to three different operations: typesetting, word processing, and data processing. In time, these operations will become integrated.

Typesetting and Graphics

Progress was made first in the typesetting area. In 1969, a "strike-on" typesetting station was purchased, and a secretary began to produce type as part of other duties. By 1972, a three-work station, computer-based phototypesetter was acquired. By today's standards, it was primitive. It required paper tape to be punched as an input to the typesetter. Operators could not see what they were typing as they worked; tables and technical symbols were difficult to set. Training was difficult as operators needed 6 months to become productive.

In the following decade, the Society gained considerable expertise in typesetting skills. The typesetting staff developed at headquarters was highly specialized in setting technical matter such as statistical tables and technical symbols—skills that trade typesetters found difficult to do because of their lack of experience.

It also became necessary to develop a graphics department with staff that could keyline, design, contract for printing, and do all of the operations required to handle the amount of type that now could be produced. The economy of setting our own type made this possible. As a result, publications were being produced in ever-increasing numbers, and all Society printed matter, from letterhead to brochures and programs, was being produced internally. Eventually, the production workers were capable of handling more work than the typesetters could supply, and the laborious, punch-tape typesetter became obsolete.

In 1978, the original typesetting equipment was replaced. A year later, it was upgraded to a more sophisticated, computer-driven composition system. With the new system, operators can see their work on a video display screen at the time of typing, and the typed material is stored on floppy diskettes. This equipment has increased the range of styles and sizes of type over that available previously. Also, all work stations are more highly computerized.

This new equipment allows tables to be set faster, permits corrections to be made electronically, allows standard or repetitive material to be set with minimal typing, and has reduced the numbers of errors. Moreover, operators become proficient in a few weeks. In short, productivity and accuracy increased, and problems virtually disappeared.

The Society was enabled to continue its publishing program (in 1981, 20,000 nonjournal publications were sold and distributed) and even add a new monthly journal, PLANT DISEASE, with only a modest increase in personnel. Costs for these projects would have been prohibitive if type had been set by an outside agency or if the newer technologies had not been available internally.

Word Processing

In 1980, a word processing facility was acquired at headquarters. It consists of a word processor that can "talk" to the typesetting equipment, an optical character recognition (OCR) device that enables typescripts to be "read" into either the word processor or the typesetting equipment without any retyping, and an interface that enables material stored in the Society's main business computer to be printed out on either the word processor or the typesetting equipment.

The word processor is especially useful for documents (such as letters and promotional brochures) that the staff writes and then revises in the office. Multiples of these documents can then either be printed or set in type without being retyped on a standard typewriter. Use of OCR typeface enables the relatively inexpensive typewriter to serve as an input to either word processing or typesetting. Thus, in the last half of 1981, all of the type set for PHYTOPATHOLOGY, except for the tables, was processed at headquarters, but the actual typing was done inexpensively by freelance typists in their own homes.

The computer interface enables names and addresses, or any other kind of information, to be taken directly from the computer to the word processor for marketing and other purposes. Personalized letters can be written automatically on the word processor. Also, it gives APS the capability of typesetting projects, such as directories, without ever having to manually type out the data stored in the computer.

The increased availability of word processing equipment—not only to headquarters' staff but to authors and scientific editors—has allowed the Society to disseminate some types of information very rapidly by producing camera-ready books. These books can be produced inexpensively and quickly because the editing and typing are done by authors and only the physical production is handled at headquarters. Camera-ready material has been used occasionally in the past; in 1951 some tables were printed camera-ready for PHYTOPATHOLOGY on a trial basis because of the high cost of linotype at that time, and since 1979, abstracts of papers presented at national and divisional meetings have been published as camera-ready copy in PHYTOPATHOLOGY. However, the physical appearance of camera-ready material that is not typed with the same typing element is distracting to the reader and such material is not met with universal acclaim. By using word processors, editing can be done by visual display terminal operators and, for multiple-authored works, the typeface is uniform. In 1982, an APS symposium was published using camera-ready material and a common typeface. More such publications are planned.

Data Processing

An IBM System 34 computer, installed in 1977, has been up-graded since its purchase. It maintains member, subscriber, and publication-buyer information; generates mailing labels and invoices; keeps a master roster of those who serve on committees; and issues financial statements and reports. In the future, manuscript review and manuscript handling processes may be computerized.

Printing Developments

APS has depended to an increasing extent on emerging technologies in printing. These affect all aspects of the headquarters' operation as a means to lower expenses, increase income, improve quality, increase speed of publication, and give control of publication to the Society instead of the printer. The change from hot metal composition and letterpress to offset printing alone avoided a probable increase in printing costs that would have made expenses 25–50% greater than now. Today, offset is as common as letterpress was in the 1960s. Even printing is being computerized to an increasing degree. Also stapling has been replaced by soft bindings or by end-stapling. Although binding has been the most labor-intensive part of the production effort, costs have been reduced substantially in recent years.

ALTERNATE FORMS OF PUBLISHING

Microform

Microform is a technique for photographically recording printed matter on film or cards at a reduced scale that can be enlarged for reading. Reductions range from 8 to 40 times, occasionally to 60, and even to 150 times. At a reduction of 1:150, 3,000 pages can be condensed to the size of a postcard. Microform includes microfilm (a continuous spool of film), microcards (sensitized cards, 7.5 × 15 cm, on which printed matter is reproduced photographically in greatly reduced form, and microfiche (a photographic film 10 × 15 cm in size on which printed material is reproduced). The Society thought that microfiche had possibilities for publication, and for a 3-year period (1973–1975), APS offered PHYTOPATHOLOGY in microfiche as well as in conventional journal form; however, only about 30 members chose this version of the journal and it was discontinued. An entire issue of the journal appeared on two microfiches. It became apparent that microfiche or other microforms are cumbersome for frequently consulted sources of information and lend themselves mainly to publications that are referred to only occasionally or to archival types of material.

Miniprint

Another variation in publication is the use of "miniprint," in which type is reduced to a size that is barely legible to the unaided eye or in which lenses must be used for reading. Even though the use of miniprint would save on paper and mailing costs, it is not a likely choice except possibly for sections of a paper, such as the materials and methods or literature cited sections.

Synoptic Publication

The advent of the word processor makes possible the production of relatively error-free copy, which then makes possible the production of two versions of a typescript. One version is a summary or synopsis that is published in a journal on a single page (plus illustrations), and the second version is a camera-ready copy of the complete text that is available on demand as an annual volume for libraries or for purchase by individuals; this second version is the author's original version and is not edited. Such a procedure may reduce printing costs but may result in publication of papers of lower quality because complete versions do not appear in the journal. Also it requires the author to prepare two versions of each paper. This has been suggested but not tried by APS.

Magnetic Tape

Still another variation in publication is the use of cassette or video tape, which is suitable for brief updates of disease situations or to announce and describe new discoveries. Oral history or interviews with notable plant pathologists may be recorded on magnetic tape for the education of future generations. For example, tapes are available at APS from the Golden Jubilee meeting in 1958.

ALTERNATIVE WAYS OF HANDLING MANUSCRIPTS

Traditional Journal

Recent technology for the publishing industry has been directed toward the use of electronics both in the publication of the traditional journal as well as in publication of the nonprinted (electronic) journal. A currently possible scenario that enables the author to fully utilize new technology starts with a standard typewriter fitted with a $20 OCR typing element. After the manuscript is typed, it is sent to the copyeditor, who arranges for peer review by selecting from a computerized list of reviewers. About 130,000 persons are listed in the reference *American Men and Women of Science*; their names and areas of expertise, as key words, are in computers and are retrievable. Also, computerized citation data are being stored by the Institute for Scientific Information. Thus reviewers can be selected for peer review from these sources.

The reviewed manuscript is then sent by the editor-in-chief to the technical editor, who edits it on a visual display terminal and transfers the edited copy onto diskettes. The copy is then typeset, or, by telephone, these diskettes can be transmitted between computers, which "talk" to each other at the rate of 300–10,000 characters per second.

If the author's computer is not compatible with the typesetter's machine, the author can find a service agency that specializes in all kinds of media conversion. In Minneapolis, for example, at least a dozen service firms can convert diskettes to tape overnight for a modest fee. The author could improve this situation by submitting to the publisher a diskette or the magnetic tape instead of a manuscript.

Electronic digitizing in which black-and-white or color photographs and line drawings can be scanned is being developed. Lasers are being used to scan and at the same time direct images onto the printing plate.

Electronic Journal

The scenario for the operation of an electronic journal is that the writer composes the work on the computer (word processor), submits the work to an editor by computer conference, and has the work reviewed by computer conference. If the work is accepted, the information is made available to the public through a search and retrieval system.

An electronic journal has distinct advantages over conventional publication. Valuable data can be retained in the system while ephemeral data can be deleted; all of the data can be published with no pressure for brevity; there are no printing, binding, or mailing delays; storage problems are eliminated; referee time is reduced because mail is bypassed; and the revision is done on word processors.

Prospects

Even though computers and electronic devices have an enormous capacity to store, retrieve, and disseminate information, many experts in the trade are not convinced that books will be replaced. More likely, computers will aid in the publication and production of better books. At present, information data bases contain information artificially organized to locate knowledge, not to disgorge knowledge. Better computers and communication systems will probably become available to make possible storage, retrieval, transmission, and display of large volumes of information. So far, the quality of information on computer terminals is inferior to the quality of printed materials. Terminal screens lack the quality graphics, color capabilities, and amount of digital data equal to that presently available on full-color printed pages.

A disadvantage of electronic journals is lack of portability due primarily to size. However, it is possible, even feasible, to reduce terminal size, especially the video display unit, to about the size of a book. Such a book-size machine could hold the annual output of 20 or more searchable journals.

Thus, we predict that the printed page will hold its appeal for the scientific audience for the next two decades at least and that electronic or computerized developments will focus on making the printed page as readable and inexpensive to produce as possible. However there will be a phasing in of electronic publication so that some kinds of information will be available only by retrieval from a computerized source. Both will coexist for a time; convenience and economy will dictate whether the printed page or electronic publication will prevail. The fact that the electronic journal is cheaper to produce than the printed page will favor its continued development.

REFERENCES

1. Burke, T. J. M., and Lehman, M. 1981. Communication Technologies and Information Flow. Pergamon Press, New York. 151 pp.
2. Langlois, E. G., ed. 1981. Scholarly Publication in an Era of Change. Soc. Scholarly Publ., Washington, DC. 86 pp.
3. O'Connor, M. 1979. The Scientist as Editor. John Wiley and Sons, New York. 218 pp.

CHAPTER 4

HEALTHY PLANTS—A THREAT TO CIVILIZATION (AND A CHALLENGE TO THE APS)

WILLIAM C. PADDOCK[1]

"Scientists have known sin," Robert Oppenheimer said after the bomb was dropped (41). If that is so then so also is it true for plant pathologists, for their science is contributing to the deaths of a larger number than were killed in the bombing of Japan. However, most plant pathologists are unaware of their blame and, unlike the physicist, their blindness has no wartime emergency to mollify their guilt.

Why the blame? Because:

1) plant pathologists during the past 40 years have played a key role in helping the third world produce more food;
2) they, more than any other agricultural scientists, should be conscious of the consequences of this action in an area of rapid population growth;
3) plant pathologists as a group have made no effort to warn of the dangers of increased food production in areas where the rate of population growth is uncontrolled; and
4) thus plant pathologists are contributing to a situation that ultimately means that more people will starve, more environmental stress and degradation will result, and the very continuation of civilization as we know it will be threatened.

WHY BLAME THE PLANT PATHOLOGIST?

We were all taught in Plant Pathology 101 that our science grew out of the Irish famine of the 1830–1840s, a famine that killed two million Irish. The famine, we were taught, was caused by a disease (*Phytophthora infestans*).

We were taught incorrectly!

The famine was not caused by a disease but by the *person* who introduced the potato into Ireland in the 1700s. Before that introduction, the Irish had come into

[1]This is hardly a pleasant message to bring to the 75th birthday celebration of the American Phytopathological Society (APS), a society proud of its accomplishments. But in praise of the APS, it is a message solicited by the editors of this publication.

a Malthusian balance with their environment, where two million Irish lived in abject poverty. But when the new technology arrived on that island in the form of the potato, food production increased markedly—and so did the Irish. By 1830, eight million Irish lived in abject poverty.

Then *Phytophthora infestans* arrived. The potato crop failed and two million Irish starved to death, two million emigrated, and four million were left in the same miserable state as before the introduction of the new technology.

The person who introduced the potato thought he was doing a good deed. Yet by increasing food production in a society that could not or would not limit its birth rate, the do-gooder assured a population explosion as surely as does one who takes a bacterial culture from a nutrient-poor to a nutrient-rich medium.

Thus, of all agricultural scientific societies, ours should best understand the Malthusian dictum: if the only check on population growth is starvation and misery, then no matter how favorable the environment or how advanced the technology, the population will grow until it is miserable and starves.

Unfortunately, our foreign aid programs throughout the world have worked diligently to repeat the Irish famine scenario. On this anniversary occasion, a brief review of this history might be appropriate to establish the extent of involvement by plant pathologists from the very beginning, even before President Truman's fourth point in his 1949 inaugural address.

The two most prominent were Dr. I. E. Melhus and a former student of his, Dr. J. G. Harrar. Melhus (who, interestingly, had made a permanent name for himself in plant pathology by discovering how *Phytophthora infestans* overwinters) had read a paper by the Russian geneticist Nikolai Ivanovich Vavilov, which said that the United States should study the indigenous food plants of Central America and Peru as sources for genetic material to improve U.S. agriculture. Consequently, Melhus wrangled funds from Earl May, an Iowa nurseryman and seed producer, and in 1941 Iowa State University set up an experimental station in Guatemala, becoming the only Land Grant College to support agricultural research outside the United States.

A year earlier another Iowan, Henry Wallace, as U.S. Vice-President-Elect, attended the inauguration of the Mexican president, Manuel Avila Camacho. Wallace was then America's symbol of dynamic scientific agriculture, having sparked an agricultural revolution with the first commercially produced hybrid corn seed. After his return to the United States, Wallace told Raymond Fosdick, then president of the Rockefeller Foundation, "If anyone could increase the yield per acre of corn and beans in Mexico, it would contribute more effectively to the welfare of the country and the happiness of its people than any other plan that could be devised" (17).

The Foundation then sent to Mexico, to determine the wisdom of starting an agricultural program, a three-man team headed by plant pathologist E. C. Stakman, who recommended Harrar to head the program. He, in turn, quickly hired two more plant pathologists: Norman Borlaug and John Niederhauser.

Also in 1941, before U.S. entry in World War II, Washington sent four teams of agricultural scientists throughout Latin America to determine how the area's agriculture could be strengthened both for the benefit of Latin America and for our own expected war effort. Pathologists were well represented on these teams, not only because of a keen interest in rubber, the supply of which was about to be cut off by the Japanese and its production in this hemisphere curtailed by disease (*Microcyclus ulei*), but also because the head of the Latin American Division of the Office of Foreign Agricultural Research (OFAR) was Dr. Ross E. (Dinty) Moore. Moore, a soil scientist, had been recruited from the United Fruit Company's banana division in Central America and had lived through the battle United Fruit waged against Sigatoka (*Cercospora musae*), a battle that, if lost, would have destroyed the company. Thus, the importance of plant pathology in the tropics was well ingrained in OFAR's planning.

As a result of the teams' recommendations, OFAR established experimental stations in four locations: Tingo Maria (the first) in the eastern montaña of Peru (headed by Benjamin J. Birdsall, a soil scientist also recruited from the United Fruit Company); Pichilinque on the eastern littoral of Ecuador on a cacao plantation abandoned because of Witches' Broom (*Grinipellis perniciosa*) and headed by plant pathologist Lee Hines; El Recreo in eastern Nicaragua, headed by soil scientist Robert Pendleton, located on an abandoned (because of Panama disease [*Fusarium oxysporum* f. sp. *cubense*]) banana plantation; and at La Aurora in the Guatemalan highlands, headed by plant pathologist Roland Lorenz.

Thus, the plant pathologists dominated, or were well established in, all overseas efforts when the initial foreign aid program was being fleshed out; they were consulted and were a major factor in the early planning stages of what was to become a $200 billion U.S. effort. In the beginning, agricultural programs received heavy emphasis and eventually experiment stations were either established or supported throughout the noncommunist third world countries.

EARLY HOPES FOR FOREIGN AID

During the first 20 years of our foreign aid effort, there was great confidence that science could solve the food problem of the hungry nations. Unquestionably, the most acclaimed program during this period was that of the Rockefeller Foundation in Mexico (Iowa State's program folded after Earl May died, and the stations set up by OFAR were always on the brink of collapse for any of several reasons). For instance, by 1952, it was being reported that the Foundation had helped Mexico achieve self-sufficiency in corn during "normal" years (34), and later Harrar, who had become president of the Foundation, was to tell Congressional committees, "Mexico has become self-sufficient in corn production, even though its population has risen some 60 percent" since work had begun in that country (21) and that Mexico before the Foundation's arrival had been (22):

> spending some very difficult to come by, hard won, foreign currency to buy what might be called bread from abroad. I mean wheat and corn....[The Foundation] by identifying the problems, selecting the priorities, and attacking the problems right at their roots...[changed things so that] by 1955 Mexico no longer had to import...wheat....

Yet, from the beginning there was concern about the effectiveness of foreign aid, a concern not expressed by the agricultural scientist and scientific societies but by the taxpayer footing the bill.

One evidence of concern could be seen in the continual changing, reorganizing, and renaming of the U.S. foreign aid agency. During its first 14 years of existence it had seven structural changes, was reviewed by eight presidential committees, and had 12 different directors. Quite a record!

WHY FOREIGN AID EFFORTS FAIL

During this time there was minimal interest in the population growth of the hungry nations. Typical was this writer who, in 1952, concluded that agricultural sciences could feed the world (22) and in 1964 wrote a book (35) about the problems of the third world, hardly mentioning population growth. In 1965 plant pathologist Norman Borlaug stated, "It seems likely that through a combination of improvements in conventional and non-conventional food production, man can feed the world's mushrooming human population for the next 100–200 years." (7).

But we were all wrong, back in those days. The optimism was either a reflection of our blindness to the facts or enthusiasm for the job we were doing. For population growth was destroying the efforts being made to feed and clothe the hungry of the world. Unfortunately, by the time this was realized so much had been invested in, and so many careers dedicated to, the ongoing programs that the whole corps of development experts (a new profession grown to over 7,000) was unwilling to pause, reassess what had been done, and determine whether programs were or were not on course.

All that was needed to keep up the effort was a shot in the arm of good American optimism. In 1968 William Gaud, then head of the U.S. foreign aid program, gave a fund-raising speech in which he coined a new term, "Green Revolution," and the foreign aid team was off with another spurt of enthusiasm. In 1970 the Nobel Peace Prize was awarded to Norman Borlaug as the "Father of the Green Revolution." In making the announcement, the chairperson of the Nobel Peace Prize Committee stated that the Green Revolution (31): 1) was "a technological breakthrough which makes it possible to abolish hunger in the developing countries in the course of a few years"; 2) had "contributed to the solution . . . of the population explosion"; and 3) "in short, we do not any longer have to be pessimistic about the economic future of the developing countries." And nowhere did a group of agricultural scientists raise its voice to caution the world that such claims were gross exaggerations of fact; if anything, we each basked in the reflected glory of that Nobel Peace Prize (not unlike scientists after the first atomic explosion; C. P. Snow reports that "the scientists were jubilant, and they wouldn't have been human if they hadn't been" [42]).

For we knew that it was not "possible to abolish hunger . . . in a few years"; that not yet was there a "solution . . . of the population explosion"; nor was the economic future of the hungry nations something we should not "be pessimistic about."

PROGRESS IN INCREASING AGRICULTURAL PRODUCTION

If one simply looks at what world farmers have done, helped by technology (and by farming farther and farther up the hillsides), one must conclude that they have achieved remarkable increases in production. World grain production (which is the best single measure of agricultural production) rose from 631 million metric tons in 1950 to 1,432 million in 1980 (11), and 1981 production is expected to be still higher.

However, during this same period nearly every country "revolutionized by the Green Revolution" became a net food importer dependent on a half dozen grain exporters, notably the United States (45). In fact, today, only North America,

Australia, and New Zealand are net exporters. Since World War II there has not appeared one significant new net food exporter in the entire world (11). Demand for U.S. exports is now growing so fast that they are expected to triple within 20 years (4). The reasons for this disturbing swing to dependence on foreign food supplies are several. For instance, the last 20 years have seen an upgrading of diets within some sectors of the third world economy while their agriculture has been unable to provide for the resulting increased demand. Furthermore, several countries that were said to be self-sufficient in the past really never were. This writer once wrote about the Mexican self-sufficiency in wheat, attributing it to improved technology (36) and, as a result, received a letter from the late William Vogt, author of *Road to Survival* (a classic that should be required reading for all agriculturalists), saying

> something [about your book] *does* dismay me: your acceptance of the Rockefeller line that "Mexico was self-sufficient in wheat." You mean there were no more people who wanted it? Or do you mean there were no more people able to pay the price? Malnutrition is wide spread . . . I'm afraid you've been had, me lad (37,46).

Indeed, Mexico, in an effort to increase wheat production, had priced wheat out of the reach of many. In the 1960s with a per capita gross national product 87% lower than that in the United States, Mexico paid a wheat subsidy 40% higher than that paid in the United States (37). Looking at today's figures, we see that Mexico is far from self-sufficient:

Mexican Imports in Metric Tons (19)

	Wheat	Corn
1975/76	1,000	1,450,000
1977/78	625,000	1,690,000
1979/80	1,020,000	3,870,000
1980/81	1,100,000	4,000,000

One reason for this growth has been Mexico's growing oil-export earnings. But the dominating, overriding reason has been, as elsewhere in the hungry nations, population growth. The advantages of increased food production have been destroyed by population growth, a growth made possible in turn by increased agricultural production. It's a vicious circle.

GROWTH IN WORLD POPULATION

Look what has happened to population growth during the first 75 years of APS (15,40) and what is projected for its 100th birthday:

Population of Third World Countries in Millions

	Year	Population	Percent Increase
Founding of APS	1908	1,169	
75th birthday	1983	3,622	209
100th birthday	2008	6,124	423

When APS was founded, the world had only 11 cities with a million or more inhabitants. By its 75th birthday, there will be over 270; 17 will have a population of 10 million or more (49).

Such growth in human numbers causes one to stand in awe of the hungry nations' abilities to feed themselves as well as they do.

The battle to feed the world, however, is being lost, as seen in the total world grain inventory figures (11,47): the number of days of grain supply *on hand* in 1961 was 102, in 1971 was 73, and in 1981 was 40.

How hungry is the hungry world? No one really knows. Cornell's Thomas T. Poleman recently said, "We simply don't have sufficient evidence to estimate the numbers of hungry people. There is no basis for coming up with concrete estimates" (32).

The United Nations says that there are more than one-half billion (probably not a concrete estimate) "suffering today from severe under-nutrition—hunger" (16).

UNICEF in 1978 said 30 million children under the age of five or 22% of all children born that year starved to death. "At current rates 900 million children will . . . die of starvation in the next 30 years alone" (29).

There are those who take heart in some reports that the population growth rate in recent years has fallen, such as reports of rates dropping from 1.9% per year to 1.64%. If true, it is not evidence that the crisis has passed. An annual growth rate of 1.64% will do everything that 1.9% will do—it just takes a bit longer. For example, the world population would increase by one billion in 13.6 years instead of 11.7 (6).

Only in Asia and the Far East has there been any upward trend in levels of "food self-sufficiency" in the last 15 years, while all other regions and groups show a steady downward trend (3). Alas, even the glimmer of hope in Asia is passing. India, which with much acclaim had achieved a modicum of "self-sufficiency" in the 1970s now has, by one estimate, 300–400 million who lack enough to eat because of poverty, and that country is now expected to import 2.5 million extra tons of wheat in 1981 (43).

India was the first country in the world to have a government family planning program, which began in 1951 and has continued through the years (at a cost of $875 million during the last 10 years) (50). Thus, India's census figures announced in March 1981 were a shock when they showed the exact same population growth rate as in 1971 and 1961 (18). During that period, the birthrate dropped but the death rate dropped just as fast, and in numbers the population is now growing by 13 million per year (instead of six million as when the family planning program began).

Theodore Hesburgh, Chairman of the Rockefeller Foundation's Board of Trustees, testifying before a joint Senate and House Subcommittee recently put the consequence of such figures in perspective, saying

> I believe that the whole problem of migration of people will be one of the great specters of the future, that I can easily foresee . . . hundreds of millions of people from the poor parts of the world such as India, marching on Europe, if they have three or four bad harvests . . . (23).

At current rates, the world population will double in 40 years. That means that the world must produce as much food in the next 40 years as has been produced in the past 11,000 (27,45).

WHEN WILL WORLD POPULATION STABILIZE?

Stabilization is a long way off. Where birthrates have declined (e.g., Costa Rica, China, Brazil, Mexico, and India), there remains a powerful momentum that guarantees continued growth—much like a car that keeps going after you take your foot off the accelerator. China, because of past high birthrates, now has increasing numbers of people coming into the parental age, which means that in spite of lower birthrates, more people will be having children and, thus, that country's continued growth is ensured.

Ainsly Coale, Princeton demographer, says "no one can rationally predict world population stabilizing at less than 8 billion. A more realistic figure is closer to 15 billion, or nearly 4 times what it is now" (18).

Leon Tabah, Director, U.N. Population Division, sees stabilization at about 12 billion around the year 2045 (12).

In 1977 President Carter directed the Council on Environmental Quality and the Department of State to jointly produce a study, finally issued in 1980, entitled GLOBAL 2000, which states, "At present and projected growth rates, the world's population would reach 10 billion by 2030 and would approach 30 billion by the end of the 21st Century" (5).

There are few who believe the world's population will ever attain such an enormous size. Even Coale doesn't believe his 15 billion figure is likely. Why? Because something will happen to prevent it, and that something is most likely to be one or more of the Four Horsemen of the Apocalypse. Consider, for instance, the stress placed on agriculture alone if Leon Tabah's "optimistic" 12 billion figure were to be reached.

By 2045, there would have to be produced nearly 300% more food than today. To do this, assuming that no grain was fed to livestock and that everyone in the world was permitted just enough to sustain a minimum level of activity (250 kilos per person per year), the world grain crop would then be three billion metric tons (vs 1.4 billion today). Every bit of arable land would be planted, using the most advanced farming technologies. What would happen when world weather was less than optimal, like 1974–75 when the world's crops were reduced by 2.2%? In 2045 it would result in a food shortage, causing the deaths in a single year of 264 million people. Added to the normal death rate, this would be more than one million deaths a day. The deaths would be as much the result of the agricultural scientist, whose technology stretched the carrying capacity of the world beyond its limit, as were the Irish famine deaths due to the person who introduced the potato.

GOVERNMENT EFFORTS TO SOLVE THE WORLD FOOD PROBLEM

Those who feel that a solution to the food problem has not materialized because of a lack of effort or interest by concerned governments cannot be aware of existing activity in this area. The U.N.'s Food and Agricultural Organization (FAO), which once could be housed in a private residence off Washington's Dupont Circle, now overflows into what must be one of the world's largest buildings (in Rome). Presidents Truman, Eisenhower, Kennedy, Johnson, Nixon, Ford, and Carter all went through the ritual of establishing some type of study group, commission, or program to solve or learn how to solve the world's food problem. The effort continues. At the end of 1980 there were no less than three studies issued within a few months (The Presidential Commission on World Hunger, The Brandt Commission Report, and Global 2000), all dire reading for those concerned with hunger.

The Presidential Commission on World Hunger is a typical example of why such studies fail: they look at the result and not the cause. The 251-page report (plus a 29-page summary) cost $3 million and was the work of a 20-member Commission. The directive given the Commission began, "To determine the basic causes of, and relationships between, domestic and international hunger and malnutrition . . . ." The word "population" is mentioned twice; the most complete statement on the subject is found in the supporting appendix and, in its entirety, says:

> The steady increase in world population, expected to nearly double by the year 2000, *complicates* [emphasis added] the problem of food production and distribution, the encouragement of self reliance.

The report further states that

> there are no physical or natural reasons why all the men, women and children of the world cannot have enough food to eat...[and in short] the prospects for eliminating world hunger will depend in large measure on the attainment of greater economic growth and increased food production in the developing countries (26).

Former Secretary of State Henry Kissinger said it just as foolishly before the World Food Conference in Rome in 1974: "We must proclaim a bold objective—that within a decade no child will go to bed hungry," and our government had posters with the statement printed and sent to those who would support U.S. foreign aid efforts.

Such statements are tragically misleading, yet where is the society of agricultural scientists that stands up and says so?

Instead, we have such recent contributions as the National Academy of Sciences looking for panaceas in "underutilized plants" and the Office of Technical Assessment holding a workshop on "innovative biological technologies for lesser developed countries (LDC)" in an effort to find "sustainable LDC agriculture" (24).

Any leader of a hungry nation would be encouraged after reading one of the National Academy of Science's contributions on the subject:

> A groundswell of support for the development of new species could lead to a cornucopia of new foods, fuels and industrial feedstocks. It may help extend productive agriculture to vast regions that today are not arable. It may help raise from despair the ever increasing numbers of humans in developing countries who waste their lives away in malnourished poverty . . . it's a challenge (24).

Occasionally someone will say that there is no solution to the world food problem unless population growth is stopped, but the words are lost within the welter of efforts to grow more food. The Green Revolution was supposed to buy 30 years' (8) or 20 years' (48) or 15 years' time (10) in which to solve the population problem. Now, it is exactly 15 years and the population problem remains.

CAN WE EXPECT ANOTHER GREEN REVOLUTION?

At a recent Congressional hearing, Congressman Romano L. Mazzoli (D-Kentucky) asked this writer, "Is there hope for another Green Revolution?" and received the answer, "I hope not!" (39). What would be accomplished by a scientific breakthrough of any of the currently discussed panaceas (genetic engineering, nitrogen-fixing bacteria on corn, the current wonder plant: the winged bean, triticale's living up to its advance billing, etc., etc.) as long as population growth continues?

We know that the world's population will be growing faster, in sheer numbers, on the 100th anniversary of the APS than on its 75th (over 100 million per year vs 79 million today).

Add more food to the world, and the population will grow still faster.

Add more food, and the denuding of the world's hillsides will be speeded up; wood for fuel will fall short of needs; water shortages will become more acute; agricultural soils will deteriorate more rapidly due to erosion, desalinization, alkalinization, and waterlogging; desertification will spread faster than currently—and it is now destroying annually an area the size of Maine; the concentrations of carbon dioxide and ozone-depleting chemicals can be expected to increase to the point where the world's climate will be significantly altered; acid rains from increased combustion of fossil fuels will threaten crops as well as accelerate the extinction of plant and animal species (already it is estimated that between half a million and two million species—15 to 20% of all plant and animal species on earth—could be extinct by the year 2000) (5,14).

We know that past civilizations have disappeared because of the destruction of their resource base. The Fertile Crescent, the Carthagenian farm land, the Mayan soils were all destroyed as their populations grew and placed greater and greater demands on their environment. Lester Brown, Worldwatch President, wonders,

> if environmental stresses undermined earlier civilizations whose population doubling times were measured in centuries, what is their impact now when population doubling time is measured in decades? (53)

Increase food production, and the hungry nations, whose population today will double in 37 years, will double in even less time. Another Green Revolution, then, would be the ultimate Pyrrhic victory!

THE OTHER SIDE

With a mote of fairness to alternate views, it should be noted that there are those who do not agree with the premise of this paper. Some take the view that science can conquer all, or that population figures show encouraging trends, or that there is actually a need for more, not fewer people ("the greatest shortage is brainpower"). For those wishing to pursue further this other side, there are a variety of publications to consult, such as those of Herman Kahn's Hudson Institute—e.g., *The Next 200 Years* (1976), John Maddox's *The Doomsday Syndrome* (1972), Wilfred Beckerman's *In Defense of Economic Growth* (1974), and most recently Professor Julian L. Simon's *The Ultimate Resource* (1981). (In reviewing the latter, biologist Garrett Hardin says the conclusions are "highly palatable to budget evaders, car salesmen, realtors, advertisers, land speculators and optimists in general; scientists find them appalling" [20].)

PRESIDENT REAGAN'S RESPONSE TO THE NEED TO DO SOMETHING

Presidents, like plant pathologists, are no different from anyone else when faced by the stark realities in the hungry world. All want to do something. But what?

As this is written, President Reagan has just met in Cancún, Mexico, with the leaders of 14 hungry nations to discuss how they might climb out of their poverty. They held a special session on their food problem, where one delegate, Nigerian President Shehu Shagari, stressing Africa's mounting food crisis, said that by 2000 Africa's food import bill would be $8.6 billion for cereals alone and that already "Africa's food bill presently equals its bill for oil" (51). (Probably forgotten by most today was the statement to Congress by our foreign aid people at the height of the Green Revolution: "By and large, Africa is not now or likely to be a chronic food deficit area" [1].)

The President told the leaders of these countries, "If a man is hungry and you give him a fish, he will eat it and will be hungry tomorrow, but if he is taught to fish, he will never be hungry again" and offered to send task forces of American agricultural experts to any nation requesting them (44).

What an original offer! How many such teams has the United States already sent out since 1940? 500? 5,000? The number blurs!

But he followed precisely what his advisors advocated. The 1980 Presidential Commission on World Hunger said "lack of political will" is the major problem (26), and that same year a group of 13 international agricultural research institutions prefaced its glowing report by saying

> the basic problems affecting food supply result from decisions made by governments and individuals . . . It is plain that the highest priority must be given to achieving a marked increase in food production in the less-developed countries themselves (13).

And on the eve of his departure, the New York Times editorialized, "The President ought to say . . . at Cancún . . . avoiding food/population disasters requires better farming, genuine land reform, fairer income distribution and—not least—real incentives for development" (33).

The President should have ignored his advisors and told his listeners:

> During the 5 days you spend traveling and meeting here at Cancún, the world's population will grow by one million, and by simply increasing either food aid or agricultural technology without a commensurate reduction in your population growth rate, your problems of poverty will only be exacerbated; every day that your populations increase, your resources become less able to meet your needs, and if you believe you are poor now, you can only face greater poverty in the future; while condoms and pills and abortion clinics financed by the industrialized nations might help, fundamentally each of your nations must develop its own single-minded ethic, preached at every level, directed towards *motivating* your people to *want* fewer children.

And if any one of you lead a nation with such an ethic and/or has achieved a falling population growth rate, then ask me for help and I will eagerly send agricultural teams to work at increasing your food production.

NATIONAL SECURITY

Some may suggest that President Reagan could hardly give such a message for fear of jeopardizing U.S. national security. With mounting U.S. dependence on third world countries for raw materials and military bases and with its current vulnerability to massive illegal immigration, a President must, of course, choose his words carefully. But the population issue is a national security issue.

Sol Linowitz's covering letter for the Commission on World Hunger told President Carter, "A hungry world is an unstable world," and the report itself added, "world starvation can be as great a threat to U.S. national security as enemy forces" (52).

Among those who have made similar statements are Robert McNamara ("Population growth is the greatest issue that the world faces"), former Director of the Central Intelligence Agency William Colby, former National Security Advisor Zbigniew Brzezinski, retired Army General Maxwell Taylor, and former Assistant Secretary of State George W. Ball (29).

Thus, *if for no other reason* than national security, President Reagan's advisors should have urged him to use Cancún to drive home the need for the have-not nations to squarely face their population problem.

Why has no professional agricultural society tried to alert governments to the inability of science and technology to continually feed a world whose population is doubling every few decades? One cynical reason might be an unwillingness to bite the hand (of various forms of foreign aid) that pays for a large percentage of the travel agricultural scientists now enjoy abroad. But in this writer's view, the reason primarily is a belief that while it may be proper for the *individual* to speak out, "it's not the *society's* business."

This may be a satisfying excuse for other societies, but not for the APS, which on its official seal (2) carries conidia of the Irish Famine's *Phytophthora infestans*!

HOW DOES THE FOOD/POPULATION PROBLEM CHALLENGE THE AMERICAN PHYTOPATHOLOGICAL SOCIETY?

The stated objective of APS in its constitution is to "promote the increase and diffusion of *all* (emphasis added) aspects of knowledge relating to plant diseases and their control" (2).

If plant disease control in a country that has a growing population means ultimately more deaths and misery than without disease control, is not the APS, by the terms of its constitution, *obligated* to promote and diffuse that knowledge?

But how?

Abraham Maslow says in *The Psychology of Science* that "it is tempting, if the only tool you have is a hammer, to treat everything as if it were a nail" (28), which is essentially what most agricultural scientists have done when confronted with problems of the hungry world: either hammer in more technology to boost food production—or hold a symposium. In 1975 APS held such a symposium, where the general tenor of the speakers could be summed up in the statement: "It should be evident to all that the relentless advance of the population monster threatens all nations and must be dealt with expeditiously and humanely to avoid disaster" (9).

Then nothing happened. The only symposium recommendation calling for action (38) died quietly after some discussion in the APS Council, with no effort, apparently, being made by the Council to find an alternate, more effective course to follow. Thus the consequences of that symposium total zero.

When the nuclear physicists reflected on what they had wrought with their Manhattan Project, little time was wasted trying to alert the public to the danger facing civilization. Not only did some band together to publish a journal with this specific task in mind (*The Atomic Scientist*), but others shouldered the responsibility of informing Congress as well as political candidates of the danger. Unless the plant pathologists do the same, they fall into the category in which the following ditty puts the scientist responsible for Germany's V-2 rocket (25):

> Don't say that he's hypocritical;
> Say rather that he's apolitical.
> Once the rockets are up
> Who cares where they come down;
> That's not my department,
> Says Wernher von Braun.

The APS is now entering its 76th year. Should it continue to approach the world food/population problem as it has in the past, a problem which the Society members helped create? Of course not!

> The ethical considerations of the uncontrolled population growth compel us to intervene on a massive scale in mankind's current demographic course as soon as humanely possible. To do otherwise would be to deny all human values. Civilization is clearly at stake. If you fail, it will perish—and our morality, more than 10,000 years in the making, will perish with it (30).

Obviously the response of the APS to such a massive, ethical challenge cannot be undertaken without inputs from the full membership. When the membership is solicited for ideas, it must be told that to be effective, no half-hearted, queasy, Pollyanna recommendations can be expected to jar public opinion; something is needed other than a simple, face-saving public relations pat-on-the-back statement. The problem is serious: it merits forceful action.

With that caveat in mind, the Society's President, with Council approval, should solicit the APS membership for suggestions on how to complete the following statement:

> WHEREAS we, as plant pathologists whose science is unique for having been born out of famine, are aware and rightfully fearful of the ominous consequences of the growing world food/population crisis;
> WHEREAS the consequences of continued population growth will be increased human misery;
> WHEREAS some nations have failed to institute measures that could lower their population growth due, in part, to confidence in the potential of agricultural research, a confidence sometimes generated by excessive claims from the scientific community;
> WHEREAS increasing food production, by lowering the death rate without

influencing the birthrate, accelerates population growth; and
WHEREAS agricultural research cannot continually keep pace with a population that doubles every 20 to 35 years:
NOW THEREFORE BE IT RESOLVED that it is the intention of the American Phytopathological Society to . . . ? . . .

To do what?

The membership, by doing nothing, is capable of speeding the destruction of our way of life, of giving civilization a helping hand on its way to the end; or the membership can take on the challenge—one worthy of a Diamond Jubilee—to help prevent it.

Which will it be?

LITERATURE CITED

1. Agency for International Development. 1969. Congressional presentation, May 29. Washington, DC. pp. 1-3, B-4, H-9.
2. American Phytopathological Society. 1979. Constitution. Page 99 in: Directory of Members. The Society: St. Paul, MN. 120 pp.
3. Anonymous. 1979. Self-sufficiency. Ceres Jan./Feb. 12:19-21.
4. Anonymous. 1981. National Agricultural Lands Study. Final report of an inter-agency study. Washington, DC. 108 pp; p. 55.
5. Barney, G. O. 1980. The Global 2000 Report to the President. Council on Environmental Quality and the Department of State, Washington, DC. 1:1-47.
6. Bartlett, A. A. 1981. Forgotten fundamentals of the energy crisis. Monograph. Environmental Fund, Washington, DC. p. 2.
7. Borlaug, N. E. 1965. Wheat, rust and people. Phytopathology 55:1097.
8. Borlaug, N. E. 1972. CIMMYT reprint and translation, series 3, Jan. International Maize and Wheat Improvement Center,
9. Borlaug, N. E. 1976. The Green Revolution: Can we make it meet expectations? Proc. Am. Phytopathol. Soc. 3:20.
10. Brown, L. 1969. House Foreign Affairs Subcommittee hearings, Dec. 5.
11. Brown, L. 1981. Building a Sustainable Society. Worldwatch Inst., Washington, DC. 433 pp.; pp. 91-93, 96.
12. Christian Science Monitor. 1976. November 8.
13. Consultative Group on International Agricultural Research. 1980. CGIAR, Washington, DC. 150 pp.; page V.
14. Eisner, T., Eisner, H., Meinwald, J., Sagan, C., Walcott, C., Mayr, E., Wilson, E. O., Raven, P. H., Ehrlich, A. and P. R., Carr, A., Odum, E. P., and Gans, C. 1981. Conservation of tropical forests. Science 213:1314.
15. Environmental Fund. 1981. Data sheet. Environmental Fund, Washington, DC.
16. Food and Agriculture Organization. 1981. Food day announcement, Oct. 16. FAO, Rome.
17. Fosdick, R. B. 1952. The Story of the Rockefeller Foundation. Harper Brothers, New York. p. 184.
18. Gall, N. 1981. The good news and the bad. Forbes Magazine, Oct. 12. pp. 194-195.
19. Grain and Feed Division, Commodity Programs FAS/USDA. 1981. Foreign Agriculture Circular (unnumbered), January 28. Washington, DC.
20. Hardin, G. 1981. Dr. Pangloss meets Cassandra. The New Republic, October 28.
21. Harrar, J. G. 1969. House Committee on Ways and Means hearings, February 19. Washington, DC. p. 248.
22. Harrar, J. G. 1969. House Subcommittee on Inter-American Affairs hearings, April 22. p. 255.
23. Hesburgh, T. 1981. Joint hearings of the Senate Subcommittee on Immigration and Refugee Policy and the House Subcommittee on Immigration, Refugees and International Law, May 5.
24. House Foreign Affairs Committee. 1981. Background papers for innovative biological technologies for lesser developed countries. Washington, DC. 511 pp.; p. 486.
25. Lehrer, T. Wernher von Braun. In: That Was the Year That Was. Reprise record number 6179.
26. Linowitz, S. M. (Chairman). 1980. Overcoming World Hunger: The Challenge Ahead. Report of the Presidential Commission on World Hunger. June. 251 pp.; appendix, 29 pp.
27. McMahon, T. 1981. Memorandum to

author. Environmental Fund, Washington, DC.

28. Maslow, A. 1966. The Psychology of Science. Harper and Row, New York. pp. 15-16.
29. Mumford, S. D. 1981. Population growth and global security. Humanist Jan./Feb. pp. 6-25; p. 11.
30. Mumford, S. D. 1981. Population growth control: The next move is America's. The Philosophical Library, New York. 167 pp.
31. New York Times. 1970. Oct. 22.
32. New York Times. 1981. Oct. 5.
33. New York Times. 1981. Editorial, Oct. 16.
34. Paddock, W. C. 1952. Can we make the world feed us all? Sat. Eve. Post, Oct. 18.
35. Paddock, W., and Paddock, P. 1964. Hungry Nations. Little, Brown, Boston. 344 pp.
36. Paddock, W., and Paddock, P. 1967. Famine—1975. Little, Brown, Boston. 286 pp.; p. 74.
37. Paddock, W., and Paddock, E. 1973. We Don't Know How. Iowa State University Press, Ames. 331 pp.; p. 217.
38. Paddock, W. C. 1976. A humanitarian response by scientists to the food/population equation: A moratorium on agricultural research and a "watchdog committee" to monitor claims of agricultural "break-throughs." Proc. Amer. Phytopathol. Soc. 3:40-46.
39. Paddock, W. C. 1981. House Subcommittee on Immigration, Refugees, and International Law hearings, Oct. 26.
40. Petersen, W. 1969. Population, 2nd ed. MacMillan, New York. 735 pp.; p. 327.
41. Sagan, C. 1981. Washington Star, June 6.
42. Snow, C. P. 1981. How the bomb was born. Discover. April, p. 71.
43. Strout, R. L. 1981. World food distribution: An unsolved puzzle. Christian Science Monitor, October 6.
44. Time Magazine. 1981. October 2, p. 19.
45. Time Magazine. 1981. October 19, p. 90.
46. Vogt, W. 1967. Letter to author.
47. Wall St. Journal. 1981. May 6.
48. Ward, B. 1970. Jacket quote on: Seeds of Change, Lester Brown, author. Praeger, New York.
49. Ward, B. 1976. The Home of Man. W. Norton, New York. 297 pp.
50. Washington Post. 1981. May 12.
51. Washington Post. 1981. October 23, p. A20.
52. Wilber, V. 1980. GAO, Hunger Commission recommendations differ. Food Monitor. July/Aug. pp. 14-16.
53. Winegrade, G. 1981. Interview. ZPG Reporter, Washington, DC. October.

CHAPTER 5

GOAL FOR PLANT HEALTH IN THE AGE OF PLANTS: A NATIONAL PLANT HEALTH SYSTEM

J. ARTIE BROWNING

My charge from the editors of this volume was to "highlight emerging problems" in plant health "that serve as a challenge to plant scientists, to indicate where we are now, if we are doing a good job," and to stress what we should be striving for. . .in 2000 AD.

First, the time frame; the year 2000 is not magic and it results in too brief a time frame for meaningful planning. The American Phytopathological Society (APS) must consider really long-range goals—not only for phytopathology, but for the entire agricultural community and the world. As the objects and beneficiaries of plant health are people, we should attempt to chart our course minimally until the time predicted for stabilization of the world's runaway population growth. Predicted by the United Nations to stabilize at anywhere from 9 to 16 billion people (18), population growth may stabilize before our Society is 150 years old (in 2058). There are many viable options after population stabilization—such as the option to seek a truly better quality of life for all of God's children; meantime, the options are few indeed while people must concentrate on feeding and clothing an ever increasing quantity of life. That "*there is not a shortage of food; there is a longage of people*" (20) is a reality against which the APS, at its Diamond Jubilee Celebration, must discuss plant health. Realizing the dramatically deceptive nature of growth by doubling (15), phytopathologists must realize that it is mathematically impossible to achieve realistic long-range goals for plant health unless their moving target—population growth—is slowed dramatically. Paddock (20, and Chapter 4 of this book) has emphasized this and justifiably challenged phytopathologists to play a heroic role in this drama. Adkisson (1), too, has stressed increasing population as the underlying cause of "the approaching crisis in sustaining high-yielding agricultural systems." In this chapter, I stress that our *minimum* response to an expanding world population must be to build sustainable agricultural systems—which then become the carrying capacity of the planet—and plant health institutions to maintain them.

Second, we must define plant health. To me, plant health is far more than the opposite of "plant disease," as used in phytopathology. Plant health describes the relative freedom of the green plant and its ecosystem from biotic and abiotic stresses that limit its producing to the maximum of its genetic potential over time.

Where are we now? Are we doing a good job? As U.S. agriculture is the most productive in the world, we must have been doing many things right. Yet I assess plant health in the United States as "insecure" and U.S. agriculture, which is based on the presumption of healthy, productive green plants, as being conducted at an unsustainable rate.

Plant pathologists must accept their proportionate share of responsibility for this situation and contribute their share, or more, toward reversing it. Reversing this situation, which is what we should be striving for, is the greatest challenge in plant health!

Before proposing a way to reverse the situation that causes my indictment of the state of plant health, let me attempt to justify my claims by considering genetic vulnerability, excessive soil erosion, and excessive energy consumption. I will use examples primarily from Iowa and the corn belt for I know them best; a quarter century of my professional life was spent there.

SOME CONSTRAINTS ON THE STABILITY OF CORN BELT AGRICULTURE

Genetic Vulnerability of Corn Belt Agriculture

Iowa—and corn belt—agriculture is Big Business, all-important to the region, our nation, and the world. But by no stretch of the imagination can it be considered a reasonably safe, secure business. Iowa's agriculture, in fact, is a very vulnerable business because it is based on a very few cultivars or resistance genotypes of two crops, corn and soybeans, and often each corn plant and each soybean plant is genetically identical to its neighbor from field to field across millions of acres. This is analogous to a tinder-dry prairie awaiting a spark to ignite it. In 1969, *before Helminthosporium maydis* race T "ignited" the 1970 southern corn leaf blight epidemic, I published (5) this statement:

> . . .diversity now largely is lacking in Corn Belt agriculture. This might not be of concern if there were assurances that the status quo could be maintained, that there were not unknown pestilences or other dangers beyond the horizon. But can the status quo be maintained?

My question was answered dramatically the following year by southern corn leaf blight that destroyed more biomass than any other plant disease epidemic in history. The *cause* of the epidemic is commonly considered to have been *Helminthosporium maydis* race T, but this is not true. This was an epidemic of our own making; the real cause was excessive homogeneity of the nation's corn crop (6). The National Academy of Science (19) tome spawned by this epidemic stated, "If uniformity be the crux of genetic vulnerability, then diversity is the best insurance against it." And Ullstrup (23) added,

> Never again should a major cultivated species be molded into such uniformity that it is so universally vulnerable to attack by a pathogen, an insect, or environmental stress. Diversity must be maintained....

The consequence to be expected from a dearth of diversity is described by Harlan (12) in an article appropriately entitled "The Genetics of Disaster."

In spite of such well-justified warnings, corn and soybeans in Iowa today are more genetically homogeneous than they were in 1970. Is it fair to ask what, if anything, we learned from the 1970 corn extravaganza? Or from similar examples on other crops, especially wheat, oats, and potatoes that, if heeded, should have prevented the 1970 epidemic? Obviously, corn belt agriculture gambles for what it hopes to gain, not what it—or humankind—can afford to lose. Is this any way to respond to a hungry world that demands not just food, but food from stable sources? Should *responsible* scientists, politicians, and other leaders allow this situation to continue?

These are not theoretical questions, for biological warnings are sounded frequently as microorganisms continually probe corn and soybeans, always seeking a more favorable (for the microorganism, in the short term) niche in the corn-soybean ecosystem. I have not space to enumerate the danger signals here. Suffice it to say that disease problems and threats of new disease problems have greatly intensified on corn, soybeans, and other crops as production has become more extensive and intensive and, especially, *as the genetic base has been made more narrow*. As the National Academy of Science (19) study on genetic vulnerability emphasized, "it is possibly not accidental that all of our serious widespread corn epidemics have occurred since 1960" when the corn industry shifted from double to single cross hybrids. Again, the solution is plain: diversify! Since we cannot divine the future and diversity is the only protection against the unknown, such as against a *future* disease-risk situation, diversity of cultivars grown in a well buffered cultural system is the *only* response that should be acceptable in a hungry world.

Excessive Soil Erosion

Much corn belt agriculture is without adequate regard for the needs of tomorrow's children. Frequently, fence-row-to-fence-row farming in response to pressure to produce exacerbates soil erosion and means that agriculture is, in effect, "eating the seed corn." This production pressure is of two types: materialistic, for today's profits, and humanistic, to help feed a hungry world. But one result, soil erosion, surely will be counterproductive.

How serious is soil erosion? Brink et al (4) reported that in 70% of 93 quarter-sections sampled in Wisconsin, estimated *soil losses averaged more than twice that "considered compatible with permanent agriculture"* (my emphasis). In Iowa, sloping land planted to corn to maximize short-term profit lost about 2 bu of top soil for each bushel of corn harvested on that land. And some soil losses exceeded that amount by three to five times! Obviously, loss of topsoil at a rate beyond the agroecosystem's ability to replenish it is mining the soil, and soil loss is incompatible with a permanent agriculture (4). Yet many corn farmers advocate massive export of corn to the USSR and corn fermentation for gasohol; they thereby advocate indirectly exporting top soil to the USSR and putting it into gasoline tanks. Small wonder the leading Iowa newspaper crusades to ban the moldboard plow. We scientists should lead in this matter by developing alternative cultural systems.

Lest soil erosion be judged an exclusive problem of the corn belt, Miller (17) wrote recently that 90% of land used for row crops and small grains in the United States was largely without soil conservation practices and that "the erosion problem on agricultural land is worse today than it was during the Dust Bowl days of the 1930s."

Excessive Energy Consumption

Typical corn belt agriculture is energy intensive. It maximizes returns per year per person, but *not* return per unit of energy input or per unit of land (24). It is doubtful whether the corn belt-type of agriculture can continue as we know it today if it is confronted with the major energy constraints many anticipate. Obviously, research on energy efficient alternatives for the corn belt is indicated and is imperative.

PROMISE OF INTEGRATED PEST MANAGEMENT IN THE CORN BELT

Integrated pest management (IPM) is a powerful concept appropriately pioneered by the entomologists because of problems experienced from excessive use of pesticides and from the development of resistance to pesticides by arthropod populations (2). It shows great promise of keeping plant pathogens and arthropod pests below an economic threshold, reducing pesticide use, protecting people and the environment, and increasing economic return to growers of multiple-pesticide-use crops such as cotton, apples, alfalfa, etc. But does it hold similar promise for corn belt crops?

In Iowa, farmers annually grow more than 20 million acres (8.1 million ha) of corn and soybeans. Herbicides are used on about 95% of the acreage for both crops. Insecticides are applied as a soil treatment to about 50% of the corn acreage, primarily for root worm control; none is applied to soybean fields. Seed treatment fungicides are applied to about 99% of the corn and 15% of the soybeans. A trace amount of each crop, primarily seed increase fields, may receive foliar fungicides. Retail value of these pesticides will exceed $300 million; thus, pesticides, too, are Big Business in Iowa and the corn belt. As might be expected, it is estimated that there are more than 4,000 pesticide salesmen in Iowa versus 119 persons, largely county extension directors, who might suggest alternative programs *if* they were available. Unfortunately, such programs generally are not available (2).

Genetic host plant resistance is the first line of defense against plant pathogens and most insects, including the first brood corn borer. In general, pesticides are the first line of defense only against weeds and certain soilborne insects, especially the root worm. Thus, the pesticides used on corn and soybeans in Iowa involve a single application, and much research will be required if alternate pest-control strategies are to be developed to circumvent that single application. IPM, then, does not promise to reduce pesticide use on corn belt corn and soybeans as on multiple-application crops like, say, cotton or apples, or corn or soybeans in the South. Several years of difficult, expensive, intensive multidisciplinary field research will be necessary before superior, integrated, energy conserving, environmentally safe cultural systems—that may even require the breeding of a different type of cultivar—are ready for extension personnel to deliver to farmers. In Iowa, I know of no one who does any real integrating short of the farmer himself, and I suspect that, again, Iowa is characteristic of the corn belt in this respect.

Thus, corn belt agriculture remains a vulnerable form of agriculture, operated by and large to maximize today's profits. It depends on the widespread use of pesticides and high yielding but genetically homogeneous cultivars.

PROPOSAL FOR A HOLISTIC NATIONAL PLANT HEALTH SYSTEM

This critical situation, as exemplified in the corn belt, demands that a long-term change be made in the philosophy by which goals are assessed and problems are approached and responded to by the agricultural community and responsible national leaders.

President Jimmy Carter, in his Environmental Message to the Congress 23 May 1977, called for "a long-term change in the way we approach this problem." I respectfully submit that no less a change than the establishment of a *national plant health system* (NPHS) is indicated.

The preciousness of human life gave impetus to the perfection, over a long period of time, of health care research, teaching, delivery, and reporting systems for people. The monetary and/or sentimental value of domesticated animals, including pets, fostered the development of a parallel system for animals. It seems that, with the increasing pressure on plants to meet human needs, it is high time we developed an NPHS for research, education, delivery (extension), and reporting to parallel those that have proven successful for our animals. That is my proposal.[1]

In the Golden Jubilee volume celebrating the 50th anniversary of the APS, Horsfall (14) wrote a provocative chapter on "the status of plant pathology in biology and agriculture." He also examined the teaching of plant pathology and proposed two graduate degrees. Those who continue to develop the *science* of plant pathology would continue to earn the Ph.D. degree. But he suggested that plant pathology participate in a new degree, possibly called Doctor of Agriculture, for those "who will practice that *art* professionally" (my emphasis).

Tammen and Wood (22) and others accepted Horsfall's (14) suggestion and wondered why the second-degree idea for the art or practice of plant pathology and agriculture was never effected. They concluded that the primary reason was "because there was no great need," due to full granaries and the low economic value of plants and their products, and that

> the need for practitioners in plant pathology will not increase until the value of the sick plant they treat increases and until the technology for treating them has been developed. This time, in fact, has arrived.

Indeed it has! Until the 1970s, plants always occurred in excess of our need and we could take them for granted, never giving them their due. But this changed strikingly in response to population pressure. In the 1970s, with no fanfare other than depleted grain stocks, rising food prices, and mountainsides depleted of firewood, we slipped into what I call the "Age of Plants." There have been many ages: the Iron Age, the Bronze Age, the Atomic Age, etc. Now we are in the most critical age of all, the Plant Age. This is because, in the Plant Age, for the first

[1] My proposal for an NPHS was presented previously to the Council for Environmental Quality, Executive Office of the President, as part of the Huffaker IPM Project Review, 15 September 1977; in testimony to the United States Senate Committee on Agriculture, Nutrition, and Forestry, Ninety-Fifth Congress, 31 October 1977; and to the Third International Congress of Plant Pathology (7).

time in history, virtually every plant has actual or potential value for food, feed, fiber, firewood, or esthetic purposes. Hence, it no longer is safe just to assume that there will be plants to meet our needs. Leaders must recognize that the basis of *any* agriculture, and therefore of civilization itself as we know it, is the healthy, productive green plant, the primary photosynthetic factory. Then, we must develop cultivars and cultural systems in which each green plant is able to yield to the maximum of its agronomic, horticultural, or silvicultural potential *over time* commensurate with energy input and impact on the environment. In the superior, sustainable cultural systems we must develop,

> the protection a given cultivar enjoys should not rest just on *each* individual plant in the population, but be characteristic of the whole population and cultural system (8).

GOAL OF THE PROPOSED NATIONAL PLANT HEALTH SYSTEM

The worthy goal of IPM—to ensure responsible pesticide usage through superior, integrated pest management and thereby protect the environment—is too narrow for the proposed NPHS. The goal of the NPHS, indeed of the United States and world agricultural communities, must be to develop sustained- (or sustainable-) yield agroecosystems (SYAEs). IPM will be a necessary strategy toward the larger goal. I will discuss the SYAE.

First, the ecosystem must be the unit. In Iowa and northern Illinois, the corn-soybean ecosystem constitutes the minimum ecosystem unit for study; corn and soybeans compete for the same land, and rotating them is an elementary but classical means of weed, insect, and disease control. The ecosystem unit in southern Illinois probably would be corn, soybeans, and winter wheat, and in Nebraska, where the *ecofallow* system was developed (9), it would be grain sorghum and winter wheat. In the Central Texas Blacklands, the ecosystem unit would be cotton, grain sorghum, winter wheat, and cattle. When and whether grazing cattle are removed from winter wheat pasture will depend on crop conditions and the anticipated relative value of beef and wheat grain. But in each case, the yield per acre per year must be a reflection of the entire ecosystem unit. That, in turn, must allow for deleterious (disease-causing) microorganisms and arthropods as well as organisms that are beneficial. All organisms belong to the ecosystem and lend themselves to management to the benefit of the SYAE.

I emphasized *sustained*-yield agroecosystems. With increasing pressure on the healthy, productive green plant to meet our needs, we plant scientists have the responsibility to make agricultural production as stable as possible—immune, if possible, from *all* sources of instability except the weather, and to do that, over time, with maximum protection of the environment and conservation of soil and energy. The world demands, and we must try to deliver, no less. An SYAE means, essentially, that a given agroecosystem can be sustained, over time, within the carrying capacity of that land and that environment. To allow the population of a country or planet, or the economy of a community, or the profit of a given farmer, to be dependent on agriculture conducted beyond the long-range carrying capacity of the land is to be irresponsible as plant scientists and government leaders and grossly unfair to the hungry (whether for food or for monetary profit) victims of such a system. Hence, the SYAE must be our goal.

For an SYAE,

> it is necessary (i) that the chemical nutrients removed by crops are replenished in the soil, (ii) that the physical condition of the soil suited to land utilization type be

maintained, which means that the humus level in the soil is constant or increasing, (iii) that there is no buildup of weeds, insects, and pathogens, (iv) that there is no increase in soil acidity, or of toxic elements, and (v) that soil erosion is controlled. All of these are essential (10).

The proposed NPHS must satisfy all of the above in its SYAE. But it is appropriate that, through IPM, immediate attention be given to the third item, buildup of weeds, insects, and pathogens. Of all of the above factors, it is plant pests that can change genetically and build up rapidly, sometimes with catastrophic results—as with southern corn leaf blight in 1970, new races of wheat or oat rusts, or pesticide-resistant mites on apple or cotton. Thus, pests are a major cause of instability of crop yields.

The strategy that stabilizes pests and contributes to an SYAE will be a stable agricultural strategy, the agroecosystem equivalent of an evolutionarily stable strategy in ecology and population genetics. Use of a parallel of the powerful, well-recognized ecological concept of an evolutionarily stable strategy to build a stable agricultural strategy—and through it an SYAE that, itself, parallels the powerful ecological concept of carrying capacity—emphasizes that the SYAE is a scientifically sound concept, a worthy goal of the NPHS, and necessary to provide relief for a hungry world.

Thus, pest management is necessary for an SYAE. But for the proposed NPHS to concentrate on pest management and ignore other causes of yield instability would be a major mistake. As emphasized earlier, it is theoretically possible to stabilize pests at a low population in an agroecosystem that cannot be sustained because of soil erosion. Obviously, loss of topsoil at a rate beyond the agroecosystem's ability to replenish it is incompatible with an SYAE.

My examples should not cause one to conclude that the SYAE need is peculiar to the corn belt, however. Greenland (10) recognized the SYAE need in developing countries. Certainly it is needed in other cropping areas of the United States. For instance, cotton yields have been decreasing in the Mississippi River Delta since 1965 even though the genetic potential for yield increase has continued its upward trend (3). Seemingly, management of the cotton *ecosystem*, not the cotton genotype, limits achievement of yield potential of the cotton cultivar. Possibly the comments of Cotton Extension Specialist James Supak, cited by Spencer (21), relative to the Texas High Plains, apply:

> I think a lot of the declining yield problem in this area goes back to the fact that soil productivity has been reduced by continuous cropping to cotton. Because cotton is a low residue crop, the physical condition of soils continuously cropped to cotton has deteriorated. As a result, we see more problems with hardpans, poor water filtration, and wind and water erosion. Also, continuous cotton production has aggravated our disease problems especially with *Verticillium* wilt and the *Fusarium* wilt-root knot nematode complex. We're rapidly getting to the point where continuous cotton is jeopardizing our soil resources....These soils can't take too many more years of continuous cotton and still maintain a reasonable level of productivity.

Obviously, SYAEs are as imperative for the cotton belt as for the corn belt.

PROPOSED INSTITUTIONAL CHANGES

With the additional pressure on plants to meet the ever-growing needs of an expanding world population, leaders must examine the institutions that deal with plant problems to ensure that an adequate plant health research, teaching, delivery, and reporting system can be developed and maintained.

I assess the present institutional system as being adequate to *evolve* into the proposed NPHS. The several relevant federal agencies are, in general, excellent. The National Science Foundation is the best means yet devised to fund scientific research, especially basic research. And the United States Department of Agriculture-Cooperative States Research Service (USDA-CSRS) has added an excellent granting capability to its formula funding of long-term agricultural research projects. But crops grow and are managed in the states, and students grow up and are educated there. For over a century, the nation's land grant university system has been a world model for research and education in agriculture. Thus, backstopped by federal funding, research, and regulatory agencies, I see the nucleus of the NPHS as being the present land grant university system, slightly modified.

THE PROPOSED NATIONAL PLANT HEALTH SYSTEM

A USDA assistant secretary of agriculture for plant health should oversee or direct the entire NPHS. USDA plant-health-related activities would come under the jurisdiction of this office. Also, through the USDA's CSRS, Agricultural Research Service, and Extension Service, the assistant secretary would exert a strong influence on the land grant university agricultural experiment stations and the extension services. The singular importance of plants to our well-being more than justifies their receiving this special, but belated, emphasis.

I visualize the NPHS as operating in four modes: research, delivery (extension), education, and reporting.

The Research Mode

In each state, this arm or division of the state agricultural experiment station (AES), in cooperation with the CSRS, the Agricultural Research Service, the National Science Foundation, the Environmental Protection Agency, etc., should fund, coordinate, and oversee all research by AES scientists in the area of plant health—or protection from pests (plant pathogens, arthropods, nematodes, weeds), fertility imbalance, water stress, etc.—including a major effort to build SYAEs suited to the soil type, climate, state, and region.

The division would require a director for plant health, who could be an associate director of the AES responsible to the AES director.

Staff of the division would include all plant pathologists, all plant entomologists, and all weed scientists in the AES. Scientists in plant breeding, crop production, soil science, field crops, horticulture, forestry, genetics, botany, biochemistry, agricultural engineering, meteorology, statistics, systems science, computer science, economics, etc., must be integral parts of the Plant Health Division team and would be included in proportion to the amount of time they devoted to plant health and the goals of the division. I think that departmental units should remain intact but that transdisciplinary efforts should be integrated and coordinated at the director's level.

The goals of this AES division would be 1) to develop cultivars and cultural systems for growing them so that the cultivar and its ecosystem would yield the maximum of their genetic potential over time, commensurate with energy input and impact on the environment—in short, to develop SYAEs, and 2) to generate basic knowledge of plant-pest-environment interactions to place and keep plant health on a sound scientific footing.

The Education Mode

Many universities currently offer curricula in pest management leading to the B.S. or M. S. degrees. This is good; it should be continued to meet the need for people educated at these levels. There will be increasing demand for such people as pest management becomes operational on more crops and areas. A corps of scout- or observer-reporters such as I propose in the next section—now used on only a few crops and in a limited way—will require many B.S. graduates from pest management curricula if the corps is utilized as I visualize and predict.

To really meet the increased demand for health care of cultivated plants, however, colleges of agriculture must plan to establish degree programs to parallel those that now lead to the degree Doctor of Veterinary Medicine. The new *professional* degree might be called "Doctor of Plant Health" (D.P.H.) or, simply, "Plant Doctor."

Persons aspiring to become plant doctors and deal as general practitioners with the *total* needs of the crop would earn a B.S. degree in one of the plant sciences, much as in the preveterinary program. Then they would enter a fairly uniform, highly structured and rigorous curriculum designed to train and educate them to become *professional* plant doctors. The structured curriculum would be rich in clinical experience, which would give it diversity; a final internship would be a necessary, integral part of the program to launch plant doctors as professional practitioners.

Professional plant doctors could go into private or socialized practice. I visualize that all privately employed general practitioners, most plant agriculture extension specialists, many area and county agents, many agribusiness personnel, and not a few growers, will want—need—the D.P.H. degree.

Some plant doctors will want to go into teaching and/or research, and for this they should obtain *academic* degrees in addition to their *professional* D.P.H. degrees. Thus, after earning the B.S. and D.P.H. degrees in the college of agriculture, they would enter the graduate college to pursue M.S. or Ph.D. degree programs in any of several relevant academic areas, much as animal scientists do today in the academic disciplines that backstop veterinary medicine.

Others have made similar proposals. Horsfall (14) clearly saw the need to stop giving the same treatment to those who would pursue the science and those who would practice the art of agriculture. He stated that

> All who survive. . .efforts to turn out professional artists and professional scientsts. . . obtain a Ph.D., and, having been squeezed through the same template, all look alike.

Also, he visualized the Doctor of Agriculture degree as preparing one to deal with the general problems of production agriculture, including those of applied plant pathology.

Tammen and Wood (22) also visualized a general agricultural degree at the M.S. level; but for the highest level of professional advancement for the aspiring practitioner, they proposed a Doctor of Plant Medicine degree, which would prepare the practitioner to deal only with diseases. Parallel degree programs would be required for entomology, soil science, agronomy, horticulture, etc. They considered that problems associated with plant agriculture are more complex than those encountered by M.D.s or D.V.M.s and too complex for any one person to prepare to deal with their whole breadth. Plant agriculture *is* very complex, but farmers deal with that complexity every day! The specializations of Tammen and Wood's (22) series of doctor's degrees would leave only the farmer to integrate pest management, just as today. I am convinced that, with superior curricula and instruction, capable students can be educated effectively as general practitioners in plant health. Such Doctors of Plant Health probably should specialize only to the extent that they would be better prepared to deal with agronomic crops, horticultural crops, or glasshouse crops. Also, climatic regions require them to deal with, say, citrus belt, cotton belt, or corn belt crops. Otherwise, they should be educated as generalists, parallel to the general practitioners among medical doctors or veterinarians, and able to deal with most plant health problems. It goes without saying, however, that such general practitioners would be backstopped by specialists, just as in human and veterinary medicine.

The concept of "phytomedicine" is a relatively old one in Germany. Grossmann (11) gave its history and a well-reasoned argument to justify it. He called for "the plant doctor with a full training in all fields of phytomedicine and its neighboring disciplines." Merrill (16) also called for a Doctor of Plant Medicine degree. Clearly, Horsfall's Doctor of Agriculture, Grossmann's Plant Doctor, Merrill's Doctor of Plant Medicine, and my Doctor of Plant Health degrees are all proposals for interdisciplinary training of general practitioners. My only objection to the reasoning of Grossmann (11) and Merrill (16) is in the title. For people and their animals, "medicine" suggests the prescription of medicinal drugs as the control of choice for disease. But the controls of choice for plant diseases are use of host plant resistance and cultural systems to maximize natural control processes. Therefore, plants are different. By manipulation of the plant's genetic system and components of the ecosystem through cultural practices, I am convinced that SYAEs with a high level of plant health can be developed with a minimum dependence on chemical pesticide "medications" as alternatives. Hence, the name I prefer is Doctor of Plant Health. That stresses the positive result desired by the farmer and a hungry world, not any one group of causal agents or any single dominant means of treatment.

The Delivery Mode

As Horne (13) has emphasized, extension is "the face of plant pathology." Unfortunately, extension is the face also for each of the other agricultural sciences and there is no general agricultural practitioner to whom a farmer can turn for initial help in choosing among them or for reference to a specialist. If a member of a farmer's family becomes ill, he can go first to a family doctor. If there is an animal problem, a local veterinarian, a generalist, is available. Either can help directly or make reference to specialists as indicated.

But if the farmer has a plant problem, the first contact normally is with a county agricultural agent. The probability is that this person has a B.S. degree in animal science or agricultural economics (generally the two most popular curricula in colleges of agriculture) and is ill equipped to deal with plant problems. Certainly the agent is not prepared professionally as are the farmer's contacts in human and animal health. I know of only two states that have a single, unified clinic to diagnose all plant problems! Thus, the farmer or the county agricultural agent must first decide whether the problem probably involves a plant disease, insect, weed, or other cause to know which clinic and specialist to contact. Meantime, valuable time is lost and frustration is experienced.

This system served American agriculture well until the 1960s. Then, interactions of crop, environment, pest, and *multiple* pesticides became common; few specialists could deal with them; and generalists were unavailable. It seems to me that, in the Age of Plants, this is a less than satisfactory way to organize the delivery of information vital to the health of the nation's multibillion dollar agricultural industry, which is based on the healthy, productive green plant!

The viable alternative is to educate *professional* plant doctors who will be well prepared as generalists and who, backstopped by specialists, can deal with any plant problem.

The analogy to human and animal doctors describes roughly the way the new plant doctors might work into extension programs in many states. In other states with more specialized crops, they would be part of more advanced, computer-linked on-line pest management programs. Central to the superior plant health delivery system would be the state's central staff of plant health extension and management specialists. Coordinated by a plant health delivery system director (who probably would be an associate director of the Cooperative Extension Service), specialists in all plant health disciplines would coordinate their work and operate a unified plant stress diagnostic laboratory. They would cooperate closely with the AES Plant Health Division research staff. Field information and problems would be given to them by county agents and by a corps of trained scout- or observer-reporters. When anyone in the system found some pest believed to be "new," it would be reported to the USDA's Animal and Plant Health Inspection Service (APHIS) personnel. Otherwise, the observer-reporters would report pest intensity, plant stress, and other crop condition information to plant health extension and management specialists and to the USDA's statistical specialists, described in the next section. Aided by their diagnostic laboratory and by appropriate check plots about the state, they, pest management (PM) consultants, or the farmer would "write the prescription" for that field or crop. The better educated farmers probably would choose to write their own prescriptions, seeing them as basic to their management decisions. It probably would be preferable for the publicly supported extension and management specialists to restrict themselves to advising on probable consequences of taking different management alternatives and for the farmers or their consultants to actually write the prescriptions.

The observer-reporter corps, the county agents, the central staff of extension and management specialists, and anyone else who "writes the prescription" should be *independent* of any commercial interests and beholden to no one. They must follow only high professional ethics. It is no accident that medical doctors write prescriptions and pharmacists sell drugs!

The Reporting Mode

The USDA's Statistical Reporting Service, Crop Reporting Board (SRS-CRB), in cooperation with each state's Crop and Livestock Reporting Service, now estimates *actual* crop yield, both during the growing season and after harvest. Thus, the SRS-CRB considers reductions from some *potential* yield when they estimate *actual* yield. Estimating *loss* from potential yield and partitioning the *cause* of that loss must be handled with precision and credibility equal to or greater than that of estimating *actual* yield.

The science of crop loss assessment is in its infancy. A USDA-CSRS ad hoc committee is working to develop a national crop loss assessment system. This effort involves stimulating research to develop crop loss assessment as a science and to improve delivery of the information.

As reliable methods of estimating losses from crop-stress interactions are developed, they will have to be made operational, starting with a corps of trained observer-reporters who actually cover the state and see the crops. The corps should be educated in PM and/or be closely overseen by the state's plant health extension specialists, who work closely with the plant health research teams. The state's unified diagnostic laboratory would help partition causes of losses. Check plots in different environments in each state (*not* commonly used at present) should be used to help determine 1) identity, intensity, and prevalence of pests and other stresses for management purposes and 2) amount of loss partitioned by cause for the SRS-CRB.

The corps of observer-reporters will require training and have responsibilities that go well beyond that of persons used by the SRS-CRB now. It is nonsensical to have a corps of trained observers for the SRS-CRB and not make use of their observations in management and vice versa. The time may come when the law says that certain pesticides may not be applied on a large scale until and unless such trained, independent observers confirm their need. Thus, the *same* trained, competent observers will feed information they generate 1) to the extension staff, PM specialists, and farmers themselves for management purposes; 2) to USDA-APHIS if something new is discovered; and 3) to USDA-SRS-CRB for yield and loss estimation both during the growing season and after harvest. Reporting would be done through county, state, and national computer linkages. The estimated *biological* yield of the crops and of *biological* losses therefrom will be reported by the SRS-CRB. The USDA Economics, Statistics, and Cooperatives Service will interpret this *economically*.

As emphasized earlier, the observer corps must be independent and have adequate checks and balances. It must report losses and yield estimates through the SRS-CRB subject to the usual SRS-CRB "lock-up" system. The commodities market, the farmer's income, and pesticide sales are far too sensitive for reporting to be done otherwise. Thus, as for the "prescription writer," the corps of reporters must be beholden to no one and subject to a strong sense of professional ethics.

CONCLUSIONS

The world is in "The Age of Plants," in which, for the first time in history, there may not be sufficient plants to meet the food-feed-fiber-firewood-esthetic needs of humankind. Leaders must recognize, finally, that the basis of any agriculture is the healthy, productive green plant, the primary photosynthetic factory, and give plants their due. A major change in the philosophy of assessing goals and approaching and responding to plant problems is indicated for the entire agricultural community. For the United States, I propose a holistic national plant health system (NPHS) for research, education, delivery, and reporting. The goal of the NPHS will be to develop sustained-yield agroecosystems (SYAEs); its focus will be the healthy, productive green plant; IPM will be a major strategy. Research will develop SYAEs. Education will produce subprofessionals and a new professional Doctor of Plant Health for delivery (extension). For reporting, a new corps of trained observer-reporters will cover the crop and report crop condition and pest incidence for management, new pests for USDA-APHIS, and data for USDA-SRS-CRB to make crop yield and loss estimates, partitioned by cause. The federal government and the land grant university system should take cooperative action to initiate this proposed NPHS as the umbrella program for plant health-related activity in the United States.

LITERATURE CITED

1. Adkisson, P. L. 1981. The approaching crisis in sustaining high-yielding agricultural systems. Plant Dis. 65:940-942.
2. Bottrell, D. R. 1979. Integrated Pest Management. Council on Environmental Quality, Washington, DC. 120 pp.
3. Bridge, R. R. 1982. Rates of gain in cotton yields in Mississippi. Proc. Beltwide Cotton Prod. Res. Conf. Nat. Cotton Council, Memphis. p. 76. (abstr.)
4. Brink, R. A., Densmore, J. W., and Hill, G. A. 1977. Soil deterioration and the growing world demand for food. Science 197:625-630.
5. Browning, J. A. 1969. Disease consequences of intensive and extensive culture of field crops. Spec. Rep. 64. Iowa Agric. Home Econ. Exp. Stn., Ames. 56 pp.
6. Browning, J. A. 1972. Corn, wheat, rice, man: Endangered species. J. Environ. Qual. 1:209-211.
7. Browning, J. A. 1978. Responsibilities of plant scientists in "the age of plants." Page 398 in: Abstracts of Papers. 3rd Int. Cong. Plant Pathol., 16–23 August 1978. Am. Phytopathol. Soc., St. Paul, MN. 435 pp.
8. Browning, J. A., Simons, M. D., and Torres, E. 1977. Managing host genes: Epidemiologic and genetic concepts. Pages 191–212 in: Plant Disease: An Advanced Treatise, Vol. I. J. G. Horsfall and E. B. Cowling, eds. Academic Press, New York. 465 pp.
9. Doupnik, B., Jr., Boosalis, M. G., Wicks, G., and Smika, D. 1975. Ecofallow reduces stalk rot in grain sorghum. Phytopathology 65:1021-1022.
10. Greenland, D. J. 1975. Bringing the green revolution to the shifting cultivator. Science 190:841-844.
11. Grossmann, F. 1971. The concept of phytomedicine. Indian Phytopathol. 24:247-257.
12. Harlan, J. R. 1972. Genetics of disaster. J. Environ. Qual. 1:212-215.
13. Horne, C. W. 1981. Extension: The face of plant pathology. Annu. Rev. Phytopathol. 19:51-67.
14. Horsfall, J. G. 1959. A look to the future—The status of plant pathology in biology and agriculture. Pages 63–70 in: Plant Pathology: Problems and Progress—1908–1958. C. S. Holton, G. W. Fischer, R. W. Fulton, H. Hart, and S. E. A. McCallan, eds. Univ. Wis. Press, Madison. 588 pp.
15. Meadows, D. H., Meadows, D. L., Randers, J., and Behrens, W. W., III. 1972. The Limits to Growth. Universe Books, New York. 205 pp.
16. Merrill, W. 1981. The doctor of plant medicine. Pages 385–387 in: Proc. Symp., 9th Int. Congr. Plant Prot., Washington, DC, 5–11 August, 1979. Vol. I, Plant Protection: Fundamental Aspects. T. Kommedahl, ed. Burgess Publ. Co., Minneapolis. 411 pp.
17. Miller, F. P. 1981. Soil erosion: Is there a way to stop it? Crops Soils 34(3):5-7.
18. Muller, R. 1976. World trends in population, resources, and relevant political leadership. Proc. Am. Phytopathol. Soc. 3:2-5.
19. National Academy of Sciences. 1972. Genetic Vulnerability of Major Crops. Washington, DC. 307 pp.
20. Paddock, W. C. 1976. A humanitarian response by scientists to the food/population equation: A moratorium on agricultural research and a "watchdog committee" to monitor claims of agricultural "breakthroughs." Proc. Am. Phytopathol. Soc. 3:40-46.
21. Spencer, W. 1982. We're wearing out our land. Cotton Grower 18(1):40.
22. Tammen, J. F., and Wood, F. A. 1977. Education for the practitioner. Pages 393–410 in: Plant Disease: An Advanced Treatise, Vol. I. J. G. Horsfall and E. B. Cowling, eds. Academic Press, New York. 465 pp.
23. Ullstrup, A. J. 1972. The impacts of the southern corn leaf blight epidemics of 1970–71. Annu. Rev. Phytopathol. 10:37-50.
24. Wortman, S. 1976. Food and agriculture. Sci. Am. 235(3):31-39.

PART II

Nature of Losses

...if we are to advance the science of plant pathology and the art of treating disease, we must be able to express the amount of sickness in quantitative terms.

—K. Starr Chester, 1959

Plant diseases were initially studied because of the losses they cause, yet today it is paradoxical that there are only a few reliable estimates of loss.

—W. Clive James, 1974

Most losses, whether prior to or after harvesting, remain hidden or poorly quantified.

—Georg Borgstrom, 1981

CHAPTER 6

NATURE OF CROP LOSSES: AN OVERVIEW

C. E. MAIN

NEED FOR DATA ON CROP LOSSES

Crop Losses

Farmers and scientists undoubtedly recognized the phenomenon of crop loss long before the science of plant pathology was recognized late in the 19th century. Today we still are attempting to understand, define, and measure the impact of such losses upon crop production and agricultural productivity. Crop losses from infectious plant diseases are a result of botanical epidemics of one type or another. Since epidemics are necessarily complex, involving host, pathogen, weather, and human activities in time and space, an understanding of the resulting loss in food, fiber, or other plant products as an accumulative process is also complex. Perhaps one of the greatest impediments to quantifying losses has been the predilection for end-of-season counts and yield measurements. The processes resulting in crop losses are dynamic, just like the epidemics that bring them about. Since 1960 great strides have been made in quantitative epidemiology. It is only natural that there now follows an emphasis upon quantifying these effects.

Renewed Interests in Crop Losses

Coincidental with the advances in botanical epidemiology, the need for accurate, reliable data on crop losses became increasingly apparent. The concept of pest management (vis-à-vis disease management) has fostered interdisciplinary concerns about the manner in which multiple pests, together with other factors, interact to affect the desired end product, i.e., the crop produce. It became apparent that timely and reliable measures of yield/quality accumulation through time (i.e., a growing season or several years) would be necessary to monitor pest constraints if decision criteria for protective actions were to be formulated (6,8). Some contend that such an approach represents disease "loss management" rather than the more restrictive "pest management" (1). More about terminology later.

World Food Shortage

The much discussed world food shortages, both realized and anticipated, made us look closely at the quantity of produce lost to pests each year. These losses have traditionally been accepted as a matter of course, and our agricultural system adjusted to them. We realize that production can be increased simply by preventing some of this loss. How did we accomplish this trade-off? Largely by the increased use, sometimes excessive, of energy-consuming chemical pesticides. This strategy worked rather well particularly in the developed countries, but at the risk of environmental pollution.

Constraints Imposed by Pesticide Restrictions

That brings us to a third factor that demanded timely quantitative data on losses. Federal regulations, mainly imposed by the U.S. Environmental Protection Agency made it mandatory, in some cases, to document production and economic benefits to obtain labeling of new pesticides. In the flurry of activity surrounding this mission, almost everyone including crop oriented scientists, federal and state agencies and private companies soon realized that the necessary concepts, principles, and methods to quantify crop response, i.e. crop losses, either did not exist or were lacking. By 1960 we had been in business as an agricultural society for more than 50 years and still we could not estimate crop loss with any degree of reliability.

WHO WILL MEET THIS CHALLENGE AND WHAT WILL MOTIVATE THEM?

Lack of Qualified Crop Loss Specialists

Plant pathology has done a good job of educating and training competent agricultural scientists. Other crop and biological disciplines have done likewise. In the 1960s a new type of practitioner began to emerge, i.e., the pest management specialist, who, by the very nature of the perspective and goals of this specialty, quickly recognized the need for quantitative data on crop losses. However, because of the breadth of knowledge and range of activities required, the pest manager became a user rather than a generator of crop loss data. A relatively small group of pest-oriented economists, in reality social scientists, also recognized the lack of good data on crop losses necessary for assisting crop consultants in formulating decision making criteria, i.e., economic thresholds, optimization parameters, etc. (1). Obviously, economists were not in a position to generate the necessary data themselves and have relied mostly on survey techniques to poll the expert opinions of crop specialists. Sometimes "hard" data were available, but more frequently only educated guesses were provided.

Who Will Take Up the Challenge?

Who will take up the challenge of developing crop loss as a subdiscipline? At most, plant pathology departments teach the concepts and methods of crop loss estimation at an introductory level involving perhaps a single lecture that includes loss statistics from the existing literature. Research projects and proposals contain loss data to justify plant pathology as an important subject matter area for funding new programs. Our European colleagues recognized the importance of teaching crop loss some years ago and they developed formal coursework and graduate programs accordingly (10). Recently, several U.S. plant pathology departments have instituted courses in quantitative epidemiology that include valuable sections on disease assessment and crop loss methodology. To my knowledge, separate courses on crop losses do not exist yet, but several are being planned.

Until such time as specific educational programs are developed to deal with this paucity, it remains for present day scientists to "retrain" themselves, using as a base several excellent references that have appeared as a result of seminars and workshops on crop losses held both in the United States and in several foreign nations (3,6,9). The Food and Agricultural Organization (FAO) of the United Nations took a bold lead in this area, starting with a very important crop loss conference in Rome in 1967. This conference brought together scientists who set the stage for an expanded, worldwide effort to stimulate interest and research on the subject. In the United States, the American Phytopathological Society and its counterparts in nematology, entomology, and weed science became active in this cause. Internationally, the International Society of Plant Pathology has become involved through its Committee on Crop Losses and Production Constraints.

The Search for Those Concerned with Crop Losses

Starting in 1977, the UDSA/SEA/Cooperative Research funded four pilot projects on crop losses. A Current Research Information System literature search of station projects in the 13 U.S. southern states revealed that only two of more than 400 projects in plant pathology, entomology, nematology, and weed science listed crop loss by title, objective, or key word. This type of accounting could be misleading, however, since many projects failed to mention crop loss per se even as a key word although many were involved in measuring crop response (yield or quality) in relation to pests, pesticide application, or pest management variables.

WHAT ARE CROP LOSSES?

The Nature of Losses

Crop losses are the result of changes in structure, organic existence, or condition of a crop to an extent that restoration of yield and/or quality is irreversible (5). Disease loss can be considered as a biological phenomenon or a social problem or both. Increasingly, development, selection, and implementation of control strategies must expand to include short- and long-term disease loss management as a component of the broader field of crop management. Zadoks and Schein (11) suggested a classification of losses (Table 1) that describes the complexity and interdependence of loss at all strata of society.

Terminology

As with any subdiscipline within plant pathology, communication about crop losses derives strength from a set of terms that are both relevant and useful. I have chosen to quote directly from Zadoks and Schein (11), who support the yield and crop loss terminology proposed by the FAO.

> A *crop* is a unit of plants grown to provide food, fiber, stimulants or other plant products. *Yield* is the measurable produce of a crop. *Crop loss* is the reduction of either quantity and/or quality of yield. *Crop damage* is the term used to indicate injuries by *harmful organisms* that collectively result in a measurable loss in yield. Activities of harmful organisms with little effect on growth or visual appearance and no consequent effects on yield are referred to as apparent *crop injury*.

The FAO has recently proposed an expanded glossary of crop loss terms that are being reviewed and debated in hallways and classrooms around the world. We need such a glossary to aid us in our understanding of crop losses.

The Concept of Yield Loss Level

Perhaps one of the most useful explanatory descriptions available relating pest effects to yield, or more specifically yield level, was proposed by the FAO and is shown in Fig. 1. This representation is surprisingly clear and teachable in its simplicity. It provides a holistic overview, which is necessary when introducing and studying the subject of crop losses, particularly in relation to control strategies and economic impact. Not everyone agrees with the details of the diagram (see Chapters 7 and 8 of this volume), but it serves very well to stimulate discussion among students and crop scientists. Zadoks and Schein (11) do a superb job of explaining the rationale of the diagram. In essence, it encompasses pest effects from the extremes of primitive to theoretical yield levels, emphasizing the differences between attainable and desirable (economic) yields. Crop loss then, by definition, is the difference between actual and attainable yield. However, these authors are careful to point out that the elimination of economic loss can be achieved only when economic boundaries—the size of losses to be prevented—are known. Thus, we come to the purpose or goal of the crop loss specialist, to quantify the size of the crop loss.

TABLE 1. Classification of actual crop losses in the economic and social sphere as affected by plant disease[a]

Direct loss[b]		
Primary[c]	Secondary[d]	Indirect loss[e]
Yield	Contamination of sowing and plant material	Farm
Quality		Rural municipality
Costs of control	Soilborne diseases	Exporters
Extra cost of harvesting	Weakening of trees by premature defoliation	Traders (wholesale and retail)
Extra cost of grading		Consumers
Costs of replanting	Costs of control	Government
Loss of income due to less profitable replacement crops		Environment

[a] After Zadoks and Schein (11).
[b] Losses sustained by the producer.
[c] Losses of yield, quality, or wages as a direct consequence of plant disease appearing before or after harvest.
[d] Loss of future production capacity.
[e] Losses in social sphere, notwithstanding more or less successful disease control.

Crop Loss Estimation

Numerous, and many times excellent, methods have been published on disease assessment. Frequently they are buried in the text of articles on other topics. Efforts are being made by FAO, USDA, and crop loss committees of the APS and other professional societies to collect and collate lists of these references. Assuming that a disease can be diagnosed properly, disease assessment is generally the determination of the amount of disease damage. To approach the above stated goal, and depending upon the magnitude of the problem (field, county, region, etc.), the process must be taken a step further. By means of a carefully designed sample survey, the incidence and intensity of disease damage can be determined. These data, together with valid, empirical yield loss models can then be used to expand crop loss estimates to larger geographical units. *Estimation* is an approximate calculation submitted by trained specialists prepared to undertake such a task. These estimates can then be used to develop a *prognosis*, i.e., a forecast or probable course to a terminal effect, rather than a *prediction,* which implies greater precision of calculation. In situtations where crop losses are estimated without attention to the above steps, i.e., opinions of authorities, the estimates represent *guesses* either at random or from admittedly uncertain evidence. We are still sorting out the terminology to accompany the evolving concepts, principles, and methods of crop losses.

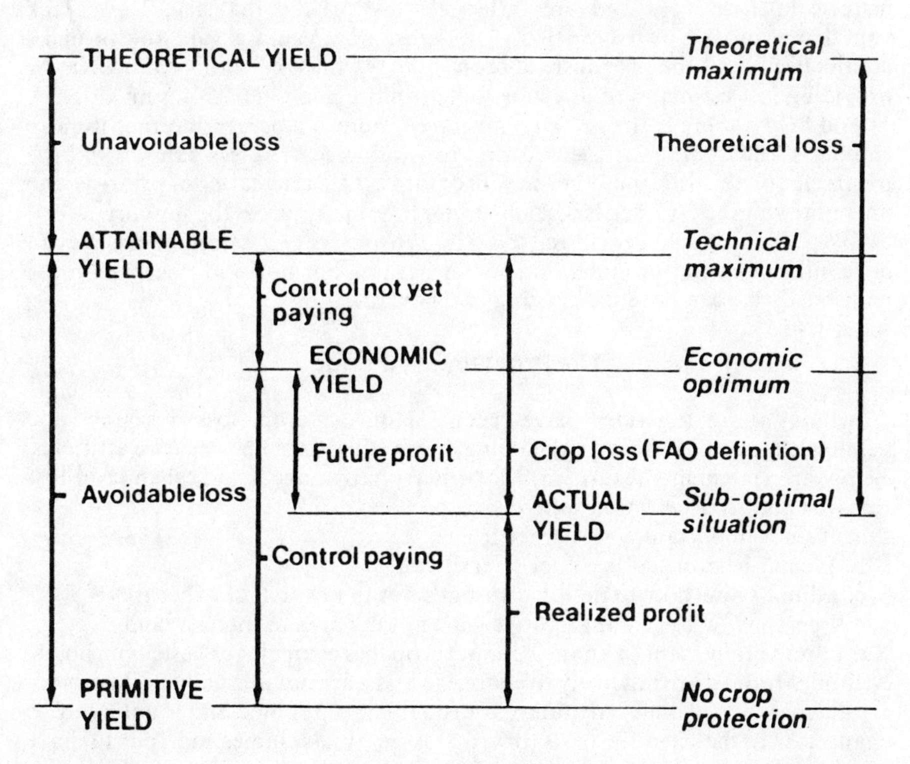

Fig. 1. Yield levels used to explain the conceptual basis for crop losses (after Zadoks and Schein, 1979).

THE SOCIOLOGY OF CROP LOSS ESTIMATION

The Problem of Overestimation

One of the most severe criticisms aimed at crop loss specialists is that of overestimation. Are we guilty as charged? In my opinion we are to some degree. There are probably several reasons, which are both sociological and technical. All too frequently overestimation is practiced to justify our vested interests for recognition and support. We may sometimes overestimate and overplay our relative importance to agriculture and society in general. This is not necessary because the real losses are sufficiently important to justify the need for plant pathology.

Although scientists are conservative, in practice they sometimes seem to assume the worst. In gaming theory this is referred to as the *mini-max principle*; play your cards in a manner expecting the greatest possible loss. Many times the rewards are found in actually playing the game rather than in emphasizing the results.

Another problem is a general lack of appreciation for the dynamics of *risk aversion* (1). We are dealing with a unique group of entrepreneurs, the farmers, who are making economic decisions based upon fewer and less accurate production records than most shopkeepers, insurance salesmen, etc. Numerous integrated pest management (IPM) studies have shown that farmers' decisions to control plant diseases vary greatly but that they are intuitively based upon sad historical experiences and are influenced by private industry. If we lack confidence in our own crop loss estimates, how can we gain the farmer's confidence about the prognosis of crop loss? We must earn this confidence by providing loss estimates of direct as well as indirect use in decision making.

And lastly, there is the *priority paradox*. There is a perception that if more emphasis and support is focused on crop loss research, less support will be available for existing and other new programs. This is a matter of priority, and only time and society's needs will tell us into how many pieces the support pie can be divided. This might be referred to as the *either-or* vs *and* syndrome (7). We can have either existing programs or new programs but not both; perhaps more realistically we can have the existing *and* new programs.

The Problem of Attitude

Although great strides have been made in crop loss research and acknowledgment has been forthcoming, there still exist some negative attitudes. Below are statements (exerpted) that I personally have heard when crop loss research and its application were discussed:

a. It's not important, at least not to me;

b. It can't be done, at least not scientifically;

c. I don't know how to do it because it's not in my area of expertise;

d. I personally don't want to do it—it's not my area of interest; and

e. I don't really want to know the exact crop loss estimate for vested purposes.

Attitudes b and d are not likely to change; a and c are not mutually exclusive but are amenable to change. Attitude e is probably deep rooted and is not likely to change. As to the scientific basis for crop loss methods, James and Teng (4) have provided the most comprehensive and detailed description of loss methodology since K. Starr Chester's famous article in 1950 (2).

Several factors have contributed to this negativism. We must address these issues if crop loss is to achieve its logical role in crop management. Let's face it; some preposterous overestimates have been collated and published in past years (e.g., >100%). We must do better. With noted exceptions, of course, loss data are usually used to fill reports and summaries loosely used for justification purposes. Far too little precise data have been directly applied to farmers' needs in making decisions. Perhaps the data demands of present day IPM programs and the relatively new area of systems agriculture with its optimization approaches will improve the situation (8). Lastly, crop loss research is difficult, tedious work not always rewarded in a manner similar to other "apparently more productive" areas of plant pathology. This problem is slowly correcting itself as increased pressures for crop loss data become more pronounced.

SUMMARY

Quantification and understanding of crop loss and its impact is important. The data demands of IPM, crop management, and the systems approach to agricultural production can only be met by increased emphasis and research on crop losses. This represents one of the most interesting and challenging areas of plant pathology for the future. Students are becoming more concerned with this topic as graduate coursework begins to include material on the concepts, principles, and methodologies of assessment and crop loss estimation. The crop loss specialist of the future will need a good background in diagnosis, quantitative epidemiology, survey sampling, statistics, data processing, and microeconomics.

This chapter has attempted to define crop loss, examine some forces that resulted in renewed emphasis on crop loss research, and describe challenges that lay ahead. The chapters that follow in this section, written by two practicing specialists, go into much greater detail on identification, assessment, and some of the factors influencing crop losses. We hope that this set of three chapters provides an overview and enough detail to stimulate your further study of this important and fascinating area of plant pathology. For those who say it cannot be done, we have provided documentation to show that it can be done. We challenge those who say they do not know how to estimate losses or that it cannot be done scientifically to join us in working together toward a common goal. To those who will support us, we accept the challenge to provide the data so needed, yet so scarce. We can only tell it like we see it.

LITERATURE CITED

1. Carlson, G. A., and Main, C. E. 1976. Economics of disease-loss management. Annu. Rev. Phytopathol. 14:381-403.
2. Chester, K. S. 1950. Plant disease losses: Their appraisal and interpretation. Plant Dis. Rep. Suppl. 193:190-362.
3. Close, R. C., Harvey, I. C., Sanderson, I. C., Gaunt, R. E., McCully, A. J., and Teng, P. S., eds. 1978. Epidemiology and Crop Loss Assessment. Proc. Aust. Plant Pathol. Soc. Workshop, Lincoln College, Canterbury, N.Z.
4. James, W. C., and Teng, P. S. 1979. The quantification of production constraints associated with plant diseases. Appl. Biol. 4:201-267.
5. Main, C. E. 1977. Crop destruction—The raison d'être of plant pathology. Pages 55–78 in: Plant Disease, An Advanced Treatise. Vol. I. How Disease is Managed. J. G. Horsfall and E. B. Cowling, eds. Academic Press, New York. 465 pp.
6. Norton, G. A. 1976. Analysis of decision making in crop protection. Agro-Ecosystems 3:27-44.
7. Salk, J. 1973. The Survival of the Wisest. Harper & Row Publ., New York. 124 pp.

8. Teng, P. S., and Gaunt, R. E. 1981. Modeling systems of disease and yield loss in cereals. Agric. Systems 6:131-154.
9. Teng, P. S., and Krupa, S. V. eds. 1980. Crop Loss Assessment. Proc. E. C. Stakman Commemorative Symp. Misc. Publ. 7. Minn. Agric. Exp. Stn., St. Paul. 327 pp.
10. Zadoks, J. C. 1974. Teaching botanical epidemiology at the Agricultural University, Wageningen. Neth. J. Plant Pathol. 80:154-164.
11. Zadoks, J. C., and Schein, R. D. 1979. Epidemiology and Plant Disease Management. Oxford Univ. Press, New York. 427 pp.

CHAPTER 7

IDENTIFICATION AND ASSESSMENT OF LOSSES

P. S. TENG and R. J. OSHIMA

Crop losses probably constitute the most significant constraint worldwide to increasing productivity and food production. Preharvest losses caused by diseases, insects, and weeds have been estimated at 35%, whereas postharvest losses are reported at 20% and consumer wastage at 10% of the total production (7). The worldwide deficit in cereal grains was estimated at 37 million metric tons for 1975. If only the preharvest and postharvest loss estimates are considered for 1975, these add up to 748 million metric tons (3). It therefore appears that, although we have insufficient cereal grains, we are at the same time, losing 20 times what we need for world consumption. These figures imply that, potentially, the world food problem can be eliminated just by reducing the amount of loss.

We have used the above illustration to show how important it is that we, as scientists, be critical of the validity of our statistics, especially when the statistics concern an issue as critical as food. Careless estimates of crop losses destroy credibility in loss values, and it is necessary to develop more accurate measurements and estimates of loss.

Information on crop losses caused by pathogens is important to justify research in some diseases; however, we still encounter controversy over whether it is possible to identify and assess losses with any accuracy. Reviews by Chester (1), Large (11), James (6), and James and Teng (9) have appeared on the subject of disease and loss assessment. Although international meetings have been convened to address losses, for instance, the 1967 Symposium on Crop Losses organized by the Food and Agriculture Organization (FAO), sessions of the 1973 and 1978 international congresses of plant pathology, and the 1980 E. C. Stakman Commemorative Symposium in Minneapolis, researchers concluded that despite the accomplishments, much more work is needed.

Crop loss implies a reduction in yield (quality or quantity) resulting from disease, in comparison to a specified yield of no loss, as defined by C. E. Main in Chapter 6 in this volume. Thus both actual and potential yields have to be determined, as well as the factor(s) accounting for the reduced yield. Loss assessment is considered to be the process of determining losses in individual plants to losses in agroecosystems.

Whether actual or potential losses are assessed will dictate the procedures used and the criteria for evaluating the validity of the assessments. Crop loss assessment is increasingly being linked to modeling. It is worthwhile questioning whether models are appropriate to meet the objectives of loss assessment, especially at regional and international levels. Questioning is singularly appropriate when the biggest units used for developing disease-loss models have been plots, and it has not been demonstrated that such models are valid when applied to regional loss assessment. The problems of scale and resolution will be important ones in the coming years, when plant pathologists will be increasingly asked to stand by the reliability of their loss estimates.

IDENTIFICATION OF LOSSES

A loss occurs when yield is less than a reference yield, and this loss can be identified with a causal agent(s). Many definitions of yield levels are available. For instance, Zadoks and Schein (25) distinguished three levels of crop losses, each depending on the reference yield value used (see also Chapter 6, this volume). Theoretical crop loss is the difference between actual and theoretical yield; economic crop loss is the difference between actual and economic yield; crop loss per se is the difference between actual and attainable yield. The FAO considers attainable yield to be yield when crops are grown under optimal conditions, i.e., the maximum potential that can be technically obtained at a site. These definitions illustrate the difficulties in identifying the meaning of loss. All the yield values defined are variable and subject to environment-genotype interactions. They are expected to differ between years and locations. Is it feasible to even attempt to do crop loss experiments when there are so many potential sources of variation?

Crop loss has commonly been expressed as a proportion term (e.g., percentage) or as crop yield units (e.g., tons). Recently, the dangers of using proportions were highlighted by MacKenzie and King (13), using data on bacterial leaf blight of rice. Significant effects of the environment on the potential yield of rice led to a determination of significantly different regression coefficients when the percent yield loss was regressed with blight severity for 2 years. When actual yield was used, the slope of the regression between yield and severity remained the same, between years, indicating a direct relation between them. MacKenzie and King stated that it is often misleading to assume that yield plus loss equals 100%. What does this mean in practical terms? If we consider the recommendations of the FAO International Collaborative Program on Crop Loss Assessment, of doing multiyear and multilocation experiments to provide data for modeling, then we have to accept procedures that reduce the effect of environmentally induced differences in yield and loss. The problem of variable reference yield in computing loss may be alleviated if the data base in the disease-loss experiments is large enough to account for most of the variation on loss. Perhaps what we should be concerned with is to increase the level at which we are willing to accept a model as being representative, for example, if the R^2 value in a regression model falls below some arbitrary values such as 0.6, then we may consider it unrepresentative of a system.

Current efforts at crop loss assessment emphasize quantitative estimates. However, changes in the nutritional quality and content and in the appearance of cereals, fruits, and vegetables can result from exposure to air pollutants. These pollutants can significantly change the market value of goods, a problem compounded by interactions between abiotic and biotic agents. Residues of heavy metals, halogens, and organic compounds from industrial sources resulted in the enactment of government regulations that affect the marketing and marketability of foodstuffs. These aspects of loss assessment have received little attention. Losses caused by one factor cannot be viewed in isolation from losses caused by other factors. Changed air quality may provide an undefined factor(s) in experiments that influence the basic disease-loss relationship. For example, oxidant air pollutants interact with the tree damaged by the Western bark beetle to cause a significant decline in some conifer species in the San Bernardino mountains of Southern California (2). If we are to recognize multiple factors in any loss program, then we need to ascertain the additive or multiplicative effects of factors in interpreting losses. It is common practice to identify regional losses additively, leading to estimates of loss that range from 12 to 24% for diseases alone in many crops (8). We must seriously question how, if these losses are actually occurring, we have been able to sustain a viable agriculture.

Because loss can occur any time from planting to harvesting and storage, the distinction is made between preharvest and postharvest losses. Few attempts have been made to use a holistic approach in loss identification and assessment; consequently objectives often are not well defined and methodology not well developed to suit the needs of a particular program. We have failed to adopt a comprehensive systems approach in our subject area, and in failing, we have probably impeded progress with our ad hoc approach to research. However, it is encouraging to see in the United States, in the past few years, the emergence of national groupings on loss assessment, such as the NCLAN (National Crop Loss Assessment Network), sponsored by the U.S. Environmental Protection Agency (EPA), and the NCLAS (National Crop Loss Assessment System), sponsored by the USDA.

ASSESSMENT OF LOSSES

Losses caused by plant disease may be estimated at many stages of the production system, from determining the effect of a single disease on a plant, to estimating regional losses of a crop from several diseases. Crop loss assessment methods may be grouped conveniently as either subjective or quantitative. The subjective approach applies mainly to loss assessment in a region using expert testimony or enquiries (25). The quantitative approach to loss assessment uses strategies in which the disease-loss relationship is characterized by experiments and modeling first, then disease intensity is determined by regional surveys, and finally, loss and intensity are integrated to give a regional or multistage loss estimate (9). There have not been studies on which approach gives the most realistic loss estimates. When the limitations of current disease-loss models are considered in conjunction with variability in cropping practice and environment, it is questionable whether quantitative methods give more reliable information than subjective ones do. It has been argued that subjective methods may be potentially more discriminating than quantitative ones (19); this is a research area that needs to be formalized.

Disease-Loss Experiments

There are many sources of variation in experiments to quantify the effect of disease on yield. All three components of the disease triangle, host, pathogen, and environment, may vary independently or interactively. The complexity is increased when we impose human intervention in the form of husbandry. Disease-loss measurements that are made primarily of host and pathogen (as disease) need to be designed to minimize these potential sources of variation. The full impact of disease on loss should be explored over all development stages of the crop.

Measurement of Disease

Since disease is the most important variable in disease-loss experiments, we should ask ourselves how accurate are the present methods for disease measurement. We further need to distinguish between measurement methods that are suitable for experimental or laboratory use and those suitable for field use. Large (11) recommended that field methods be abstractions of experimental methods for assessment, whereas James and Teng (9) suggested that the same method must be used in experiments and in regional loss assessment. Small errors in disease measurement are likely to be magnified when disease values are converted into loss figures for individual fields and when a regional loss figure is calculated. Disease measurement methods may be direct (e.g., disease assessment using keys and standard area diagrams) or indirect (e.g., remote sensing, incidence-severity relationships, and spore counts). Of these, the most commonly used method is disease assessment using field keys or standard area diagrams. Much has been written on the theory and application of disease assessment; it remains essentially a subjective method, relying on the ability of the eye to distinguish categories of diseased area. There is always the "hidden human error" in this method. However, when one considers that the amount of disease on individual plants in a population will also have a certain amount of variation, then the two together could give a misrepresentation of disease intensity. Researching disease measurement should have a high priority in crop loss assessment, as it will lead to more reliable loss estimates and possibly clarify current thinking on what constitutes disease.

The difficulties involved in assessing injury from abiotic agents, and in establishing useful relations between injury assessments and loss, have spurred research into quantitative substitutes for abiotics. Researchers have attempted to link release of ethylene and other hydrocarbons to stress induced by air pollutants. While some of these methods have been useful in the laboratory, they are not yet used in the field. In general, methods that are designed to measure some factor of physiological stress do not discriminate between the types of stress. These methods are subject to the same variation between plants as are visual assessment methods.

With the rapid development of microprocessors, accurate, portable "disease sensors" may soon be available for field use. Such sensors should be compatible with color video image analysis. Even if disease measurements can be made objective, it is doubtful whether loss estimates that depend on a model will increase in precision very much. The reason is that common biological variation, in disease per plant and yield per plant, is inherently present and cannot be reduced by increased instrument precision.

Knowledge of the spatial distribution of a disease, at various phases of the epidemic, is essential for designing a sampling procedure that will give the best estimate of the mean amount of disease in a field. Furthermore, it is often desirable that more than one disease be sampled in any field during a given visit.

We have just begun to recognize the need to develop multidisease sampling theory and methodology. With increasing emphasis on scouting in integrated pest management, multidisease sampling becomes important.

The characterization of ambient air pollution is an underdeveloped area, and most of these characterizations have been in urban locations in conjunction with human health studies to aid regulatory agencies. Despite reported crop injury, little is known about the composition and spatial distribution of ambient air pollution in agricultural areas.

With air pollutants, visible injury (which is equated with disease severity) does not appear to be a repeatable indicator of loss. Crop loss may be caused by doses of pollutants that do not produce visible symptoms. Actual monitored concentrations of air pollutant appear to be necessary in modeling the pollutant-loss relationship (22).

Measurement of the Disease-Loss Relationship Through Experiments

When attempting to quantify the relationship between disease and yield loss, we should ask whether we can hope to account for variation caused by factors other than disease on crop yield. If we cannot, then we should lower our acceptance value for models defining disease-loss relationships, especially models that have been derived from field data. However, if we strive for the ideal model, then we need to explore methods either for reducing the influence of factors other than disease on yield or for incorporating these factors into the disease-loss models. Research is needed into the relative usefulness of disease-loss models that only relate to the effect of disease per se and models that can account for the effect of factors other than disease on the disease-loss and yield relationships.

Experimental design is a major constraint in the generation of data for disease-loss modeling. A disease-loss model incorporates both an experimenter's hypotheses concerning a particular system and the empirical relationships dictated by the biological data base. We would be presumptive to say that we can improve on the precision of current models until the data base realistically reflects the interaction between disease and yield. Experimental design needs to relate to the particular kind of modeling activity, whether it is empirical, explanatory, or theoretical (19).

The relationship between disease and yield loss, as derived from field experiments, has been quantified either as a response plane or a response surface. The so-called "critical point" models (6) are, in effect, response planes, and yield loss and disease intensity are the vertical and horizontal axes, respectively. Critical point models relate loss to a single disease assessment during crop growth. Multiple point models, which relate yield loss to several sequential disease assessments during crop growth, may be viewed as response surfaces having a third axis of development stage; that is, multiple point is yield loss = f (disease intensity, development stage), whereas critical point is yield loss = f (disease intensity). If the relationship between loss and disease at various crop development stages is a response surface, then we are using inadequate designs for generating the data base needed to characterize disease losses. Literature reviews (6,9) show that, almost without exception, all disease-loss experiments have been done using designs that cater to classical analysis of variance through paired or multiple treatments with replication. Designs that have been and are currently being used, are the randomized complete block, latin square, and split-plot.

When data from conventional experimental designs have been used for modeling disease-losses, researchers have adopted two approaches: either all values from treatment plots have been used for modeling, ignoring the blocking effect, or only the mean values of each treatment have been used. Neither approach is satisfactory. The first concedes that blocking has failed to produce comparable results within replicates of the same treatment; the second discounts inherent biological variability in diseased populations and in yield response to disease. Furthermore, in any disease-loss experiment, a wide range of epidemics needs to be produced and often a compromise is made that reduces the number of epidemic treatments so that the number of replicates is satisfactory to produce significance between treatments. If the purpose of an experiment is to provide information on a disease-loss response, then it is more important to have more treatments than to have fewer treatments with replication. Fewer treatments mean that the full effect of disease on loss will not be explored adequately, especially in crop response at different development stages to different disease intensities.

The argument that replication is necessary at low levels of disease or air pollution exposure is not valid because this is the most sensitive area for disease-environment and disease-disease interactions. High variability due to interactions and the low magnitude of plant response may make the allocation of resources for treatment replication a poor cost/benefit decision.

Research is needed on the use of response surface methodology as an alternative to experimental designs that have been commonly used and accepted by plant pathologists.

A related issue when considering experimental design in disease-loss experiments is the size of the host unit. Researchers have used various units: single tillers, single plants, microplots, and large plots. Using single tillers allows a wide range of disease intensities to be explored during a single cropping season, thereby leading to a fuller exploration of the effect of disease on loss (15). Each tiller can be treated as one datum point in modeling, conforming to the requirement in regression analysis that each independent observation be without variance. More resources are required when size of host unit is increased for the same number of treatment points. An argument against using small host units in disease-loss experiments is that the amount of variability between units is increased with reduction in size, and that the influence of factors other than disease will be so great to make it difficult to characterize the relationship. One distinguishing feature of the empirical models developed using single tillers is the low accounting of variation. New methods may remove some of the variation due solely to host-environment interactions (5).

Crop loss assessment is a practical activity that must relate to farming systems. Yet, when we review the literature, we find that most of the disease-loss relationships cannot account for differences in husbandry, cultivars, and environment. Furthermore, even with a single disease, little research has been done on the effect on yield of factors such as pathotype symptoms, plant part on which symptom occurs, and even intercultivar reaction. Vanderplank (23) noted the phenomenon of "interplot interference" in epidemiology experiments, where treatment plots with different amounts of disease may influence each other. Although there is an awareness of the representational error created by this phenomenon when extrapolating results to the farm, much remains to be done on the quantification, reduction, and correction of representational error.

Disease-loss experiments must consider the epidemiological properties of the system. For example, plot size influences the rate at which epidemics develop (23). A disease-loss experiment conducted with epidemics generated in small plots bears few similarities to the dynamic interaction of disease epidemics and yield in the field. We are faced again with the problem of scale and resolution.

Although much has been written on the need for formal experiments to define a disease-loss relationship, questions remain on the appropriateness of examining single diseases in isolation and on the use of the experimental approach. Stynes (17) and Wiese (24), recognizing the need to account for multiple factors in the total production system, used factor analysis techniques to partition sources contributing to yield loss. Data for analyses were obtained by intensively monitoring selected fields. With more research this approach may lead to a better understanding of disease-loss relationships.

Modeling the Disease-Loss Relationship

The modeling and experimental phases of a crop loss assessment program are interrelated. The type of model to be developed influences the experimental design to be used. In turn the kind of experiment done dictates what model may be developed. The term "model" is used in this chapter with a general meaning: a model is any mathematical representation of a disease-loss system (19). Categories of models are: empirical-regression models, where disease is the only independent variable(s); empirical-regression models that account for environment; explanatory-simulation models, where disease effect on yield is determined through changes in physiological processes; and theoretical-simulation models, where the model is based on theoretical relationships. Categorization of models offers a means for comparison and discussion. Not all models will fit into a particular category.

The complexity of summarizing and interpreting disease-loss systems lends itself to modeling. Aided by computers, modeling has developed increasingly into a practical tool for plant pathologists.

We are faced in disease-loss modeling, vis-a-vis the explanatory-simulation models, with a situation in which the methodology and technology for modeling is available but the biological research has not kept pace. For many disease-loss systems, we know how to relate the effect of pathogen on crop yield on a physiological basis, but are unable to do so for lack of data. There is a need for systems-directed, physiological research, where modeling can identify areas for basic research that will lead to a better understanding of the whole system.

Early efforts at presenting the interrelationships between disease intensity, yield loss, and crop development stage led Kirby and Archer (10) and Greaney (4) to develop matrix tables linking all three. This simple, although clumsy and inefficient, way of using data from disease-loss experiments was an elementary modeling effort. An improvement on tables was linear regression, which led to critical-point models. The multiple-point and area-under-disease-progress-curve models are refinements to incorporate epidemiological and physiological properties of the host. All of these empirical-regression models are inherently linear by nature of the technique used, and it is appropriate to question the applicability of the technique. Tammes (18) conceptualized the relationship between crop yield and a biological stress as a curve with upper and lower thresholds of injury and a linear response between these thresholds. Linear regression models may apply to the linear portion of this conceptual response. When attempting to evaluate the appropriateness of linear regression for disease-loss models, we need to consider the potential use of a model. For example, epidemiological theory states that the most effective disease control is obtained when disease is less than 0.05, i.e., during the exponential phase of multiplication. If a disease-loss model is to be used for estimating the effectiveness of control at disease ≤ 0.05, then it may very well have to be a nonlinear model. However, given the large standard deviations in disease intensities of field epidemics when disease is <0.05, it may be difficult to derive any empirical model with precision. Furthermore, with many diseases, the detection threshold in field sampling may be above 0.05. Despite their limitations, empirical regression models may remain the most applicable of all the types of disease-loss models for practical disease management.

Regression does not imply cause and effect, and recently an attempt was made by Madden et al (14) to quantify the Tammes (18) relationship, using a modified Weibull model. The Weibull model is a very versatile one, and these authors (14) have made a significant contribution by developing a model that facilitates estimation of the thresholds worked out by Tammes (18). The next step should be the development of an unbiased theoretical model, using what is known of epidemiology and yield physiology.

It is necessary to distinguish the different system levels of modeling disease-loss. Empirical-regression disease-loss models are mainly plot-level models, addressing a mean response of a defined number of plants to disease. Explanatory-simulation models that are currently being researched address both the physiological process and plot levels. The theoretical models aim at process, plot, and region. We referred earlier to the potential of a representational error being incurred when results from one level are used to interpret phenomena at another level. It may well be that the loss estimates given in our introduction are a result of this scale problem and that we are magnifying our loss estimates. The same error may be incurred in model application. A disease-loss model developed using plot-level data should be recognized as being pertinent only to a plot-level phenomenon, yet we are often faced with having to make regional loss assessments. Similarly, explanatory-simulation models are mainly process-level models, and it is a quantum jump from process-level simulation to simulation of losses occurring over a region. We need to resolve this issue of how to make loss assessments at a higher level using models developed at a lower level.

There has been much progress in developing crop physiological models, and models are currently available for wheat, soybean, corn, sorghum, and cotton (16). A potentially powerful tool is the coupling of disease epidemic models with these crop models to provide a model that can be used both for research on disease-loss and for loss assessment. A major obstacle to this coupling effort is the lack of data on the functional relationships between pathogen and host processes, e.g., the relationship between fungal mass per unit host mass and rate of transpiration. Research effort needs to be directed to defining these functional relationships; the payoffs will be rewarding. With a coupled model, it will be possible to account not only for biological factor interactions, but also for environment and husbandry.

A novel approach has been proposed by the developers of SIMED, an alfalfa crop simulator at Purdue University, which addresses the issue of scale (Don Holt, *personal communication*). In addressing the need to forecast large-scale crop production as affected by various factors, a model was developed that may be described as being intermediate in sophistication between the empirical-regression and explanatory-simulation models. Model structure may be represented by the equation:

$$Y = K * f(X1) * f(X2) \ldots\ldots * f(Xn),$$

where Y = yield, K = photosynthetically active solar radiation, and X is any factor that influences yield. When used to estimate yields, the model was reduced to:

$$Y = M * W * E * B \ldots\ldots$$

where M = maximum yield (calculated using historical data), W = a weather factor, E = pest factor and B = management factor. The magnitude of each factor is determined either empirically or by using another simulation model. This approach minimizes the problem of scale and accounts for the influence of other factors on overall yield at the regional level. The model is routinely used for crop production forecasting by Control Data Corporation. It may be the approach needed to integrate the many ad hoc efforts in the two national loss assessment programs, NCLAN and NCLAS. The approach has conceptual elegance and helps to bridge not only the problem of scale, but also the gap between static and dynamic models.

Model validity is a central issue in systems research (19). Much remains to be done in relating model validity to application. Often so much effort is put into the activities of experimentation and modeling that resources become inadequate for validation.

APPLYING ASSESSMENTS OF DISEASE-LOSS

The experimental and modeling phases in a crop loss program should result ideally in some algorithm for estimating disease-loss in a given situation. While some argue that crop loss assessment is in its infancy (24), others have pointed out the huge resources already expended on loss research (20). In the following section, we will attempt to show where the gaps occur in potential and current efforts at applying disease-loss assessment.

Integrated Pest Management (Diseases)

Crop loss assessment should provide the objective data base for IPM decisions that impinge at different levels of organization: single fields, farms, or a complete agroecosystem. In this regard, the estimation of loss caused by disease must be predicted using a model appropriate to the level of organization.

Although many predictive systems have been developed for decision making in IPM (see Chapter 32, this volume), few use potential loss as a criterion for control. On the individual field, predictive systems have focused generally on identifying critical periods when weather has favored infection. Control is based on the premise that if such critical periods occur, then disease will develop to intensities causing a significant crop loss. This procedure should be made more quantitative by linking the potential loss to prediction of favorable infection conditions. Some researchers have proposed that a crop loss model be the basis for making control decisions and that each decision be evaluated on the extra amount of profit generated (21).

To make better use of crop loss assessment in IPM, there needs to be research on integrating the operational components of IPM: biological monitoring, environmental monitoring, epidemic prediction and projection, loss estimation, and benefit/cost analysis. Research on the integration of these components is entirely legitimate and should not be considered merely adaptive and left to agricultural extension personnel.

Perhaps a major obstacle to our advancing the science of loss assessment is the relative inattention given to application. Little research has been done on the precision and size of models needed for on-line disease management. Advances in microprocessor technology will likely force us to reexamine our approaches in modeling disease-loss systems. With increasing use of microcomputers and disease monitoring by farmers and scouts, future disease-loss models will have to consider changes in disease measurement and sampling.

Resource Allocation for Research and Extension

Crop protection policy is often set on perceptions of what macrolevel behavior is. Regional crop loss data should provide a further basis for setting research and extension priorities. This, presumably, is one of the objectives of national systems designed to assess loss due to biotic or abiotic factors.

Many aspects of loss assessment methodology need to be addressed before it will be possible routinely to use regional loss information for policy decisions. A discussion of the USEPA/NCLAN system, designed to estimate the effect nationally of air pollutants on crop yield, will illustrate some of the areas that need research. The Clean Air Act of 1977 requires that the EPA review National Ambient Air Quality Standards every 5 years and provide supporting information for their retention or revision. NCLAN was planned to determine crop losses in each of five defined regions in the United States and then arrive at a national estimate. One of the most difficult questions the researchers had to address was how to develop models that could be validated at the regional level and used for making these regional estimates. Since air pollutant concentrations do not remain constant over the whole region, losses needed to be estimated with a smaller spatial unit of resolution, e.g., a county. Aggregating loss estimates from plot to county to region is a procedure susceptible to error at any stage, with no way of validating or estimating the confidence limit of the regional loss estimate. In view of this, any change in policy made will have to recognize the weaknesses in the information. Even at the county level, the use of plot data leads to an averaging effect of pollutants on yield.

Another problem area in regional loss assessment is the lack of reliable crop yield and weather data. Regardless of the type of model used for loss assessment, a reference yield has to be established. Commonly the yield data are provided by state crop reporting services. Lack of specific data on crop cultivar and cropping practices is a major obstacle in regional loss assessment, one that needs to be addressed by state and federal agencies.

To make regional crop productivity and loss information available in a timely manner requires the development of computerized delivery systems. There are encouraging signs that regional systems will be developed by linking currently operational state surveillance systems such as those of the Michigan Cooperative Crop Monitoring System and the Minnesota Cooperative Pest and Disease Surveillance System. Although these systems are relatively sophisticated in their computer software, much needs to be done to improve basic procedures in data gathering, e.g., disease assessment and sampling.

Regional loss assessment may be considered as a process of collecting data from a predefined geographical unit (county, state, etc.) dependent on the available model. A region may consist of clusters of geographical units with common characteristics, with loss being estimated for each cluster. Little research has been done on techniques to define regions into such clusters (e.g., counties), using criteria that will not bias the loss assessment. For example, it is conceptually wrong to average factors linearly if these factors have a nonlinear effect on yield. Clustering may also reduce some of the present computational problems encountered in using crop physiological models for large-scale simulation. However, computer technology may enhance application of such models for many small geographical units, to predict loss for a great variety of situations in the same region. Concern that big models may need to be condensed so they can be used for regional loss assessment may be unfounded.

Socioeconomic Impact of Diseases as Production Constraints

The impact of crop loss caused by disease, although experienced directly by the producer, filters through eventually to the consumer. The social and economic effects of substantial crop loss are more apparent than the marginal losses that agriculture sustains yearly.

We should increase our efforts to study the effect of crop losses, beyond the farm into the community and beyond. Losses in the field may well result in benefits if crop prices are increased.

Crop loss must be viewed as more than just loss in productivity; there is the issue of energy efficiency of production since losses that occur in spite of control measures have increased impact.

Expertise from other disciplines needs to be tapped to enable evaluation of the socioeconomic implications of crop loss. Plant pathologists should be encouraged to acquire skills in economic analysis. Currently, several projects in the United States are attempting to define economic criteria for evaluating the effects of loss. Results are not yet available.

Crop losses that occur in subsistence agriculture may have an effect on the social systems concerned in a manner entirely different from the effect of losses in technological agriculture. C. A. J. Putter of Natal University, South Africa (*unpublished data*, 1979) noted that in some primitive tribes, reactions to crop loss are built into local customs and tradition. Although losses occur in crops grown under subsistence agriculture, it is generally observed that most of these wild pathosystems are inherently more stable than crop pathosystems. A study of the social aspects of crop loss in primitive systems is important, since loss has a more immediate impact on the viability of such societies. Perhaps indications may be found of how the developed world will react to another major crop failure.

CONCLUDING REMARKS

During this Diamond Jubilee year, the American Phytopathological Society (APS) should not only take stock of its achievements but be prepared to criticize itself and ask whether American phytopathology is directing sufficient effort in increasing world food supply by decreasing crop loss. Lyman's (12) criticism in 1918 that plant pathologists were not making sufficient efforts to quantify loss is certainly applicable today. We have seen recognition and respectability conferred on crop loss assessment activities only recently in the United States, in spite of this being the bread and butter of plant pathology. With the important leadership role that U.S. agriculture and agricultural research plays internationally, it is important that crop loss assessment become formalized as an active tool for crop improvement and protection.

International organizations can do much to promote information exchange that will lead to advances in science. The crop loss program of the FAO is to be credited for its role in imparting current knowledge of loss assessment methodology throughout the developing world through regional workshops. Similarly, the Disease-Loss and Production Constraints Committee of the International Society of Plant Pathology has played a vital role in this respect. We hope that the APS will continue to support the efforts of these organizations.

Crop losses will continue to be major constraints to crop production and improvement unless significant breakthroughs are made in breeding for resistance. We would like to conclude this chapter with the thought that crop loss assessment has developed as a result of failures in plant breeding; perhaps it is now time for loss assessment to influence the manner in which breeding is done.

LITERATURE CITED

1. Chester, K. S. 1950. Plant disease losses: Their appraisal and interpretation. Plant Dis. Rep., Suppl. 193:189-362.
2. Cobb, F. W., Jr., Wood, D. L., Stark, R. W., and Parmeter, J. R. 1968. Theory on the relationship between oxidant injury and bark beetle infestation. Hilgardia 39:141-152.
3. Food and Agriculture Organization, United Nations. 1976. Production Yearbook. F.A.O., Rome. 296 pp.
4. Greaney, F. J. 1935. Method of estimating losses from cereal rusts. Proc. World's Grain Exch. and Conf. Regina, Canada. 1933 2:224-235.
5. Hau, B., Kranz, J., Dengel, H. J., and Hamelink, J. 1980. On the development of loss assessment methods in the tropics. Pages 254–261 in: Crop Loss Assessment. E. C. Stakman Commemorative Symp. Misc. Publ. 7. Agric. Exp. Stn., Univ. Minn., St. Paul. 327 pp.
6. James, W. C. 1974. Assessment of plant diseases and losses. Annu. Rev. Phytopathol. 12:27-48.
7. James, W. C. 1980. Economic, social and political implications of crop losses; A holistic framework for loss assessment in agricultural systems. Pages 10–16 in: Crop Loss Assessment. E. C. Stakman Commemorative Symp. Misc. Publ. 7. Agric. Exp. Stn., Univ. Minn., St. Paul. 327 pp.
8. James, W. C. 1981. Estimated losses of crops from plant pathogens. Pages 79–94 in: Handbook of Pest Management in Agriculture, Vol. 1. D. Pimentel, ed. CRC Press, Boca Raton, FL. 597 pp.
9. James, W. C., and Teng, P. S. 1979. The quantification of production constraints associated with plant disease. Appl. Biol. 4:201-267.
10. Kirby, R. S., and Archer, W. A. 1927. Diseases of cereal and forage crops in the United States in 1926. Plant Dis. Rep., Suppl. 53:110-208.
11. Large, E. C. 1966. Measuring plant disease. Annu. Rev. Phytopathol. 4:9-28.
12. Lyman, G. R. 1918. The relation of phytopathologists to plant disease survey work. Phytopathology 8:219-228.
13. MacKenzie, D. R., and King, E. 1980. Developing realistic crop loss models for plant diseases. Pages 85–89 in: Crop Loss Assessment. E. C. Stakman Commemorative Symp. Misc. Publ. 7. Minn. Agric. Exp.

Stn., St. Paul. 327 pp.
14. Madden, L. V., Pennypacker, S. P., Antle, C. E., and Kingsolver, C. H. 1981. A loss model for crops. Phytopathology 71:685-689.
15. Richardson, M. J., Jacks, M., and Smith, S. 1975. Assessment of loss caused by barley mildew using single tillers. Plant Pathol. 24:21-26.
16. Ruesink, W. G. 1981. Environmental inputs for crop loss models. Pages 131–134 in: Plant Protection: Fundamental Aspects, Vol. 1. Proc. Symp. Int. Congr. Plant Protection, 9th. T. Kommedahl, ed. Burgess Publishing Co., Minneapolis. 411 pp.
17. Stynes, B. A. 1975. A synoptic study of wheat. Ph.D. thesis, University of Adelaide, South Australia. 291 pp.
18. Tammes, P. M. L. 1961. Studies of yield losses. II. Injury as a limiting factor of yield. Neth. J. Plant Pathol. 67:257-263.
19. Teng, P. S. 1981. Validation of computer models of plant disease epidemics: A review of philosophy and methodology. Z. Pflanzenkr. Pflanzenschutz 88:49-63.
20. Teng, P. S., and Gaunt, R. E. 1981. Modeling systems of disease and yield loss in cereals. Agric. Syst. 6:131-154.
21. Teng, P. S., Blackie, M. J., and Close, R. C. 1978. Simulation modelling of plant diseases to rationalize fungicide use. Outlook Agric. 9:273-277.
22. Teng, P. S., Krupa, S. V., and Kromroy, K. R. 1981. Assessing crop losses due to ozone and sulfur dioxide in Minnesota. (Abstr.) Phytopathology 71:908.
23. Vanderplank, J. E. 1963. Plant diseases: Epidemics and control. Academic Press, New York. 349 pp.
24. Wiese, M. V. 1980. Comprehensive and systematic assessment of crop yield determinants. Pages 262–269 in: Crop Loss Assessment. E. C. Stakman Commemorative Symp. Misc. Publ. 7. Agric. Exp. Stn., Univ. Minn., St. Paul. 327 pp.
25. Zadoks, J. C., and Schein, R. D. 1979. Epidemiology and Plant Disease Management. Oxford University Press, New York. 427 pp.

CHAPTER 8

TOWARD THE MANAGEMENT OF CROP LOSSES

D. R. MacKENZIE

Famine. It's a sobering thought. The projections for food production to the year 2000 indicate that the world's population will soon face serious trouble. One fear is the fear of famine.

Sociopolitical polemics has helped to create factions. Each group embraces singular solutions to a very complex problem—the world's food crisis.

Neo-Malthusians argue that the only way to escape the "dismal theorem" is through population control. Short-term solutions to pending disaster include concepts such as triage (withholding food shipments from the helpless) (6) to a proposal to declare a moratorium on the transfer of technology to those countries unwilling or unable to control their population explosion (2).

Neo-Marxists, on the other hand, see the need to redistribute the world's wealth as the solution to the world's food crisis. They divide the estimated total calories available in the world by the number of people as one proof that the redistribution of the world's food, land, capital, etc., through "a new economic order," would be a just and complete solution to the world's food crisis (5).

Technology, considered by some to be a third ideology, offers hope in the form of more Green Revolutions that will supply more food to the ever-increasing number of mouths that will need to be fed in the years to come. Genetic engineering that will put *Rhizobium* nodules on wheat plants and the breeding of highly nutritious crops to feed the malnourished are often stated dreams repeated in the popular press as solutions to the world's food crisis.

Confusion over the causes and hence the solutions to the world's food crisis continues. I have found it helpful to recognize that there is not one food crisis, but indeed there are three food crises. Once the complexity of the world's food crisis is recognized, it is easier to deal with the combinations of solutions that will undoubtedly be required if significant progress is to be made in the short time that remains.

THE THREE CRISES

World Food Crisis I—Chronic Malnutrition

People have undoubtedly long experienced the continuing unavailability of sufficient food to feed all of their number. The present situation is undoubtedly worse than ever before because of the magnitude of the number involved. Estimates vary, but there is general agreement that more than 500 million people fail to get sufficient food on a day-to-day basis. The solution to this food crisis can come from several activities, including producing more food, redistributing available foods more effectively, and improving existing food crops through genetics. Everyone should reflect on the 10 years of progress that might have been if Henry Kissinger's "bold objective" declared at the 1974 World Food Conference that "no child should go to bed hungry" had been faced squarely. It was not.

World Food Crisis II—Malthusian Inevitability

The exponential growth of the human population on a global scale has shown some recent signs of tapering off, but the projections for the next few decades are causes for serious concern. The most rapid growth is expected to occur in the countries that can least afford additional problems from ever larger populations. Existing stresses on food supplies will not permit the added burden of more people—especially for those populations with the potential to nearly double by the turn of the century.

The expectation that the First World (the haves) can provide the needy Third World (the have nots) through grain shipments is unrealistic. The magnitude of the problem is greater than the First World's capacity to ship grain half-way around the world. The solution to this world food crisis will come from several efforts directed at both a reduction in the explosion of seemingly unbounded populations *and* increases in the capacity of regions to produce food for local consumption. To do less will be a demonstration of Malthus' dismal theorem in unparalleled proportion.

World Food Crisis III—The Prospect of Famine

Short-term shortfalls in food production are often manifested as famine. When food reserves prove inadequate to compensate for crop losses some individuals must do with less. In many parts of the world, those blessed with an abundance of production, reserves are sufficient so that few or none will suffer. In other production regions, unexpected production shortfalls can have catastrophic consequences on the lives of many. One solution offered a decade ago for famine prevention was the establishment of an international grain reserve to be drawn on in times of famine. Apparently nothing has been done yet on this proposal.

I have had special reason to reflect on the causes and consequences of famine, not only as a plant pathologist, but as a guest in a country in a time of famine. In the early summer of 1972 I traveled with my family to a new position in the Philippines. Our arrival in the Philippines was met with a tropical storm of fury not often seen. After the typhoon crisscrossed the northern island of Luzon, the "rice basket" of the Philippines was left in ruins. Many portions of central Luzon were as much as 3 m under water. Residents had little comfort and no security. The sanctuary of their home offered little protection from the many rats and poisonous snakes seeking shelter from the rising floods. Uncounted numbers of water buffalo, the pride and wealth of the Philippine rice farmer, were drowned, exhausted after swimming endless hours and finding no place to stand.

When the flood waters finally receded, the rice crop was in total destruction and the Philippines was faced with the prospect of famine. Grain reserves, which teeter on the brink of insufficiency in most Third World countries, proved insufficient in the face of this devastating typhoon. The supermarkets of Manila and sari-sari stores of the village were soon out of food. Consider the prospect of endless empty shelves with nothing to buy, except highly priced leaves of sweet potato plants, one of the few crops able to survive the heavy rains.

There is little or no documentation on the suffering experienced by the Filipino population. But the situation became so severe that the International Rice Research Institute stripped its research program of all available rice seed for distribution to the hungry. Valuable genetic stocks were reduced to the lowest quantity possible in an effort to make food available for the famished.

Several years later I looked up the records of rice production in the Philippines to reflect on those hard times. Imagine my surprise when I found that the production figures showed little deviation and certainly not the precipitous drop I expected to see for 1972. Figure 1 presents the rice production figures for the Republic of the Philippines for the last two decades (8). The story the data tell is important in developing an understanding of what is crop loss. Although the short-term shortfall in rice production in the Philippines in 1972 had catastrophic consequences for several million people, the rice production statistics fail to recognize that impact on their lives.

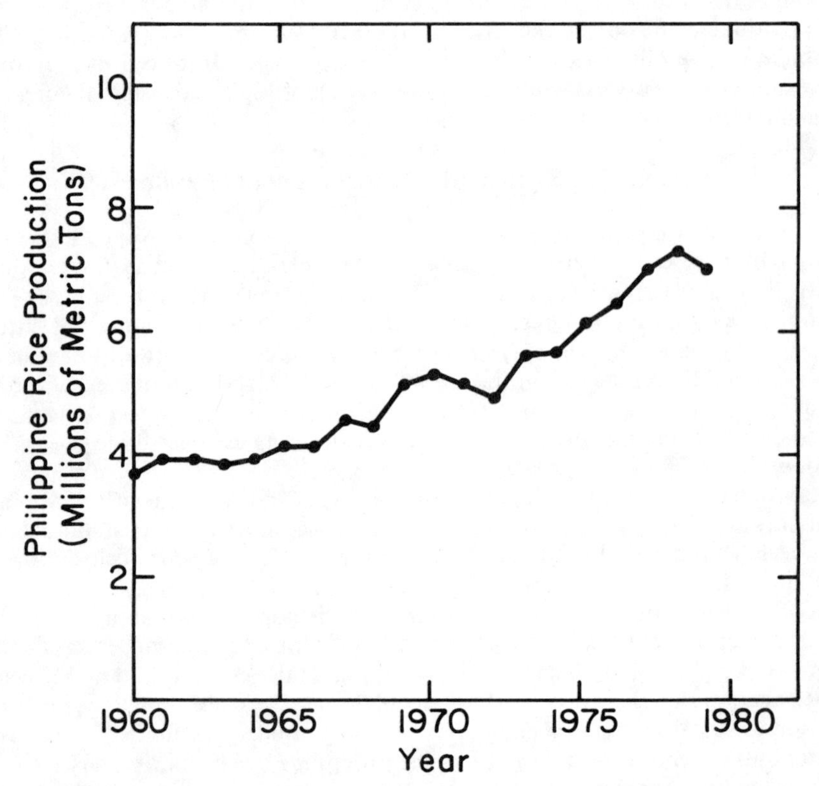

Fig. 1. Philippine rice production from 1960 to 1980 in millions of metric tons. Based on FAO production statistics (2).

WHAT IS CROP LOSS?

There have been many theoretical treatments given to the concept of crop loss. To use one as an example, Zadoks and Schein (10) developed the concepts of theoretical yield, attainable yield, and actual yield.

They defined theoretical yield as the yield to be obtained under the best of the conditions according to calculations based on considerations of crop physiology. The difference between theoretical yields and actual yield (yield obtained under current crop husbandry practices) was termed by Zadoks and Schein (10) "theoretical loss."

The high yields from crops grown under optimal conditions using available technology, as, for example, in experimental plots, were termed "attainable yield" (10). The difference between actual yield and attainable yield was termed by Zadoks and Schein (10) "crop loss."

YIELD LOSS AND CROP LOSS

There is a growing awareness that this common definition of crop loss is wrong. What Zadoks and Schein (10) were defining is really yield loss. And yield loss is not synonymous with crop loss! Regional estimates of crop losses through yield loss estimates are difficult to make and are not the product of the average yield loss per hectare and the hectares of production. New methods of crop loss estimation for crop regions are needed that can account for errors and biases of the estimates.

To demonstrate another problem encountered in crop loss estimation, consider the information in Table 1. The itemized yield loss percentages for each problem were obtained from one source (9). The sum of these items gives the incredible total of 41.4% loss for wheat. Who would accept this as a valid estimate for the U.S. wheat crop?

The interpretation of information for several pests requires more sophisticated methods rather than just simple addition. Moreover, projection of unit area yield losses to give regional crop losses is a topic that has yet to receive adequate attention.

In years past, in times of plenty, or today in those countries lucky enough to be categorized as a "have" nation, crop losses caused by plant disease, insects, weeds, drought, floods, etc., are not measured in terms of human suffering but in hours or dollars lost. In the less fortunate "have not" countries, low food reserves contribute to human suffering when even slight shortfalls occur in food production. As the world's food crisis intensifies and food reserves continue to be drawn down, it seems inevitable that for many people crop losses will shift from being an item of economic importance to one of vital interest.

The importance of the crop failure in central Luzon in 1972 was apparent to all who witnessed it. But its impact was lost on those who only look at the rice production statistics. The ability of a region to respond to disaster through crop replanting and food redistribution permits annual production statistics to be deceptive and, more importantly, to be poor measures of human suffering. The lesson to be learned from experiencing famine first-hand is that the definitions and the statistical summaries we use are inadequate measures of crop losses.

CROP LOSS MANAGEMENT

The challenge of the problems of crop loss estimation and prediction are immense. But the benefits of a crop loss management system could be spectacular. What would a crop loss assessment system 10 to 20 years in the future look like? I hesitate to speculate so far into the future, as the technology that will be useful for understanding crop losses is evolving at a rate described by some as explosive (7). However, if famine prevention is to be attained, it will come through a better understanding of crop losses coupled with an ability to forecast and manage these losses. Plant pathology can make many important contributions in this area. And this is the aspect of the world's food crisis that is appropriate for our attention.

Crop losses can be split into two categories, crop losses that are expected and those that are unexpected. People learned to cope with expected crop losses a long time ago. Low expectations from land areas that are unproductive for reasons of pest attack, poor climate, poor soil, etc., are compensated for through the use of additional land area. More recently, research directed at specific problems has been significantly effective in increasing production expectations. The upward trend in Figure 1 is a consequence of dealing with expected losses through new technology. This is the real benefit of technologies that led to the Green Revolution.

Unexpected losses have proved to be much more difficult to deal with. In many ways, it is the unexpected losses that can be catastrophic and consequently produce the most human suffering.

Developing methods to predict unexpected losses would seem to be a desirable objective. Accomplishing this will require much research and a considerable

Table 1. Estimated losses of United States wheat for all causes

Cause	Percent Loss[a]
Disease	
Stem rust	4.0
Leaf rust	2.5
Root rots	1.0
Septoria	1.0
Wheat streak mosaic	1.0
Loose smut	0.8
Cercospora	0.6
Scab	0.5
Soilborne mosaics	0.5
Common bunt	0.4
Powdery mildew	0.4
Take-all	0.2
Bacterial	0.1
Dwarf bunt	0.1
Leaf and head blights	0.1
Miscellaneous virus diseases	0.1
Snow molds	0.1
Stripe rust	0.1
Others	0.5
Subtotal	14.0
Insects	6.0
Weeds	12.0
Harvest losses	5.0
Storage losses	4.4
Total	41.4

[a]United States Department of Agriculture (9).

financial investment for the design and maintenance of a crop loss assessment system.

I, too, declare a bold objective. By 1993 we should have in place a functioning global crop loss assessment system for the purpose of managing crop losses on the important world food crops. Toward this objective, I offer the following proposed system as a model to be argued in the design of research working toward that end.

Identification of Benefactors

Understanding yield loss is a key element in the management of pests at the production unit (farm) level. Cost-benefit decisions on strategies aimed at specific pests need to include the relationship between pest quantities and their consequences on the yield of that crop. The present practice of arbitrarily assigning thresholds of tolerance is, for the most part, inadequate and frequently unrealistic. Crops in production are commonly under attack from several pests that may be operating as interacting forces on that crop's yield. Pest threshold numbers would then be expected to vary with specific situations. These relationships must be understood as we implement integrated pest management programs.

The yield response of crops to pest attack undoubtedly varies by cultivar. Plant pathologists have long termed this phenomenon tolerance, but little or no quantitative information is available to help interpret these differences as part of economic thresholds for pest attack.

In addition to production unit managers (farmers) intraregional production specialists (e.g., county agents, Cooperative Extension specialists, etc.) would benefit from regional perspectives on crop situation reports to assist them in recommendations to contiguous areas.

State planners and interstate (regional) planners could benefit from continuously updated crop production statistics to help keep them abreast of changes in pest forecasts and to make recommendation for alternative pest management strategies.

Finally, national and international planners could benefit from broad scale crop production information for several purposes. Administrative decisions and the redistribution of resources and personnel would become more effective with additional knowledge on crop losses. Justification of action taken could be more easily made with a functioning crop loss information network. Some have asked if there are research projects that cost more than the crop losses sustained? Who knows! National or international documentation on crop losses would be a valuable asset to scientists and administrative planners for assigning priorities for research endeavors. The consequences of changes (e.g., loss of a pesticide registration or the "breakdown" of a cultivar's resistance) could be helpful in planning and establishing priorities in research projects.

One obvious benefactor of a crop loss assessment system would be national crop reporting services. Presently, many nations expend considerable amounts of money to forecast crop production. In the United States, for example, the Department of Agriculture's (USDA) Crop Reporting Service provides updates on expected production for the major crops. Weekly, throughout the growing season, information is compiled during the "lock up." Once the report is completed, the room is "unlocked" and the participants are then free to use the information in whatever way they desire.

The commodity exchanges are financial devices designed to promote crop price stability. Crop production forecasts are a major factor in commodity exchange activities, and buyers are said to benefit from the USDA Crop Reporting Service forecast through buying or selling contracts for future delivery of major crops. Presently, little or no pest-caused crop losses are reflected in these production forecasts.

Crop loss forecasting methods would add considerably to the USDA's Crop Reporting Service forecasts. Additionally, private endeavors such as Agserve (a division of Control Data Corporation) (1) may find such services useful for advising international grain dealers on the desirability of buying or selling commodities on the international market.

Realistic crop loss assessment systems would, therefore, have the potential of reaching from production unit management to global considerations as part of a total effort to forecast crop production, stabilize food prices, and hopefully prevent famines.

Data Collection

Weather information is an obvious imperative for a realistic crop loss forecasting system. For plant diseases, items such as temperature, rainfall, leaf wetness, and solar radiation could prove to be key factors in forecasting many potentially important plant diseases. The collection of these data will undoubtedly prove costly, as the number of sites required will be many. Much needs to be learned for many plant disease problems.

Fortunately, semiconductor electronics technology has advanced to the point where today it is possible to deploy miniaturized data loggers to monitor several sensors for extended periods while operating on dry cell batteries. Presently the costs are reasonable, given the level of technology. There are reasons to believe that prices will continue to decline as the technology advances.

Large questions remain as to the number of sites (and the density of sites) that will be required to draw appropriate conclusions for an effective system. The collection of weather data and the potential for telemetry via satellite to receiving stations offers prospects once thought impossible. More difficult will be the realization of the need to collect biological information such as the incidence of various pests and specifics on the cultivar of the crop being grown.

The earlier euphoria over satellite reconnaissance has not proved to be as adaptable to pest detection as once thought. The resolution of most satellites now available for agriculture is limited to about 80 m. Plans for orbiting more sophisticated satellites capable of about 30-m resolution offer the promise of applications in agriculture but little hope of detecting pest problems until they are well out of hand (see Chapter 18). The limitations of satellite reconnaissance may require a substantial number of scouts visiting production areas to sample pest incidence and severity for interpretive purposes.

The experience of the U.S. National Weather Service in staffing an adequate number of local weather experts to collect the necessary information for weather forecasting has forced them to use volunteers. Criticism of the volunteer system as being inadequate and fraught with error seems unfair given the unlikely funding of a paid staff sufficient to meet the needs of the U.S. National Weather Service's forecasting system. The prospect of sufficient funds for a U.S. National

Crop Loss Assessment System with paid staff seems, in view of past experience, slim but not hopeless. In less wealthy nations it seems ludicrous to propose that scarce financial resources will be used to staff a comprehensive crop loss assessment system. Volunteers will be essential for the collection of those data.

Immediate access to satellite information will be an absolute necessity for any crop loss assessment system. Presently, several would-be applications of satellite reconnaissance data have been dropped because individuals have experienced delays of up to 9 months in delivery of data. Crop loss assessment systems with 9-month delays would, of course, be useless.

AgRISTARS is a U.S. research program that may be successful in overcoming previous shortcomings of remotely sensed data from space. In a statement to the U.S. House of Representatives USDA's H. W. Hjort (4) stated that

> We recognized the need for coordination between agencies involved in complex space related technology, and we took a lead role in establishing a multiagency program ultimately to be known as AgRISTARS (Agriculture and Resource Investing Surveys Through Aerospace Remote Sensing). The (United States) Department (of Agriculture) is now engaged with other agencies in this broad-based program of research and development to determine how aerospace technology can be applied to agriculture and renewable resources as an information source.

Later in the statement he gave the objectives of AgRISTARS as improvements in "objectivity, reliability, timeliness. . .and reliable crop production forecasts. . . ." The fiscal 1980 budget of AgRISTARS was $32.3 million—a mere glimpse of the beginning costs for a crop loss assessment system budget.

One of the appeals of the AgRISTARS is its potential for worldwide applications. As an example, Hjort (4) cites the success of the Crop Condition Assessment System. Developed by the USDA's Foreign Agricultural Service, this system analyzed Landsat satellite weather data and other information in 1979 and pinpointed problems with both winter and spring wheat production in the Soviet Union. This information caused the USDA to reduce its midseason estimates of expected USSR wheat production.

Satellite information may well prove to be valuable when coupled with ground truth provided by scouts. That information combined with on-site data loggers may prove sufficient to represent the existing conditions and project the likelihood of crop losses. But to extend this potential to reality will require the development of biologically valid crop loss models for multiple pest complexes.

Data Interpretation

Before anyone can begin to account for the impact of pests and to project crop losses, there is a need to develop the ability to understand and predict crop production in the absence of pests. Crop growth models have received a lot of attention from crop physiologists in the past decade. Considerable advances have been made, but much more remains to be done. The key to successful crop loss forecasting systems will be the ability to project accurately pest-absent crop yields and crop production. For it is from those estimates that yield loss and crop loss predictions can be subtracted to obtain realistic forecasts of crop production.

The pest disciplines of entomology, plant pathology, and weed science have side-stepped this important consideration for a long time. Most yield loss equations of the past have expressed crop yields as a percentage of the unaffected check plots. This manipulation of crop response to pests, expressed as a percentage, covers up our inability to project the real impact of pest attack on yield (as tons per hectare) and on production (as tons per year).

Undertaking the design of a workable crop loss assessment system on a regional, national, or international scale is a considerable effort. In the United States this effort has been underway for a decade. The leadership for this project has been provided by John Fulkerson (see Chapter 46 of this volume) of the USDA's Cooperative State Research Service (CSRS), and significant progress has been made. However, much remains to be done.

Using pilot research projects, the CSRS National Crop Loss Assessment System has identified directions for research and areas of difficulty that need attention. One major difficulty repeatedly encountered by the CSRS National Crop Loss Assessment System Design Committee and the participating scientists is the conflicting mandates of government agencies. Some have referred to these differences as expressions of territoriality. For instance, the Animal and Plant Health Inspection Service (APHIS) of the USDA lays claim to the detection of pests. This is, of course, a legitimate activity of APHIS and much more effort is needed. However, the needs of a National Crop Loss Assessment System may differ significantly from the efforts of APHIS, requiring what would appear to critics as duplication of effort. Other examples of these overlapping needs can be foreseen with the USDA's Agricultural Research Service, the Crop Reporting Service, and many more.

The solution would seem to be a coordinated and directed effort that recognizes the commonality of interests and requirements and makes plans to avoid duplication. Moreover, there must be a recognition that some tasks that would appear to be duplication by name are, in fact, quite different tasks. The pest detection system proposed by Epstein (3) for APHIS is not the pest detection system needed for the National Crop Loss Assessment System, although there are elements in common.

A second area of concern for the National Crop Loss Assessment System is the need to recognize that predicting and managing yield losses at the farm level is a task quite different from crop loss management. Although the data sets may be common to both activities, the approach to the problems will undoubtedly prove to be different. This fact has been lost on those who persist in equating yield loss with crop loss.

Information Delivery

The interpretation of weather and biological data will require sophisticated models that will no doubt force us to depend on large computers. The mass of information required and the need for speed in interpretation all indicate a centralized computer system or better yet, a network of powerful computers all capable of intercommunication.

Regional, national, and international planners stand to benefit in many ways from a rapid, large-scale computerized crop loss assessment system. But such a system may exclude many potential users at the production unit level, albeit, the downward flow of information may well be equally important as a pest management tool for the individual production unit. Crop production units (farms) will have two needs as participants in a crop loss assessment system.

First, they will need to process local information for recommendation on action to be taken as part of a pest management program. Second, crop production units will need access to updated production expectations and the consequent changes in crop value. In many farm operations these computing needs can now be met with home computing systems available for U.S. $2,000. Justifying such an expenditure would be difficult if these were the only tasks to be performed on the home computing system. Fortunately, many other tasks such as accounting, inventory, tax reports, etc., are easily accomplished on these machines and help justify such a large expenditure.

This technological solution will undoubtedly have drawbacks in that many individuals will be hesitant to participate. They may well be overwhelmed with the requirements of operating such a sophisticated information system. Great opportunities for research exist in discovering the types of technology that are acceptable to farmers of different backgrounds and cultures. Such psychological or human factor research has been neglected in agriculture but is becoming increasingly important as we work toward more sophisticated technology and more precise and quantitative solutions to complex food production problems.

Many farmers may choose not to spend money on home computing systems, and consideration must be given as to how information will be delivered to such individuals. Government subsidized programs and/or hardware is one alternative worthy of consideration. The problems become even more complex in Third World nations.

Intraregional participants will have other needs for information delivery. Individuals such as county agents can make good use of home computing systems connected to an information delivery network with crop loss information. Under some national conditions it may prove valuable to terminate the regional crop loss assessment system at the county agent's office and have this office provide face-to-face services for individual production units. This may be particularly true in Third World nations where farmers could not be expected to buy expensive communications equipment. Researchers, administrators, and planners at both the state and regional levels can share in the information delivery system through networks of computers. Technology at this level of sophistication is already available in many North American and European countries and can only continue to improve with new technology. For instance, with the use of telephone lines it is now possible within the United States to transmit, at relatively high speed, tremendous amounts of information at very low cost. Commercially available data banks such as Radio Shack's CompuServe demonstrates this potential with many other applications. For the cost of a phone call and a very low charge per hour, subscribers to CompuServe have access to information on the business world, airline schedules, weather, stock market, professional football scores, and even the "pro" football injury list. CompuServe's electronic mail service provides opportunities for the transmittal of information to other subscribers. Additionally, computer "games" and interactive computer programming advice provide attractive inducements to subscribers. And if that is not enough, 64 K bytes of computer work space is available to each subscriber for personal activity.

This one system is described as an example of how computers have become and will continue to become, an important part of daily living in much of the world. The challenge of developing crop loss forecasting systems and strategies for crop loss management will depend very much on our ability to use Toffler's

(7) "Third Wave" to its fullest advantage. Computer technology will undoubtedly be an important part of any functioning crop loss assessment system. The requirements for large data bases of both biological and meteorological information and a need for rapid evaluation of that information will require computer technology.

The challenge is immense, but then so are the potential benefits. We should dream of a future wherein production shortfalls in one region could be forecast and met with the orderly transfer of food reserves from another. In Third World regions this prospect is surely better than resigning ourselves to the prospect of famine. The management of an international grain reserve with advanced warning of crop production problems would aid greatly in benefiting the most people when resources are limited. Other contributions to be made would be in recognizing potential losses before they occur and taking appropriate actions to avoid them. Together these efforts make up the overall strategy we can call crop loss management.

The contributions that plant pathology can make as its part of solving the world's food crisis will be significant and worthy of our efforts. We've come a long way in our brief 75-year history. The challenges of the coming decades will test our abilities even further.

ACKNOWLEDGMENTS

I thank C. Main and P. Teng, authors of chapters on crop losses found in this volume, for sharing with me many of their thoughts, ideas, and philosophy on crop loss. I am specifically grateful to Dr. Main, who first verbalized to me that the real objective is not to record crop losses but to manage them.

I also thank the many members of the National Crop Loss Assessment System Design Committee, who have worked for so long to establish the present base of knowledge from which I have borrowed so freely.

LITERATURE CITED

1. American Society of Agronomy. 1981. Industry News. Crops Soils. p. 25.
2. Aycock, R. 1976. Paddock-Niederhauser proposals. Phytopathol. News 10(9):3-4.
3. Epstein, A. H. 1981. Pest information needs (editorial). Plant Dis. 65:703.
4. Hjort, H. W. 1980. Statement prepared for the Subcommittee on Space Science and Applications of the Committee on Science and Technology, United States House of Representatives. (Mimeographed) 7 pp.
5. Lappe, F. M., and Collins, J. 1979. Food First, Beyond the Myth of Scarcity. Random House, Inc., New York. 619 pp.
6. Paddock, W., and Paddock, P. 1965. Famine—1975! Little, Brown and Co., Boston. 276 pp.
7. Toffler, A. 1980. The Third Wave. Bantam Books, New York. 537 pp.
8. United Nations, Food and Agricultural Organization. 1960–1980. Commodity Reports (Yearly). Reports 12–29. FAO, Washington, DC.
9. U.S. Department of Agriculture. 1965. Losses in Agriculture. USDA Handbook No. 291. 120 pp.
10. Zadoks, J. C., and Schein, R. D. 1979. Epidemiology and Plant Disease Management. Oxford University Press, New York. 427 pp.

PART III

Biological and Physical Constraints to Plant Health

The principal differences between the two cases [health and disease] is, that whereas the normal healthy plant varies more or less regularly and rhythmically about a mean, the diseased one[tends]to vary too suddenly or too far in some particular directions from the mean....

—H. Marshall Ward, 1901

If we interpret disease as any apparently abnormal condition of an organism or its parts or functions, it is evident that the diseases of plants, like those of other living things, include morphological and physiological disturbances which may be induced by a variety of environmental factors, living or nonliving.

—Benjamin M. Duggar, 1909

There is no need for a model to look like the real thing; it only has to function like it.

—S. D. Garrett, 1975

The development of a crop, and therefore its health, is always the result of interplay between biological and environmental factors, as influenced by human agency. In other words, crop health is a highly complex affair.

—Josef Palti, 1981

CHAPTER 9

THE ROLE OF GENETICS IN ETIOLOGICAL PATHOLOGY AND MAINTENANCE OF PLANT HEALTH

R. G. GROGAN

In most publications on plant diseases, etiology refers only to a specific pathogenic microorganism if an infectious disease is under discussion or to an external factor of the environment when an abiotic disease is the object of inquiry. In contrast, etiological plant pathology, as defined by Ling (20,21), is concerned with *all* factors and their interactions that are involved in the causation of plant disease.

Webster's Third New International Dictionary defines disease as follows:

> Disease is an impairment of the normal state of the living plant body or of any of its components that interrupts or modifies the performance of the vital functions, being a response to environmental factors, to specific infective agents, to inherent defects of the organism, or to combinations of these factors.

Thus, plant diseases can have a relatively simple or a complicated etiology, but in all instances the plant genotype and environmental factors (biotic and abiotic) and interactions between them are the key elements.

Major emphasis in plant pathology, however, has been placed on diseases caused by biotic pathogens, and discussions of the maintenance of plant health usually emphasize protection from the so-called pests (weeds, insects, and other biotic pathogens). Moreover, discussions involving genetics usually pertain to inheritance of resistance to biotic pathogens or to host-parasite interactions; the same bias for biotic pathogens usually is evident also when abiotic factors of the environment are under consideration with regard to disposition to disease (6,30,44).

It is not my intention to belittle the importance of biotic pathogens and other pests for, indeed, they can cause catastrophic losses and crop failure. I will, however, emphasize the necessity for a broader concept of disease etiology, especially in relation to genetics, disease diagnosis (13,39), and maintenance of plant health.

The concept of etiology still generally accepted is similar to that proposed by Whetzel (41): "The study or discussion of a disease which deals with the chief causal factor, its nature, classification and relations with the suscept." However, if etiological pathology is to be complete it must include a consideration of genetics because a plant's genetic constitution influences or controls its reaction to all biotic and abiotic factors of the environment.

LINK'S CONCEPT OF ETIOLOGY

Such a broad concept was described by Link (20,21) as a "thorough-going etiology." One of his main objectives was to convince "germists" that diseases result from a complex of causes instead of a "chief causal factor," usually meaning a biotic pathogen. Link dubbed this narrower concept "pseudoetiology." His broader concept included all antecedents of the "pathic state or condition" which developed as a result of one or several "pathic events," defined as any occurrence, either external or internal, potentially capable of causing injury. Many kinds of occurrences are possible, inasmuch as any factor of the internal or external environment affecting plant growth and development is potentially injurious. In fact, Link defined Plant Pathology as the science devoted to the study of plant injury. An injury caused by one or several pathic events may be minor and within the plant's capacity for adaptation and ready repair. If so, the plant would remain healthy. If, however, the injury were so severe that the plant is unable to adapt readily and its range of easy tolerance is exceeded, permanent injury (disease) would result. Accordingly, disease usually results from the interaction of causal complexes comprising internal factors (genetic and cytoplasmic components and their interaction), correlative influences among cells that normally result in tissue differentiation, and all external factors of the environment that influence plant development and its composition and disposition at any time of reference (past history).

Although Link stated that diseases usually develop as the result of a complex of causal or pathic events, he recognized that the relative complexity of etiology is variable. A defective gene or an incompatible gene-cytoplasm combination could result in severe, permanent injury and, in this case, external factors of the environment would have little or no influence. In another instance a gene or gene interaction could condition susceptibility to a potentially injurious factor such as chilling temperature and the plant would remain healthy unless subjected to a time-temperature dose in excess of its ability to adapt readily and repair the damage. Similarly a plant with a genotype conditioning susceptibility to a primary biotic pathogen would remain healthy unless inoculum were present and weather factors were favorable for infection and disease development; the same weather factors in absence of inoculum, however, usually would not be injurious. In a still more complicated etiological situation, a plant might have a genotype that conditioned susceptibility to a potential biotic pathogen, but only when abiotic or biotic factors of the external environment changed the plant's disposition to greater susceptibility.

This latter situation demonstrates the fallacy of defining "predisposition by the environment" as being due to nongenetic factors (6,44) because it is axiomatic that the effect of any factor of the external environment is always conditioned by the plant's genotype. This concept is discussed in greater detail in a subsequent section.

CONCEPTS OF STRESS AND STRAIN

Link's concept of etiology is similar to and compatible with the more recent concepts of environmental *stress* and *strain* described by Levitt (19). A biological stress is any environmental factor (biotic or abiotic) capable of inducing a potentially injurious strain in a plant, and stress resistance is the capability of the plant to survive the unfavorable stress by preventing the strain or by reparation of the damage through the expenditure of metabolic energy. If a strain is relieved when the stress is removed, it is called *elastic*; if not relieved it is called *plastic*. When a stress factor acts on a plant, it may produce injury in several different ways: 1) it may induce a *direct stress injury*; 2) it may produce an elastic strain that is not injurious if applied only for a short time, but when applied for a longer time, may produce an *indirect stress injury* (as an example, exposing a genetically susceptible plant to a chilling temperature stress usually is not injurious if the exposure is of short duration. If the exposure is prolonged, however, the indirect stress injury may become so severe that irreparable damage is done); 3) a stress may not injure a plant by the strain it produces, but by the production of a secondary stress that may result in *secondary stress injury*. As an example, a high temperature stress may not be injurious, but may result in a water deficit that is injurious.

Plants may escape injury either by avoidance or tolerance of stress. Avoidance can be accomplished in various ways. For example, crop plants usually are grown at a time or place when the stress is not likely to occur. Specifically, frost- or chilling-sensitive cultivars are planted when warmer weather is expected; or drought may be avoided in some cereal cultivars by early planting and subsequently earlier maturity. A plant would be more tolerant to injury if there were an increase in the time and intensity of exposure to stress necessary to produce injury or if there were a decrease in the amount of injury produced by a certain stress exposure. Inasmuch as these escape or tolerance characteristics are conditioned by both the plant's genotype and its past history of exposure to various environmental factors, the occurrence of a diseased or healthy phenotype also is dependent upon these interactions.

Although Levitt (19) did not discuss biotic factors of the environment in detail, the same concepts apply both to biotic and abiotic stress factors. Thus, the concepts of Levitt and Link are compatible and can be combined into a single comprehensive scheme of etiology in which predisposing factors of the environment and inherent susceptibility are not treated as separate categories coordinate with etiology as is often done in current plant pathology literature.

PREDISPOSITION OR DISPOSITION?

Colhoun (6), in a review of "predisposition by the environment," defined the term as "the effects of nongenetic conditions on susceptibility of plants to pathogens when these conditions act solely on the plant before infection occurs." As discussed previously, this definition seems too restrictive and, in fact, defines an impossibility, i.e., changes of a host-plant's susceptibility to disease by factors of the environment without influence or control of the change by the plant genotype. I can think of several examples, but will describe only one to illustrate the fallacy of this concept. A fruit such as tomato that is genetically susceptible to chilling stress can be predisposed (or disposed) to infection by a facultative parasite such as *Alternaria alternata*, but a chilling-resistant tomato, if such were available, would not become susceptible if subjected to the same treatment. The sensitive tomato, if subjected to prolonged chilling, could suffer severe injury per se (necrosis and failure to ripen), but the chilling-resistant tomato would not be affected by the same treatment. Thus, genotypes of the two exert a major influence, and predisposition results from the interaction of the suscept's genotype and environment.

The environmental factors that have been associated with predisposition to biotic pathogens are essentially the same as those that are recognized to be potential stress factors (deficient or excess water, low or high temperature, defoliation, fruit set, transplanting shock, wounding by various agents, excess or deficiency of mineral elements, high or low light intensity, toxic chemicals, atmospheric pollutants, and primary and secondary pathogenic biotic agents).

Schoeneweiss (30) recognized the role of environmental stress factors in predisposition to biotic pathogens, and also that stresses alone may injure plants. However, he down-played the role of stresses per se in the etiology of disease as indicated by the following statement:

> Although extreme stresses may injure plants through direct irreversible plastic strains, in most cases plants tolerate or adapt to stresses without permanent injury in the absence of disease organisms.

This statement is highly questionable and I regard it as further evidence of biased emphasis in plant pathology on biotic-pathogen diseases. Nevertheless, the hypothesis that stress injury induces predisposition by reducing the plant's capability to resist infection or retard disease development, especially of diseases involving weak, nonaggressive, facultative parasites, seems plausible.

The practical utility of the term predisposition is impaired, however, by the definitional restriction that its induction must occur prior to infection so that the effect supposedly is solely on susceptibility of the host and not to an effect on the pathogen. Inasmuch as we usually are interested in the final outcome, i.e., disease incidence and severity and its effect on yield, it seems that we should consider stress and other factors present both before and after infection because disease severity is conditioned by the host-parasite-genetic-environmental interactions after infection as well as the host-genetic-environmental interactions prior to infection. In fact, in most instances the former interaction is the more important controlling factor. Therefore, I suggest that the term "disposition," meaning inclination or tendency, that can be applied throughout the whole process of disease development, has greater utility than "predisposition." The foregoing discussion illustrates the validity of Link's concept of a thorough-going etiology (20,21) in which the diseased plant is studied as a whole, and disposing factors and inherent susceptibility are not considered as separate categories coordinate with etiology.

YIELD POTENTIAL AND CONSTRAINTS ON YIELD

Yield can be measured as total biomass produced, but usually refers to the amount of economically valuable components harvested. In either case, it results from the interaction and integration of all physiological processes involved in growth and reproduction. Thus, it is inherited as a complex quantitative character that is influenced by the effects of many genes, but their individual role usually is not identified. There are various physiological and morphological constraints on yield (see chapters 10 and 11) that are subject to co-control by the plant's genotype and environment, and it often is difficult to determine whether environment or genotype is more influential.

Wallace et al (38) indicate, however, that frequently more than half of the yield variation in F_2 populations derived from crosses between parents with high and low yields is genetically controlled. The remainder of the yield variation is associated with factors of the "normal" environment that occur haphazardly when crop cultivars and selections are compared in field test plots. Obviously, however, if extreme fluctuations in the environment result in severe stress and permanent irreparable injury (disease), environment may exert a much greater effect than usual on yield. Thus, for maintenance of plant health, and especially for prevention of catastrophic losses, plant cultivars need to have genetic constitutions that confer as much resistance or tolerance as possible to various biotic and abiotic stress factors. This fact has been exploited with much publicized success for biotic diseases, but resistance to abiotic stress factors, for the most part, has been used inadvertently as an unidentified component of polygenically inherited yield potential.

Wallace et al (38) presented examples in several crop plants wherein genotypes within species exhibit variation in the many physiological components of yield (relative growth rate, net assimilation rate, net CO_2 exchange rate, photosynthetic enzyme activity, and harvest index, i.e., the ratio between total and economic yield). Most of these yield components are complex, quantitatively inherited characters that result from the interaction of many subcomponent processes, as well as environmental influences on the individual processes and the genes controlling them. They concluded, nevertheless, that genetic variability exists for all the yield components and that breeding for higher yield requires identification of limiting components and an understanding of their genetic control. Similarly, to breed for resistance to stress factors requires identification of the factors most likely to be encountered where a potential cultivar will be used and determination of genetic variability and genetic control of resistance to the stress factors.

Wallace (39) and Wiese (43) have suggested the use of repeated surveys and multiple regression analysis of the data to assess the relative impact of various biotic and abiotic determinants of disease and yield losses, respectively. Similar methodology probably would prove useful for decisions on which stress factors are likely to be most important and, therefore, prime targets for resistance breeding. Whether the stresses are biotic or abiotic, there is considerable evidence that genetic variability is available for exploitation. Some examples illustrating this fact will be discussed in the sections that follow.

RESISTANCE TO BIOTIC PATHOGENS

The genetics of host-parasite interactions and the use of genetic resistance for disease control has been the subject of voluminous literature that has been reviewed by many authors (7,8,15,16,24,27). Thus, I will comment only briefly on several concepts and on some of the controversy still ongoing. Much of the controversy and confusion was started by the introduction of the terms *vertical resistance* (VR) and *horizontal resistance* (HR) by Vanderplank (36).

He defined VR as being effective against some but not all races of a pathogen, whereas HR is effective against all races; VR is simply inherited whereas HR was said to be polygenically inherited. Vanderplank theorized that VR acts epidemiologically by decreasing the amount of effective incoming inoculum whereas HR reduces the rate of epidemic development. Another difference is that VR often is short-lived whereas HR is long-lived and durable. Thus, HR is the preferable type of resistance and recommendations have been made for the elimination of vertical genes in disease-resistance breeding programs (29). Detailed studies involving genetic analysis indicate, however, that both HR and VR may be controlled by a small number of genes that are race specific, and Ellingboe (8) has suggested that "horizontal resistance is resistance that has not yet been shown to be vertical." Johnson (18) argued, however, that it is not possible to determine whether all resistance is race-specific or whether any resistance is race-nonspecific. He concluded, therefore, that

> durable resistance could only be identified in cultivars that remained adequately resistant during widespread and prolonged commercial use, and that cultivars possessing such resistance were the most obvious parents in breeding for durable resistance.

Others (12,24) have proposed that durable resistance results when VR genes are pyramided in a single cultivar, but this result is not consistent for all combinations of VR genes and, therefore, may be due to other unrecognized genes in the genomes (18). The use of multiline cultivars is another approach to gene management which results in better balance among populations of host and pathogen somewhat similar to that of natural ecosystems. Obviously much more information is needed. Doubtless, however, the use of genetic resistance will continue to be one of our most effective methods for biotic disease control. However, we should avoid putting all of our eggs in one basket. To quote Browning et al (4):

> The challenge...is to develop agroecosystems that utilize specific and general *genetic* resistance to create cultivars with dilatory *epidemiologic* resistance or tolerance, and to grow these in a well-buffered cultural system that encourages natural antagonists and homeostatic tendencies in the interacting host, pathogen, and antagonist populations.

RESISTANCE TO ABIOTIC STRESS FACTORS

Although limited effort has been directed toward the development of cultivars with improved genetic resistance to abiotic stress factors, there is ample evidence that heritable variability is available for breeding cultivars with improved resistance. The following are some examples.

Epstein (9) cited several examples in which absorption, transport, and efficient utilization of iron, magnesium, boron, phosphate, and nitrogen by various crops are genetically controlled, and in some instances simply inherited. Bangerth (1) also reported differences in calcium uptake and distribution and in the incidence and severity of Ca-deficiency disorders among cultivars of tomatoes, lettuce, peanuts, brussels sprouts, cabbage, and rootstocks of several fruit species. Furthermore, cultivars of various crops (beet, barley, grape, and almond) differ significantly in tolerance to total salts (2,10) and to toxicity of specific ions such as manganese and aluminum (11,23,28,37,40). Unfortunately, the opportunity for breeding cultivars with improved nutritional characteristics has not been widely utilized (9).

Similarly, little effort has been made to develop cultivars of chilling-sensitive crops with increased resistance, and results attained have not been spectacular, although some progress has been made in increasing seed germination of cotton, soybean, beans, corn, and tomato at low temperature (35). Also, Lyons and Breidenbach (22) reported that attempts have been made to use resistance from *Passiflora* spp. and *Lycopersicon hirsutum* for breeding chilling-resistant passion fruit and tomato.

Most work on cold resistance has been done with winter wheat, but according to Gusta and Fowler (14), little improvement has been made since the introduction of Crimean wheats in the late 1800s. They suggested that the failure to make progress may be due to lack of effective screening methods for individual plants, limited information on the genetic control of cold tolerance, and the limited genetic variability for this character in the germ plasm that has been screened.

Considerably more effort has been made to develop drought resistance, especially in cereal crops. This doubtless is due to the fact that drought, which often is associated with high-temperature stress, is the most important environmental stress factor that limits crop production in a large portion of the world's arable land; some 36% of the land classified as arid or semiarid receives only 12–75 cm of rainfall annually, and the other 64% often undergoes temporary drought during the growing season (17).

Traditionally, early maturity coupled with earlier planting has been used for drought avoidance in many crops. Otherwise, selection for drought escape or tolerance usually has been done empirically and as a component of total yield. More recently, increased emphasis has been given to the identification and utilization of specific plant characteristics associated with drought avoidance and tolerance, i.e., quicker response of stomates to moisture stress, increased cuticular thickness, and leaf shape, orientation, and hairiness. The extent, density, and depth of rooting also are important characteristics related to the ability of plants to avoid or decrease drought stress (33,34).

Two characteristics used as indicators of relative drought tolerance are the recovery of stomatal function after severe stress and the relative capacity for photosynthesis during and following moisture stress (34).

The measurement of physiological characteristics is time-consuming and usually is not feasible for the screening of large, segregating populations. Furthermore, the ultimate test of drought resistance is a yield comparison in the field under natural drought conditions. Thus for breeding drought-resistant wheat, Townley-Smith and Hurd (33) recommended that crosses be made between only a few parents carefully selected by a combination of tests to insure that they have desiccation tolerance, capacity to maintain low-water deficit under stress, extensive root systems, and high-yield potential. In the F_2 generation, selections were made for highly heritable combinations of desired characteristics such as awnedness, maturity, height, leaf shape, and orientation. These selections were increased in the F_3 and large populations of the F_4 were screened for yield under drought conditions. As a result of this breeding system, Hurd and his colleagues have produced two high-yielding, drought-resistant cultivars of durum wheat that are now commonly grown in the Canadian provinces (17,33). This success story should encourage more intensive work specifically intended to improve drought resistance of other crops. Indeed, there is evidence of increased effort to develop drought-resistant cultivars of a number of other important food crops (3,26,31–33,42).

Hopefully, much more effort will be devoted in the future to the development of resistance to the various abiotic stress factors that periodically contribute to impaired plant health and, thus, to lower and variable yields. For maximum effectiveness in this endeavor, there doubtless is great need for closer cooperation among plant breeders, physiologists, and biochemists.

CONCLUSIONS

I have emphasized the necessity for a concept of etiology that includes *all* factors both internal and external involved in plant-disease causation. I also have emphasized the great importance of abiotic stress factors in etiology and the maintenance of plant health, which have received far too little attention and research effort from plant pathologists relative to that given to biotic pathogens and diseases.

The role of plant genotype is crucial in etiology and plant-health maintenance. It can contribute directly to disease development when mutant genes are defective. Moreover, we now have an example of "genetic parasitism" with the complex of genes associated with the T-DNA of *Agrobacterium tumefaciens* (25). But of even more importance is the role of suscept genotype in abiotic diseases and the interaction of the host and parasite genomes in biotic diseases. In fact, the role of a cultivar's genotype is so crucial in etiology that it essentially controls what problems are likely to occur and their severity. Thus, in effect, the plant breeder who makes selections that eventually are released as cultivars influences what plant pathologists will be doing in subsequent years. Consider, for example, the epidemics caused by *Helminthosporium victoriae* on oats and *H. maydis* on corn; both epidemics were the result of the release and widespread use of germ plasm with high susceptibility to two pathogens that otherwise would have remained of little consequence (4). Nevertheless, many plant pathologists and breeders had to become involved in long-term research efforts on various aspects of these diseases. Perhaps this was unavoidable, but it illustrates the dangers of widespread use of a narrow germ plasm base and the need for close cooperation between various disciplines involved in the maintenance of plant health. To quote Buddenhagen (5), the objectives for crop improvement should be "stable high farm yield and high quality . . . which must include all necessary aspects of parasite/crop/environment interaction."

LITERATURE CITED

1. Bangerth, F. 1979. Calcium-related physiological disorders of plants. Annu. Rev. Phytopathol. 17:97-122.
2. Bernstein, L. 1963. Salt tolerance of plants and the potential use of saline waters for irrigation. Pages 273–283 in: Desalination Research Conference. Publ. 942. Nat. Acad. Sci. Nat. Res. Council, Washington, DC. 451 pp.
3. Blum, A. 1979. Genetic improvement of drought resistance in crop plants: A case for sorghum. Pages 429–445 in: Stress Physiology in Crop Plants. H. Mussell and R. C. Staples, eds. John Wiley, New York. 510 pp.
4. Browning, J. A., Simons, M. D., and Torres, E. 1977. Managing host genes: Epidemiologic and genetic concepts. Pages 191–212 in: Plant Disease, an Advanced Treatise, Vol. 1. J. G. Horsfall and E. B. Cowling, eds. Academic Press, New York. 465 pp.

5. Buddenhagen, I. W. 1981. Conceptual and practical considerations when breeding for tolerance or resistance. Pages 221–234 in: Plant Disease Control—Resistance and Susceptibility. R. C. Staples and G. H. Toenniessen, eds. John Wiley, New York. 339 pp.
6. Colhoun, J. 1979. Predisposition by the environment. Pages 75–96 in: Plant Disease, an Advanced Treatise, Vol. 4. J. G. Horsfall and E. B. Cowling, eds. Academic Press, New York. 466 pp.
7. Day, P. R. 1973. Genetic variability of crops. Annu. Rev. Phytopathol. 11:293-312.
8. Ellingboe, A. H. 1981. Changing concepts in host-pathogen genetics. Annu. Rev. Phytopathol. 19:125-143.
9. Epstein, E. 1972. Mineral Nutrition of Plants: Principles and Perspectives. John Wiley, New York. 412 pp.
10. Epstein, E., and Jefferies, R. L. 1964. The genetic basis of selective ion transport in plants. Annu. Rev. Plant Physiol. 15:169-184.
11. Foy, C. D., Arminger, W. H., Fleming, A. L., and Lewis, C. F. 1967. Differential tolerance of cotton varieties to an acid soil high in exchangeable aluminum. Agron. J. 59:415-418.
12. Green, G. J., and Campbell, A. B. 1979. Wheat cultivars resistant to *Puccinia graminis tritici* in Western Canada: Their development, performance, and economic value. Can. J. Plant Pathol. 1:3-11.
13. Grogan, R. G. 1981. The science and art of plant-disease diagnosis. Annu. Rev. Phytopathol. 19:333-351.
14. Gusta, L. V., and Fowler, D. B. 1979. Cold resistance and injury in winter cereals. Pages 159–178 in: Stress Physiology in Crop Plants. H. Mussell and R. C. Staples, eds. John Wiley, New York. 510 pp.
15. Harlan, J. R. 1976. Diseases as a factor in plant evolution. Annu. Rev. Phytopathol. 14:31-51.
16. Hooker, A. L. 1967. The genetics and expression of resistance in plants to rusts of the genus *Puccinia*. Annu. Rev. Phytopathol. 5:163-182.
17. Hurd, E. A. 1976. Plant breeding for drought resistance. Pages 317–353 in: Water Deficits and Plant Growth. T. T. Kozlowski, ed. Academic Press, New York. 383 pp.
18. Johnson, R. 1981. Durable resistance: Definition of, genetic control, and attainment in plant breeding. Phytopathology 71:567-568.
19. Levitt, J. 1972. Responses of Plants to Environmental Stresses. Academic Press, New York. 697 pp.
20. Link, G. K. K. 1932. The role of genetics in etiological pathology. Q. Rev. Biol. 7:127-171.
21. Link, G. K. K. 1933. Etiological phytopathology. Phytopathology 23:843-862.
22. Lyons, J. M., and Breidenbach, R. W. 1979. Strategies for altering chilling sensitivity as a limiting factor in crop production. Pages 179–196 in: Stress Physiology in Crop Plants. H. Mussell and R. C. Staples, eds. John Wiley, New York. 510 pp.
23. Munns, D. N., Johnson, C. M., and Jacobson, L. 1963. Uptake and distribution of manganese in oat plants. I. Varietal variation. Plant and Soil 19:115-126.
24. Nelson, R. R. 1978. Genetics of horizontal resistance to plant disease. Annu. Rev. Phytopathol. 16:359-378.
25. Nester, E. W., and Kosuge, T. 1981. Plasmids specifying plant hyperplasias. Annu. Rev. Microbiol. 35:531-565.
26. O'Toole, J. C., and Chan, T. T. 1979. Drought resistance in cereals—Rice: A case study. Pages 373–405 in: Stress Physiology in Crop Plants. H. Mussell and R. C. Staples, eds. John Wiley, New York. 510 pp.
27. Parlevliet, J. E. 1979. Components of resistance that reduce the rate of epidemic development. Annu. Rev. Phytopathol. 17:203-222.
28. Rhue, R. D. 1979. Differential aluminum tolerance in crop plants. Pages 61–80 in: Stress Physiology in Crop Plants. H. Mussell and R. C. Staples, eds. John Wiley, New York. 510 pp.
29. Robinson, R. A. 1980. New concepts in breeding for disease resistance. Annu. Rev. Phytopathol. 18:189-210.
30. Schoeneweiss, D. F. 1975. Predisposition, stress, and plant disease. Annu. Rev. Phytopathol. 13:193-211.
31. Sheldrake, A. R., and Saxena, N. P. 1979. Growth and development of chickpeas under progressive moisture stress. Pages 465–483 in: Stress Physiology in Crop Plants. H. Mussell and R. C. Staples, eds. John Wiley, New York. 510 pp.
32. Sullivan, C. Y., and Ross, W. M. 1979. Selecting for drought and heat resistance in grain sorghum. Pages 263–281 in: Stress Physiology in Crop Plants. H. Mussell and R. C. Staples, eds. John Wiley, New York. 510 pp.
33. Townley-Smith, T. F., and Hurd, E. A. 1979. Testing and selecting for drought resistance in wheat. Pages 447–464 in: Stress Physiology in Crop Plants. H. Mussell and R. C. Staples, eds. John Wiley, New York. 510 pp.
34. Turner, N. C. 1979. Drought resistance and adaptation to water deficits in crop plants. Pages 341–372 in: Stress Physiology in Crop Plants. H. Mussell and R. C. Staples, eds. John Wiley, New York. 510 pp.
35. Vallejos, C. E. 1979. Genetic diversity of

plants for response to low temperatures and its potential use in crop plants. Pages 473–489 in: Low Temperature Stress in Crop Plants. J. M. Lyons, D. Graham, and J. K. Raison, eds. Academic Press, New York. 565 pp.

36. Vanderplank, J. E. 1963. Plant Diseases: Epidemics and Control. Academic Press, New York. 349 pp.

37. Vose, P. B., and Griffiths, D. J. 1962. Resistance to aluminum and manganese toxicities in plants related to variety and cation-exchange capacity. Nature 196:85-86.

38. Wallace, D. H., Ozbun, J. L., and Munger, H. M. 1972. Physiological genetics of crop yield. Adv. Agron. 24:97-146.

39. Wallace, H. R. 1978. The diagnosis of plant diseases of complex etiology. Annu. Rev. Phytopathol. 16:379-402.

40. Weiss, M. G. 1943. Inheritance and physiology of efficiency in iron utilization in soybeans. Genetics 28:253-268.

41. Whetzel, H. H. 1929. The terminology of phytopathology. Proc. Int. Cong. Plant Sci., Ithaca, 1926. 2:1204-1215.

42. Wien, H. C., Littleton, E. J., and Ayanaba, A. 1979. Drought stress of cowpea and soybean under tropical conditions. Pages 283–301 in: Stress Physiology in Crop Plants. H. Mussell and R. C. Staples, eds. John Wiley, New York. 510 pp.

43. Wiese, M. V. 1982. Comprehensive and systematic assessment of crop yield determinants. Pages 262–269 in: Crop Loss Assessment. Proc. E. C. Stakman Commemorative Symp. Misc. Publ. 7. Minn. Agric. Exp. Stn., St. Paul, 327 pp.

44. Yarwood, C. E. 1959. Predisposition. Pages 521–562 in: Plant Pathology, Vol. 1. J. G. Horsfall and A. E. Dimond, eds. Academic Press, New York. 674 pp.

CHAPTER 10

PHYSIOLOGICAL CONSTRAINTS TO MAXIMUM YIELD POTENTIAL

G. E. EDWARDS, S. B. KU, and J. G. FOSTER

> Plant Physiology is the study of plant function at the macroscopic, cellular and molecular levels of organization. It is not an isolated discipline, but rather it merges with other branches of science. One cannot consider the effect of light intensity, or water stress, or temperature on plant growth, in isolation from the physics of the air surrounding the plant, or from soil science, or plant breeding.... Plant Physiology is much more experimental today and will, I suggest, become more predictive in the future.
>
> —B. J. Forde[1]

It is possible to discuss physiological differences among plants and the differential effects of various environmental conditions upon plant growth. To extrapolate from such discussions how physiological phenomena may influence yield is obviously complex. To some extent, the physiological responses of a species must be considered in the context of a given environment.

A number of reviews and meetings in the last few years have addressed in some detail the physiological principles related to crop yield (4, 22, 25, 29). At an international conference held in 1975 at Michigan State University, several working groups developed research imperatives for improving crop productivity. One aspect of these recommendations was the identification of physiological problems requiring concerted study (4). The following is our personal perspective of how physiological constraints in selected areas may limit crop productivity and how these constraints must be evaluated in the future in relation to environmental conditions.

CARBON INPUT: INFLUENCE OF LEVELS OF CO_2 AND O_2

Most of the world's economically important crop species are C_3 plants. Fixation of atmospheric CO_2 by these plants occurs in the chloroplast, where ribulose bisphosphate carboxylase/oxygenase (RuBPC/O) catalyzes the carboxylation step of the reductive pentose phosphate pathway (Calvin cycle).

[1]B. J. Forde. 1972. The Banks Lecture, Predicting the Growth of Plants. Journal of the Royal New Zealand Institute of Horticulture. pp. 19-24.

RuBPC/O, however, also exhibits oxygenase activity, and competition between CO_2 and O_2 for the active site of the enzyme results in lowered photosynthetic efficiency. C_4 plants, including maize, sugarcane, and sorghum, possess two photosynthetic cell types, whose functions are integrated to increase photosynthetic efficiency. Initial fixation of CO_2 into C_4 acids occurs in mesophyll cells and is catalyzed by phosphoenolpyruvate carboxylase (PEPC), which is insensitive to O_2. Subsequent decarboxylation of the C_4 acids in bundle sheath cells results in a mechanism whereby CO_2 is concentrated at the site of RuBPC/O. Temporal separation of the two carboxylation reactions in the same cell allows plants like pineapple that perform Crassulacean acid metabolism (CAM) to function effectively under unfavorable environmental conditions such as drought. In these plants initial fixation of CO_2 into C_4 acids via PEPC occurs in the dark; decarboxylation and refixation of the released CO_2 by RuBPC/O takes place in the light, at which time stomatal closure prohibits loss of water and CO_2 from the leaf (1, 5–7).

Under their optimal conditions of temperature and light, C_4 plants fix CO_2 at rates twice those of C_3 plants and are markedly superior to C_3 plants in utilizing CO_2 at low concentrations (atmospheric or less). Photosynthesis by C_3 plants does not saturate until the CO_2 concentration in the leaf reaches 600 μl/l; C_4 photosynthesis saturates at about 100 μl/l (Fig. 1). Under current atmospheric concentrations of CO_2 (about 320 μl/l), the intercellular CO_2 concentration during photosynthesis is about 200 μl/l in C_3 plants and 150 μl/l in C_4 plants; thus, atmospheric CO_2 levels are usually limiting for photosynthesis by C_3 plants but not for photosynthesis by C_4 plants. Further, as a consequence of the competition between O_2 and CO_2 for the active site of RuBPC/O, O_2 inhibits C_3 photosynthesis, the ratio of carboxylation to oxygenation being determined by the relative concentrations of the two gases and by the kinetic properties of RuBPC/O for CO_2 and O_2. Reduction of net photosynthesis by photorespiration (the oxidation of glycolate to CO_2) lowers the photosynthetic potential of C_3 plants even more. When the O_2 concentration is low or if the CO_2 concentration is elevated, photorespiration is eliminated and net photosynthesis is enhanced. In C_4 plants, photosynthesis is insensitive to the atmospheric O_2 concentration; photorespiration is not apparent; and the CO_2 compensation point is low (0–5 μl/l) compared to that characteristic of C_3 plants (50 μl/l). It is believed that the C_4 pathway of photosynthesis serves as a CO_2-concentrating mechanism and raises the CO_2 concentration at the site of RuBPC/O to values much higher than those solely achieved by equilibration with atmospheric CO_2 levels. This mechanism would permit carboxylation to compete more favorably with oxygenation and thus suppress photorespiration. Biochemical similarities between C_4 photosynthesis and CAM suggest that a comparable mechanism ensures that carbon is assimilated under CO_2-saturated conditions.

It is not understood why the apparently wasteful process of photorespiration has persisted in C_3 plants throughout evolution. Functions for photorespiration such as providing amino acids (glycine and serine) for protein synthesis and dissipating excess photochemical energy have been proposed, but the existence of C_4 and CAM plants, the enhanced growth and yield of C_3 plants under elevated CO_2 or lowered O_2, and the presence of RuBP oxygenase activity in nonphotosynthetic organisms and photosynthetic anaerobes argues against these theories. That is, so far no obligatory function for photorespiration has been determined.

The sensitivity of C_3 photosynthesis to O_2 represents a substantial reduction in carbon assimilation. Therefore, control of O_2 inhibition of photosynthesis represents an enormous potential for increasing net carbon assimilation and thus, presumably, crop productivity. Since O_2 inhibition of photosynthesis may occur as an inevitable consequence of the reactions at the active site of RuBPC/O, an altered enzyme with differential regulation of the carboxylation and oxygenation reactions may be futile. However, genotypic variation in specific activity of RuBP carboxylase that correlates with photosynthetic capacity has been reported in tall fescue (23). Genetic variations in net photosynthesis and stomatal properties also exist in a number of crop species. These observations form the basis for searching for lines with higher photosynthetic efficiency. Chemical control of photorespiration may also be a reasonable approach. Several chemicals, such as L-hydroxy-2-pyridinemethane sulfate, 2-hydroxy-3-butynoic acid, aminoacetonitrile, and 2,3-epoxypropionic acid have been used to control glycolate metabolism in the photorespiratory process (5,28). Although these chemicals block glycolate oxidation, no sustained

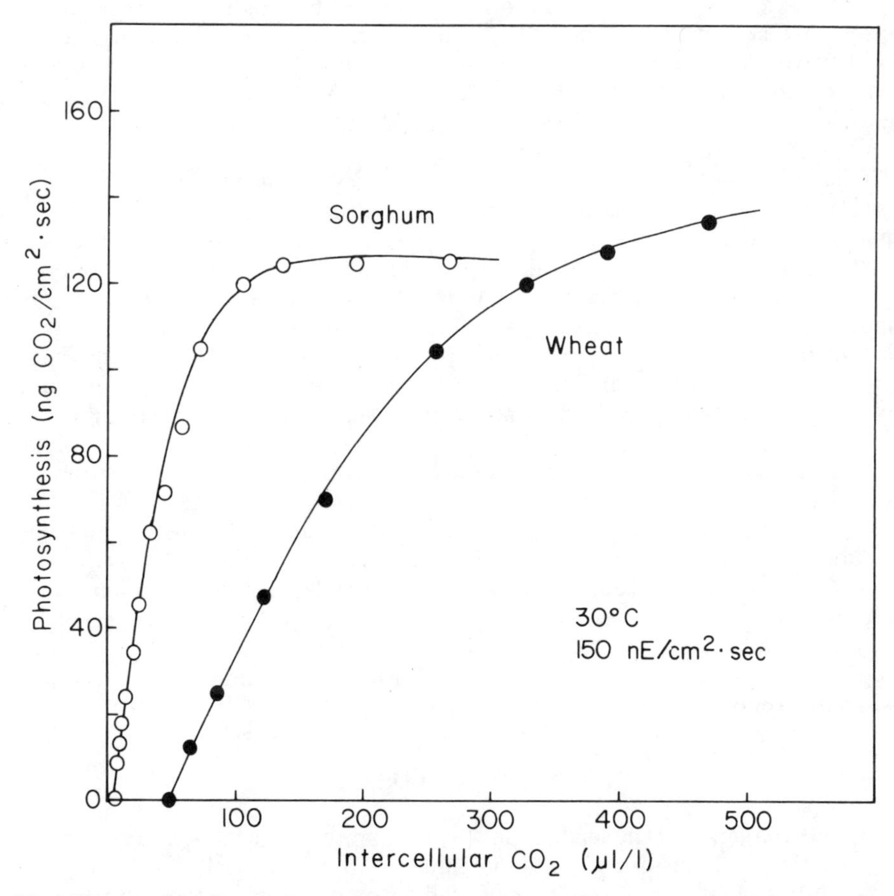

Fig. 1. Photosynthesis by a C_4 plant such as sorghum saturates at CO_2 concentration in the leaf of 100 μl/l, whereas photosynthesis by a C_3 plant such as wheat does not saturate until the CO_2 concentration in the leaf is about 600 μl/l (Ku and Edwards, *unpublished data*).

enhancement of net CO_2 uptake has been observed. Nonetheless, since net photosynthesis and crop yield in agriculturally important crops are substantially increased in elevated CO_2 and low O_2 conditions, research on chemical control of photorespiration must be continued.

From 1860 to 1977 the average levels of atmospheric CO_2 increased exponentially from about 285 to 335 μl / l (32), presumably due to deforestation and burning of fossil fuels. The annual rate of increase has accelerated in recent years. By the middle of the next century, if present trends continue, the amount of CO_2 in the atmosphere could approach twice the current value and could be six-fold higher by the year 2300. There is also an annual oscillation in atmospheric CO_2 concentration that is apparently related to the annual course of photosynthesis. This oscillation follows the change in growing season in both southern and northern hemispheres (local CO_2 level reaches its annual minimum during the growing season) but is reduced at higher elevations and diminishes toward the tropics (33).

Continued elevation of the CO_2 concentration in the atmosphere will certainly change the agricultural environment and, based on the sensitivity of photosynthesis to O_2, the impact on C_3 crops will be greater than that on C_4 crops. In fact, some improvement in yield of C_3 crops in the past may well have been due to the increased CO_2 in the atmosphere. For some years horticulturists have practiced CO_2 enrichment of enclosed crops, and increased yields by CO_2 enrichment above 600 μl/l have been reported in a number of C_3 species of economic importance, including tomato, potato, soybean, and wheat (16,31). Similar attempts to improve yields from C_4 species such as maize and sorghum (16,31) have not been successful. The CO_2-dependent increases in yield by C_3 plants, ranging from about 30 to 75%, result from an increase in the net assimilation rate. Elevated CO_2 also increases water use efficiency and symbiotic nitrogen fixation in many crops (13). Improved water use efficiency may enable many traditional crop species to extend into areas that are presently arid or may lengthen the growing season in regions with seasonal drought. The effect on crop yield of interaction between CO_2-enriched atmospheres and other environmental factors such as light, water, and temperature awaits further investigation.

Recently, it has been found that O_2 plays an important role in controlling sink activity of reproductive structures. Photosynthesis and dry matter production in soybean, rice, and sorghum have been substantially increased under low O_2, but seed production was impaired. At least 21% O_2 is required for maximum seed production in both C_3 and C_4 plants (13). The role of O_2 in reproductive growth is independent of CO_2 concentration and, thus, is not related to the O_2 inhibition of photosynthesis in C_3 plants. Hopefully, further studies will lead to a chemical or physical explanation for this role of O_2 and reveal methods to regulate reproductive growth and harvest indices (ratio of economic growth to total biomass produced).

NITROGEN USE

Crop productivity is dependent on the availability of nutrients, and continual fertilization is necessary if yield potential is to be realized. Since limiting availability of nutrients can be expected in many cases in the future, it is important to identify differences in the ability of various crops and weeds to competitively extract nutrients from the soil. This will be dependent on the development not only of crops that are productive under high fertilization but also of crops that can compete effectively in environments where leaching or irreversible adsorption of nutrients by soils diminish the efficacy of fertilization (22).

With the rising cost of fertilizer, there is increasing need for crops that show maximum growth per unit of nutrient applied, especially in developing countries. Thus nitrogen, which is the most expensive component of fertilizer, is attracting considerable attention in the plant sciences, particularly with respect to improving biological N_2 fixation in existing crops that fix N_2 symbiotically (e.g., soybeans), developing the capacity for biological N_2 fixation in crops lacking it, and evaluating the efficiency of growth on inorganic nitrogen within and among species. The main food crops of the world, maize, potatoes, and the small grains, lack symbiotic N_2 fixation through *Rhizobium*. However, it has been shown that some diazotrophs may associate with various cereals and grasses, providing potential for biological N_2 fixation. Enhancement of both photosynthesis and the capacity for N_2 fixation in legumes (e.g., soybeans) grown in CO_2-enriched environments has demonstrated that biological N_2 fixation is limited by photosynthesis and the availability of photosynthate to the microorganisms (13). Since C_4 plants, which lack symbiotic N_2 fixation, are known to have a high capacity for carbon assimilation, introduction of the capacity for N_2 fixation conceivably could greatly improve productivity of these species. Slow progress in achieving this goal, however, emphasizes the need for continued selection of crops having high efficiency in using inorganic nitrogen (nitrate and ammonia) in the immediate future.

The nitrogen content of leaves ranges from about 1.5 to 7% of the dry weight and is dependent, in part, on the availability of nitrogen through fertilization. Photosynthesis tends to increase linearly with increasing total nitrogen content of the leaf, but in C_4 plants like maize, photosynthesis per unit of leaf nitrogen is substantially higher than that of C_3 plants such as rice and wheat (3,24; Fig. 2). The greater nitrogen use efficiency in C_4 plants (expressed as biomass produced or carbon assimilated per unit of plant nitrogen) may be due to their CO_2-concentrating mechanism, which provides near saturating levels of CO_2 for RuBP carboxylase, resulting in a more efficient use of the photosynthetic machinery. As a consequence, C_4 plants may be more competitive in areas where the nitrogen content of the soil is low. Further research is needed to determine whether certain plants (including many C_4 species) that evolved in tropical or arid soils low in nitrogen content have an adaptive advantage for uptake and utilization of inorganic nitrogen or other nutrients.

TEMPERATURE

Many aspects of physiology and biochemistry could be considered in analyzing possible effects of low or high temperature on plant growth. Over a certain temperature range, activity of a process will be related to its intrinsic dependence on temperature while, outside that range, irreversible or near irreversible damage may occur. The effects of temperature on photosynthesis vary among species and ecotypes within species. A given species has a temperature optimum for photosynthesis that depends on both inherent genetic factors and environmental conditions. In some species, this optimum may be more or less fixed, whereas, in others, there is acclimation (showing phenotypic plasticity) to seasonal changes in temperature. This ability to adapt to a change in temperature has been recognized in some plants (2), but it has not been studied in most crops. Large fluctuations in temperature may occur during the course of a growing season. For a plant such as maize, which experiences temperatures as high as 30–35 C during much of the growing season, inability to adapt to low temperatures in the initial stages of growth could represent a physiological constraint on yield.

Various species are recognized as being suited for growing best in either tropical, subtropical, temperate, or arctic environments. Most plants of economic importance are not able to tolerate extremes in temperature and have a rather narrow temperature optimum for maximum growth. Although there are exceptions, C_4 and CAM plants are generally more tolerant of high temperatures and more sensitive to low temperatures than are C_3 plants, as illustrated for temperature dependence of photosynthesis in Fig. 3. CAM species (e.g., cacti) that evolved in arid climates may experience day-night extremes in leaf temperature of 45 C and 15 C. Stomatal closure in these plants during the day as a means to conserve water can result in extremely high leaf temperatures when the heat load is high and transpirational leaf cooling is low. Survival under these conditions requires enzyme and membrane stability over a wide temperature range (11), and the inherent ability of CAM plants, in contrast to C_3 and C_4 species, to tolerate such extremes of temperature and conserve water may

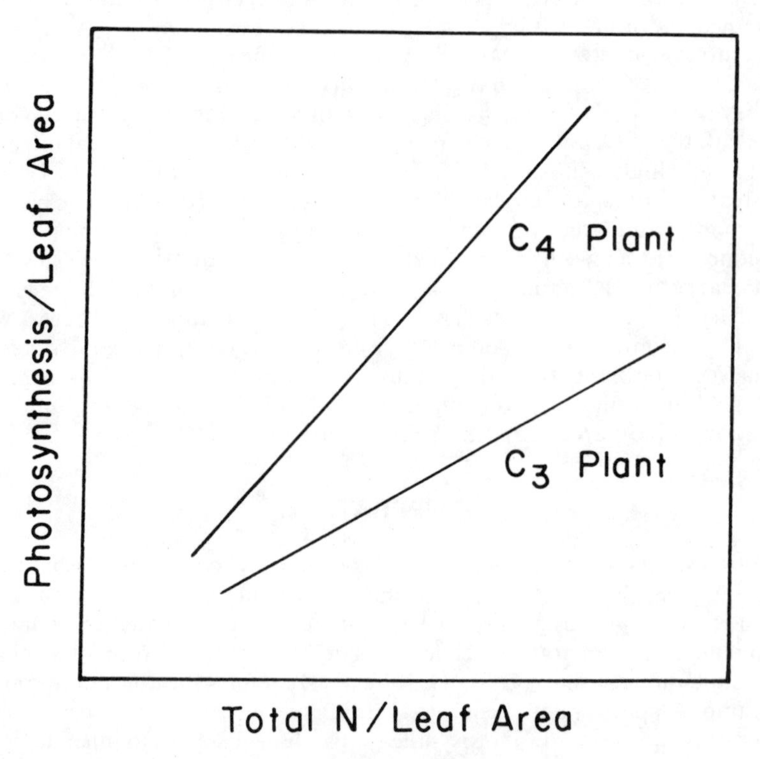

Fig. 2. There is some evidence, particularly among the monocotyledons, that C_4 plants like maize have a greater nitrogen use efficiency than C_3 plants like wheat (3,33).

become more important in the future as the need for making use of marginal, hot, semiarid land increases.

Many C_4 plants like corn and sugarcane are better able to grow under high temperatures than are C_3 plants such as wheat and barley. In part, this may be due to the fact that many of the C_4 plants evolved under tropical or subtropical conditions. However, it is recognized that the mechanism of photosynthesis in C_4 plants itself provides a high internal concentration of CO_2 for carbon assimilation. Photosynthesis in C_3 plants is limited at 25–40 C due to the low concentration of CO_2 in the atmosphere and the decreasing solubility of CO_2 with increasing temperature (17,18). Such plants also have an increased level of photorespiration at high temperatures. Because of the doubling of the atmospheric level of CO_2, as is projected to occur in 50 years (32), the capacity for growth of these plants could increase dramatically at the higher temperatures—two- to threefold depending on the species and other conditions such as water and nutrients (Fig. 4). There would be much less effect of increased atmospheric levels of CO_2 on C_4 and CAM plants since these plants already photosynthesize at relatively high internal concentrations of CO_2. In CAM plants the CO_2 concentration in the leaf can be an order of magnitude higher than the external level during fixation of internal CO_2 derived from malate decarboxylation in the light.

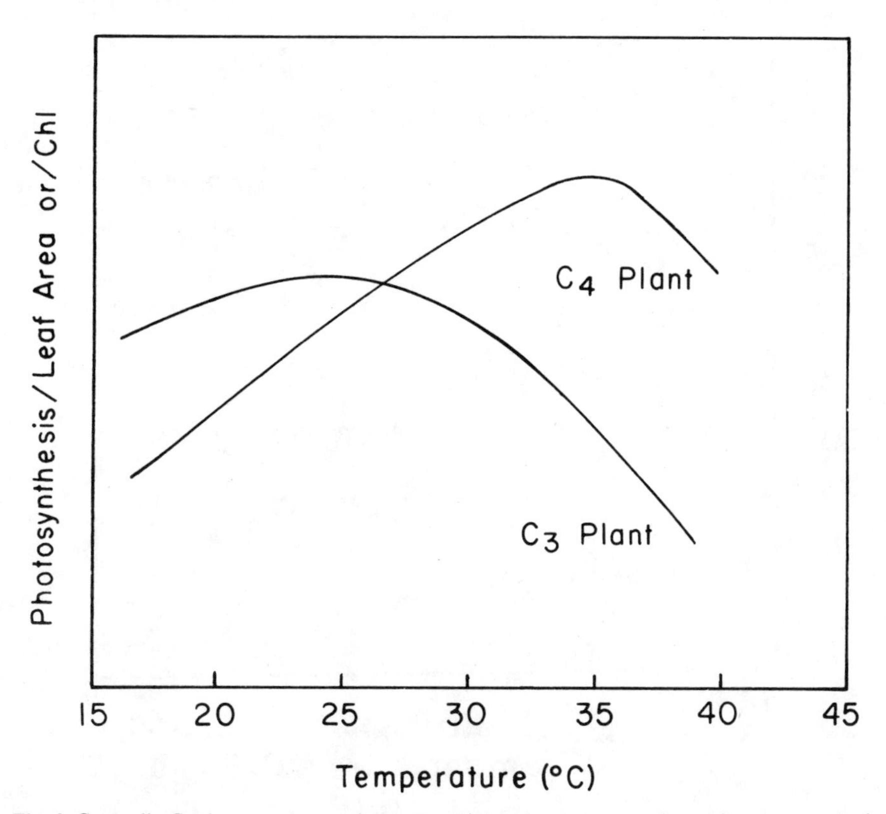

Fig. 3. Generally C_4 plants, such as maize, have a higher temperature optimum for photosynthesis than C_3 plants like wheat.

In plants having a low temperature optimum for growth, a number of processes, in addition to photosynthesis, may be affected by high temperatures. Such plants generally have rather high rates of dark respiration at high temperatures. Irreversible damage to the photochemical electron transport system (photosystem II), denaturation of proteins, reduction in the capacity to synthesize protein, and loss of semipermeability properties of membranes at 40–50 C, depending on the species, impose extensive limitations on growth and development. In C_3 species, the first major effect is observed at moderately high temperatures (25–35 C) and is due to increased photorespiration and the limited availability of CO_2 for photosynthesis. Therefore either elimination of photorespiration in C_3 plants or the increase in atmospheric CO_2 in the future may allow these species to perform better at high temperatures. Extremely high leaf temperatures (35–45 C), which may become more common if the earth gets warmer due to the greenhouse effect from increasing concentrations of atmospheric CO_2, will necessitate cultivation of species (likely C_4 or CAM plants) that are particularly adapted to tolerate such temperatures.

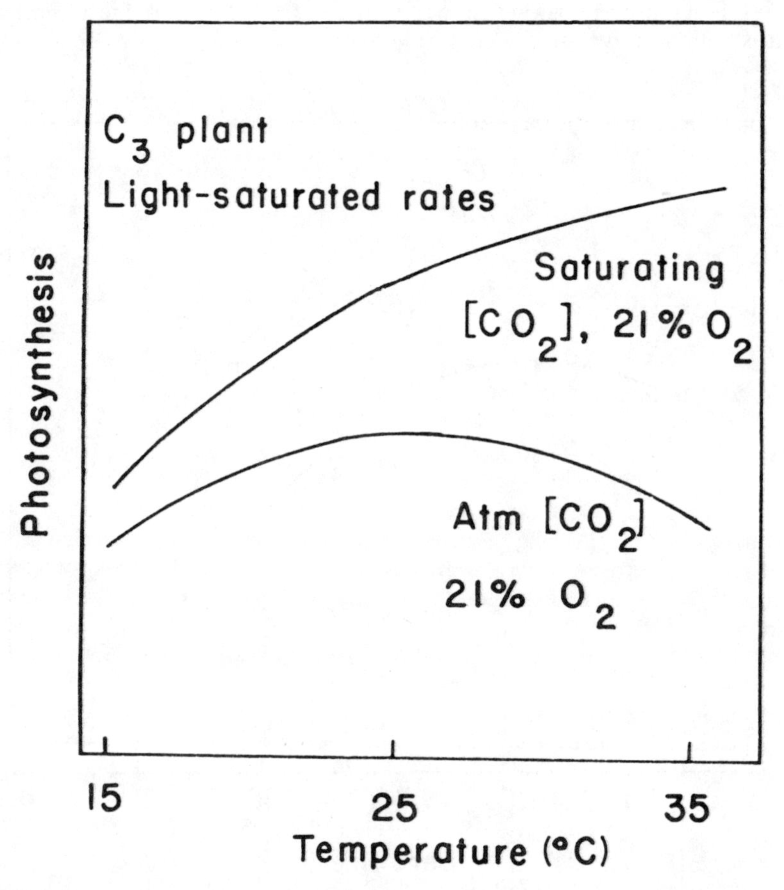

Fig. 4. The influence of temperature on photosynthesis in a C_3 plant under nonphotorespiring conditions with varying levels of CO_2 and light.

Tolerance of C_3 plants to low temperatures and their generally better photosynthetic performance under these conditions, compared to that of C_4 plants, may be due, in part, to differences in the levels of certain photosynthetic enzymes, the cold lability of some key enzymes of the C_4 pathway (pyruvate, P_i dikinase, PEPC) and differences in the phase transition of lipids in membranes. The degree of saturation of lipids in membranes affects their interaction with proteins and the temperature at which the membrane solidifies as it converts from a fluid to a gel. The latter will influence metabolite transport across membranes as well as membrane-associated electron transport systems of the chloroplasts and mitochondria.

At very low temperatures the catalytic capacity of RuBP carboxylase may be more rate limiting in C_4 than in C_3 plants since the level of the enzyme is severalfold higher in C_3 than in C_4 species.

It is interesting that differences occur in the cold-lability of pyruvate, P_i dikinase between species, between varieties within a species, and during the course of a growing season. Cultivars of maize grown in the northern part of Japan have a more cold-tolerant pyruvate, P_i dikinase than cultivars grown in the southern part of Japan. At low temperatures the enzyme dissociates and becomes inactive (6). Species in which pyruvate, P_i dikinase is stable at low temperatures apparently have an additional protein factor that stabilizes the enzyme (T. Sugiyama, *personal communication*).

Knowledge of why some species can adapt to varying temperatures and others cannot may eventually be of use in selecting crops that are better able to cope with broad fluctuations in temperature during a growing season. Physiologically based means of improvement of crop growth in relation to temperature should be pursued but not considered a substitute for the empirical and common-sense approaches that have been used in the past.

LIGHT

Light is an important environmental factor controlling plant growth and development, but physiological constraints may limit the ability of a plant to effectively utilize available light energy. Measurements of the quantum efficiency, expressed as moles of CO_2 fixed per mole of quanta absorbed, are made at low light (up to 5% of full sunlight) where light, not CO_2, is a limiting factor. Under this condition, carbon assimilation increases linearly with increasing levels of light. The quantum efficiency is remarkably similar for C_3 and C_4 plants at atmospheric conditions (1 CO_2 fixed per approximately 15 quanta absorbed) even though there are striking differences in the biochemistry of carbon assimilation.

However, in C_3 plants, the quantum efficiency decreases with increasing temperature and falls below that of C_4 plants due to the high levels of photorespiration (photorespiration utilizes energy that could otherwise be used in CO_2 assimilation). In these plants the efficiency of light utilization is increased by about 50% when photorespiration is eliminated (by increasing CO_2 concentrations or lowering the O_2 concentrations). Therefore, increases in the atmospheric amounts of CO_2 in the future will be beneficial to C_3 plants, which includes most of our crop species, as their quantum efficiency for photosynthesis will increase.

The efficiency of light utilization at high light intensities varies widely among species. Measurements on single leaves show that photosynthesis of most C_3 plants growing in direct light is saturated at about 25–40% of full sunlight, whereas photosynthesis in C_4 plants increases up to full sunlight. In plants adapted to grow in shade (C_3 or C_4), photosynthesis becomes light saturated at a relatively low light intensity. Analysis of the physiological basis for such responses may contribute to improving the efficiency of light utilization among crops.

In a closed canopy, photosynthesis of C_3 plants usually continues to increase up toward full sunlight, although generally not at a linear rate. Therefore, a canopy may use light more efficiently than single leaves can. Considering the angle of the sun and canopy structure, most leaves in the canopy receive less than direct full sunlight. As a crop develops in the early part of the growing season, the canopy will be open and the light response of the crop may be closer to that of single leaves. Particularly during this period, C_4 plants, which use much light with greater efficiency, may grow faster than C_3 plants (if other conditions are optimal). For this reason some of the worst weeds are C_4 plants and may be very competitive in early stages of their growth.

In the future, as physiological processes become better understood, growth can be modeled, using a mechanistic rather than empirical approach, by considering physiological and environmental factors. Some potential for this is illustrated in Fig. 5 by a computer-generated plot of the response of photosynthesis to varying light and CO_2 concentrations based on equations derived from considerations of the mechanism of photosynthesis.

WATER

Water availability is probably one of the most critical factors limiting terrestrial plant productivity. Crop yield can be diminished by moisture excess or deficit. Flooding reduces soil aeration and thus the supply of O_2 available to roots. With poor aeration, beneficial microbial activity and water and nutrient uptake by plants are seriously inhibited. On the other hand, severe drought can cause stomata in the leaf to close, reducing photosynthesis. Since water acts as a transporting agent and participates as a reactant in many biochemical processes, water deficit also leads to reduction of a number of other physiological activities such as nutrient uptake, carbohydrate and protein metabolism, transportation of ions and metabolites, and general plant growth and development.

Plants vary widely in their efficiency of water usage (grams of H_2O transpired per gram of CO_2 assimilated). Generally, C_4 plants are about twice as efficient as C_3 plants in utilizing water, as can be seen in retrospect in the report of Shantz and Piemeisel (26) early in this century. Efficient use of available water by these plants is due to a combination of a slightly higher stomatal resistance to gas exchange and a more efficient assimilation of CO_2 through the C_4 pathway of photosynthesis. These strategies for efficient water utilization, which entail both anatomical and physiological features, have been considered as a mechanism for adaptation and evolution of C_4 plants in xeric environments. However, under semiarid conditions, there are adaptations to water limitations among C_3 plants that are yet to be exploited in crop production. For example, very high rates of photosynthesis have been identified in C_3 herbaceous desert annual species (20). The water use efficiency of CAM plants, which is even higher than that of C_4 plants, is consistent with the theory that CAM plants are best adapted to survival in arid lands.

In addition to some morphological adaptations (e.g., leaf succulence, sunken stomata, and tap roots), plants that are tolerant or adapted to water stress may possess some unusual physiological and metabolic characteristics. The identification and measurement of such features are very important in a selection program for crop improvement. One approach that may prove successful is to follow the sequence of metabolic events in tolerant and susceptible lines or species exposed to water stress. For example, under excess water conditions, flood-intolerant species show a rapid increase in the rate of glycolysis and accumulate toxic amounts of alcohol (10), whereas some flood-tolerant species tend to accumulate nontoxic organic acids, especially malate, and are able to make effective use of nitrate as an alternative electron acceptor to O_2 during anaerobiosis.

Proline has been identified as a metabolite that may be beneficial to plants under drought, salt, and cold stress. Under drought conditions, total free amino

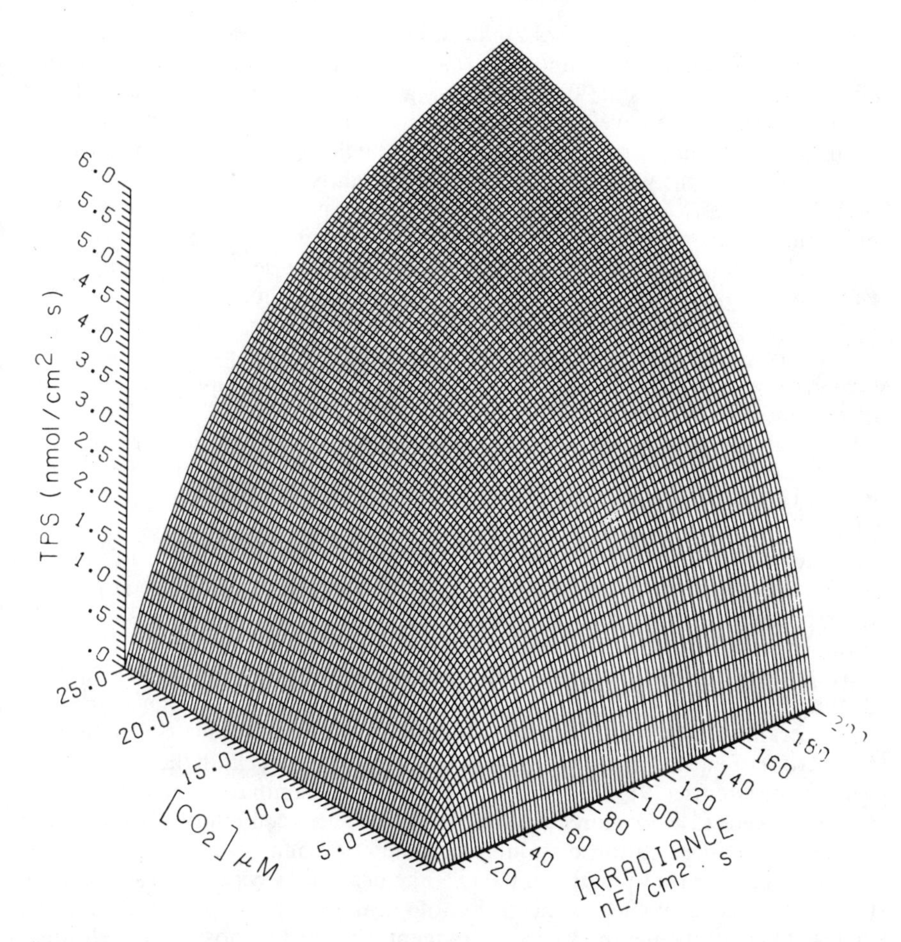

Fig. 5. Computer-generated plot of the true rate of photosynthesis (TPS) in a C_3 plant under nonphotorespiring conditions with varying levels of CO_2 and light. The model incorporates values of the maximum carboxylation potential, maximum capacity for generation of energy, carboxylation efficiency, and quantum efficiency (H. Farazdaghi and G. Edwards, *unpublished data*).

acids in leaves of many species (both C_3 and C_4) often increase, but the increase in proline is most pronounced. In barley, the extent of proline accumulation by seedlings of 10 cultivars exposed to water stress was positively correlated with their drought resistance ratings (27). Proline may serve as a storage compound for reduced carbon and nitrogen during stress. The relationship between proline metabolism and drought tolerance needs to be investigated in a wide range of species and cultivars.

In the future, in order to overcome the physiological constraints imposed on crop productivity by water availability, experimental approaches should be directed toward a better understanding of the physiological and metabolic responses of plants to water stress, the range of genetic variation, and the basic mechanism of plant adaptation to water stress.

SALINITY

Crop yield from the earth's 3.2 billion ha of land surface that is potentially suitable for cultivation (4) is limited, to a large extent, by the availability of water and the salinity of the soil (9). Through irrigation, water can be supplied during critical stages of the developmental process, and many marginal regions have been converted into cropland. However, as much as two thirds of the United States currently has water that is considered slightly saline, and saline water is prevalent in many other countries as well (21). Irrigation, therefore, is a mixed blessing. Deserts offer fertile soil, an abundance of sunlight, and a long growing season. Irrigation provides the appropriate quantity and seasonal distribution of water necessary for plant growth and development, but, at the same time, increases the salinity of the soil and, in many cases, exacerbates the detrimental effects of existing salt concentrations. Irrigation for other purposes, including seed germination, crop cooling, and frost protection, contributes to salinization of soil not routinely irrigated to relieve drought.

Virtually all of the world's economically important crop species are sensitive to salt. Because the metabolic machinery of cells (30), including enzyme activity (9,12,14,15) and membrane permeability (8,9), are adversely affected by salt, plants utilize energy derived from photosynthesis to exclude salt during uptake of water and nutrients from the soil. The osmotic stress that accrues as salt accumulates at the root surface causes tissue desiccation. Diminished growth, interrupted developmental processes, and plant death result in unrealized production potential.

Halophytes not only exclude salt, but also possess internal mechanisms for dealing with saline environments. Dynamic cytoplasmic activity, even in these salt-tolerant species, is not adapted to high concentrations of ions. Salt, admitted by the roots, is stored in cell vacuoles or in special storage cells of the epidermis of leaves and stems (19). In addition, salt may be eliminated through special excretory cells (21). Osmoregulation can also prevent cell injury. By increasing the cellular concentration of soluble organic compounds such as proline and glycerol, salt-tolerant plants can balance the osmotic effect of high soil salinity (9,14). As a consequence of these physiological and biochemical mechanisms, yields from salt-tolerant species frequently surpass those of traditional agricultural species. However, energy used to exclude, sequester, or excrete salt is energy that could be used more profitably for vegetative and reproductive growth. Moreover, germination, even of salt-tolerant species, is generally inhibited by saline conditions, and subsequent effects of salinity vary with plant age, humidity, day length, and light intensity (9). It is of interest that some CAM plants that evolved in regions subject to fluctuations in water availability also tolerate relatively high amounts of salt.

Improving productivity from soil that is experiencing a continual rise in salinity may be accomplished by: 1) altering the environment, 2) biochemically or genetically manipulating traditional crop species toward salt tolerance, or 3) utilizing more tolerant species. Until recently, better management of irrigated land to minimize salt accretion and reclamation of marginal lands were viewed as the most practical approach. However, the problem of increasing soil salinity conceivably can be circumvented by breeding or engineering salt tolerance. Although time-consuming, identification of individuals possessing desirable characteristics through selection and breeding studies has had limited success. The process has been shortened by the use of cell cultures to select for tolerance and the development of procedures for generation of plants from cultured cells, but single cells and intact plants frequently respond differently to environmental factors.

Variations in plant responses to soil salinity suggest that adaptability may be controlled genetically. Before genetic engineering can provide a solution to the problem of low productivity by saline soils, the factors that confer salt tolerance on plants must be identified. Certainly anatomy plays a critical role in the ability of some plants to perform well under conditions of osmotic stress. Metabolic regulation may be equally important. Enzymes of halophytes do not require salt and are adversely affected by salt, but the accumulation of amino acids, polyols, and betaine under stress clearly indicates a biochemical response to the environment (14). The presence of elevated levels of organic compounds, however, has not proved to be a good marker for tolerance to salt stress. An intensive effort is needed to determine the degree to which the structure and function of cellular membranes are related to adaptability to osmotic stress. Cross-adaptation, the ability of a plant exposed to one adverse environment to modify its response to a second stressful situation, and the utilization of hormones to ameliorate the effects of environmental stress need further investigation to identify the mechanism(s) involved and to evaluate the feasibility of incorporating these processes into nontolerant species.

Recently, research has shown that a number of halophytes are more productive in terms of biomass and have a nutritional value that is equivalent, or superior, to many of our staple crops (21). Cultivation of species such as these would permit the extension of agriculture into regions considered marginal by current standards and would help supply the world's increasing demands for food, fiber, and fuel. Success of halophytic agriculture, however, depends upon the development of techniques for desalinizing tissue that does not exclude salt and for utilizing tissue having a high salt content.

LITERATURE CITED

1. Andrews, T. J., and Lorimer, G. H. 1978. Photorespiration—Still unavoidable? FEBS Lett. 90:1-9.
2. Berry, J., and Björkman, O. 1980. Photosynthetic response and adaptation to temperature in higher plants. Annu. Rev. Plant Physiol. 31:491-543.
3. Brown, R. H. 1978. A difference in nitrogen use efficiency in C_3 and C_4 plants and its implications in adaptation and evolution. Crop Sci. 18:93-98.
4. Brown, A. W. A., Byerley, T. C., Gibbs, M., and San Pietro, A., eds. 1975. Crop Productivity-Research Imperatives. Int.

Conf. sponsored by Mich. State Univ. Agric. Exp. Stn. and the Charles F. Kettering Foundation. Mich. Agric. Exp. Stn., East Lansing.

5. Chollet, R. 1977. The biochemistry of photorespiration. Trends Biochem. Sci. 2:155-159.
6. Edwards, G. E., and Huber, S. C. 1981. The C_4 pathway. Pages 237–281 in: The Biochemistry of Plants. A Comprehensive Treatise, Vol. 8, Photosynthesis, M. D. Hatch and N. I. Boardman, eds. Academic Press, New York.
7. Edwards, G. E., and Walker, D. A. 1982. C_3, C_4: Mechanism, Cellular and Environmental Regulation of Photosynthesis. Blackwell, Oxford, U.K. 488 pp.
8. Ellouze, M., Ghorsalli, M., and Cherif, A. 1980. Action du chlorure de sodium sur la composition lipidique des feuilles due tournesol (*Helianthus annuus* L.) et de la "Lime Rangpur" (*Citrus limonia* Osbeck). Physiol. Veg. 18:1-10.
9. Flowers, T. J., Troke, P. F., and Yeo, A. R. 1977. The mechanism of salt tolerance in halophytes. Annu. Rev. Plant Physiol. 28:89-121.
10. Garcia-Novo, F., and Crowford, R. M. M. 1973. Soil aeration, nitrate reduction and flooding tolerance in higher plants. New Phytol. 72:1031-1039.
11. Gerwick, B. C., Williams, G. J., Spalding, M. H., and Edwards, G. E. 1978. Temperature response of CO_2 fixation in isolated *Opuntia* cells. Plant Sci. Lett. 13:389-396.
12. Greenway, H., and Osmond, C. B. 1972. Salt responses of enzymes of species differing in salt tolerance. Plant Physiol. 49:256-259.
13. Hardy, R. W. F., Havelka, U. D., and Quebedeaux, B. 1976. Opportunities for improved seed yield and protein production: N_2 fixation, CO_2 fixation, and O_2 control of reproductive growth. In: Genetic Improvement of Seed Proteins. National Acad. Sci. Washington, DC. 394 pp.
14. Hellebust, J. A. 1980. Reactions to water and salt stress in plants. Pages 147–156 in: Plant Membrane Transport: Current Conceptual Issues. R. M. Spanswick, W. J. Lucas, and J. Dainty, eds. Elsevier/North Holland Biomedical Press, Amsterdam.
15. Kalir, A., and Poljakoff-Mayber, A. 1981. Changes in activity of malate dehydrogenase, catalase, peroxidase, and superoxide dismutase in leaves of *Halimione portulacoides* (L.) Aellen exposed to high sodium chloride concentrations. Ann. Bot. 47:75-85.
16. Kramer, P. J. 1981. Carbon dioxide concentration, photosynthesis, and dry matter production. BioScience 31:29-33.
17. Ku, S. B., and Edwards, G. E. 1977. Oxygen inhibition of photosynthesis. I. Temperature dependence and relation to O_2:CO_2 solubility ratio. Plant Physiol. 59:986-990.
18. Ku, S. B., Edwards, G. E., and Tanner, C. B. 1977. Effects of light, CO_2, and temperature on photosynthesis and transpiration in *Solanum tuberosum*. Plant Physiol. 59:868-872.
19. Lüttge, U. 1975. Salt glands. Pages 335–336 in: Ion Transport in Plant Cells and Tissues. D. A. Baker and J. L. Hall, eds. Elsevier/North Holland Biomedical Press, Amsterdam.
20. Mooney, H. A., Ehleringer, J., and Barry, J. A. 1976. High photosynthetic capacity of a winter annual in Death Valley. Science 194:332-334.
21. Neary, J. 1981. Pickleweed, palmer's grass, and saltwort. Can we grow tomorrow's food with today's salt water? Science81 2(5):38-43.
22. Proceedings of the symposium increasing the biological efficiency of vegetable crops. 1978. HortScience. 13:672-686.
23. Randall, D. D., Nelson, C. J., and Asay, K. H. 1977. Ribulose bisphosphate carboxylase. Altered genetic expression in tall fescue. Plant Physiol. 59:38-41.
24. Schmitt, M., and Edwards, G. E. 1981. Photosynthetic capacity and nitrogen use efficiency of maize, wheat, and rice: A comparison between C_3 and C_4 photosynthesis. J. Exp. Bot. 32:459-466.
25. Schrader, L. E. 1980. Contributions from biochemistry and plant physiology. Pages 25–43 in: Moving Up the Yield Curve—Advances and Obstacles. L. S. Murphy, E. Doll, and F. Welch, eds. Am. Soc. Agron. and Soil Sci. Soc. Am., Madison, WI.
26. Shantz, H. L., and Piemeisel, L. N. 1927. The water requirement of plants at Akron, Colo. J. Agric. Res. 34:1093-1190.
27. Singh, T. N., Paleg, L. G., and Aspinall, D. 1972. Proline accumulation and varietal adaptability to drought in barley: A potential metabolic measure of drought resistance. Nature New Biol. 236:188-190.
28. Usuda, H., Arron, G. P., and Edwards, G. E. 1980. Inhibition of glycine decarboxylation by aminoacetonitrile and its effect on photosynthesis in wheat. J. Exp. Bot. 31:1477-1483.
29. Wallace, D. H., Ozbun, J. L., and Munger, H. M. 1972. Physiological genetics of crop yield. Adv. Agron. 24:97-146.
30. Wignarajah, K., and Baker, N. R. 1981. Salt induced responses of chloroplast activities

in species of differing salt tolerance. Photosynthetic electron transport in *Aster tripolium* and *Pisum sativum*. Physiol. Plant. 51:387-393.

31. Wong, S. C. 1979. Elevated atmospheric partial pressure of CO_2 and plant growth. I. Interactions of nitrogen nutrition and photosynthetic capacity in C_3 and C_4 plants. Oecology 44:68-74.
32. Woodwell, G. M. 1978. The carbon dioxide question. Sci. Am. 238:34-43.
33. Woodwell, G. M., Whittaker, R. H., Reiners, W. A., Likens, G. E., Delwiche, C. C., and Botkin, D. B. 1978. The biota and the world carbon budget. Science 199:141-146.

CHAPTER 11

MORPHOLOGICAL CONSTRAINTS ON MAXIMUM YIELD POTENTIAL

R. M. SACHS

Form and function are ultimately, if not always obviously, related. Hence, it is inevitable that physiological constraints to yield overlap with morphological ones. But there is a practical, research-related reason for dissecting out morphological from physiological constraints, namely, the fact that topics such as canopy structure, phenology, sink strength, assimilate partitioning, and meristematic activity are poorly defined at the physiological systems level. A geneticist can readily get a measure of, for instance, assimilate partitioning (in terms of harvest index) without referring to underlying physiological and/or biochemical systems. An indication of confusion of terms is found in a paper on physiological aspects of peanut yield improvement (17) in which the three major physiological processes studied were harvest index, rate of fruit set, and duration of fruit filling, none of which are mentioned in Chapter 10 (this volume). They are reviewed here because in the author's view they are best "understood" in morphological terms.

MERISTEMS—DETERMINANTS OF FORM AND CENTERS OF ASSIMILATE UTILIZATION (SINKS) AND SOURCE TISSUE DEVELOPMENT

From germination through fruit growth, the placement and functioning of meristems, and the control of meristematic activity, regulate the size, shape, and time of formation of plant organs. Active meristems are regions of intense metabolic activity and are, thus, the major mobilization centers.

Concentration gradients of assimilates from photosynthesizing source leaves to meristematic "sink" tissues are the parameters accounting for assimilate partitioning. Relative steepness of the gradients is a function of the metabolic storage activities of competing sinks and their nearness to source leaves (11).

Harvestable products, such as expanding fruit, storage roots and tubers, and stem tissues (e.g., sugarcane) accumulate photosynthetic assimilates and nutrients at relatively rapid rates. Other meristematic tissues, particularly the shoot and root apical meristems and cambium in woody perennials, compete with these centers for assimilates. Consciously or not, distribution of assimilates between the harvested product and competing tissues is selected by breeders in crop improvement programs.

Development of source tissue, namely photosynthetically active leaves, is dependent upon leaf initiation at the shoot apical meristem and subsequent leaf expansion resulting from continued meristematic activity in the leaf primordia. For some time, young, expanding leaves are net sinks and their transition to the net source stage varies, according to species and cultural conditions (34), undoubtedly as a function of assimilate demands for growth. Rapid leaf initiation and expansion are the two key factors ensuring early canopy closure for most efficient light interception.

Spring vigor, i.e., the capacity for relatively rapid growth at low temperatures, is a particularly important yield-limiting factor for temperate zone crops. According to Troyer (49), spring vigor in corn seedlings is a heritable trait that can be incorporated into otherwise desirable cultivars and it extends the growing season as effectively as does delayed leaf senescence at grain filling. Vigor is a term that refers to the relative growth potential of meristematic tissues.

CANOPY ARCHITECTURE

For the last 20 years it has been generally recognized that for high leaf area indices (LAI), photosynthetic efficiency is greater with more erect leaves (16,18). Since high LAI are sought to ensure complete interception of incident irradiation—(and experimental data indicate that yield in most crops is a direct function of LAI up to values of 4)—it is likely that more erect canopies will be selected. Most of the work on canopy architecture has been limited to grain crops; thus, there may be special features of those crops that are linked to leaf position. In corn, the leaf subtending the developing ear, the most important one for kernel development, is relatively deep in the canopy, and upright leaves may indeed let more light fall on the critical ear leaves (15). The uppermost (flag) leaves of barley, wheat, rice, and other grain crops play the principal role for grain filling and are unlikely to be shaded to any great extent. Erect architecture in cereal crops may provide some other as yet unrecognized benefits—such as reduced water stress in the flag leaf owing to less incident radiation at mid-day when atmospheric water deficit may exceed the plant's water transport and transpiration capacity. In some leguminous species, leaves move to a more erect position in response to water deficit; in these species a more upright architecture confers greater drought resistance (14,28). Drought avoidance is influenced by canopy architecture in another important way. Light penetration into the canopy may promote root growth simply by promoting greater photosynthesis in the lower leaves, which are the primary sources of photosynthates to the root system (18). Andries et al (2,3) found greater yields in more open foliar architecture in okra-leafed cotton cultivars, which permitted greater penetration of light to developing bolls and more air circulation within the canopy. There were concomitant reductions in boll rot and boll weevil reproduction, resulting probably from lower humidities within the canopies.

EVOLUTION IN WHEAT: A CASE STUDY IN YIELD INCREASE THROUGH CHANGED ASSIMILATE DISTRIBUTION

Evans and Dunstone (19) examined several physiological and morphological traits of lines of wheat in greenhouse, pot plant studies with representatives from wild, diploid progenitors and improved tetraploid and hexaploid cultivars (Table 1). The plants were grown under noncompetitive conditions with optimal moisture and nutrition, temperature (including vernalization where required), light intensity, and daylength. Under these noncompetitive conditions, evolutionary advance from diploid to hexaploid cultivars was noted by the increased size of leaf and grain and the *redistribution of dry matter* rather than by increased efficiency of photosynthesis (Table 1). Note that grain yield per plant was relatively high in some of the diploid lines, but in highly competitive (high population density) field conditions, yield per unit area may be very different. Grain weight was greater 1) in the cultivated than in the wild species and 2) in species with higher ploidy. Leaf area, total per plant and of individual leaves, increased in proportion to grain weight, but photosynthesis rate decreased in proportion to leaf area. Nevertheless, the larger area of the flag leaves more than offset the lower rates of photosynthesis—flag leaf photosynthesis, together with that of the ear, accounts for essentially all of the assimilates in wheat grain (20). Large leaf area in dense planting may be a disadvantage owing to mutual shading by neighboring plants, but if the flag leaves tend to be more upright, this problem is reduced.

With high soil moisture contents, large leaf areas (and, thus, large transpirational surfaces) can be supported. When water is scarce, particularly during the grain-filling period, the small-leaved, wild wheats, with greater drought resistance, outperform the improved wheats. Modern agronomic practices, depending on more ample water supplies than found in the centers of origin of the wild types, utilize the large-leaved cultivars to great advantage.

A significant advance in the cultivated wheats is seen in the larger proportion of their dry weight found in the grain; the harvest index of the cultivated hexaploid wheat was close to 50%, whereas that of a wild diploid was only 34%. *Aegilops speltoides* continued to tiller at a high rate up to maturity, resulting in a large proportion of vegetative, nonbearing tissues. Since harvest data were obtained when the main ear on the seedling axis was mature, many tillers on the wild species were still immature or vegetative.

Evans et al (21) also found a close correlation between phloem cross-sectional area at the top of the main stem (inflorescence axis) and maximum rate of import of assimilates by the ears. Faster grain growth in the cultivated wheats was accompanied by a larger transport system; thus, the specific mass transfer rate (g hr^{-1} mm^{-2}) for assimilates was about the same for all wheat cultivars; phloem cross-sectional areas varied considerably within a cultivar depending upon cultural conditions and was not considered a yield-limiting factor.

SIGNIFICANCE OF HARVEST INDEX

Evans and Dunstone's study (19) underlines the importance of harvest index (HI) in wheat improvement programs, and this index seems to be applicable in nearly all cereals. Sims (43) found that improvement in oat yield in Australian cultivars is almost entirely accounted for by increased HI. He suggested that further improvements could be achieved by breeding and selection for this character. The mean HI and range of variability in cereals, based on data from Singh and Stoskopf (44), in percent for each crop, is as follows: winter wheat, 39 (28–46); winter barley, 45 (44–47); winter rye, 27 (27–29); spring barley, 51 (35–52); and oats, 41 (42–50). They concluded that yield advances were

TABLE 1. Some characteristics of wheat species in *Aegilops* and *Triticum*[a]

Character	*Aegilops*		*Triticum*						
	speltoides	*squar*	*boeticum*	*monococcum*	*dicoccoides*	*dicoccum*	*durum*	*spelta*	*aestivum*
Wild (W) or cultivated (C)	W	W	W	C	W	C	C	C	C
Genome	B	D	A	A	AB	AB	AB	ABD	ABD
Ploidy	2	2	2	2	4	4	4	6	6
Tillers per plant (no.)	71	43	34	46	20	18	11	16	14
Grains per ear (no.)	12	20	30	21	27	37	47	36	36
Weight per grain (mg)	5	10	14	32	33	36	35	56	49
Grain yield per plant (g)	4	8	14	32	17	25	18	32	25
Harvest index (%)	10	32	34	40	36	39	38	36	49
Area of largest leaf (dm^2)	5	8	9	19	35	29	31	41	29
Photosynthesis rate									
Flag leaf[b]	34	36	46	35	35	30	29	27	31
Ear[c]	1	2	3	2	6	4	6	4	3

[a] Data adapted from Evans and Dunstone (19). Each value is the average of all lines of the species computed for individual plants in pots.
[b] CO_2, mg/dm^2 per hour.
[c] Gross photosynthetic rate: CO_2 in milligrams per ear per hour.

correlated with increased HI and not with increased total productivity per unit area (total biomass yield). Murata and Matsushima (39) found this to be so in rice, and Duncan et al (17) concluded that this has occurred in peanuts.

A harvest index of 50% appears to be the upper limit achieved in breeding programs with several crops. Is this likely to be the upper limit? Can it be achieved for all crops? Are low values the result of genetic or environmental (cultural) limitation or some combination of the two? In peach trees, HI increases from 30% in young bearing trees to nearly 70% in mature trees (7). Keep in mind that the redistribution of assimilates required for increased HI is the result of changed competition among all sinks, a concept that is well illustrated in the study with peach trees (Fig. 1). Dwarf wheat cultivars have reduced straw (stem) weight, but Mackey (35) found that the root system of one dwarf cultivar was

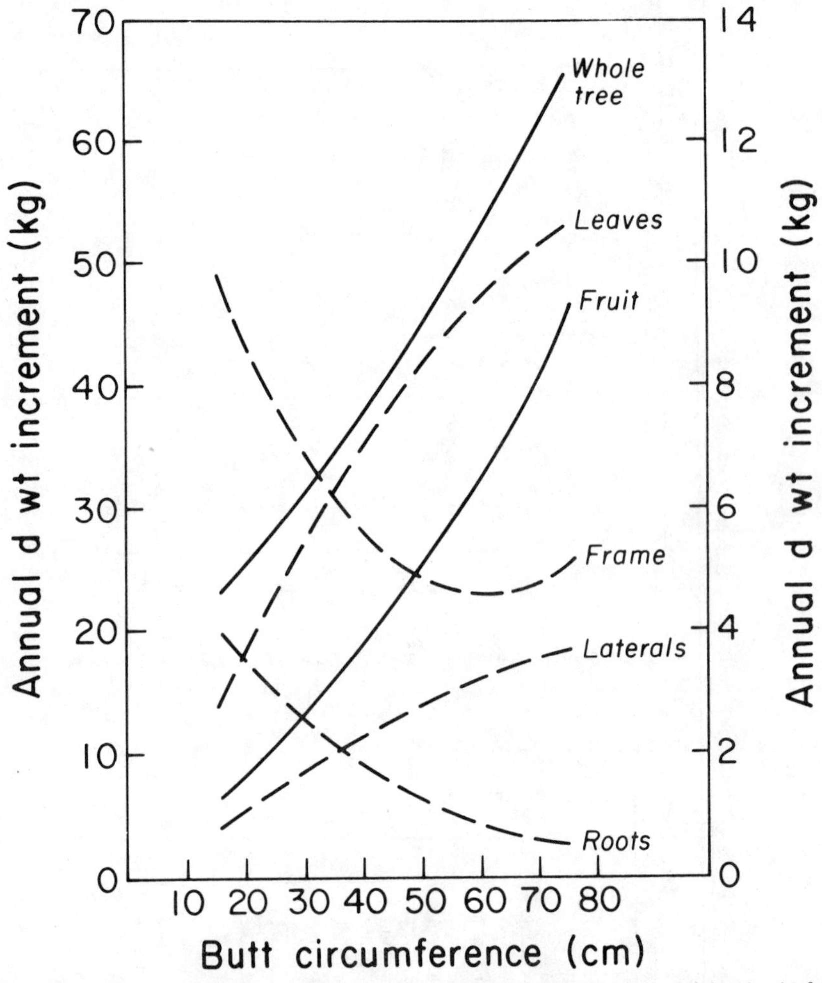

Fig. 1. Effect of tree size (butt-circumference) on the annual increment of dry weight (d wt) of the whole tree and its components. Solid lines refer to left hand ordinate, broken lines refer to right hand ordinate. Reprinted, by permission, from Chalmers and Van den Ende (7).

also reduced. Such reductions in root systems may be tolerated in some crops under certain conditions where rainfall (or irrigation) is ample. Reductions in other nonreproductive components may pose other problems. Donald (13) presented a conceptual model of a wheat plant of potentially greater yield than any commercially available cultivars. His plants would be short statured, single stemmed (uniculm), sown to give high densities and rapid canopy closure as well as large numbers of fertile ears. Leaves would be small, encroaching minimally upon neighboring plants; increased photosynthetic tissue would be achieved by awns situated ideally for grain filling. The inflorescence would be large with many florets so that sink-capacity did not limit yield. With few (or no) nonbearing culms, there would be a net conservation of water. For better exploration of soil water and nutrients, deep root systems would be derived predominantly from the embryo root axis and not from the internode system. Mackey (35) found that, in wheat, deep-rooting characteristics could be selected independently of shoot stature or inflorescence structure. The major lesson is that short statured, otherwise reduced cultivars need not necessarily have shallow root systems. The goal of increasing HI to some optimal value should be the goal for other agronomic and horticultural crops.

SOURCE VS SINK LIMITATIONS TO YIELD

A key question for this chapter on morphological constraints is whether yield is limited by storage tissue or photosynthetic tissue. For most crop communities, light, CO_2, duration of green area, etc., *appear* to limit productivity, but (as reviewed above) improved utilization and storage of assimilates have accounted for major yield advances. Productivity increases may be achieved by breeding species that have more efficient mechanisms for utilizing light energy either in the primary photoreaction in the CO_2 assimilation step or by decreased respiratory losses (Edwards et al, Chapter 10, this volume), yet these have been difficult areas in which to make progress. We can expect further advances by continuing to improve sink strength, but rational progress requires that we assess the need for sink strength. Can we determine for each situation whether assimilation (source) or assimilate utilization, distribution, and storage limit productivity and, if so, take advantage of this knowledge in crop research programs? Is there perhaps some sequence in crop development where first assimilation, then sink capacity, then assimilation, etc., takes its turn as the limiting factor? The answers are not clear cut and certainly not the same for all species in all environments (18,52).

In sugar beets, net assimilation rate is reduced if storage tissue is not adequate, as shown by Thorne and Evans (47) for beet tops on chard roots, but there is no clear evidence for sink limitations to yield in a field-grown beet crop. Indeed the model for the sugar beet crop presented by Fick et al (22) assumes assimilate limitation to beet development and sugar storage. But increased beet storage capacity or beet growth rate may have been a factor in sugar beet breeding programs (25), permitting exploitation of climates with longer or more favorable growing seasons. Sink limitations to overall yield in sugar cane are indicated by a few studies with gibberellin-induced increased growth rate (6), but such treatments appear to have little or no agronomic value. Historically there has been a selection for cane cultivars with thicker stems and greater leaf area in the upper canopy (25). Enhanced sugar storage in cane seems to follow diminished stalk elongation, but this is not the result of overall increases in sugar per unit crop area. It is due to an increased proportion of mature, sugar-storing to immature, assimilate-consuming tissue (6).

Sink limitations to overall yield, not merely yield of storage tissues, were examined in early studies by Chandler and Heinicke (9,10) with grapes, apples, and gooseberries. More recent studies were done with apples by Maggs (36,37). Plants with fruit accumulate dry weight more rapidly than nonfruiting plants do (Tables 2 and 3). There was a 40–80% difference in cumulative dry weight gain between the fruiting and nonfruiting plants in the 5- to 10-year interval from planting to harvest. Leaf dry weight could not have accounted for the differences observed; thus the data indicate that neither the roots nor the tops were equivalent to developing fruit as alternative sinks for assimilates. Since the data for Wealthy apple and the grape cultivars are normalized for leaf area, net photosynthetic rate (per unit leaf area) was apparently reduced in the nonfruiting plants. There are other explanations for apparent reductions in photosynthesis in defruited plants. Maintenance costs may have increased (40) and/or biomass composition may have been altered such that the glucose value increased (38). Maggs (36,37) and Hansen (29) found that photosynthesis in apple leaves is

TABLE 2. Effect of fruiting in pruned and unpruned grape vines of two cultivars planted in 1918 or 1920 on the amount of leaf area available during the life of grape when harvested in spring 1925[a]

Cultivar and plant date	Treatment		Dry matter (lb/1,000 ft^2 of leaf area)					
	Pruned	Fruited	Roots	Tops	Prunings	Fruits	Blossoms	Total
Concord								
1918	+	−	3.67	5.75	4.31	0.09	0.02	18.83
	+	−	3.49	5.06	4.48	8.96	...	21.98
1920	+	−	5.87	7.44	3.11	0.06	0.02	16.50
	+	−	5.51	6.74	3.77	10.84	...	26.86
Delaware								
1920	−	−	4.26	7.25	...	1.02	0.08	12.60
	−	+	3.25	5.74	...	16.86	...	25.85
	+	−	5.83	6.77	1.98	0.06	0.01	14.66
	+	+	4.39	6.53	1.79	7.68	...	20.39

[a]Data from Chandler and Heinicke (9). Multiply values by 0.049 to get t/ha of leaf area.

TABLE 3. Dry weight distribution and total dry weight gain in fruiting and nonfruiting apple trees[a]

Cultivar	Plant parts	Dry matter (lb/1,000 ft^2 of leaf area)	
		Fruiting	Nonfruiting
Oldenburg[b]	Roots	38.2	41.1
	Shoots	52.6	52.8
	Fruits	29.5	0.1
	Total	120.3	94.0
Wealthy[c]	Roots	29.4	21.3
	Shoots	47.4	33.4
	Fruits	17.5	0.6
	Total	94.7	55.4

[a]Trees planted in 1917 and harvested in 1926, based on data from Chandler and Heinicke (10). Multiply values by 0.049 to get t/ha of leaf area.
[b]Values are average dry weights of near uniform leaf areas of 12 fruiting and 8 nonfruiting trees.
[c]Data based on 5 fruiting and 3 nonfruiting trees.

higher with subtending fruit than when fruit are removed, and Chalmers et al (8) obtained similar results with peach trees. Readers are referred to Barden (4) and Lakso and Seeley (33) for fuller treatments of the physiological systems, and their interactions with environmental parameters, that are likely to be involved in reducing photosynthesis in leaves of defruited plants. Feedback inhibition of photosynthesis, which one expects in sink-limited plants, has not been found in all species (24). In the author's view, this is one of the central research problems for crop scientists interested in yield limitations.

An important practice, dependent upon sink limitations to yield, is found in table grape production. Yields of Thompson Seedless grapes are often doubled as a result of gibberellin application to the young berries (Table 4; 50,51) a treatment known to increase cell number and cell expansion in berries. The cluster-thinned vines are extraordinarily under-sinked and, as expected, the largest gibberellin-induced yield increment is seen here. Although the sugar content per fruit is decreased, the overall sugar yield per unit area is increased significantly.

Adams (1) analyzed the problem of assimilate supply and yield by studying correlations among the components of yield in crops with reproductive structures. In these crops, economic yield is usually defined by the equation:

$$\text{Yield} = \text{avg. seed wt.} \times \text{seeds/fruit} \times \text{fruit/area}.$$

The common relationship that seed weight *decreases* as the number of fruit per unit area *increases* is an example of yield component compensation. In such cases overall yield may remain unchanged. In dense plantings of navy beans and corn, Adams found negative correlations only. In widely spaced plantings, correlations were close to zero or positive. Agronomists assume that negative correlations indicate source rather than sink limitations to yield. In widely spaced plantings, light penetration into the canopy permits greater photosynthesis, reduced source limitations, and introduces sink limitations to yield (and hence, positive or no correlations among the components of yield). Can a case be made for sink limitations in densely planted communities? Duncan (14) suggests that yield increases in corn up to the theoretical maximum, even in high density plantings, may depend in part upon increasing sink size by increasing the rate of grain filling, the duration of grain filling, or both. Both notions are supported in studies by Daynard et al (12), Johnson and Tanner (31),

TABLE 4. Influence of gibberellic acid on yield and sugar content of Thompson Seedless grapes[a]

		Fruit yield[a]		Sugar content[a]	
Cluster per vine (no.)	Gibberellic acid	Per vine (lb)	Per acre (lb)	Per fruit (%)	Per acre (lb)
16[b]	−	19.5	8,775	22.2	1,948
	+	44.3	19,935	17.3	3,448
48[c]	−	70.7	31,815	15.1	4,804
	+	82.5	37,125	14.3	5,308

[a]Data from Weaver and McCune (50,51). To get yield per vine in kg, multiply values by 0.454; to get kg/ha, multiply values by 1.12.
[b]Thinned.
[c]Unthinned.

and, in terms of exploiting the growing season, by Troyer (49). Increased HI by assimilate redistribution is not suggested here. Advanced cultivars of corn would be characterized by high rates of cell division and expansion in the ovaries of ear inflorescences and by additional florets on the ears and increased longevity of assimilatory tissue. Spiertz et al (45) found that increased longevity of the flag leaf is associated with higher grain yields in some wheat cultivars. Citations for quantitative analyses of negative correlations among yield components, such that the degree of source or sink limitation can be estimated, would be very valuable. Gifford et al (26), however, analyzed changes in HI as a function of cultural conditions (shading and CO_2 enrichment applied at different stages of crop development). They found both source and sink limitations in barley; e.g., poor light conditions before and at the time of inflorescence initiation resulted in relatively small inflorescences and, hence, reduced sink strength at the time of grain filling. Analysis of other published data leads to the conclusion that sink limitations *only* may exist in some field crops (26). Fischer and Wilson (23) found only sink limitations during the period of grain filling in a sorghum crop. Tollenaar (48) concludes that both source and sink limitations occur in maize; sink limitations arise with adverse conditions during early stages of ear development, at the time of silking following spikelet initiation. Poor light, moisture stress, or low nitrogen at this time can result in poor ear development. Below et al (5) found that at early stages of grain filling in densely planted field corn, the supply of reduced N and not carbohydrate may limit grain development, results suggested by earlier studies of Hageman et al (27). Since supply of N to a plant depends in part upon root development and soil exploration, we must look to preflowering stages of crop development for subsequent failure to develop adequate sink strength in the fruiting structures. But there are good reasons for looking at reduced N fertilizers whenever solar radiation, or source tissue, is limiting yield. Nitrate reduction is costly of carbon-skeletons, reduced adenine dinucleotide, and of adenosine triphosphate and for this reason is closely tied to photosynthesis itself (30,46).

SUMMARY

1. It has been established that there are morphological constraints to yield; these are now explained as source and/or sink limitations.

2. For diagnostic purposes references to the morphological structures responsible for assimilation and storage should begin when they are at the meristematic stage.

3. Increased harvest index in many crops denotes changes in assimilate partitioning favoring development of the economically valuable tissue at the expense of other tissues. For example, suppression of axillary branching (tillering), nodal root systems, and shoot axis elongation in cereals has accompanied increased inflorescence size, floret number, ovary growth rate, and fruit size.

4. Crop growth rate (biomass produced per unit time) may not have been increased in any major crop as a result of breeding programs, but extended formation of assimilatory and sink tissues, resulting in a longer growing season may have increased *annual* biomass yields. Since they have the potential of a longer growing season, perennial crops should be capable of greater yields than those obtained from annual crops. In temperate zones, cultivars that commence

growth in the cool early spring and continue through late fall will best exploit the total annual solar radiation.

5. A key issue for crop morphologists, physiologists, and field managers is to keep source and sink tissues functionally coordinated in agricultural systems. Kohl (32) has recently proposed a model to facilitate coordination. The reader is referred to Gifford and Evans (25) for a penetrating review of the source-sink problem and its significance for future research in the physiological/ morphological realms.

LITERATURE CITED

1. Adams, M. W. 1967. Basis of yield component compensation in crop plants with special reference to field bean, *Phaseolus vulgaris*. Crop Sci. 7:505-510.
2. Andries, J. A., Jones, J. E., Sloane, L. W., and Marshall, J. G. 1969. Effects of okra leaf shape on boll rot, yield and other important characters of upland cotton, *Gossypium hirsutum* L. Crop Sci. 9:705-710.
3. Andries, J. A., Jones, J. E., Sloane, L. W., and Marshall, J. G. 1970. Effects of super okra leaf shape on boll rot, yield and other important characters of upland cotton, *Gossypium hirsutum* L. Crop Sci. 10:403-407.
4. Barden, J. A. 1978. Apple leaves, their morphology and photosynthetic potential. HortScience 13:644-646.
5. Below, F. E., Christensen, L. E., Reed, A. J., and Hageman, R. H. 1981. Availability of reduced N and carbohydrates for ear development of maize. Plant Physiol. 68:1186-1190.
6. Bull, T. A., and Glaziou, K. T. 1975. Sugar cane. Pages 51–75 in: Crop Physiology: Some Case Histories. L. T. Evans, ed. Cambridge Univ. Press. 374 pp.
7. Chalmers, D. J., and Van den Ende, B. 1975. Productivity of peach trees: Factors affecting dry weight distribution during tree growth. Ann. Bot. 39:423-432.
8. Chalmers, D. J., Canterford, R. L., Jerie, P. H., Jones, T. R., and Ugalde, T. D. 1975. Photosynthesis in relation to growth and distribution of fruit in peach trees. Aust. J. Plant Physiol. 2:635-645.
9. Chandler, W. H., and Heinicke, A. J. 1925. Some effects of fruiting on growth of grape vines. Proc. Am. Soc. Hortic. Sci. 22:74-80.
10. Chandler, W. H., and Heinicke, A. J. 1926. The effect of fruiting on the growth of Oldenburg apple trees. Proc. Am. Soc. Hortic. Sci. 23:36-46.
11. Cook, M. J., and Evans, L. T. 1978. Effect of relative size and distance of competing sinks on the distribution of photosynthetic assimilates in wheat. Aust. J. Plant Physiol. 5:495-507.
12. Daynard, T. B., Tanner, J. W., and Duncan, W. G. 1971. Duration of the grain filling period and its relation to grain yield in corn, *Zea mays* L. Crop Sci. 11:45-48.
13. Donald, C. M. 1968. The design of the wheat ideotypes. Pages 377–387 in: Proc. Intl. Wheat Genetics Symp. 3rd. K. W. Finley and K. W. Shepherd, eds. Butterworth Sci. Publ., Sydney (Australia).
14. Duncan, W G. 1967. Corn yields to meet the challenge. In: Maximum Crop Yields—The Challenge. Spec. Publ. 9. Am. Soc. Agron., Madison, WI.
15. Duncan, W. G. 1975. Maize. Pages 23–50 in: Crop Physiology: Some Case Histories. L. T. Evans, ed. Cambridge Univ. Press. 374 pp.
16. Duncan, W. G., Loomis, R. S., and Williams, W. A. 1967. A model for simulating photosynthesis in plant corn varieties. Hilgardia 38:181-205.
17. Duncan, W. G., McCloud, D. E., McGraw, R. C., and Boote, K. J. 1978. Physiological aspects of peanut yield improvement. Crop Sci. 18:1015-1020.
18. Evans, L. T. 1975. The physiological basis of crop yield. Pages 327–355 in: Crop Physiology: Some Case Histories. L. T. Evans, ed. Cambridge Univ. Press. 374 pp.
19. Evans, L. T., and Dunstone, R. S. 1970. Some physiological aspects of evolution in wheat. Aust. J. Biol. Sci. 23:725-741.
20. Evans, L. T., and Rawson, H. M. 1970. Photosynthesis and respiration by the flag leaf and components of the ear during grain development in wheat. Aust. J. Biol. Sci. 23:245-254.
21. Evans, L. T., Dunstone, R. L., Rawson, H. M., and Williams, R. F. 1970. The phloem of the wheat stem in relation to requirements for assimilate by the ear. Aust. J. Biol. Sci. 23:743-752.
22. Fick, G. W., Loomis, R. S., and Williams, W. A. 1975. Sugar beet. Pages 259–295 in: Crop Physiology: Some Case Histories. L. T. Evans, ed. Cambridge Univ. Press. 374 pp.
23. Fischer, K. S., and Wilson, G. L. 1975. Studies of grain production in *Sorghum*

bicolor (L. Moench). III. The relative importance of assimilate supply, grain growth capacity and transport system. Aust. J. Agric. Res. 26:11-23.

24. Geiger, D. R. 1976. Effects of translocation and assimilate demand on photosynthesis. Can. J. Bot. 54:2337-2345.
25. Gifford, R. M., and Evans, L. T. 1981. Photosynthesis, carbon partitioning and yield. Annu. Rev. Plant Physiol. 32:485-509.
26. Gifford, R. M., Bremner, P. M., and Jones, D. B. 1973. Assessing photosynthetic limitation to grain yield in a field crop. Aust. J. Agric. Res. 24:297-307.
27. Hageman, R. H., Leng, E. R., and Dudley, J. W. 1967. A biochemical approach to corn breeding. Adv. Agron. 19:45-86.
28. Hall, A. E. 1981. Adaptation of annual plants to drought in relation to improvements in cultivars. HortScience 16:37-38.
29. Hansen, P. 1970. ^{14}C studies on apple trees. VI. The influence of the fruit on the photosynthesis of the leaves and the relative photosynthesis yields of fruits and leaves. Physiol. Plant. 23:805-810.
30. Hewitt, E. J. 1975. Assimilatory nitrate-nitrite reduction. Annu. Rev. Plant Physiol. 26:73-100.
31. Johnson, D. R., and Tanner, J. W. 1972. Calculation of the rate and duration of grain filling in corn (*Zea mays* L.) Crop Sci. 12:485-486.
32. Kohl, H. C., Jr. 1981. A Crop Productivity System. Dept. Env. Hort., Univ. of Calif., Davis.
33. Lakso, A. N., and Seeley, E. J. 1978. Environmentally induced responses of apple tree photosynthesis. HortScience 13:646-650.
34. Larson, P. R., Isebrands, J. G., and Dickson, R. E. 1980. Sink to source transition of *Populus* leaves. Ber. Deutsch. Bot. Gesellschaft. 93:79-90.
35. Mackey, J. 1973. The wheat root. Pages 827–842 in: Proc. Int. Wheat Genetics Symp. 3rd. Mo. Agric. Exp. Stn., Columbia.
36. Maggs, D. H. 1963. The reduction in growth of apple trees brought about by fruiting. J. Hortic. Sci. 38:119-128.
37. Maggs, D. H. 1964. Growth rates in relation to assimilate supply and demand. I. Leaves and roots as limiting regions. J. Exp. Bot. 15:574-583.
38. McDermitt, D. K., and Loomis, R. S. 1981. Elemental composition of biomass and relation to energy content, growth efficiency, and growth yield. Ann. Bot. 48:275-290.
39. Murata, Y., and Matsushima, S. 1975. Rice. Pages 73–99 in: Crop Physiology. Some Case Histories. L. T. Evans, ed. Cambridge Univ. Press, New York. 374 pp.
40. Penning DeVries, F. W. T. 1975. The cost of maintenance processes in plant cells. Ann. Bot. 39:77-92.
41. Sachs, R. M., and Weaver, R. J. 1968. Gibberellin and auxin-induced berry enlargement in *Vitis vinifera* L. J. Hortic. Sci. 43:185-195.
42. Shackel, K. S., and Hall, A. E. 1979. Reversible leaflet movements in relation to drought adaptation of cowpeas, *Vigna unguiculata* (L.) Walp. Aust. J. Plant Physiol. 6:265-276.
43. Sims, H. T. 1963. Changes in hay production and the harvest index of Australian oat varieties. Aust. J. Exp. Agric. Anim. Husb. 3:198-202.
44. Singh, I. D., and Stoskopf, N. C. 1971. Harvest index in cereals. Agron. J. 63:224-226.
45. Spiertz, J. H. J., Tentag, B. A., and Dupers, L. J. P. 1971. Relation between green area duration and grain yield in some varieties of spring wheat. Neth. J. Agric. Sci. 19:211-222.
46. Steer, B. T. 1978. Integration of photosynthetic carbon metabolism and nitrogen metabolism on a daily basis. Pages 309–320 in: Photosynthesis and Plant Development. R. Marcelle, H. Clijsters, and M. Van Poucke, eds. Dr. W. Jan K. Publ., The Hague.
47. Thorne, G. N., and Evans, A. F. 1964. Influence of tops and roots on net assimilation rate of sugar beet and spinach beet and grafts between them. Ann. Bot. 28:499-508.
48. Tollenaar, M. 1977. Sink-source relationships during reproductive development in maize. A review. Maydica 22:49-75.
49. Troyer, A. F. 1978. Corn yields as influenced by flowering date and season. Pages 1-15 in: Annual Corn and Sorghum Research Conference, 33rd. Am. Seed Trade Assoc., St. Louis, MO.
50. Weaver, R. J., and McCune, S. B. 1959. Effect of gibberellin on seedless *Vitis vinifera*. Hilgardia 29:247-275.
51. Weaver, R. J., and McCune, S. B. 1961. Effect of gibberellin on vine behavior and crop production in seedless *Vitis vinifera*. Hilgardia 30:425-444.
52. Yoshida, S. 1972. Physiological aspects for grain yield. Annu. Rev. Plant Physiol. 23:437-464.

CHAPTER 12

CONSERVATION OF PLANT AND SYMBIONT GERM PLASM

PAUL H. WILLIAMS

Energy, knowledge, and germ plasm represent fundamental resources with which humankind has evolved. The degree to which these resources are understood, preserved, and wisely exploited will in great measure determine the course of the future. Largely through an increased understanding of energy in all its forms and complexities, as mediated through the physical and biological sciences, humanity has developed vastly diverse environments enriched with the social, philosophical, and religious traditions that form the fabric of civilization. Within civilization exist institutions and implements for the acquisition, preservation, dissemination, and creation of knowledge.

Germ plasm is the physical and chemical encapsulation of life's evolutionary heritage and its link to the future. Sequestered in the genetic material of each living entity are the encoded residues of its evolutionary past. Through ceaseless sifting and winnowing of the primary genetic material, coupled with continuous selection for adaptation and fitness, evolution of the 5–10 million species estimated to exist has gone on (13). Over time, each species has become uniquely adapted to its own particular setting in the environment, and many species have become highly interdependent for survival.

In the prehistoric past, humanity's influence on the evolution of most species was relatively minor. However, with the onset of civilization, population growth and the development of agriculture have played increasingly important roles in shaping the evolution of many species. Few regions of the earth have not felt the impact of civilization through the perturbations of the plow, the saw, or the drag line. The essentiality of carbon in the chemical skeletons of organic molecules and as a combustible source of energy has resulted in rapid alterations in the chemical composition of the atmosphere, water, and soil over vast regions. These environmental alterations are now occurring at such a pace as to overtax the adaptive flexibility of many species, with the result that extinctions are estimated to be occurring at the rate of one species per day (13). This alarming fact only occasionally comes to the public's attention when epidemics expose the vulnerability associated with widespread genetic homogeneity within a species. An important issue in species extinction is the interdependence of species within ecosytems. Ravin noted that the extinction of each plant species may be accompanied by the loss of 10–30 other species (13).

Central to the issue of species interdependence is that of the vulnerability of the human species itself. Our future may depend on the wisdom exercised in managing the biological diversity within altered ecosystems supporting civilization's needs. Humankind is morally bound not only to its own future generations but also to the future of other species to maintain an evolutionary continuum of *all* species through careful planning and management of altered ecosystems.

In this chapter I will address a number of general and specific issues related to the maintenance of the evolutionary continuum in crop plants and their symbionts within earth's rapidly changing biosphere. Whether one is considering plants, animals, or fungi, prokaryotes or eukaryotes, parasites or commensals, predators or prey, the fundamental issues are essentially the same.

Throughout the chapter I refer to higher plants and their *symbionts* and have found de Bary's broad definition of symbiosis to be useful, i.e., that of sharing a *common life* without specific reference to levels of interdependence. Many symbioses are beneficial to one or both species; others are of transitory or of lasting detriment to one. Particularly where important trophic dependencies exist between the symbionts, the interdependent phenotypes have been fixed in the genotypes of each partner through coevolution. Person and Christ describe this evolutionary relationship in Chapter 34.

Although plant pathologists have specialized in parasitic symbioses largely between higher plants and microbial pathogens, increasing attention is being given to mutualistic symbioses of mycorrhizal fungi and rhizobacteria (Chapters 30 and 31).

DIVERSITY, VULNERABILITY, EROSION, AND EXTINCTION

Diversity, vulnerability, and genetic erosion are interrelated concepts that must command an enormous effort in the future on the part of biologists. If, as current evolutionary theory would have us believe, the maintenance of diversity is necessary for our future survival, then at issue is the question of how much and what kinds of diversity are essential. In the face of continued extinctions, the question of how little species diversity can be tolerated over protracted periods of time will become crucial. Species diversity implies the maintenance of unique genetic materials, normally perpetuated through mating systems, that confer a stable balance between genetic fitness and flexibility. Species extinction results in the irretrievable loss of these unique genetic combinations.

The question of species extinction and species interdependence should be one of great importance to plant health specialists. Loss of what today may be considered insignificant wild plant species that serve as hosts for crop symbionts could shift dramatically the evolutionary balance of these symbionts to the detriment of the crop species. Although such ideas are largely speculative, they point to the need for more information on the ecology and evolution of interrelated species in both natural and crop ecosystems.

In the absence of our ability to predict reliably the course of change in the environment and the rate of extinction, biologists have generally assumed that species extinctions are important losses that must be minimized at almost any cost. Whether such a position is defensible is open to question and seriously in need of continued research. Certainly in the absence of adequate documentation on the questions of how much species diversity is essential to provide a sustainable evolutionary path to the future, we are obligated to maintain existing diversity as fully as possible.

Relating directly to species diversity is the question of intraspecific diversity. The degree to which a species can tolerate the erosion of unique subpopulations, ecotypes, morphotypes, or pathotypes and still maintain the potential to regain these special phenotypes is unknown. Such information will vary for different species but will be of importance in establishing priorities governing the long-term conservation of cultivated crops and their wild relatives.

CONSERVATION, MANAGEMENT, AND PRESERVATION

The conservation ethic has grown directly from our recognition of the need to maintain an optimized diversity among all species. Many complex issues focus on germ plasm conservation, including the notions of maintenance and preservation. Maintenance implies the conservation of germ plasm in a dynamic and evolving condition within a particular ecosystem. Ecosystems of varying complexity and agroenergy inputs provide a wide range of environments, each of which imposes varying selection pressures on wild and crop species. The homogenized high-energy-input cropping systems of developed nations supported by sophisticated breeding and agronomic practices foster rapid evolution among crops and their symbionts, whereas greater evolutionary stability is associated with traditional forms of agriculture. Within natural ecosystems, one finds relative evolutionary stability coupled to species diversity and genetic heterogeneity.

Ecosystem management must be at the center of any germ plasm conservation strategy since only within each particular ecosystem can one maintain the fully evolving range of adapted diversity. Conservation efforts must be directed not only to agroecosystems, but also to the preservation of natural reserves. Natural preserves will sustain unique pools of genetic material that could not be maintained under conventional systems of cultivated cropping. Intensive study of natural preserves will provide insight into the fundamentals underlying complex diversity in natural ecosystems. From basic research in ecology, population genetics, and physiology will come information on adaptation and evolution. Such information will also be of value in maintaining greater evolutionary stability in agroecosystems.

The establishment of wild preserves raises numerous national and geopolitical issues that are now gaining worldwide attention. It is imperative that plant health scientists lend their support to the establishment of international and national ecological reserve systems. In addition, we must be prepared to specifically document the importance of such areas as reservoirs of untapped germ plasm and information on economic plants and their symbionts.

Perhaps in need of even greater immediate attention are those regions where primitive and traditional cropping systems exists. Frequently found in centers of species diversity, these agroecosystems are intimately tied to the traditional pattern of culture. Crop species interface with wild populations creating opportunities for gene flow between wild and domesticated forms. Traditional agriculture provides an enormous world resource of crop diversity in what are known as primitive cultivars, land races, and farmer's seed. These sources of germ plasm are of great value to plant breeders in that they provide a broad array of adapted germ plasm that has passed the threshold of domestication. For the most part, such ecosystems exist in developing nations and, in the face of 20th century population, agricultural, and economic pressures, are in rapid decline.

The intimate links between traditional farming practices and the diversity of germ plasm has raised even more important and sensitive ethical and geopolitical issues than those surrounding the preservation of natural reserves. The question of how best to maintain germ plasm diversity that is dependent on traditional forms of farming will continue to be of great national and international concern. Paddock in Chapter 4, Buddenhagen in Chapter 42, and Zentmyer et al in Chapter 47 have addressed these important issues from various perspectives.

Encouragement of the practice of growing food in home gardens and small private plots could foster genetic diversity among many plant species, particularly if home gardeners were educated in how to produce seed from superior selections. In this respect, both amateur and professional horticultural organizations, together with public urban extension specialists, could make significant contributions to germ plasm conservation.

VIRULENCE SOURCE PRESERVES

Of particular concern to plant health specialists should be the maintenance and study of virulence source preserves (VSP), those unique ecosystems that support maximization of pathotypic diversity among various symbionts. In some instances, these will be natural regions where the wild progenitors of economic species exist in balance with diverse populations of pathogenic and beneficial symbionts. For example, the wild oat populations in the Middle East exist in balance with a wide range of rusts and other species. In other situations, VSPs are found within agroecosystems where sufficient host diversity and environments favorable for endemic disease exist. The genetically heterogeneous mosiac of winter and spring barley fields in Northern Europe supports a vast and continually evolving population of *Erysiphe graminis* f. sp. *hordei* with subpopulations that closely match in virulence the resistance makeup of local cultivars. Likewise in the Toluca Valley of Mexico, a favorable environment combines with the intermating populations of *Phytophthora infestans* on wild and cultivated *Solanum* species to provide an important VSP used by potato breeders throughout the world for the selection of field resistance to late blight.

Some of the most valuable sources of pathotypic diversity of soilborne pathogens exist within the disease nurseries used by plant breeders and plant pathologists for screening and selection of resistant segregants in breeding programs. Such plots as the Burkholder plot at Cornell University have supported for many years a continued cropping of many bean cultivars and breeding lines that in turn have supported an evolving diversity of bean pathogens not normally found in commercial fields. Such plots represent valuable, unique national and world resources and should be maintained and protected with the full legislative support afforded natural preserves. VSPs should be subject to continual surveillance for shifts in environment as well as for changes in virulence structure of species and populations.

Although ecosystem management will provide the long-range solution to sustained evolution and germ plasm conservation, time and resource allocation in the next few decades will be insufficient to develop a comprehensive global system of diverse ecosystem management. In the face of this reality, germ plasm collection and preservation schemes will provide an important means of stemming the rate of genetic erosion.

GERM PLASM PRESERVATION

Only within the past 10 years have there begun significant national and international efforts to develop a coordinated global system to preserve the diversity of germ plasm among economic plants. Global leadership has been assumed by the International Board for Plant Genetics Resources (IBPGR), founded in 1974 by the Consultative Groups on International Agricultural Research (5). Working in close association with the Food and Agricultural Organization of the United Nations, the IBPGR's primary objective is the development and administration of an international plant genetic resource conservation network. In the years since it was created, great progress has been made in organization of a global network of genetic resource centers, and a large number of collecting missions have been accomplished. Among the IBPGR's objectives are 1) collection, conservation, and evaluation of plant genetic resources with particular reference to species of major importance and their wild and cultivated relatives, 2) strengthening of programs of existing institutions and encouragement of new programs, particularly in areas of major genetic diversity, 3) stimulation of the availability of material for plant breeding and other scientific activities, 4) establishment of standards, methods, and procedures for exploration, evaluation, and conservation of germ plasm stocks in both seed and vegetatively propagated material, 5) establishment of a network of replicated storage centers, and 6) promotion of training and dissemination of information and materials among centers.

On the basis of their economic importance, diversity, and rate of erosion, IBPGR has established first, second, third, and fourth priority status for crops and regions of the world (6). Crop advisory committees and ad hoc working groups comprising breeders and specialists from the scientific community have provided expert advice on most major crops. As of 1981, more than 50 crops had been assigned first priority status. The IBPGR network includes national, regional, and international institutions or gene banks having adequate facilities and staffing to support long-term seed storage in what are designated *base collections.* The board assists in the establishment of gene banks in the centers of diversity where host genotypes can be evaluated or regenerated and where regional quarantines prevent dispersal of pathogens with plant germ plasm. Most of the International Agricultural Centers have accepted the designation to house base collections of their mandate crops, e.g., the International Rice Research Institute (IRRI) in the Philippines (rice), International Crops Research Institute for the Semi-Arid Tropics (ICRISAT) in India (sorghum, millets, groundnut, pigeon pea, chickpea), International Institute of Tropical Agriculture (IITA) in Nigeria (African rice, cowpea), Centro Internacional de Agricultura Tropical (CIAT) in Colombia (*Phaseolus*), and the Centro Internacional de la Papa (CIP) in Peru for potato. Regionally coordinated gene banks, such as that in Bari, Italy (Mediterranean collections) and the Centro Agronómico Tropical de Investigación y Enseñanza (CATIE) genetic resources center in Turrialba, Costa Rica (Central American collections), are foci for regional collections. A large number of national gene banks have accepted the designation of the IBPGR to maintain base collections of specific crops on a global or regional basis, e.g., the National Seed Storage Laboratory (NSSL) in the United States for rice, wheat, maize, sorghum, *Phaseolus, Amaranthus*, tomato, and eggplant. In 1980, the IBPGR expanded its efforts in supporting the long-term storage of clonally propagated crops through research on in vitro culture, and it also intends to designate "field gene banks" (previously called clonal repositories) for these crops (e.g., sweet potato, cassava, citrus, banana, etc.).

IBPGR has established criteria for collection of representative diversity, multiplication, and storage of stocks. As a matter of principle, a portion of each sample is left in the country where it is collected. A major portion is placed in long-term storage and, where possible, samples are partitioned to major working collections for evaluation and description, an initial step toward utilization by plant breeders. Detailed descriptors and information guides are being prepared by specialized committees to assist workers in the characterization of collections.

It is in this area of accession evaluation that plant pathologists and other plant health specialists can make important contributions to world germ plasm resources. For the most part, evaluative criteria with respect to disease and pest resistance are few or very crude. Virtually no standardization within crops and between disease or other stress-related phenotypic criteria exist. In the development of criteria to evaluate biologically created stress, the relationship of host genotype and ontogeny needs to be understood in relation to pathotypic variability. Such criteria need to be sufficiently precise to be usable by plant breeders.

Without question the IBPGR represents the most realistic and potentially effective channel through which global conservation of germ plasm can be maintained. As collection activity adequately samples the existing diversity throughout the world and gene banks store and maintain such diversity, the preserved germ plasm will be evaluated and continue to make its way back to agroecosystems through the efforts of breeders. In a reversed form of introgression, gene flow into natural ecosystems will occur from agroecosystems. Our most realistic long-term strategy for germ plasm, therefore, will be one of maintaining a slow but deliberate balance of managed ecosystems and gene banks in which there is a continuous flow of genetic material between the two. Over time the role of IBPGR may be broadened to guide the stewardship of evolving ecosystem preserves.

NATIONAL PLANT GERM PLASM SYSTEMS

Essential to the success of the IBPGR's mission is the effective functioning of many national plant germ plasm preservation programs. Until recently, the major activities in plant germ plasm preservation have been performed by those developed nations whose food and economic crops have originated from regions of diversity outside their national boundaries. Important collections of plant germ plasm have been maintained for many years by the Soviet Union, the Democratic Republic of Germany, The Netherlands, France, the United Kingdom, and a number of other European countries. Japan, China, and India all have substantial germ plasm collections in gene banks. Today many nations have initiated programs for the collection and maintenance of crop species of importance to their citizens. The United States is a nation that recognized early its virtual total dependence on exotic species for its food crops. Although the United States is not unique among nations in this position, it has one of the longest histories of collection, conservation, and exploitation of foreign germ plasm as a contribution to the agricultural welfare.

The U.S. National Plant Germ Plasm System

As early as 1819, Secretary of the Treasury William H. Crawford requested American Consuls to send useful plants back to the United States. From 1836 to 1862, the office of the U.S. Patents Commission directed plant introductions and established the government commitment to collect and investigate agricultural crops from around the world. With the creation of the U.S. Department of Agriculture (USDA) in 1882, plant exploration continued to be emphasized, particularly the procuring of introductions from the Orient and Europe. Not until 1898 was the section of Seed and Plant Introduction formally established. It had a budget of $2,000 per year. Since 1898, more than 400,000 plant introductions have been accessioned by the USDA.

During the early years, new accessions were introduced and maintained at one of the federal plant introduction stations located either at Chico, CA; Glendale, MD; Savannah, GA; or Miami, FL, or in research and breeding collections at Beltsville, MD, and elsewhere. Passage of the Research and Marketing Act of 1946 and the establishment of the National Coordinating Committee for New Crops provided for the development of regional plant introduction stations located at Pullman, WA; Ames, IA; Experiment, GA; and Geneva, NY, and the interregional Potato Research Station at Sturgeon Bay, WI. Recently the USDA has initiated the development of a number of centers for the maintenance of clonally propagated fruit and nut crops. The regional stations have served as centers for deposition, increased evaluation, and distribution of those species that were most suitably adapted for propagation in the particular regions. Activities of the regional stations were coordinated with the guidance of regional technical committees comprising representatives from each of the region's agricultural experiment stations.

An important addition to the U.S. germ plasm conservation effort was the establishment of long-term storage capability. In the early 1940s, the National Research Council members became disturbed by the fact that thousands of plant introductions had been lost due to inadequate storage facilities. The National Coordinating Committee for New Crops examined the feasibility of a long-term storage repository and, with the support of a number of public and private organizations such as the American Seed Trade Association, American Society of Agronomy, the National Wheat Breeders Association, established the National Seed Storage Laboratory at Colorado State University in Fort Collins. Designed to accommodate approximately one-half million seed lots, its purpose was to serve as a repository for reserve seed stocks and introductions held in the regional stations, stocks of open pollinated farmer's seed, land races, discontinued cultivars and breeding lines, as well as for newly developed cultivars and specialized stocks, including those used for genetics and pathogen virulence analysis. Seed storage in the NSSL is not distributed unless all remaining available sources have been exhausted. Germination of seed is normally tested at 5-year intervals, and before germination declines below predetermined limits required to maintain the characteristics of the stock, it is increased again under contract in the same area in which the seed was produced originally.

In June 1971, the state agricultural experiment station directors and the Agricultural Research Service of the USDA published a general progress report entitled *The National Program for Conservation of Germplasm* (11). This was designed to inform the public of the germ plasm conservation program, its history, and success. This informative and well-prepared publication documented many accomplishments in the development and release of superior crop cultivars in the past. It indicated strongly the need for continued support of the national program for germ plasm. It is ironic that as this very publication was in preparation, *Helminthosporium maydis* was establishing itself in epidemic proportion on "superior germ plasm" of the sort that the report was extolling. Perhaps we were lulled by full granaries and readily available research funds available in the 1960s.

Profile of an Awakening Giant

The 1970, *H. maydis* epidemics spawned the 1972 National Academy of Sciences report, which pointed out that vulnerability stems from genetic uniformity and that a number of important U.S. crops were highly vulnerable on the basis of uniformity (8). The report challenged leaders of agriculture interested in our national germ plasm to 1) establish a watchdog system to identify exotic pests and evaluate susceptibility of U.S. plants to these pests, 2) improve and enlarge the agricultural research pool of available germ plasm, 3) establish a national monitoring system to warn the agricultural community of hazards associated with new or widespread cropping practices, 4) continue development of germ plasm resources through plant introduction and storage, 5) develop new plant cultivars incorporating resistant genetic material, 6) collect parasites for research on improving crop resistance, and 7) devise a means for mitigating economic losses from future epidemics. Stimulated by this report, the Agricultural Research Policy Advisory Committee (ARPAC) reviewed the national program for plant germ plasm and pointed out a lack of directed national effort to effectively utilize germ plasm resources. They observed a general underuse of collections and a lack of information and basic research on patterns of genetic variation, genetic transfer, wide crosses, and cytoplasmic inheritance, concluding that

> while considerable effort had gone into genetic resource management, the effort was too haphazard, unsystematic and uncoordinated and never received the high priority it deserved.

The ARPAC report also noted that the "situation was serious, particularly dangerous to the welfare of the nation and appears to be getting worse instead of better." General recommendations in the ARPAC report advised 1) establishment of a National Plant Genetic Resources Commission, 2) development of a national plan for plant genetic resources, and 3) collaboration in international efforts to conserve world plant genetic resources (10). In response to the ARPAC report, Secretary of Agriculture Earl Butz established the National Plant Genetic Resources Board (NPGRB) in 1975 as a means of providing advice and guidance in plant genetic resources. The board comprises personnal of private companies and federal and state agencies with broad interests in plant germ plasm.

An important contribution of the NPGRB was the publication in 1979 of the report *Plant Genetic Resources: Conservation and Use* (9). This report recommended a seven-phase program encompassing broad activities associated with collection, maintenance, evaluation, and distribution of plant genetic resources, but it emphasized the utilization of these resources for the purposes of basic and applied research on plant genetics and breeding. The report recognized the need for involvement by federal and state governments as well as by private industry in the full exploitation of the national plant genetics resources.

Established in 1975, the National Plant Germplasm Committee was formed from representatives from State Agricultural Experiment Stations, the National Council of Commercial Plant Breeders, and the USDA. The Committee advises on policy, coordinates the activities, and serves as the voice of the National Plant Germplasm System (NPGS) (12). As can be seen from the foregoing discussion, the NPGS is a cumbersome giant with many of the problems associated with independent bureaucratic evolution among different lines of administrative and fiscal authority. Much of the criticism relating to the ineffectiveness of the NPGS stems from the lack of a comprehensive long-range plan for the assessment of genetic vulnerability in U.S. crops.

In a review of the NPGS during 1980, the U.S. General Accounting Office identified a number of critical problems within the system that could be resolved with adequately coordinated planning and fiscal support (10). In response to the report, the USDA Germplasm Taskforce Study developed a long-range plan for 1983–1997, in which current strengths and weaknesses of the working, management, and advisory components of the NPGS have been identified. On paper, the NPGS long-range plan looks very sound. It will indeed be an important challenge, not only for plant scientists and administrators but also for the public, to ensure the full development of this plan. The recent appointment within the USDA Agricultural Research Service of an assistant to the deputy administrator for germ plasm, who will serve as a national focal point for the NPGS, holds promise for more effective management of our national germ plasm resources.

An essential component of the NPGS to be implemented in 1982 and 1983 will be the computer-assisted Germplasm Resources Information Network (GRIN). GRIN will consist of computer-related hardware, software, and telecommunication links at various major NPGS sites. Regional plant introduction stations and specialized collection sites will be able to load, process, and retrieve plant characteristic data as determined by plant advisory committees and provide germ plasm users with other important information. Taxonomic data, geographical origin, and maintenance data will be available, as well as specialized reports on general utilization of collections and characteristics of specialized collections (1).

Thus, although the U.S. NPGS is cumbersome and imperfect, the coordinated revitalization of the basic working components of plant introduction, base collection storage, working collection activity, and user research should provide an effective national resource capable of supporting the global efforts of the IBPGR in the years to come.

PRESERVATION TECHNOLOGY

The preservation of plant germ plasm will continue to be a complex worldwide undertaking into which increasing fiscal resources will be channeled. The need for new preservation technologies will provide opportunities and challenges for young scientists. Although cryogenic storage provides preservation of most microbes for unlimited duration, more needs to be known about the potential for accumulated genetic variation over long periods of time within quiescent germ plasm. Cryogenics hold promise for the preservation of both seed and clonally propagated plant parts, tissues, and cells, although this is an area in need of much more extensive research. The potential of pollen and forms of gametes as long-term storage propagules is in need of more research. The attractiveness of gametic storage lies in the efficiency of space with which pollenlike genetic propagules might be conserved. Similarly, there is a need to examine the potential for storing cytoplasmic organelles and organellar genomes, either within whole cells or perhaps in isolated form.

The rapidly advancing technologies in microbial genetics, plant cell biology, and genetic engineering may provide new methodologies for the preservation of plant germ plasm that will obviate the need for cryogenic storage. Isolated genetic material from plants or symbiont cells could be purified, concentrated, and stored in stable form, then later introduced into cells for the production of new regenerated cells. In the light of the growing technologies of the "new biology," the generation of approaches for the utilization and exploitation of crop and microbial germ plasm seems virtually unlimited.

Underlying germ plasm preservation technology must be a body of genetically based theories that can be used to guide policy on the quantities and representation of the germ plasm to be stored, the means of increase upon depletion, and the limits of erosion to be tolerated within preserved materials. For the most part, germ plasm collections throughout the world are maintained on a relatively ad hoc basis. Many are being inappropriately increased such as to exert strong selection pressures favoring genetic drift. Many national collections have excessive or unnecessary redundancy. In light of present limited worldwide storage facilities and resources, redundancy seriously limits the acquisition of new, important collections. Virtually all collections would benefit from the development of more sound theory, based on population genetics and guided by a thorough understanding of the reproductive biology of each species. Perhaps adequate representation of most species can be maintained as relatively heterogeneous intermating gene pools, the sizes of which could be based on the limits for preserving rare alleles. Highly structured mating protocols for seed increase could be established to minimize selective advantages that might be associated with either favorable or deleterious linkages within individuals. Such schemes could reduce the number of accessions in base collections, permitting more thorough characterization of the existing collections. One of the great deficiencies of existing collections is the general lack of useful descriptive information on each accession.

SYMBIONT PRESERVATION

Despite the intimate interrelations between higher plants and many of their symbionts, no comprehensive national or global program has been developed to coordinate the conservation of symbionts in the same way that the preservation of higher plant germ plasm is coordinated. This fact is a serious commentary on the insights and farsightedness of our profession and must stand as a major challenge in the immediate years to come. In spite of this serious deficiency, various individuals and organizations have established and maintained collections of major beneficial and pathogenic plant symbionts. Recognizing the lack of coordinated attention to plant symbiont germ plasm, the American

Phytopathological Society (APS) organized, in 1980, a National Work Conference on Microbial Collections of Major Importance in Agriculture and published the report of the conference (2). With representatives from a number of the world's major microbial collections in attendance, the conference focused on the main problems associated with microbial collections and set forth recommendations to assist in the development of national and international programs for microbial germ plasm conservation.

MICROBIAL COLLECTIONS

Specialized Research Collections

Many specialized collections contain large numbers of genetic variants within a narrow variety of species. Such collections are invaluable and even indispensable in studies of genetics and taxonomy, yet they are the most vulnerable to destruction, particularly upon retirement of individuals or termination of specific projects. Such working collections are frequently the major source of representative virulence used by plant breeders in resistance breeding programs. Among the obligate biotrophs, strains with specific virulence are maintained through propagation on hosts of appropriate differential resistance. The *Phytophthora* collection at the University of California, Riverside, the *Rhizobium* collection at North American Plant Breeders, Inc., and the International Collection of Phytopathogenic Bacteria at the Unversity of California, Davis, are examples of working or research collections. Without a national policy and support on preservation, losses of such collections will not be averted.

Major Reference and Research Collections

These collections of symbionts often involve collaborative efforts of individuals, industry, and state and federal institutions and are housed at public institutions. Collections such as the *Puccinia graminis* collection at the USDA Cereal Rust Laboratory at the University of Minnesota and the *Fusarium* species collection at the Pennsylvania State University, University Park, are widely used by pathologists, breeders, and taxonomists of the world.

National Microbial Collections

A number of major national collections exist throughout the world. The American Type Culture Collection (ATCC) is the only major national collection in the United States (3). It is a private, nonprofit organization founded in 1925 and governed by a board of trustees representing 16 scientific societies including APS. The primary function of national collections such as the ATCC are the accession, preservation, authentication, and distribution of authentic reference and type cultures. The ATCC serves as a national repository for medical, industrial, and agricultural microbial stocks. Among its inventory of more than 1 million frozen or freeze-dried ampoules are more than 29,000 different strains of bacteria, fungi, chlamydiae, rickettsiae, protozoans, algae, cell lines, viruses, and antisera. The ATCC also serves as a national and international repository for patented microbes. In addition to the major collection activities of acquisition of new significant cultures, and the reproduction, preservation, and authentication of these cultures, the eight departments of the ATCC conduct considerable basic research and offer numerous technical and informational services.

Although only a small fraction of most national microbial collections comprise pathogens or beneficial symbionts, the ATCC serves an important purpose for plant pathologists by providing a permanent repository for isolates used and reported in research publications.

COMMUNICATION: THE KEY TO SUCCESSFUL CONSERVATION

As with any complex evolving enterprise of global dimensions, the key to continued positive growth and evolution of conservation efforts is effective communication. Within the effort to conserve plant and symbiont germ plasm, effective communication is essential at all levels of a system. Computer-assisted data banking and telemetrically assisted communication linkages have arrived globally. At the level of genetic resource specialists, those whose daily concerns directly focus on germ plasm maintenance, there is great need to improve communication with national networks such as the U.S. National Germplasm Systems or GRIN and to implement international networks among the IPBGR base collections. Much must be done to refine the actual language or descriptor criteria and notations that will be used in disseminating technical information internationally among specialists, and here plant biologists and microbiologists, including plant pathologists, have an important role to fill. In preparing descriptor information for crop germ plasm, it will be important to provide quantitative information on the nature of the particular symbiont-host relationship in a way that is meaningful to plant breeders and epidemiologists. Such information must be encoded precisely so as to be readily stored and handled by computerized data banks and yet must be sufficiently explicit to have value to the user.

In the hands of germ plasm users—breeders, geneticists, ecologists, microbiologists, and epidemiologists—the descriptive criteria will assume an almost infinitely diverse complexity. Indeed, in the hands of the breeder, the nuances of genetic recombinants and phenotypic expression may defy anything but the most objective form of descriptive expression—the scent of the violet, the blush of the rose!

In recent years, a number of important initiatives in communication have been undertaken among germ plasm specialists and users. The IBPGR crop descriptor list, newsletters, and reports play an important role in the communication among genetic resource specialists throughout the world. Another publication, the quarterly news journal *Diversity*, has recently been initiated by the Laboratory of Information Science in Agriculture at Colorado State University. In the first issue, *Diversity* states that its objective is to be the plant genetic resources community's link to the NPGS and to other national and international activities related to germ plasm (1). If *Diversity* continues to function with the fund of pertinent information and the perspectives found in the early issues, the plant genetic resources community, worldwide, will be well served.

WHOSE GERM PLASM?

Within the last 25 years, crop germ plasm in the form of improved and newly developed cultivars has come to be recognized as an important economic property and has gained much of the national legislative protection afforded inventions of any kind. Plant cultivar protection and breeder's rights legislation vary somewhat among nations and are likely to affect the future of crop research and germ plasm conservation. Although plant protection legislation is highly controversial, it is probably needed in the developed nations, where strongly protected economic incentives are required to stimulate development of new cultivars having more diverse genetic backgrounds than those presently being used (4). For the developing world, the benefits from plant cultivar protection are much more unsure. Critics of plant cultivar protection see the concentration of germ plasm in the hands of a few, large, multinational corporations who, through breeder's rights, conceivably could "gain control" of a major portion of the world's germ plasm used in developed agroecosystems (7). I believe this fear to be basically unfounded, because the bulk of the diversity for crop germ plasm exists in national and international repositories for use by virtually everyone.

In the United States, patent-protected cultivars must be entered in gene banks for long-term storage. Protection is afforded for 17 years on seed-propagated crops and 18 years on clonally propagated crops. For hybrids, the parental stocks are patented.

The acquisition of many of the world's major seed firms by a relatively few multinational corporations is seen by some as a major threat to the world's germ plasm preserves through the overconcentration of resources (7). Within the past 10 years, I have observed rejuvenation of many of the old U.S. and European seed firms as corporate funds for capital improvement and research were allocated to them for further growth and development. The seed companies today contain many more highly trained and qualified breeders, pathologists, and production specialists than they ever had in the past. I also believe that enlightened private corporate policy will have the same concerns with respect to the management of genetic diversity among crop species as those of the public institutions. I can envision large crop-breeding firms maintaining their own gene banks, from which their breeders could draw and deposit materials. As long as strong public support exists for national and international programs of germ plasm preservation, I cannot envision how cultivar protection will be a detriment to the long-term maintenance of germ plasm. The growing group of private germ plasm users and the custodians representing public interests in government and educational institutions should ensure the wise management of our crop germ plasm resources into the future.

LITERATURE CITED

1. Anonymous. 1982. Diversity: Your Link to the National Plant Germplasm System. Diversity 1:2-3.
2. American Phytopathology Society. 1981. National Work Conference on Microbial Collections of Major Importance to Agriculture. Am. Phytopathol. Soc., St. Paul, MN. 52 pp.
3. ATCC. 1981. American Type Culture Collection. ATCC, Rockeville, MD. 28 pp.
4. Barton, J. H. 1982. The international breeder's rights system and crop plant innovation. Science 216:1071-1075.
5. International Board for Plant Genetics Resources. 1981. Crop Genetic Resources. IBPGR, FAO. Rome, Italy. 14 pp.
6. International Board for Plant Genetics Resources. 1981. Revised Priorities Among Crops and Regions. IBPGR, FAO. Rome, Italy. 18 pp.
7. Mooney, P. R. 1979. Seeds of the earth: A Private or Public Resource? Int. Coalition for Developmental Action, London. 120 pp.
8. National Academy of Sciences. 1972. Genetic Vulnerability of Major Crops. NAS, Washington, DC. 307 pp.
9. National Plant Genetic Resources Board. 1979. Plant Genetic Resources, Conservation

and Use. U.S. Dep. Agric., Washington, DC. 20 pp.

10. U.S. Comptroller General. 1982. The Department of Agriculture Can Minimize the Risk of Potential Crop Failure. GAO, CED-81-75. 35 pp.

11. USDA. 1971. The National Program for Conservation of Crop Germplasm. U.S. Dep. Agric., Washington, DC. 73 pp.

12. USDA. 1977. The National Plant Germplasm System. U.S. Dep. Agric., Washington, DC. 12 pp.

13. USDS. 1982. Proceedings of the U.S. Strategy Conference on Biological Diversity. U.S. Dep. State, Pub. 9262. 126 pp.

CHAPTER 13

SOMATIC MODIFICATION OF GERM PLASM AND THE POTENTIAL FOR GENETIC ENGINEERING

JAMES F. SHEPARD

Modern plant breeding techniques have, over the past 50 years, contributed to spectacular gains in crop productivity. In the United States, average yields per hectare of corn have increased more than threefold since 1930 and sorghum yields more than fourfold; wheat yields have more than doubled. Despite these generally acknowledged achievements, there is justified concern that with existing technology alone, the likelihood of achieving equivalent increases in crop productivity during the next half-century is quite dim. Close inspection reveals that we may have nearly optimized cultural and fertilizer inputs and hence must rely even more on improved plant genotypes as the source of increased production. Moreover, there are numerous important crops that, from a genetic potential point of view, have been only modestly enhanced during the past 50 years. Soybeans, for example, have undergone slightly more than a 50% rise in yield per hectare (about 1% per year), and the increment of increase in new cultivars is declining. We continue to rely on antique cultivars of potato, sweet potato, citrus, banana, and numerous other vegetatively propagated crop plants. And production practices and/or constraints may change more rapidly than can be accommodated for by the plant breeder. Norman Berg, Director of the United States Department of Agriculture's Soil Conservation Service recently commented in Forbes magazine, "Yields are increasing at decreasing rates. Our technology may have reached the point of diminishing returns." This is at a time when farmland is being removed from production at an alarming pace. The land base of the United States is 891 million hectares, of which 167 million hectares is now used for cropland, with additional reserves of 51 million hectares in forest, pasture, and rangeland. The National Agricultural Lands Study recently reported that each year, 1.2 million hectares of valuable cropland is consumed by urban expansion, including 405 thousand hectares classified as "prime" farmland. Since the sacrifice of farmland is certain to continue, it remains a constant challenge to produce more food on less land, especially when we consider that the yield from one of every three hectares harvested in the United

States is shipped abroad. From this we must conclude that, despite an enviable record of prior success in crop plant improvement, there is a need for additional research strategies and technologies that can complement and extend the capabilities of conventional approaches.

One possible avenue has been popularized under the umbrella term "plant genetic engineering." While genetic engineering is often used synonymously for recombinant DNA technology, in plants, it pertains to all procedures that have the fundamental goal of directing phenotypic changes in plants through manipulation of their genetic constitution. Thus, in its broadest sense, plant genetic engineering includes conventional plant breeding technologies as well as such recent techniques as protoplast and cell culture, interspecific protoplast fusion, and genetic transformation of cells with genes either cloned in bacterial plasmids or as components of crude DNA preparations. In the following paragraphs, some recent progress in the development of new or nonconventional genetic techniques will be summarized with emphasis on those with potential application to the improvement of crop plant performance.

CLONING PLANTS FROM CULTURED CELLS

Organisms of asexual or somatic origin are termed "clones," which is derived from the Greek *clon*, meaning a twig or slip suitable for plant propagation. In 1903, Webber (32) of the U.S. Department of Agriculture proposed the adoption of *clon* to distinguish plants that are propagated asexually and, "which are simply parts of the same individual. . . ." *Clon* later became "clone," and its usage now encompasses all asexual life forms including genetic macromolecules, i.e., genes "cloned" in bacterial plasmids.

Clones are often portrayed as carbon copy replicas of the parent, and at first glance, such deductions seem entirely reasonable. Naturally occurring clones, such as potato plants grown from tubers, are generally identical in appearance and may thus be expected to possess functionally equivalent genomes. However, most science dictionaries now define "clone" as an individual asexually derived from a single cell or a common ancestor through mitosis, and the definition is not intended to imply that a population of asexually derived plants will necessarily display phenotypic or genetic homogeneity. That is, clones derived from single somatic cells of a parent plant need not, by definition, possess exactly the same genetic composition.

In recent times, plant cloning has been refined to the point where a single cell removed from the body of a plant can be cultured and then induced to reform a complete individual. This extraordinary capability of single cells, "totipotency," was first suggested from experiments with cultured carrot callus by F. C. Steward at Cornell University about 25 years ago. Within the ensuing decade, the phenomenon was firmly established for somatic cells of tobacco and several other species. In 1971, a further advance in plant cloning was made by Itaru Takebe and colleagues in Japan (31). They combined pectin-digesting (pectinase) and cell wall-degrading (cellulase) enzymes to dissociate tobacco leaves into living but "wall-less" plant cells, referred to as "protoplasts." Once isolated, protoplasts were cultured in a defined medium that promoted growth and division. In the final step, resulting cell masses called "calli" were induced to regenerate (redevelop or reform) small shoots, which eventually grew into entire plants. This ability to obtain normal appearing tobacco plants from single protoplasts was encouraging in that if similar techniques could be developed for species of major commercial interest, it might be possible to genetically alter or transform single plant cells with techniques akin to those used in microbial systems and then obtain specifically modified plants.

With the development of tissue, cell, and protoplast techniques for plant regeneration has come terminology to identify the origin of such clones. Plants regenerated from traditional callus cultures have been called "calliclones" (30) and those from protoplast-derived calli "protoclones" (28). Recently, Larkin and Scowcroft (17) proposed the more general term "somaclone" to include plants regenerated ". . .from any form of cell culture." In this article, the earlier designations, i.e., calliclones and protoclones will be retained.

Somatic Variation from Cell and Tissue Culture

For a number of years, in vitro cloning has been an important adjunct to plant propagation. Meristem or shoot tip culture has freed elite plant genotypes from viral infection, and considerable effort has also been applied to the in vitro or micropropagation of valuable species. Both of these approaches rely on retention of parental genotype in regenerated materials.

An alternative to the precise duplication of the parent in asexual progeny would be to induce and then recover improved somatic variants of a plant strain through in vitro techniques. Such an approach is a sophisticated form of mutation breeding whereby cultured cells are exposed to some mutagenic circumstance and then regenerated into a population of plants that displays extensive variability for critical traits. Recent evidence suggests that this strategy may indeed provide useful genetic changes within a cultivar or breeding line without the simultaneous loss of essential agronomic characters.

The first example of presumed genetic improvement of a crop plant through tissue culture was with sugarcane (*Saccharum officinarum*), in which plants regenerated from callus exhibited numerous phenotypic changes of potential value (14). Calliclones were identified that produced cane yields greater than yields from the parent, greater numbers of stalks, and increased concentrations of sucrose (18). Others calliclones possessed greater resistance to such important diseases as downy mildew (*Sclerospora sacchari*), eyespot (*Drechslera sacchari*), the leafhopperborne Fiji disease agent (13), and smut (*Ustilago scitaminea*) (18). One particularly interesting clone, Pindar 70-31, was resistant to both the Fiji and downy mildew diseases and still about equivalent to the parent in yield potential (20). In these studies, plants were regenerated from multicellular callus, and mutagens did not appreciably increase the frequency of resistant individuals (13). It is also significant that even without in vitro selection, the frequency of resistant calliclones was high enough to permit efficient recovery. In the data of Heinz et al (13), individual clones differed in their degree of general resistance to a specific pathogen, and there was little to suggest that a single gene was responsible for any of the reactions.

Subsequently, Larkin and Scowcroft (17) reported that in a population of sugarcane calliclones, there was a definite shift toward enhanced resistance to both the Fiji and the downy mildew diseases. Moreover, the same phenomenon occurred even when the parental source already possessed a useful resistance level. At maturity, disease-resistant clones did not suffer a compromise in yield or other vital characters. These results are both intriguing and puzzling. With neither intentional mutagenesis nor in vitro selection, it is unexpected that genetically stable disease resistance would appear in such high frequency and be so easily recoverable. The fact that it does so, and reproducibly in sugarcane, suggests definite possibilities for other plant species.

Additional examples of clonal variation arise from callus cultures. Using tissue cultures from immature embryos of Texas male sterile cytoplasm maize, Gengenbach and associates (11) obtained cell lines and then plant clones resistant to the pathotoxin of *Helminthosporium maydis* race T. Resistance to the pathogen was shown to be maternally inherited, but with a concomitant loss of cytoplasmic male sterility. Variability among regenerated maize plants is not confined to cytoplasmic factors. Green et al (12) found an extensive array of cytological abnormalities in individual clones, including instability of chromosome number and/or integrity. In 1981, Edallo et al (9) examined the frequencies of spontaneous, simply inherited mutations in plants regenerated from immature embryo callus. Virtually all clones studied had retained their normal chromosome number. However, the mean number of mutations observed after a second selfing averaged 0.9–1.2 per plant, respectively, for the two inbred lines tested. The actual frequency of mutation was undoubtedly much higher because only endosperm and morphological seedling characters were scored.

For *Hordeum* sp., Orton (21) reported a variety of cytological and phenotypic modifications in regenerated plant populations. Sibi (29) found that calliclones from cotyledonary callus of lettuce displayed numerous phenotypic alterations despite having retained a normal chromosome number. When the phenotypic variants were "selfed," sexual progeny displayed several forms of genetic modification. In some, point mutations were suggested while in others, i.e., "phenovariants or variants," the phenotypic change was fixed and thus did not segregate after six cycles of selfing. Examples included changes in leaf shape and color. Generally, these characters were uniparentally inherited in diallel crosses and were termed "cytohets" or "cytoheteroplasmics" after the terminology of Sager (23). Surprisingly, the vegetative vigor of F_1 progeny of these crosses exceeded by as much as 44% that of parental control crosses. Still other examples, appear in the recent review by Larkin and Scowcroft (17).

Extensive phenotypic variability has also been documented for clonal populations regenerated from protoplasts. In 1978, protoclones from mesophyll protoplasts of the potato (*Solanum tuberosum*) cultivar Russet Burbank were screened with the early blight fungus (*Alternaria solani*). Several clones displayed increased resistance over parental Russet Burbank, and the trait was stable over several tuber generations (19). Later, and with an additional group of 800 protoclones, a similar range of variation was recorded for sensitivity to *Phytophthora infestans* (25,28). Of these, 20 (nearly 2%) were deemed to possess a degree of general resistance to the pathogen greater than that of the parent. For some protoclones, resistance was expressed against both race 0 and race 1,2,3,4 of the fungus, while in others, response to the two races was distinctly different. Again, the relative resistance reaction was passed through tubers to succeeding generations. In no case, however, have protoclones emerged whose reactions mimic the major or "R" gene form of hypersensitivity.

Protoclones of potato cultivars also display high frequency variation for horticultural characters. Preliminary observations (28) indicated that such complex traits as tuber number, size, shape, and surface texture differed between clones, as did a range of foliar characteristics. Later (24), replicated field plots established statistically significant variation for more than 20 characters. Most protoclones were altered in more than a single trait, and never was one completely identical to the parent clone. Overall, however, the protoclones closely resembled Russet Burbank in morphology and general tuber characteristics and, therefore, had retained essential traits of the cultivar. Significantly, all of the variant protoclones in this population possess the full complement of 48 chromosomes.

These results with calliclones and protoclones of different crop species strongly suggest that in the absence of intentional mutagenic treatment, extensive somatic variation appears in plant populations regenerated from cultured cells, at least for certain species. Moreover, the phenotypic changes are stable, broad in spectrum, and include disease resistance as well as agronomic characters. The genetic origins of change in all of these examples, whether nuclear or cytoplasmic, remain to be established. It has also been debated (3,27) whether the process of in vitro cell culture is of itself sufficiently mutagenic to account for the variation, or whether genetic changes preexisted in individual cells and were merely recovered by plant regeneration. There is insufficient data to prove either hypothesis, but the fact that cultured cells can undergo gross chromosomal changes such as aberrations, deletions, additions, etc., is well documented (7). Whatever the mechanism(s) of genetic change, in vitro culture of cells and protoplasts now appears to be a useful mechanism for developing improved breeding lines for hybrid synthesis, e.g., of lettuce, as well as for the direct selection of enhanced clones among vegetatively propagated species.

Possibilities for In Vitro Selection

As explained above, phenotypic variation is commonplace in plant populations regenerated from cultured cells. Even for species that appear stable through the process of cell or protoplast culture, it is questionable whether sufficient analysis of progeny has been made to firmly establish an absence of variation. Given the presence of phenotypic change at the mature phenotype, it would be extremely valuable to select for the altered trait at some stage in the cultural process. This would permit the screening of far greater numbers of genotypes and increase the overall efficiency of the recovery process.

Numerous investigators have shown that selections may be made with cultured cells, and in some cases they have lead to plants possessing the specific phenotypic modification. For a thorough review see Chaleff (4). For disease resistance characters, host-specific toxins from plant pathogens have been useful in identifying resistant clones (27), and even crude culture filtrates from *P. infestans* have been claimed to select for cells and then plants resistant to late blight of potato (1,2). Of course, the efficacy of toxins as selective tools for disease resistance relies not only on the ability to isolate the proper compounds, but also on a close correlation between response at the cell and the mature plant levels. For the moment, at least, these conditions restrict the number of useful systems available.

In vitro selection for complex genetic traits such as yield potential will require even more ingenuity, but possibilities do exist if we remember that cell or protoplast regeneration is a multistep process and that several developmental stages are available for selection. With potato, for example, regeneration involves the initial culture of protoplasts, the growing of resultant colonies on Medium-C (26), which conditions them for subsequent stages, colony transfer to Medium-D to induce shoot morphogenesis, and a final growth phase on Medium-T. Each of these stages is available for selection, and some progress has been made for potato with Medium-T. On this composition, differences between protoclones in vegetative vigor, one component of yield potential, are manifest. Shoots that have arisen from protoplast-derived calli (p-calli) may also be induced to undergo tuberization on Medium-T under proper environmental conditions. This may permit the development of techniques for selecting genotypes that form tubers efficiently under erstwhile suboptimal environments. Certainly, numerous other possibilities will emerge as systems are developed and understood. A very useful first step, however, may be a better understanding of the physiological processes expressed during plantlet development and of how they may be identified and manipulated in vitro.

PROTOPLAST FUSION

Plant protoplasts, by definition, lack a cell wall. Consequently, when brought together under appropriate conditions, protoplasts of widely divergent species may be induced to fuse with one another and form a heterokaryocyte (for reviews see 5,8,27). Conceptually, intergeneric protoplast fusion offers a means for transferring resistance genes from a donor to a recipient plant species, even when the two are sexually incompatible.

Heterokaryocytes of distantly or unrelated plant species have given rise to plants with chromosomes of only one parent or the other, equal numbers from both parents, and a variety of intermediate combinations (8,27). Moreover, even when all chromosomes of a donor species seem to have been lost, there is evidence that some genetic function of the donor had been retained (22). These findings suggest that interspecific protoplast fusion between sexually isolated species should be useful for creating novel genetic combinations ranging from amphiploids to simple translocations to transfer of cytoplasmic genetic characters. As pertains to disease resistance, this subject was recently reviewed (27), so here I will present only subsequent findings.

Nicotiana nesophilia and *N. stocktonii* are sexually incompatible with *N. tabacum*, although interspecific hybrids have been created through the embryo rescue technique. Evans et al (10) fused mesophyll protoplasts of the first two species with ones from suspension cultures of an albino mutant of *N. tabacum* and obtained 21 somatic hybrid plants, all of which were octaploids ($2n = 8x = 96$). *N. nesophilia* × *N. tabacum* plants expressed a hypersensitive reaction to tobacco mosaic virus (from *N. nesophilia*). Progeny analysis for one of the somatic hybrid clones was suggestive of genetic recombination between the two parental genomes.

Protoplast fusion has also been recently used in attempts to transfer triazine herbicide (atrazine) resistance from *Solanum nigrum* to *Nicotiana sylvestris*. Certain *S. nigrum* strains have been identified that are resistant to triazines by virtue of a single mutation in the chloroplastic DNA, and resistance to the herbicide is inherited as a maternal or cytoplasmic trait. Jain et al (15) X-irradiated mesophyll protoplasts from *S. nigrum* to render them mitotically inactive and then performed fusions with *N. sylvestris* protoplasts. Some regenerated plants showed increased atrazine resistance, but the level was not as great as in the resistant *S. nigrum* donor, and evidence was not presented that proved the hybrid nature of resistant clones.

The studies by Evans et al (10) and the preliminary results of Jain et al (15) give encouraging indications of the potential value of protoplast fusion for introgression of genetic characters such as disease and herbicide resistance into a species. Once present, conventional plant breeding techniques may be used to utilize the gene(s) in a crop improvement program.

GENETIC TRANSFORMATION

An extension of the above procedures for modifying the genetic complement of somatic cells is the introduction and integration of foreign DNA. Conceptually, cloned gene preparations might be supplied to protoplasts via microinjection, as DNA encased in liposomes, or ligated into a vector molecule such as the Ti plasmid from *Agrobacterium tumefaciens*. One might also envision supplying crude DNA preparations from a donor source to protoplasts. Progress in these areas has been closely monitored, (for reviews see 4–6,16), but in all cases, transformational frequencies are sufficiently low as to require efficient in vitro selection techniques to identify transformants.

CONCLUSIONS

It is clear that substantial progress has been made over the past decade in our ability to manipulate and culture plant cells. Concomitantly, there has been equivalent progress in the field of plant molecular genetics. This has permitted a coordinating of technologies and the emergence of an exciting new avenue for crop plant improvement. The breadth of these developments should not be overlooked in the field of plant pathology. Work cited above has shown promise for developing enhanced general disease resistance through protoplast and cell culture without intentional mutagenesis and without the loss of essential agronomic characters. Moreover, the findings apply to more than a single plant species, and future studies will undoubtedly add numerous other plants to the list. Such examples for disease resistance and for complex agronomic characters are keys to the potential of protoplast/cell culture as a means to expand the range of genetic variability within classes of elite plant genotypes.

Progress in the field of intergeneric protoplast fusion has also been encouraging, not simply for creating amphiploid dead-ends, but for introgressing mitochondrial, chloroplastic, and nuclear genetic elements from alien donors. Hence, if, say, protoplast mutagenesis and culture proves inadequate for developing a desired improvement in a genotype, protoplast fusion and plant regeneration provides an additional mechanism for acquiring the trait(s). And, on the horizon is a technology by which specific genes or gene clusters may be added to a genome for the purpose of conferring or modifying a single character. If the next decade is as fruitful as the past one, it should be an exciting period indeed for the whole of plant biology.

ACKNOWLEDGMENTS

Some of the research described above was supported by National Science Foundation Grants PFR-7712161 and PCM-8020713. Contribution No. 82-168-B, Kansas Agricultural Experiment Station.

LITERATURE CITED

1. Behnke, M. 1979. Selection of potato callus for resistance to culture filtrates of *Phytophthora infestans* and regeneration of resistant plants. Theor. Appl. Genet. 55:69-71.
2. Behnke, M. 1980. General resistance to late blight of *Solanum tuberosum* plants regenerated from callus resistant to culture filtrates of *Phytophthora infestans*. Theor. Appl. Genet. 56:151-152.
3. Brettell, R. I. S., and Ingram, D. S. 1979. Tissue culture in the production of novel disease-resistant crop plants. Biol. Rev. 54:329-345.
4. Chaleff, R. S. 1981. Genetics of higher plants, applications of cell culture. Cambridge University Press, Cambridge. 184 pp.
5. Cocking, E. C., Davey, M. R., Pental, D., and Power, J. B. 1981. Aspects of plant genetic manipulation. Nature 293:265-270.
6. Coutts, R. H. A. 1980. The potential of protoplasts for transformation or transduction studies. Pages 231–236 in: Plant Cell Cultures: Results and Perspectives. F. Sala, B. Parisi, R. Cella, and O. Ciferri, eds. Elsevier/North Holland Biomedical Press, New York.
7. D'Amato, F. 1977. Cytogenetics of differentiation in tissue and cell cultures. Pages 343–357 in: Plant Cell, Tissue, and Organ Culture. J. Reinert and Y. P. S. Bajaj, eds. Springer-Verlag, New York.
8. Dudits, D., Hadlaczky, G., Lazar, G., and Haydu, Z. 1980. Increase in genetic variability through somatic cell hybridization of distantly related plant species. Pages 207–214 in: Plant Cell Cultures: Results and Perspectives. F. Sala, B. Parisi, R. Cella, and O. Ciferri, eds. Elsevier/North Holland Biomedical Press, New York.
9. Edallo, S., Zucchinali, C., Perenzin, M., and Salamini, F. 1981. Chromosomal variation and frequency of spontaneous mutation associated with in vitro culture and plant regeneration in maize. Maydica 26:39-56.
10. Evans, D. A., Flick, C. E., and Jensen, R. A. 1981. Disease resistance: Incorporation into sexually incompatible somatic hybrids of the genus *Nicotiana*. Science 213:907-909.
11. Gengenbach, B. G., Green, C. E., and Donovan, C. M. 1977. Inheritance of selected pathotoxin resistance in maize plants regenerated from cell cultures. Proc. Natl. Acad. Sci. U.S.A. 74:5113-5117.
12. Green, C. E., Phillips, R. L., and Wang, A. S. 1977. Cytological analysis of plants regenerated from maize tissue cultures. Maize Genet. Coop. Newslett. 51:53-54.
13. Heinz, D. J., Krishnamurthi, M., Nickell, L. G., and Maretzki, A. 1977. Cell, tissue and organ culture in sugarcane improvement. Pages 3–17 in: Plant Cell, Tissue, and Organ Culture. J. Reinert and Y. P. S. Bajaj, eds. Springer-Verlag, New York.
14. Heinz, D. J., and Mee, G. W. P. 1971. Morphologic, cytogenetic, and enzymatic variation in *Saccharum* species hybrid clones derived from callus tissue. Am. J. Bot. 58:257-262.
15. Jain, M., Aviv, D., Davis, D. G., Galun, E., and Gressel, J. 1981. Conferring herbicide tolerance on tobacco by cybridization with atrazine resistant *Solanum nigrum*. (Abstr.) Plant Physiol. (Supp) 67(4):154.
16. Kleinhofs, A., and Behki, R. 1977. Prospects for plant genome modification by nonconventional method. Annu. Rev. Genet. 11:79-101.
17. Larkin, P. J., and Scowcroft, W. R. 1981. Somaclonal variation—A novel source of variability from cell cultures for plant improvement. Theor. Appl. Genet. 60:197-214.
18. Liu, M. C., and Chen, W. H. 1978. Significant improvement in sugarcane by using tissue culture techniques. (Abstr. 1725) in: Abstracts of 4th Int. Congr. of Plant Tissue and Cell Culture. Univ. Calgary, Calgary, Alberta, Canada, August 20–25, 1978.
19. Matern, U., Strobel, G., and Shepard, J. F. 1978. Reaction to phytotoxins in a potato population derived from mesophyll protoplasts. Proc. Natl. Acad. Sci. U.S.A. 75:4935-4939.
20. Nickell, L. G. 1977. Crop improvement in sugarcane: Studies using in vitro methods. Crop Sci. 17:717-719.
21. Orton, T. J. 1980. Chromosomal variability in tissue cultures and regenerated plants of *Hordeum*. Theor. Appl. Genet. 56:101-112.
22. Power, J. B., Frearson, E. M., Hayward, C., and Cocking, E. C. 1975. Some consequences of the fusion and selective culture of *Petunia* and *Parthenocissus* protoplasts. Plant Sci. Lett. 5:197-207.
23. Sager, R. 1972. Cytoplasmic Genes and Organelles. Academic Press, New York. 152 pp.
24. Secor, G. A., and Shepard, J. F. 1981. Variability of protoplast-derived clones of potato. Crop Sci. 21:102-105.
25. Shepard, J. F. 1980. Mutant selection and plant regeneration from potato mesophyll protoplasts. Pages 185–219 in: Emergent Techniques for the Genetic Improvement of Crops. I. Rubenstein, B. Gengenbach, R. L.

Phillips, and C. E. Green, eds. Univ. Minn. Press, Minneapolis. 242 pp.

26. Shepard, J. F. 1980. Abscisic acid-enhanced shoot initiation in protoplast-derived calli of potato. Plant Sci. Lett. 18:327-333.
27. Shepard, J. F. 1981. Protoplasts as sources of disease resistance in plants. Annu. Rev. Phytopathol. 19:145-166.
28. Shepard, J. F., Bidney, D., and Shahin, E. 1980. Potato protoplasts in crop improvement. Science 208:17-24.
29. Sibi, M. 1979. Expression of cryptic genetic factors in vivo and in vitro. Pages 339–340 in: Proc. Conf. Broadening Genet. Base Crops., Wageningen, 1978. Pudoc, Wageningen.
30. Skirvin, R. M., and Janick, J. 1976. Tissue culture-induced variation in scented *Pelargonium* spp. J. Am. Soc. Hortic. Sci. 101:281-290.
31. Takebe, I., Labib, G., and Melchers, G. 1971. Regeneration of whole plants from isolated mesophyll protoplasts of tobacco. Naturwissenschaften 58:318-320.
32. Webber, H. J. 1903. New horticultural and agricultural terms. Science 18:501-503.

CHAPTER 14

AMBIENT METEOROLOGICAL FACTORS—LIGHT, TEMPERATURE, AND MOISTURE

STELLA MELUGIN COAKLEY

My particular research area, that of relating climatic variation to plant disease occurrence, predisposes me to consider first the climate that is ultimately responsible for any given set of ambient weather conditions. I am challenged by the potential for applying the increasing understanding of climate to alleviating environmental constraints of plant health and crop yields. Following the section on climate, I will assess how ambient meteorological conditions affect plant health and suggest areas of research where much more effort should be spent in the future.

CLIMATE

Climate vs Weather

Traditionally in plant pathology, researchers have been concerned more with day-to-day meteorological conditions (weather) and how these affect plant disease than with year-to-year variations in meteorological conditions (climate). Weather is defined as the current and predictable meteorological state of the atmosphere including temperature, precipitation, relative humidity, wind speed and direction, cloud cover, barometric pressure, and radiation. Ambient meteorological conditions, i.e., the encompassing or surrounding weather conditions, are readily measured and can be duplicated to some degree in controlled environment situations for the purpose of studying their effects on plant health and disease. The use of synoptic meteorology (3–5 day forecasts) is discussed by Seem and Russo in Chapter 20 of this book. The problems and progress in forecasting beyond 3–5 days are discussed later in this chapter.

In the past, *climate* has been traditionally defined as the average weather conditions at a location. In agriculture, 30-year averages of temperature and precipitation are commonly used to describe the climate of an area, a practice that obscures climate variation. However, in the last decade the concept of climate has changed from "an average of weather" to "a dynamic physical system" that produces our weather (26). *Climatic state* (or *climate* as I use it) takes into account the dynamic nature of climate and is defined as:

> the average (together with the variability and other statistics) of the complete set of atmospheric, hydrospheric, and cryospheric variables over a specified period of time in a specified domain of the earth-atmosphere system. The time interval is . . . considerably longer than the life span of individual synoptic weather systems . . . and longer than the. . .time. . .over which the behavior of the atmosphere can be locally predicted. . . .

Climatic variation is defined as "the difference between climatic states of the same kind, as between two Januaries or between two decades," whereas *climatic variability* is defined as "the variance among a number of climatic states of the same kind" (28).

Climatic Variation

Since the early 1970s when a series of weather anomalies resulted in major agricultural losses, the popular press has published numerous statements and forecasts on various global temperature warming or cooling and on impending climatic disaster from increasing atmospheric CO_2. I would like to present a synopsis of the current scientific positions on climatic change—an update on the ". . . entrancing advances in meteorology and climatology . . . " that Horsfall mentioned in 1959 (20).

There is general agreement that the world's climate has been fluctuating, although there is no agreement that it is changing in a definite direction (26). Climatic trends in one area tend to be compensated for in another on both a hemispheric and global basis (29). As indicated by 30-year averages, regional variations in the mid-latitudes have been 1–2 C for temperature and 10–20% for precipitation over the past several centuries. Weather extremes, e.g., in the USSR in 1972 and in parts of western Europe in 1976, frequently result in substantial reductions in crop yields. The global mean temperature now is almost as high today as in 1940, although in the Northern Hemisphere there was a decrease of 0.5 C between 1940 and 1965 (3,19,26).

There is concern that the burning of fossil fuels will result in an unprecedented global warming because of the increase in atmospheric carbon dioxide (CO_2) concentration. The CO_2 concentration has increased from an estimated 280–300 ppm in 1880 to 335–340 ppm in 1980 due mainly to burning of fossil fuels (19). The additional CO_2 affects the earth's heat balance because it is virtually transparent to solar radiation but absorbs outgoing infrared radiation that would otherwise escape to space. Climate models, which include a doubling of CO_2 concentration from 300 to 600 ppm, give estimates of a global surface temperature increase of 2–3.5 C (estimated uncertainty factor of 2), with changes greater above 50° latitude ($\sim$5 C in the arctic for a global change of 2 C) (19,21). Such climatic models also estimate large regional climatic variations and shifting of precipitation patterns. An increased length of growing season of about 10 days for every 1 C increase would benefit some regions, and a warmer earth might have shifts of precipitation and soil moisture patterns in some regions (19). Although most models show temperature increases as the result of increased CO_2, the current models do not accurately simulate many parts of the climatic system (e.g., oceans, clouds, polar ice, and ice sheets), and the certainty or magnitude of increased temperature or its eventual effects are questionable. My conclusion must be that although climatic variation is almost assured, its direction and the actual annual average global or regional temperatures and precipitation to be expected are unsure. Agriculturists should be aware that climate variations which are "small" to climatologists may be "large" to biological organisms. I contend that the climatic variations that have occurred in North America have been large enough to affect the pattern of plant disease occurrence. Investigating the effects of a varying climate on plant disease is a new field with many opportunities for fruitful research.

Long-Range Forecasting

Long-range forecasting attempts to predict the character of the weather to be expected in a particular month, season, or year. The "character" of the weather is expressed relative to the normal climatologically expected meteorological value for a particular time (30). Lamb (22) discussed the problem of forecasting weather or climate beyond 3–5 days and indicated that such forecasts are "concerned with foreseeing the development (and steering) of circulation systems which do not now exist," whereas shorter-range forecasts deal with extrapolation of the movement and development of existing circulation systems. Long-range forecasts must take a more statistical form than synoptic forecasts. The goals of forecasting the weather of seasons ahead are to: 1) forecast averages of temperature, rainfall, and other variables that characterize the weather, 2) indicate frequencies (or probabilities) of occurrence of days with rain, hail, frost, long droughts, etc., and 3) indicate the probable duration of temperature and rainfall beyond threshold values needed for some human purpose. Existing forecasts now fall almost entirely in category 1 (22).

Readers interested in long-range forecasting methods are referred to the excellent review by Nicholls (30). Two types of long-range forecasts are routinely available for the United States. Since 1974 the U.S. National Weather Service Climate Analysis Center has produced 30-day "outlooks" every 15 days; on the average, 30% of the temperatures will be above or below normal class limits, whereas 50% of the precipitation will be above (heavy) or below (light) the median. An outlook of 90-day average temperatures is divided into above normal (60% chance of occurrence), below normal (60% chance of occurrence), or indeterminate (50% chance of above or below normal). The "above" and "below" categories are compared with the "normal" temperature for 1941–70, and each has a natural climatic frequency of 50%. The 60% probability of occurrence is based on the verification scores of 22 years of experimental seasonal predictions. A descriptive outlook of 90-day precipitation is published only for winter and summer seasons.

How good are these forecasts? Unfortunately, the forecasts are not very good and therefore not very useful. A study concluded that for 100 U.S. cities, the concurrence of temperature forecasts was 11% greater than chance and precipitation forecasts were only 2% better than chance. Nicholls (30) reviewed available studies of the accuracy of monthly, and bimonthly temperature and rainfall forecasts prepared in the United States, the USSR, Japan, and the United Kingdom and concluded that no convincing evidence of a useful level of concurrence has been evident in the forecasts and that there was no evidence of any improvement over the past decade or even the past 30 years. Experimental seasonal forecasts may be slightly improving.

Why are the forecasts so poor and what is the outlook? First, the physical processes and chains of causation in climate are not well enough understood.

Second, there is a large amount of normal climatic "noise" that is not and probably never will be predictable because it reflects sampling variability as well as normal variability. Beyond this is a part of the variance that is assumed to be potentially predictable; part of which may now be predictable and part of which may be predictable later. And, third, attempts to make predictions for all times of the year and for all regions may obscure the accuracy of predictions that can be made for some variables in some regions during some seasons but not in others. Meteorologists have been "urged to respond positively to the need for climatic forecasts and for general climatic anomaly information" (41).

If available, long-range weather predictions could be used to: 1) avoid spraying when pathogens or pests are unlikely to be troublesome, 2) decide whether marginal crops can be grown, and 3) choose cultivars or crops most likely to thrive in the expected climate, giving consideration to the pathogens or other pests expected.

Climatic Information Available

Meteorological data collection in the United States is supervised by the National Climatic Center at Asheville, NC 28801, as part of the National Oceanic and Atmospheric Administration's (NOAA) Environmental Data and Information Service. All data collected are available to the public and may be obtained on paper, microfilm, microfiche, or magnetic tape. *Selected Guide to Climatic Data Sources* (7) is a useful publication for determining the various kinds of data sets available. What data (location, type, and length of record) are available for a particular state can be determined by looking at the station index and map provided in the back of that state's *Annual Summary of Climatological Data.*

The NOAA-National Weather Service Cooperative Observer Program records daily maximum and minimum temperature and precipitation data for approximately 9,000 locations in the United States. Of these, subsets of 3,000 and 4,000 locations have long-term data sets, usually going back to 1900 or shortly thereafter, for temperature and precipitation, respectively. For stations with long-term records, 30-year averages, recalculated every 10 years, are available. There are 300 meteorological stations located at airports (also known as first order stations), which collect extensive surface meteorological data including hourly temperature, wetbulb and dewpoint temperatures, relative humidity, windspeed, precipitation, and cloud cover. A total of 3,000 stations in this country collect hourly precipitation data.

Once the original data are obtained, the researcher is faced with reducing the data to a manageable form. If computing facilities and an able programmer are available, the possibilities for analysis are limited only by money and computer time available. Following are ideas for initial data investigation.

Climatic averages expressed as arithmetic means are usually inadequate summaries. Standard deviation, range, and frequency of extremes should be determined for the data. Frequency distributions of temperature and precipitation can also be useful. Variability at a location can be determined by calculating the standard deviation (σ). Variability between stations can be compared by calculating the coefficient of variation (CV) at each station as follows:

$$CV = \frac{\sigma}{\bar{X}} \times 100$$

where $\bar{X}$ is equal to the mean for that station.

Year-to-year variability at a location may obscure gradual climatic trends that may be affecting plant health. The technique used to remove short-term irregularities is to determine the running-mean (moving average) by calculating the mean values for successive, overlapping periods (usually for 5-, 10-, or 30-year periods) and then plotting the running means on a graph. For example, a 5-year running-mean for year 3 is determined by adding the means for years 1 through 5 and dividing by 5. The mean for succeeding years is determined by averaging years 2–6, etc. Temperature trends thus calculated for winter and spring in the Pacific Northwest were associated with the increased occurrence of stripe rust on winter wheat (12,13).

Precipitation data (including all liquid and frozen forms of water) are usually available as kind, amount, duration, and frequency of precipitation. From these data the following can be determined: 1) rainfall intensity (amount/duration), 2) areal extent of a rainstorm, and 3) frequency of rainstorms of specified intensity. Rainfall intensity is usually greater for shorter periods of rainfall, and high-intensity rain is usually associated with increased drop size (2).

Preliminary analysis should include correlation analysis between meteorological variables and disease data. Possibilities are limited mainly by one's imagination. Summarizing temperature in a degree-day by using the optimum temperature for pathogen infection as the base temperature can prove to be very revealing (13).

Measurement of the Ambient Environment

For most epidemiological studies, the meteorological data routinely collected by the National Weather Service's cooperative network do not adequately represent the microclimatic conditions that surround the plants. Micrometeorology (or microclimatology) is the study of meteorology in the boundary layer of the atmosphere where temperature and humidity can change strikingly in a short distance and where the plants modify their environment. Several references (6,31,38,42,43) give information on the principles of meteorological measurement, instruments available, sources of error in measurement, and methods for making simple instruments.

Although automation and computing capabilities facilitate processing of large volumes of data, they also encourage accumulation of unnecessary data that may make analysis more difficult than need be. Sampling intervals should be carefully selected to reflect the time scale of the organism being studied. For example, daily maximum and minimum temperature measurements are usually adequate for studies of plant growth, whereas hourly or more frequent measurements may be best for studying spore formation and release.

Pennypacker (31) discussed the rapid development of new data acquisition and analysis techniques and the pros and cons of utilizing automated systems for research studies. Automation is not always the best or least expensive option for short-term studies where new instrumentation is rapidly becoming available because more effort may be required to use a computer system than to process the data manually.

When data are recorded, the sampling error should be noted and included in research reports because measurements of the physical environment usually are interpreted as being exact. Instrument errors, however, may be large even in properly functioning equipment (6,17,31,32,42). Unless measurements of responses are accurate, they have little value to other researchers. Use of inaccurate measurements will introduce errors into mathematical models. Pennypacker (31) states that a 5 C error in a temperature measurement (probably not uncommon) may cause a 10- to 14-day shift in a disease progress curve generated by computer simulation.

We frequently see in the literature such terms as "low temperature disease," "warm summer," "cold winter," "wet conditions," "high temperature, low moisture environment," etc. These terms are rarely defined when used and there are no quantitative definitions of "high" or "low," "cold" or "warm," etc. The use of "cold" and "warm" is to be discouraged in all cases because what is warm to a penguin could be very cold to a parakeet. What is meant by high and low temperature and moisture conditions should be clearly stated in research reports, as Spilker et al (39) did in quantifying their terms, including degree of error possible. For example, high temperature conditions were 33 ± 1 C day (12 hr) and 24 ± 1 C night (12 hr), whereas low humidity was 43 ± 4% rh (ambient). All research writing should be as exact in defining conditions.

A major challenge in the area of data handling is for the researcher to be adequately prepared to analyze the data collected. In the past, graduate training in statistics, quantitative analysis, and computer programming has been largely left up to the students to select and pursue according to their own self-motivation. Such course work, including experimental applications to plant health studies, should be strongly supported, encouraged, and probably required for all graduate students. Even if researchers employ someone to do such work (for large data handling projects this is more cost-effective), they will be handicapped if they do not understand the techniques required.

Relating Macroclimate and Microclimate

Although microclimatic data are preferred for epidemiological studies, they are not routinely collected in most regions. Standard (macroclimate) data are readily and routinely available for many areas and, very importantly, available for past periods for which collection of microclimatic data is impossible (10,11). Bourke (4) and subsequently Burrage (6) have reviewed and discussed the problem of relating macroclimate to microclimate. The closest agreement between the two occurs during cloudy and wet conditions, which favor uniformity of environmental conditions in an area; these conditions are also the most favorable for fungal activity. The greatest differences in macro- and micrometeorological conditions occur on calm sunny days when temperature and humidity in the crop differ greatly from that of the macroclimate. We found temperature and windspeed to be the most highly correlated variables ($r = 0.87$ and 0.75, respectively) between a field plot at Pendleton, OR, and the Pendleton Weather Station Office located at the airport. Relative humidity was not significantly related (16).

Additional studies are needed to successfully relate microclimate to macroclimate because a recent literature search indicates that essentially no progress has been made in this area in the past 10 years.

In addition to differences between macro- and microclimate, there are differences both horizontally and vertically in the plant canopy, and differences between the climate above and below the canopy. The environment of the crop is influenced principally by the distribution, density, and orientation of leaves. This varies between crops as well as within a crop during the growing season (6). For example, temperature may vary widely through the canopy, with any extreme temperatures measured above the canopy being of such short duration that disease within the canopy is not adversely affected. Because of this, spores produced within the protection of the canopy may be very important to epidemic development (34). Quantifying the relationships between the environments of the different locations is an important new research area. Mary Ann Sall and her colleagues at the University of California, Davis, are developing "uncoupling indices" to show the difference between conditions outside the grapevine canopy and those under the canopy for disease organisms (e.g., *Botrytis cinerea*). They can then demonstrate how different canopy structures affect the "uncoupling index." Such information can be used in managing the crop in order to minimize disease (Sall, *personal communication*).

HOW CLIMATE AFFECTS PLANT HEALTH

Climate (manifested as ambient meteorological conditions) affects plant health in a wide variety of ways. Climate (primarily temperature and moisture) delimits to a great degree what species of plants can be grown in a particular area, whereas weather determines a large portion of year-to-year variation in crop yields. In temperate zones, the length of the frost-free period together with the availability of water are major limiting factors (8). The genes of a plant determine what environment the plant will thrive in and how it will respond to environmental and pathological stresses.

Disease caused by biotic pathogens is limited to those areas where the environment is simultaneously favorable for a sufficient time period for the host, the pathogen, and any necessary vector. As the climate varies, the geographical distribution of disease also will vary. How ambient meteorological factors through time affect the host, the pathogen, the interaction of the host and pathogen, and any vector is represented by the concept of a disease pyramid (5). Disease development will depend upon how frequently the environment is favorable for each part of the pyramid.

Assessment of Constraints

The yield and quality of a crop is a function of the plants' genotype (genetic potential) and the environment (25). To obtain maximum crop yields, full advantage must be taken of all positive aspects of the environment (sunlight, temperature, water, and nutrients), and negative factors (biotic, chemical, or physical, including meteorological) must be excluded. To do this, assessment of these components as they affect the crop is necessary. Based on 40 years of research, Clements (9) has published an excellent crop log of sugarcane, which records its progress from start until harvest. For Hawaii, it is made up of observational, chemical, and physical data collected at 5-week or 1-week intervals for the crop. Energy available to the crop is measured as average maximum and minimum temperature and average light intensity for the preceding 5-week period. Calculated available energy combined with tissue moisture index and crop age are used to calculate crop expectancy. Clements (9) found that maximum and minimum temperatures, soil moisture, and light were third, fourth, eighth, and ninth, respectively, in order of contribution to the partial regression for explaining growth.

Wiese (45) has been developing a comprehensive and systematic assessment of crop yield determinants for dryland spring peas with the goal of developing techniques for quantitatively assessing crop losses. More than 40 environmental and management variables were included for 100 sample sites. Weather was considered in terms of hail, frost, precipitation, and degree days, but none of these was found to be significantly correlated with yield; I am not surprised since weather variables in a five-county region would probably not normally vary enough during a single growing season to be reflected in yield variability. For multiple year studies over more widely varying climatic conditions, I would expect weather variables to contribute a significant amount to determining yields. More research such as Clements and Wiese have done is needed; if preharvest yield proponents and constraints can be identified and quantified, it would be an important first step towards managing limiting factors and maximizing yields.

Environmental Stress

Plants respond to environmental changes on both diurnal and seasonal cycles. Whenever the environment changes in an unexpected way (from the plant's viewpoint) and adversely influences normal growth by interfering with physiological processes, environmental stress occurs. This may directly injure the plant or increase its susceptibility to biotic pathogens. Biological stress is defined by Levitt (23,24) as "any environmental factor capable of inducing a potentially injurious strain [physical or chemical change] in living organisms."

Levitt (23) divides environmental stresses into biotic and physiochemical. Biotic stress factors (e.g., disease-causing organisms) are themselves subject to environmental conditions. Physiochemical stress factors include temperature, water, radiation, chemical, and wind. Numerous "abiotic" diseases are caused directly by adverse meteorological conditions, and these deserve more research attention than they receive (18). Meteorological factors that produce a direct and obvious abiotic disease are much easier to quantify and study than those that increase or decrease a plant's susceptibility to biotic pathogens. Yarwood (46) defined predisposition as the "tendency of non-genetic conditions, acting before infection, to affect the susceptibility of plants to disease." Although both Yarwood (46) and Colhoun (15) deal with predisposition as separate from genetic susceptibility, whether disease develops "depends upon the influence of environmental factors on the genetically controlled response of the host plant to the presence of the pathogen or its metabolites" (35,36). How a plant responds to stress cannot be separated from its genetic makeup (18, and Chapter 9 of this volume). Stress not only influences host susceptibility (predisposition) but also the pathogen and the host-pathogen interaction. Research is needed for many disease situations to sort out whether the environment is affecting the host, the pathogen, or the interaction of the two. Research is particularly needed on how environmental stresses modify or regulate host-pathogen interactions. Most stresses have the largest impact on disease development involving nonaggressive pathogens (35). Some of this research is best done initially under controlled environmental conditions, even though Colhoun (14) cautions researchers that such conditions are very dissimilar from those in the field. It is important to consider the interactions of the various meteorological variables as they may affect disease, and these are easier to control and repeat in a controlled environment. Schoeneweiss (35) cites references for methods and techniques developed for this purpose. Downs and Hellmers (17) reviewed both the strengths and weaknesses of available controlled-climate facilities.

In the past, most experiments were designed to test the effect of single meteorological factors on disease development. Experiments should be designed to test various levels of each factor as well as interactions of factors. The effect of a specific factor on infection may be relative to and pertain to only certain combinations of the other factors. Disease develops through time and hence data collected under set conditions cannot accurately predict quantitative disease development. Rates of processes (e.g., spore germination) under varying temperature, light, and moisture conditions are needed. Furthermore, behavior of the pathogen on artificial media under a varying environment should be confirmed on host tissue whenever possible (37, and Chapter 19 of this volume.)

Light is the meteorological parameter which has been least considered in respect to plant health. The roles of light intensity and photoperiod in host susceptibility, pathogen behavior, and host-pathogen interactions need to be clarified. The effect of the full radiation spectrum should be considered instead of immediately limiting investigations to the visible spectrum as has been done frequently in the past. It is difficult to produce normal light conditions in most growth chambers and in the field (varying light is usually associated with varying temperature); however, Shaner (37) suggests that much can be learned from light intensity experiments in the field by using shades to lower intensity. Clements (9) suggests that using a day-degree measurement may be a reasonable measure of sunlight intensity. If so, there would be many opportunities to use the relationship; therefore this concept should be investigated further even though it may be true only for some areas.

Major climatic changes, especially reduced precipitation, over a period of years have been implicated as stress factors affecting the incidence and severity of primarily tree diseases (36). However, little quantitative work on the effects of short-term deficits or seasonal variations of water supply or temperatures has been done for annual crops. Clements (9) found that any time the moisture index dropped due to a deficiency of soil moisture, the plant underwent a drought reaction and even when water became available, the moisture index did not fully recover. Considerably more attention is warranted to determine how variations in temperature and moisture predispose plants to disease as well as to directly reduce yields.

Accurate measurements of surface moisture and relative humidity (RH) near the surface of a leaf are difficult. It is extremely hard to maintain a high RH (near 100%) at the leaf surface without condensation, and for many pathogens it is not known whether free moisture or a high RH is required for infection. Numerous measurements in each experiment are necessary to get an accurate estimate of the duration of surface moisture needed for fungal activity. On a macroclimatic scale, very little is known about how the intensity or seasonal distribution of precipitation affects disease. The most effective form of moisture for disease development is often not known.

Attempts to correlate meteorological factors singly or as independent variables in a multiple regression analysis with some aspect of disease may be complicated because of correlations between the factors. Temperature and precipitation are negatively correlated with each other in some seasons and regions (12). The relationship between light intensity and temperature also needs to be determined. When independent variables are significantly correlated with each other, a linear combination of them can be used to replace the separate variables in a multiple regression (27).

Rotem (34) put forth the "hypotheses of compensation" to explain the success of pathogens in several differing climatic regions. In brief, Rotem's hypotheses are that favorable environmental factors may compensate for unfavorable factors, a pathogen's weakness may be compensated for by a strength, and a high frequency of one phase in the pathogen's life cycle may compensate for the low frequency of another phase. Readers are referred to his work for details since such pathogen compensation activities may greatly affect the validity of studies on how ambient light, temperature, and moisture affect disease.

Distribution of Disease

Reichert and Palti (33) used a "patho-geographical approach" to predict the likelihood of specific disease occurrence in a specified region based on its world distribution and origin and the ecological characteristics of the host and pathogen. "Geophytopathology" was the term used by Weltzien (44) to describe a similar concept for the study of the distribution patterns of plant disease. To develop the above approaches, deficiencies of data must be eliminated. I encourage the collection of the following data: the meteorological ranges and optima of temperature, moisture, and light intensity and the duration for the pathogen, the host, and any vector involved; information on topography, elevation, and day-length should accompany disease data. Such studies offer the possibility of forecasting potential disease occurrence on a much wider scale than is possible for prediction of disease outbreaks requiring detailed meteorological collection and analysis. Using the methods of Reichert and Palti, the potential exists for establishing more scientifically based quarantine regulations, introducing new crops and cultivars more successfully, and establishing seasonal thresholds for disease occurrence (i.e., date before which disease will not occur) (33). Weltzien (44) provided details of using maps and models to describe the areas of actual and potential occurrence of a plant pathogen. It is essential to realize that disease intensity is highly variable within the area of occurrence. Standard methods need to be developed to assess disease intensity in areas larger than single fields. Maps of disease occurrence as a function of meteorological parameters would be invaluable in predicting shifts of disease occurrence as the result of variable climatic conditions. Computer graphics can greatly aid development of such maps.

Control of Disease

The last considerations for this chapter are how the ambient factors considered affect control of disease and how control could be modified by a variable climate.

Management

There is a trend towards increasing production by multiple cropping. Usually, however, the cultivars are being grown under suboptimum growing conditions at least part of the time in order to fit two crops into one growing season. Cultivars that yield well under stress should be used in such instances. Minimum tillage is another cultural practice that will probably increase biotic stress by favoring survival of plant pathogens and insects (see Chapter 15, this volume). Cultivars should be selected under appropriate selection pressure to meet such needs. Readers are referred to Sumner et al (40) for a recent review of these topics.

Light, temperature, and particularly moisture can be managed in some cases to control disease. Temperature can be managed more easily in controlled environment houses and for postharvest conditions; moisture can be managed best when the crop is irrigated in the field or grown under controlled conditions. Areas with climatic conditions unfavorable to disease should always be selected to produce pathogen-free seed and planting material. Selecting time of planting to optimize crop growth and avoid conditions that favor the pathogen or its vector is another method. Interrupting the timing of the host and pathogen by planting early-maturing cultivars that mature before the pathogen reaches epidemic levels or by utilizing a host-free period to reduce disease carried over from year to year are other methods (47).

Environmental factors can be selected that will promote escape from disease because of the pathogen-host relationships discussed earlier. For example, low temperatures decrease insect vector mobility, and high temperatures decrease fungus and nematode vector mobility (1). A variable climate may greatly modify success in biological control.

Chemical

Light, temperature, and moisture, alone and together, affect the biological activity and selectivity of chemicals applied to control pests. Unfavorable conditions may inactivate the chemical, cause toxicity, or increase the risk of resistant pests emerging. Existing herbicide application methods are not adequate in variable weather and require improvement (29).

Resistance

Resistance is the most common method for preventing or decreasing the damage from disease. Resistance may increase or decrease as the temperature varies. Temperature-sensitive, adult-plant resistance to stripe rust in the wheat cultivars Gaines and Nugaines is an excellent example. Resistance increases with host development and is greatest under greater, late-season temperatures. Below-normal spring and summer temperatures lower host resistance and facilitate epidemic development. Trends since 1961 in the Pacific Northwest have been toward temperatures that favor stripe rust development on certain cultivars (12,13).

Environmental stresses may result in disease development in normally resistant plants, and hence research with pathogens on stressed and nonstressed genetically uniform hosts should yield useful information. When one attempts to fulfill Koch's postulates in the absence of stress, it is probable that weak pathogens will fail to produce disease symptoms in vigorous plants and hence may be discarded as the causative agents.

In addition to disease resistance, we should strive to select cultivars that have the capacity to thrive under variable meteorological conditions. Lewis and Christiansen (25) provide an excellent review of the principles for breeding plants for stress environments. It is easier to select for some stress factors such as soil problems than to select for atmospheric temperature and moisture, which vary not only from location to location but also from year to year in an unpredictable way. Controlled environment chambers can greatly facilitate initial selection of material because temperature, moisture, and light can be consistently repeated and varied at the will of the researcher. This may also allow selection for a specific stress factor when highly variable and uncontrolled field conditions make results difficult to interpret. Selection material should also be grown under high biotic stress whenever possible; controlled environments can be used to vary climatic conditions in the presence of biotic pathogens. Particular needs include improving crop response to high and low temperature and water stresses and being sure genetic diversity is sufficient.

In the past, plant photoperiod requirements as well as cold tolerance limited the testing of new cultivars under stress environments. The release of semidwarf wheat cultivars that are photoperiod insensitive has allowed extensive testing of these under stress environments. Researchers are urged to find ways around conditions that limit testing of cultivars under new environments.

Relative stability of polygenic and monogenic resistance, as well as sensitivity to environment, should be considered when selecting for resistant cultivars to be grown under variable climates. A variable climate may also affect anatomical and biochemical resistance mechanisms. Additional research also is needed to establish how predisposing stress factors influence phytoalexin synthesis and effectiveness and the development of mechanical barriers as disease resistance mechanisms (35).

In this chapter I have covered somewhat superficially a very wide range of ways that ambient light, temperature, and moisture may affect plant health. I hope that some of the ideas presented will lead to research that will provide a new understanding on this subject. A goal for us all should be to understand the climate-weather-disease development interactions sufficiently so that we can use this information to maintain plant health under variable climatic conditions, which in time, may become predictable.

ACKNOWLEDGMENTS

Research was supported by the Climate Dynamics Section, Atmospheric Science Division, National Science Foundation (Grant 80-17604).

LITERATURE CITED

1. Agrios, G. N. 1980. Escape from disease. Pages 17–37 in: Plant Disease. Vol. 5, How Plants Defend Themselves. J. G. Horsfall and E. B. Cowling, eds. Academic Press, New York.
2. Barry, R. G., and Chorley, R. J. 1971. Atmosphere, Weather, and Climate. Methuen and Co. Ltd., London. 379 pp.
3. Bolin, B. 1980. Climatic changes and their effects on the biosphere. No. 542. World Meteorol. Org., Geneva. 49 pp.
4. Bourke, P. M. 1970. Use of weather information in the prediction of plant disease epiphytotics. Ann. Rev. Phytopathol. 8:345-370.
5. Browning, J. A., Simons, M. D., and Torres, E. 1977. Managing host genes: Epidemiologic and genetic concepts. Pages 191–212 in: Plant Disease. Vol. 1, How Disease is Managed. J. G. Horsfall and E. B. Cowling, eds. Academic Press, New York.
6. Burrage, S. W. 1978. Monitoring the environment in relation to epidemiology. Pages 93–101 in: Plant Disease Epidemiology. P. R. Scott and A. Bainbridge, eds. Blackwell Scientific Publ., Oxford.
7. Butson, K. D., and Hatch, W. L. 1979. Selective Guide to Climatic Data Sources. Key to Meteorological Records Documentation No. 4.11. NOAA-U.S. Dep. Commerce Environ. Data Info. Serv., Washington, DC. 142 pp.
8. Christiansen, M. N. 1979. Organization and conduct of plant stress research to increase

agricultural productivity. Pages 1–14 in: Stress Physiology in Crop Plants. H. Mussell and R. C. Staples, eds. Wiley-Interscience, New York.
9. Clements, H. F. 1980. Sugarcane Crop Logging and Crop Control: Principles and Practice. Pittman Publ. Ltd., London. 520 pp.
10. Coakley, S. M. 1978. The effect of climate variability on stripe rust of wheat in the Pacific Northwest. Phytopathology 68:207-212.
11. Coakley, S. M. 1979. Climate variability in the Pacific Northwest and its effect on stripe rust disease of winter wheat. Climatic Change 2:33-51.
12. Coakley, S. M., and Line, R. F. 1981. Climatic variables that control development of stripe rust disease on winter wheat. Climatic Change 3:303-315.
13. Coakley, S. M., and Line, R. F. 1981. Quantitative relationships between climatic variables and stripe rust epidemics on winter wheat. Phytopathology 71:461-467.
14. Colhoun, J. 1973. Effects of environmental factors on plant disease. Ann. Rev. Phytopathol. 11:343-364.
15. Colhoun, J. 1979. Predisposition by the environment. Pages 75–96 in: Plant Disease. Vol. 4, How Pathogens Induce Disease. J. G. Horsfall and E. B. Cowling, eds. Academic Press, New York.
16. Crowe, M. J., Coakley, S. M., and Emge, R. G. 1978. Forecasting dew duration at Pendleton, Oregon, using simple weather observations. J. Appl. Meteorol. 17:1482-1487.
17. Downs, R. J., and Hellmers, H. 1976. Controlled Climate and Plant Research. Tech. Note 148. No. 436. World Meteorol. Org., Geneva. 60 pp.
18. Grogan, R. G. 1981. The science and art of plant-disease diagnosis. Ann. Rev. Phytopathol. 19:333-351.
19. Hansen, J., Johnson, D., Lacis, A., Lebedeff, S., Lee, P., Rind, D., and Russell, G. 1981. Climate impact of increasing atmospheric carbon dioxide. Science 213:957-966.
20. Horsfall, J. G. 1959. A look to the future—The status of plant pathology in biology and agriculture. Pages 63–70 in: Plant Pathology, Problems and Progress, 1908–1958. C. S. Holton, G. W. Fischer, R. W. Fulton, H. Hart, and S. E. A. McCallan, eds. Univ. Wis. Press, Madison.
21. Kellogg, W. W. 1979. Influences of mankind on climate. Annu. Rev. Earth Planet. Sci. 7:63-92.
22. Lamb, H. H. 1977. Climate, Present, Past and Future. Vol. 2, Climatic History and the Future. Methuen and Co. Ltd., London. 835 pp.
23. Levitt, J. 1980. Responses of Plants to Environmental Stresses. Vol. 1, Chilling, Freezing, and High Temperature Stresses. Academic Press, New York. 497 pp.
24. Levitt, J. 1980. Responses of Plants to Environmental Stresses. Vol. 2, Water, Radiation, Salt, and other Stresses. Academic Press, New York. 607 pp.
25. Lewis, C. F., and Christiansen, M. N. 1981. Breeding plants for stress environments. Pages 151–177 in: Plant Breeding, Vol. 2. K. J. Frey, ed. Iowa State Univ. Press, Ames.
26. Mitchell, J. M., Jr. 1980. History and mechanisms of climate. Pages 31–42 in: Das Klima—Analysen and Modelle, Geschichte und Zukunft. H. Oeschger, B. Messerli, and M. Svilar, eds. (In English) Springer-Verlag, Berlin.
27. Mosteller, F., and Tukey, J. W. 1977. Data Analysis and Regression: A Second Course in Statistics. Addison-Wesley, Reading, MA. 558 pp.
28. National Research Council. 1975. Understanding Climatic Change: A Program for Action. Natl. Acad. Sci. U.S.A., Washington, DC. 239 pp.
29. National Research Council. 1976. Climate and Food: Climatic Fluctuation and U.S. Agricultural Production. Natl. Acad. Sci. U.S.A., Washington, DC. 212 pp.
30. Nicholls, N. 1980. Long-range weather forecasting: Value, status, and prospects. Rev. Geophys. Space Phys. 18:771-788.
31. Pennypacker, S. P. 1978. Instrumentation for epidemiology. Pages 97–118 in: Plant Disease. Vol. 2, How Disease Develops in Populations. J. G. Horsfall and E. B. Cowling, eds. Academic Press, New York.
32. Pennypacker, S. P., and Stevenson, R. E. 1980. Validity of quantitative ambient temperature-response relationships derived from in vitro investigations. Prot. Ecol. 2:189-198.
33. Reichert, I., and Palti, J. 1967. Prediction of plant disease occurrence. A pathogeographical approach. Mycopathol. Mycol. Appl. 32:337-355.
34. Rotem, J. 1978. Climatic and weather influences on epidemics. Pages 317–337 in: Plant Disease. Vol. 2, How Disease Develops in Populations. J. G. Horsfall and E. B. Cowling, eds. Academic Press, New York.
35. Schoeneweiss, D. F. 1975. Predisposition, stress and plant disease. Ann. Rev. Phytopathol. 13:193-211.
36. Schoeneweiss, D. F. 1978. Water stress as a predisposing factor in plant disease. Pages 61–99 in: Water Deficits and Plant Growth.

Vol. 5, Water and Plant Disease. T. T. Kozlowski, ed. Academic Press, New York.
37. Shaner, G. 1981. Effect of environment on fungal leaf blight of small grains. Ann. Rev. Phytopathol. 19:273-296.
38. Sheehy, J. E., and Tearle, A. M. 1975. The construction of instruments for measuring and manipulating the plant environment. Tech. Rep. 16. March. Grassland Research Institute, Hurley, U.K. SL6 5LR. 36 pp.
39. Spilker, D. A., Schmitthenner, A. F., and Ellett, C. W. 1981. Effect of humidity, temperature, fertility, and cultivar on the reduction of soybean seed quality by *Phomopsis* sp. Phytopathology 71:1027-1029.
40. Sumner, D. R., Doupnik, B., Jr., and Boosalis, M. G. 1981. Effects of reduced tillage and multiple cropping on plant diseases. Ann. Rev. Phytopathol. 19:167-187.
41. Thomas, M. K. 1979. Climatic Forecasts—The Value of Climatic Information and Predictions. World Meteorol. Org. Bull. 28(2):92-96.
42. Unwin, D. M. 1980. Microclimate Measurement for Ecologists. Academic Press, New York. 97 pp.
43. Van der Wal, A. F. 1978. Moisture as a factor in epidemiology and forecasting. Pages 253–295 in: Water Deficits and Plant Growth. Vol. 5, Water and Plant Disease. T. T. Kozlowski, ed. Academic Press, New York.
44. Weltzien, H. C. 1978. Geophytopathology. Pages 339–360 in: Plant Disease. Vol. 2, How Disease Develops in Populations. J. G. Horsfall and E. B. Cowling, eds. Academic Press, New York.
45. Wiese, M. V. 1980. Comprehensive and systematic assessment of crop yield determinants. Pages 262–269 in: Assessment of Losses Which Constrain Production and Crop Improvement in Agriculture and Forestry. Misc. Publ. 7-1980. Univ. Minn. Agric. Exp. Stn., St. Paul, MN. 327 pp.
46. Yarwood, C. E. 1959. Predisposition. Pages 521–562 in: Plant Pathology. Vol. 1, The Diseased Plant. J. G. Horsfall and A. E. Dimond, eds. Academic Press, New York.
47. Zentmyer, G. A., and Bald, J. G. 1977. Management of the environment. Pages 121–144 in: Plant Disease. Vol. 1, How Disease is Managed. J. G. Horsfall and E. B. Cowling, eds. Academic Press, New York.

CHAPTER 15

SOIL PHYSICAL FACTORS THAT AFFECT PLANT HEALTH

R. I. PAPENDICK and L. F. ELLIOTT

The physical condition of the soil plays a key role in plant health. Successful seed germination, seedling emergence, root growth, and ultimately crop yields are highly dependent on a favorable soil-air-water-temperature regime provided by a good soil environment. Physical factors combined with the amount and state of organic matter in the soil affect plant nutrient uptake and availability, root colonization by harmful or beneficial microorganisms, and physiological processes within the plant. For example, poor root growth in compact soil may be the single or combined result of physical impedance, lack or excess water, or oxygen deficiency that restricts root proliferation and nutrient absorption. Many times poor physical conditions of soil can increase root diseases and soil insect damage so that plants die or grow poorly because of these secondary effects. Pathogen and insect pest response to changes in the soil environment are well documented and must be considered when dealing with the effects of soil physical factors on plant growth and crop production.

PHYSICAL EDAPHIC FACTORS

Soil physical factors of greatest importance to plant health include water status, aeration, temperature, strength, and porosity characteristics. While not a strict soil physical property, soil organic matter will affect soil physical factors—some dramatically. None of these factors are independent variables; a change in one may cause a change in others. For example, a high soil water content may result in lowered soil temperatures or a deficiency of oxygen for plants and microorganisms.

The soil physical factors are influenced to a large extent by the soil structure or aggregation of the primary particles. The characteristics of soil structure that are most important to plants are the pore-size distribution (relative numbers of pores in different size classes) that results from aggregation, and aggregate stability (the hardness of the individual aggregates and their ability to hold their shape upon wetting or reform upon drying). Many physical problems of soils relate to poor structure. Good structure is needed to maintain adequate aeration and ready infiltration and to provide an optimum root environment.

Soil Water

The state of the soil water is probably the most important soil physical factor affecting plant health. Soil water is not only essential for life processes of plants and microorganisms, but it also has many interacting components that individually or in combination affect biological systems. These effects may arise from the potential energy of the water per se or from the indirect effects of water potential or water content on such things as gas or solute diffusion or soil strength.

Water relations of plants and microorganisms are best described in terms of the potential energy of water (water potential) because, normally, work must be expended to obtain water from the plant surroundings (21). The work required depends mainly on the amount of water in the system, solute concentration, and to a lesser extent on water temperature. In most soil and plant systems, only the matric and osmotic components are considered because they exert the greatest effects on water flow and availability for physiological processes.

Water potential affects plant growth in at least three ways, all of which are interrelated. First, as the cell water potential decreases, plant growth generally decreases. Second, the conductivity of the soil for water depends on the water potential; as the soil water potential decreases, soil water conductivity decreases and flow to plant roots is slowed, which may lower the plant water potential and, hence, reduce growth. Third, a decrease in soil water potential, especially in soils with high bulk density, can lead to marked increases in soil strength and, hence, increased mechanical impedance to root elongation.

The water potential of microorganisms in soil varies with the water potential of the surrounding soil and is not directly subject, like that of higher plants, to atmospheric desiccation. Besides affecting growth, the matric potential, through control of the size distribution of water-filled pores, determines the ability of microorganisms of various sizes to disperse in soil and move through water-filled channels.

Diffusion of gases and solutes for all textural classes of soils is directly related to water content and not to the water potential. Therefore, water content, not soil water potential, is the proper variable to use in studies of reactions of plant roots and microorganisms in soil that are diffusion limited (16).

Soil Aeration and Oxygen Supply

Soil aeration consists of exchanging atmospheric O_2 consumed by respiring plant roots and microorganisms for the respiratory end product CO_2. Restricted aeration results in a reduction of O_2 and a buildup of CO_2 in soil, which can reduce water absorption and nutrient uptake by plants and can slow growth processes. Also, healthy roots subjected to O_2 stress generally undergo marked morphological and physiological changes; they can become short and stubby with fewer root hairs, become concentrated near the soil surface, and, under extreme stress, die. Often it is difficult to determine if the response by plants or microorganisms is caused by O_2 deficiency or CO_2 toxicity.

For most plants, there appears to be a critical soil O_2 concentration below which aeration will limit growth and crop yields. This value varies considerably for different plant species and is dependent on physical characteristics such as temperature, water status, compaction, and on certain chemical and biological factors operating in the soil. Some plants such as tomato (*Lycopersicon esculentum*) are very sensitive to soil O_2 content and grow best at O_2 levels approaching or exceeding atmospheric concentrations. Others, such as barley (*Hordeum vulgare*) are much less sensitive, and certain hydrophytes such as rice (*Oryza sativa*) can grow in the absence of O_2 in the soil air. Respiration by microorganisms in soil is generally not limited by O_2 concentration until it is down to about 1–3% by volume (19). However, growth of some organisms may be reduced by increasing the CO_2 concentration, while others may be stimulated, particularly when O_2 is not limiting (3,7,14). Soil aeration status places selective ecological pressures on microbial competition and ultimately root colonization, which in turn affects plant health and vigor.

Soil Temperature

Temperature has a pronounced influence on most biophysical and biochemical processes associated with growth of plants and microorganisms. Processes such as assimilation, respiration, transpiration, organic matter decomposition, and photosynthesis are strongly temperature dependent and, in the growth range for plants or microorganisms, increase with increasing temperature until some limit is reached. Similarly, soil physical processes such as gas and solute diffusion, water flow in soil, and evaporation are temperature dependent and all processes accelerate as temperatures increase.

Plant growth response to soil temperature is usually related to the mean temperature over a period, obtained by averaging the daily maximums and minimums or by averaging them singly. Sometimes, mean air temperatures can be used to approximate mean soil temperatures. However, averages give no indication of temperature extremes that can be harmful to plants or microorganisms if they approach or exceed critical values. Sometimes, cumulative or integrated soil temperature above a specified base temperature is more closely related to plant growth than is average temperature.

Soil Strength

This soil property is of most significance in mechanical impedance to seedling emergence and root penetration. Seedling emergence is most affected by soil crusts that develop when a soil dries after rain or irrigation that breaks down the surface structure and washes the fines into the voids. Crust strength varies with soil water and organic matter content, texture, mineralogy, aggregate stability, and field environmental conditions and, therefore, cannot be readily predicted from laboratory measurements. The critical crust strength for seedling emergence varies considerably for different plant species. For example, the limiting crust strength for wheat (*Triticum aestivum*) at an intermediate soil water content is at least twice that for grain sorghum (*Sorghum vulgare*) (9). Another consideration is that the emergence force of seedlings is a function of the time after planting. The emergence force of corn (*Zea mays*) seedlings at the optimum environment reaches a peak 120 hr after planting, and that of cotton 100–120 hr after planting (5,15). Thus, after planting, if drying of the soil increases the strength of the soil rapidly to a strength greater than the force generated by the seedling, the seedling may not emerge even though conditions at planting time were suitable for emergence.

Soil compaction in contrast to crusting is not confined to the surface layers but may extend to a depth of 30 cm or more and is largely the result of equipment traffic. The direct effects of compaction on the soil are loss of total pore space, a decrease in average pore size, and an increase in soil bulk density. Compaction of soils is often an insidious process and its rate of increase varies with soil properties and management. Soils that are low in organic matter with a high content of silt or fine sand and that are tilled wet are very susceptible to compaction. However, with any soil, the primary cause of this problem is cultivation when the soil is too wet.

Porosity and Size Distribution of Pores

A medium-textured soil in its natural state will average about 50% solid material and 50% pore space. The porosity may vary from 40% for sands to 60% for clays, with the range within a textural class being dependent on organic matter content. The pore spaces provide channels for air, water, plant roots, and microorganisms and are an important factor controlling biological processes in the soil. Low aeration porosities, i.e., below about 10%, may be restrictive to root growth and can lead to anaerobic soil conditions.

Equally or more important to biological processes than total porosity is the size distribution of pores. An ideal soil is described as having an equal division between large and small pores. This provides adequate aeration (large pores) and water retention (small pores). Aeration porosity is generally taken as the fraction of the pore space that is air filled at −1/3 bar matric water potential.

Soil porosity and pore-size distribution are most readily altered in the field by tillage management. Cultivation of relatively dry soil can increase the total porosity more than 50%, but the increase becomes less as the soil is tilled under wet conditions. Following the initial tillage operation, repeated tillage usually decreases the total porosity, but not to the level of the untilled soil. Similarly, soils tend to settle after cultivation, and as a result the porosity will decline over the season along with a concomitant shift from large to small pore sizes.

CURRENT AND EMERGING PROBLEMS WITH PHYSICAL EDAPHIC FACTORS

There are numerous indications that many cropland soils in the world are suffering from declining productivity. Much of the decline in soil productivity can be attributed to the deterioration of soil physical properties due to mismanagement or severe land abuse. Soil that loses its productivity often responds less to improved management, and if the productivity loss is severe enough, the soil must be abandoned for agricultural purposes.

Soil Erosion

Loss of topsoil by wind and water erosion caused by poor soil management is by far the largest single factor contributing to deterioration of physical properties and decline in productivity of most cropland soils. The U.S. Department of Agriculture (USDA) has assigned a soil loss tolerance value (T-value) for most cultivated soils. This value defines the maximum rate of soil loss that will permit sustained crop production. These T-values never exceed 11.2 mT/ha/yr, and some are less, depending on the soil depth and other factors (10).

In 1977, erosion rates were above 22.4 mT/ha/yr on 32% of the lands used for row crops in the South, 9% in the Northeast, and 19% in the corn belt (10). Nationally, water erosion alone exceeded the T-value on more than 45.3 million hectares or on 27% of the cropland (10). Soil losses are generally largest with row crops and are considerably less with meadow crops or rotations that include meadow crops. Some reports over the past several years indicate that erosion losses are increasing. One cause for the increase in erosion is the decreased use of crop rotations that specifically include a hay crop and a concomitant increase in hectarages of row crops like corn and soybeans. More row cropping with conventional practices means more intensive tillage and increased erosion. One estimate is that under a "high export scenario" with unrestricted land use and practices for all-out production, soil erosion would be increased between 50 and 80% in the corn belt and 21% nationwide (4). In this situation, the nation would lose 2 mT of topsoil for each mT of corn exported.

Topsoil on much of the world's croplands is only a few centimeters deep, usually less than 30 cm, and is underlain by subsoil that is considerably less productive. As topsoil is lost, subsoil becomes part of the cultivated layer, thereby reducing the soil's organic matter, nutrient content, and structural characteristics that make the soil ideal for seedling establishment and root growth. The direct effect of topsoil loss on crop yields is difficult to assess because increased use of technological inputs such as fertilizer, cultivation, improved crop cultivars, and pest and disease control can mask the reduced capabilities of the soil to produce crops. Evidence suggests that corn yields are reduced annually by an average of 80–106 kg/ha for each centimeter of topsoil lost (8). Thus, at a 27 mT/ha annual soil loss rate, corn yields would decline 80–106 kg/ha every 4.9 years, assuming all production factors remained constant (assuming 132.5 mT/ha per centimeter of topsoil).

Compaction

A trend in farming practices characteristic of many areas of developed countries is increased size and weight of farm tractors and implements and increased speed of tillage operations. Larger equipment is a part of larger farm size, increased hectarage of cultivated crops, and decreased use of meadow in the rotation. These practices are associated with increased compaction, which is becoming more severe on many soils. The greater traction of the equipment enables earlier tillage in the spring when the soil is too wet and is subject to compaction by wheel traffic. Besides adverse effects on root penetration and water and air movement, the hard or compacted layers require more energy for tillage. For example, tractor wheel slippage increased from 14 to 25% when the percentage of land previously wheel-tracked increased from 20 to 40% (23).

Data, although limited, show that crop yield losses from compaction can be substantial. Cotton yields in trafficked plots were 6.4 bales/ha compared with 7.7 bales in untrafficked areas (6). Farmers surveyed in southcentral Minnesota identified soil compaction as their most serious crop production problem (20). Yield losses, according to their best estimates during wet years when compaction was the worst, averaged 7.5% for corn, 10% for wheat, 13% for sugar beets, and 54% for potatoes. Loss of crop value in the United States as a result of soil compaction was estimated to be $1.18 billion in 1971 and $3 billion in 1980 (20).

These values do indicate that soil compaction is becoming an increasingly adverse element in crop production and should be given considerable attention in the years ahead.

Interestingly, compaction may be masking the effect of declining soil organic matter contents on aggregation and stability of aggregates of some soils. Normally, any loss in soil organic matter associated with increased intensity of row cropping will lead to less soil aggregation and a decrease in water stability of aggregates. However, wheel-induced compaction may over time cause a slight but continuous increase in aggregate size and stability, thus appearing to compensate for any loss of organic matter (23). This effect of compaction on aggregation should not be construed as being beneficial.

Loss of Organic Matter

Soil organic matter content is perhaps the best single indicator of soil quality and the main constituent that distinguishes topsoil from subsoil. Loss of organic matter results in loss of soil tilth and productivity. Most mineral soils contain only 2–4% organic matter by weight, and desert soils may contain less than 0.5% organic matter. Though the amount of organic matter in most soils is relatively small, its influence on soil properties can be very large. Organic matter serves as the chief granulating agent in soils and provides structural stability and optimum air and water relations. The effect of organic matter (degraded straw) on soil and ash aggregation and soil water retention is illustrated in Table 1 (13). A greater aggregate stability is indicated by a lower amount of solids in suspension. The degraded straw significantly increased the water stability of the Palouse silt loam and volcanic ash. The water retention of the soil was unaffected by the organic matter addition, but the water retention of the ash was increased significantly. Organic matter also darkens the soil, causing it to absorb heat readily and warm up rapidly in the spring.

Native amounts of organic matter usually decline rapidly once the soil is cultivated. The equilibrium organic matter content in the soil depends on the amount of tillage, the amount and type of organic material being added, and the soil environment. For example, in Iowa, an annual addition of 5 t/ha corn stover was required to maintain the soil organic carbon at 1.8% with continuous corn cropping; greater additions increased the soil organic matter content (11). Organic matter losses averaging about 50% below virgin levels appear inevitable for many soils after cultivation for 50–60 years; however, this loss value will vary considerably depending on the crop rotation and tillage management. Declines are greatest with intensive row cropping and least with meadow in the rotation.

It appears that organic matter contents of most cropland soils are continuing

TABLE 1. Water stability and water retention of Mt. St. Helens' volcanic ash and soil aggregates

	Percentage solid in suspension[a]		Water retention (grams per gram of solid)[a]	
Treatment	Soil	Ash	Soil	Ash
Degraded straw	8.5 a	25.0 c	0.379 a	0.461 b
None (control)	13.6 b	40.5 d	0.378 a	0.449 d

[a] Figures not followed by the same letter are significantly different ($P = 0.05$) using Duncan's multiple range test for a variable sample.

to decline slowly. There is a need for crop production practices that stabilize the organic matter contents at some point to maintain soil productivity and to prevent further reduction. Continuing declines in soil organic matter content will lead to subtle deterioration of soil structure and tilth and in turn create seedbed preparation and seedling emergence problems and adversely affect root growth. This deterioration may also adversely affect plant-root-microbial associations.

Soil Water Concerns

Competition for available fresh water supplies for uses other than agricultural has increased markedly in recent years, and it appears certain that in the future agriculture will lose some of its water. Irrigated agriculture, practiced mainly in arid and semiarid areas, will be most seriously affected by reduced water supplies. Less water for irrigation coupled with a growing demand for food will place greater pressure on nonirrigated crop-producing areas where soil water is often the main factor limiting production. Most experts agree that water shortage problems in the future will be solved by technologies that conserve water supplies, reduce crop water requirements, and increase plant tolerance or resistance to soil water stress.

Traditionally, the primary problem in the U.S. dryland regions is the lack of sufficient water for annual cropping. Fallowing is a water conservation practice plagued by low storage efficiency—the technique at best saves only 25–35% of the precipitation that falls over a 2-year period. In some areas, e.g., the northern Great Plains, fallowing promotes the development of saline seeps and has ruined significant areas for future crop production. There is increasing pressure in some dryland areas to crop the land annually, which in most years will force crops to grow under increased water stress. Thus, improving crop water-use efficiencies, being able to predict where and when precipitation is adequate for annual cropping, and managing crops to reduce water stress will undoubtedly be topics of increased emphasis in future years.

Plant water stress is also a major problem in humid regions. It is not uncommon for areas receiving more than 1,000 mm of precipitation annually to undergo extended dry periods during critical crop growth stages in early to midsummer. Many times, poor rain distribution is coupled with poor soil physical conditions and low organic matter. Management of the soil water to reduce plant water stress may be just as crucial under these conditions as in the dryland regions, although the frequency of stress may be less.

A major problem with soil water management under irrigation is improper timing and amount of water application. Unfortunately, efficient irrigation often requires more effort or is more costly than less efficient methods that are wasteful of both water and energy.

STRATEGIES TO IMPROVE SOIL PHYSICAL CONDITIONS

Poor plant health indicates that something is wrong with the plant's soil or aerial environment and is interfering with growth. If the problem is related to deterioration of some physical property of the soil, one should investigate further to see if the cause was from excessive soil loss by erosion, excess water, salt buildup, water stress, low organic matter content, or other factors. However, these factors are symptoms or external signs that something has gone wrong with the soil system. Rather than to treat the symptom, it is best usually to treat the underlying cause. For example, where soil compaction has become a serious problem, it is probably better for the long term to treat the cause of compaction rather than to devise mechanical ways to break up the hard soil, knowing that the solution is only temporary. To accomplish this, it may be necessary to design a different cropping or tillage system, apply organic manures, grow legumes in the rotation, or manage crop residues in a different way.

Fortunately, improvement of most soil physical properties affecting plant health is within the capability of present technology or can be achieved through technology development. There are, for example, workable practices for maintaining or increasing soil organic matter, which in turn will reduce compaction and improve other soil properties such as infiltration, water retention, and aeration. Conservation tillage systems under development offer tremendous potential for erosion control as well as for conserving water and increasing the organic matter content of some soils.

Conservation Tillage

Conservation tillage is defined as any "tillage sequence which reduces loss of soil and water relative to conventional tillage." It is currently viewed by conservationists as the most promising and economical approach for erosion control and is gaining acceptance by farmers in many parts of the United States and in many developed countries in the world. The USDA projects that most farming operations in the United States will be using some form of conservation tillage by the year 2000.

The main features of conservation tillage are increased soil surface roughness and maintenance of crop residues on the soil surface. These practices not only reduce runoff and erosion but affect the radiation or heat balance at the soil surface. Usually conservation tillage involves "reduced tillage," which means less field traffic, less fuel consumed, lower machinery investment, and sometimes, but not always, more chemicals for weed and pest control. The extreme case of reduced tillage is "no-till," where the crop is planted with just enough tillage to place and cover the seed in soil. Many studies indicate that no-till agriculture reduces soil erosion to almost zero. Since decreased expenditures for fuel with reduced tillage or no-till systems usually more than compensate for the increased use of chemicals for weed and pest control, there is a reduced cost incentive with these practices as well as a soil conservation incentive.

Success with reduced tillage in crop production depends on soil type, drainage, climate, and management practices. In the corn belt, yields of either rotation or continuous corn with reduced tillage often equal yields from conventional tillage methods on well-drained soils. Yields for the two methods on poorly drained soil are about equal when corn is grown in rotation with other crops. In the Northwest, yields of small grain crops with no-till and heavy surface residues are often 20–30% less than those from conventional tillage planting because of poor seed placement, inadequate weed control, rodent damage, increased disease, phytotoxicity, and possibly some yet unknown factors.

The presence of surface residues and minimum disturbance of the soil with reduced tillage systems may change drastically the soil-plant environment compared with that of conventional tillage. With reduced tillage, the soil is wetter and cooler in the spring, which may delay germination and slow early growth of some crop plants. Surface residues may also reduce the number of freeze-thaw cycles and result in improved soil aggregation at the surface. No-tillage usually results in an increase in the soil bulk density in the shallow layers and a reduction in aeration porosity and a less oxidative environment. This alteration of the physical environment should lead to increases in organic carbon and nitrogen contents in the soil and thus give long-term improvement in the soil structure.

The incidence and importance of certain biological agents such as insects, beneficial microorganisms, and plant pathogens that affect crop growth may be changed by reduced tillage management. For example, reduced tillage with increased surface residues would predictably favor root diseases that are problems in cool, moist soil (e.g., those caused by *Pythium*) and suppress those that are favored by warm, dry soil (e.g., stalk rot of corn or sorghum). Changes in tillage and residue management can affect plant root colonization, both by harmful and beneficial microorganisms. Overall, root diseases might be decreased by reduced tillage because increases in the surface soil organic matter content is generally suppressive to soilborne plant pathogens.

Surface residues associated with conservation tillage systems may serve as an important primary source of inoculum for diseases. Some examples for corn include the fungus pathogens for southern corn blight, northern leaf blight, and yellow leaf blight. Some wheat diseases caused by pathogens that are known to survive or increase in surface residues include tan spot, take-all, and Septoria leaf blotch.

Farmers in many areas of the world are finding that the advantages of reduced tillage systems, particularly no-till, often far outweigh the disadvantages. Many agricultural leaders and scientists agree that a more sustainable type of agriculture can be achieved with a reduced tillage system. Its rate of acceptance will depend to a large degree on how rapidly technology can be developed to overcome associated crop production problems.

Organic Matter Maintenance

Organic matter is of most benefit to the soil physical properties if it is concentrated in the upper few centimeters of soil. There it is most beneficial for such things as infiltration, stabilizing the soil against erosion, and improving the structure and tilth for seedbed preparation or planting, seedling establishment, and any subsequent cultivation. Thus, practices used to manage organic materials applied to soil should be designed to incorporate or mix these materials in the shallow soil layers.

Experiments generally show that soil organic matter levels can be maintained through proper crop rotations and tillage, residue management, and through regular applications of animal manures or other organic materials. Legumes or other forage crops in the rotation are especially effective for slowing the decline of organic matter or increasing its equilibrium level in the soil. Organic materials added to the soil at regular intervals as green manures (e.g., rye, oats, sweet clover, vetch), barnyard manure, sewage sludge, garbage, or other organic materials will also aid in maintaining the soil organic matter content.

The most important available source of organic material on most farms is crop residues. Stubble, stover, and roots are the main materials. The amount of material available for return to the soil depends largely on the crop yields. Soil organic matter levels can actually be increased by high crop yields, which increase the amount of residues for return to the soil (11). Moreover, as indicated in the previous section, surface residue management and less soil disturbance with conservation tillage should aid in maintaining organic matter at higher levels than those resulting from conventional tillage, which is another advantage for this system.

Control of Soil Water Potential

Irrigation combined with tillage practices provides the most direct means for controlling the soil water potential in the field. Frequent watering with limited amounts applied will maintain high water potentials in the upper layers. Infrequent, heavy irrigations allow surface layers to dry for longer periods. Maintaining crop residues on the soil surface would tend to keep the soil water potential relatively constant over extended periods between irrigations or rains. Raised soil beds is a management tool to maintain upper portions of the crop root zone in dry soil in furrow irrigation systems. This appears effective for protection against certain root pathogens (e.g., *Pythium* sp.) that are found in shallow parts of the root zone but need a high water potential for infection.

Special management techniques can be used to control plant water stress. For example, subsoiling will temporarily break hard soil layers and enable better root penetration. This practice appears to relieve water stress of irrigated beans in the Columbia Basin of Washington and thereby increase the plant's resistance to *Fusarium* root rot (1). Cultivation can be used to hasten drying of the tillage layer by disrupting continuity of the capillaries from the deeper layers to the surface. Once a dry layer is formed, moisture loss from below is markedly impeded and soil water is conserved (17). Other methods for controlling the soil water regime include 1) control of the wetting characteristics of soils through chemical modification of hydrophobic and hydrophilic properties, 2) use of plants with different transpiration characteristics, 3) optimal sequencing of deep-rooted and shallow-rooted plants, 4) improved drainage and tillage methods, and 5) amendments to change the surface color characteristics of the soil to alter the radiation of heat balance at the soil surface (18).

An interesting approach for increasing a plant's resistance against disease is by osmotic adjustment of the plant's water potential through additions of fertilizers or other salts, or possibly organic amendments to soil. Christensen et al (2) reported that applications of chloride salts to soil reduced the severity of take-all of wheat. They postulated that chloride lowered the osmotic potential of plant cells, which slowed the rate of root colonization by the pathogen, *Gaeumannomyces graminis* but did not adversely affect plant growth.

Organic Farming

There are several thousand farmers in the United States operating commercial farms profitably with minimum or no use of chemical fertilizers or pesticides. They are referred to as organic farmers, and although there is considerable diversity in their individual farming approaches, they collectively advocate that sustainable and successful farming must be based first and foremost on care and protection of the soil. If this is done, economical crop production on a sustainable basis is more likely assured.

Studies made by the USDA (22) and Washington University (12) of commercial-scale organic farming show that the farmers rely on recycling of organic materials and use of green manure crops and legumes in the rotation to supply nutrients and to maintain nutrient balance and soil organic matter. As a group, organic farmers are highly committed to protecting the soil resource, and their basic approach to this end is to maintain the topsoil in good physical condition through regular use of soil-building crops and animal manures. Weeds, insect pests, and plant diseases are controlled through use of crop rotations, timely tillage, and biological controls.

It is clear that the basic tenets and potential benefits of organic farming address many of the central agricultural policy issues of today and of coming years, such as environmental protection, less energy-intensive modes of production, soil and water conservation, and improved food quality and safety. For this reason, it appears likely that more farmers will at least attempt to adopt some of the organic techniques, especially for soil and water conservation, as they become better understood and more economically advantageous. The challenge for plant and soil scientists today is to understand why and where organic techniques can be successful and to develop this technology to achieve sustainable agricultural systems in general.

FUTURE RESEARCH

The design of sustainable, productive cropping systems depends on a thorough understanding of the soil physical environment and its relationship to plant health. For this reason, soil physics research cannot be separated from plant or microbiological research; it must be an integral part of it. Moreover, a systems approach is needed to study soil-plant health relationships. Too often researchers isolate certain soil factors such as water or temperature and investigate their individual effects on plant growth—and mainly in the laboratory rather than in the field. Comprehensive studies on the interaction of various soil physical factors and how they influence plant growth and health are often lacking. This information is needed if we are to develop management systems to control the soil environment for maximum crop production. In addition to a systems approach, future research relative to soil physical factors and plant health should focus on the following general goals:

1. To determine the changes in relationships of soil microorganisms (including plant pathogens) and plant root microflora that are associated with surface residue systems.
2. To develop methods and management practices, including irrigation management, for controlling the soil water potential in the field as a means of optimizing plant growth and as a selective factor in the control of root pathogens.
3. To determine the levels of organic matter required to maintain the productivity of different soils at satisfactory or sustainable levels for indefinite periods.
4. To investigate causes for increased root diseases under compact soil conditions and to define the individual and interactive roles of water stress, soil aeration, and soil strength on disease severity.
5. To develop improved techniques for water conservation to increase soil-water storage during noncrop periods for the purpose of developing cropping systems that use water more efficiently than conventional crop-fallow systems do.

6. To develop plants with greater tolerance or resistance to water stress in limited rainfall environments.

7. To determine plant resistance to disease and insect pests when plants are grown under organic or conventional farming methods.

8. To assess the potential beneficial or harmful effects of microbe-root associations as affected by changes in soil physical conditions and on ways that these associations can be managed to improve plant health.

LITERATURE CITED

1. Burke, D. W., Miller, D. E., Holmes, L. D., and Barker, A. W. 1972. Counteracting bean root rot by loosening the soil. Phytopathology 62:306-309.
2. Christensen, N. W., Taylor, R. G., and Jackson, T. L. 1980. Relationships between chloride mediated reductions in plant water potential and severity of take-all root rot of wheat. Page 77 in: Agronomy Abstracts. Am. Soc. Agron., Madison, WI.
3. Clark, F. E., and Kemper, W. D. 1967. Microbial activity in relation to soil water and soil aeration. Pages 472–479 in: Irrigation of Agricultural Lands. R. M. Hagan, H. R. Haise, and T. W. Edminster, eds. Am. Soc. Agron., Madison, WI.
4. Cory, D. C., and Timmons, J. F. 1978. Responsiveness of soil erosion losses in the corn belt to increased demands for agricultural products. J. Soil Water Conserv. 33:221-226.
5. Drew, L. O., and Garner, T. H. 1967. Seedling thrust force measurements vs. soil strength. Paper No. 67-612. Am. Soc. Agric. Eng., St. Joseph, MI.
6. Gill, W. R., and Trouse, A. C., Jr. 1972. Results from controlled traffic studies and their implications in tillage systems. Pages 126–131 in: Proc. No-Tillage Systems Symposium. Ohio State Univ., Columbus.
7. Griffin, D. M., and Nair, N. G. 1968. Growth of *Sclerotium rolfsii* at different concentrations of oxygen and carbon dioxide. J. Exp. Bot. 19:812-816.
8. Hagan, L. L., and Dyke, P. T. 1980. Yield-soil loss relationships. In: Proc. Workshop on Influence of Soil Erosion on Soil Productivity. U.S. Dep. Agric., Agric. Res., Washington, DC.
9. Hanks, R. J. 1961. Soil crusting and seedling emergence. Trans., Int. Cong. Soil Sci. 7th. 1960. Madison, WI. 1:340-346.
10. Larson, W. E. 1981. Protecting the soil resource base. J. Soil Water Conserv. 36:13-16.
11. Larson, W. E., Clapp, C. E., Pierre, W. H., and Morachan, Y. B. 1972. Effect of increasing amounts of organic residues on continuous corn: Organic carbon, nitrogen, phosphorus, and sulfur. Agron. J. 64:204-208.
12. Locheretz, W., Shearer, G., and Kohl, D. H. 1981. Organic farming in the corn belt. Science 211:540-547.
13. Lynch, J. M., and Elliott, L. F. 1983. Aggregate stabilization of volcanic ash and soil during microbial degradation of straw. Appl. Environ. Microbiol. In press.
14. Macauley, B. J., and Griffin, D. M. 1969. Effects of carbon dioxide and oxygen on the activity of some soil fungi. Trans Br. Mycol. Soc. 53:53-62.
15. Miles, G. E., and Matthes, R. K. 1969. Emergence force measurement of seedlings. Paper No. 69-507. Am. Soc. Agric. Eng., St. Joseph, MI.
16. Papendick, R. I., and Campbell, G. S. 1981. Theory and measurement of water potential. Pages 1–22 in: Water Potential Relations in Soil Microbiology. J. F. Parr, W. R. Gardner, and L. F. Elliot, eds. Spec. Publ. No. 9. Soil Sci. Soc. Am., Madison, WI.
17. Papendick, R. I., Lindstrom, M. J., Cochran, V. L. 1973. Soil mulch effects on seedbed temperature and water during fallow in eastern Washington. Soil Sci. Soc. Am. Proc. 37:307-314.
18. Parr, J. F., and Papendick, R. I. 1978. Factors affecting the decomposition of crop residues by microorganisms. In: Crop Residue Management Systems. W. R. Oschwald, ed. Spec. Publ. No. 31. Am. Soc. Agron., Madison, WI.
19. Parr, J. F., and Reuszer, H. W. 1959. Organic matter decomposition as influenced by oxygen level and method of application to soil. Soil Sci. Soc. Am. Proc. 28:214-216.
20. Sampson, R. N. 1981. Farmland or Wasteland—A Time to Choose. Rodale Press, Emmaus, PA. 422 pp.
21. Taylor, S. A. 1972. Physical Edaphology. W. H. Freeman and Co., San Francisco. 531 pp.
22. U.S. Department of Agriculture. 1980. Report and Recommendations on Organic Farming. Spec. rep. prepared for the

Secretary of Agriculture. U.S. Govt. Printing Off., Washington, DC. 164 pp.
23. Voorhees, W. B. 1979. Soil tilth deterioration under row cropping in the northern corn belt: Influence of tillage and wheel traffic. J. Soil Water Conserv. 34:184-186.

CHAPTER 16

AIR QUALITY AND PLANT HEALTH

EILEEN BRENNAN and IDA LEONE

If "past is prologue," then a glance backward should provide some guidance for the future concerning air quality and plant health. About 15 years ago (1967) the American Phytopathological Society sponsored a symposium titled "Trends in Air Pollution Damage to Plants" (7). A series of papers portrayed the prevailing state of affairs concerning pollution and proffered insights into possible developments in the near future. With those papers as a basis, we shall try to project the air pollution challenge to plant health into the next two decades.

Darley sounded the keynote of the symposium when he declared that it was time for all of us to become more acutely aware of the "insidious alterations" in the quality of the ambient air and of the serious consequences of those alterations. In the first of the formal presentations, Wood specified sulfur dioxide, fluoride, ozone, and peroxyacetyl nitrate (PAN) as the most important of the plant pathogenic air pollutants and predicted that the concentration of these last two gases would increase in the next two decades with the anticipated increase in population and in energy demand. Judiciously, Wood intimated that the accuracy of his prediction depended on future attitudes and technology. Further along in the symposium, Heggestad asserted that damage to crop and ornamental plants from air pollution was already a serious problem and that he was not optimistic about any reversal. He reviewed the acute effects of the phytotoxic contaminants, especially ozone, on the foliage of sensitive plant species and averred that if resistant cultivars had not already been substituted for the more sensitive ones, the problem at the time would indeed have been even more serious. Heggestad ended his presentation by citing a list of research needs, ranging from basic biochemical studies to an evaluation of the economic losses due to pollution. Hepting, another participant in the symposium from the Forest Service, declared that it still remained for forest pathologists to evaluate the air pollution sensitivity of tree species, select trees to monitor and ameliorate pollution, and determine if trees themselves contribute to air pollution. Treshow, in his turn, espoused the cause of the total environment, encouraging studies beyond those of agricultural crops and dominant forest species. Although the national losses to agriculture were estimated at more than $500 million, he suggested that the figure cited might actually be minor compared with potential losses to the far greater acreage of nonagronomic forest and range vegetation constituting the natural ecosystems of the country. California researchers

Dugger and Ting were at the point of speculating whether PAN injury could be avoided by dark treatment of plants or by genetic manipulation of chlorophyll in PAN-sensitive plants; and finally, Smith, a meteorologist, called for closer cooperation among the various disciplines to gain a better understanding of the complex interactions involving plants and pollutants.

Fifteen years later, the symposium deliberations still embody some of the basic problems and needs concerning air quality and plant health, but some major changes have occurred in the perception, if not the reality, of ambient air pollution and its effect on vegetation. In this review we will first reflect on what has come to pass since the symposium took place and then contemplate possibilities for the future.

AMBIENT AIR QUALITY, PRESENT AND PREDICTED

The quality of ambient air in the United States has not deteriorated to the extent predicted 15 years ago, mainly because of the Clean Air Act Amendments of 1970. Congress passed a set of regulations that called for the development and attainment of national ambient air quality standards. The act required the federal Environmental Protection Agency to establish air quality control regions throughout the nation and to determine the pollutants to be regulated and the levels permissible in ambient air. The Agency subsequently set ambient air quality standards for the following six pollutants: carbon monoxide, sulfur dioxide, nitrogen dioxide, total suspended particulate matter, hydrocarbons, and photochemical oxidant (now regulated as ozone). As a result of this legislation, between 1970 and 1976 the average national level for SO_2 declined by 27%, CO levels dropped 20%, and those for NO_2 decreased by 12% (13). One should not be totally comforted by these statistics, however, inasmuch as peak rather than average concentrations may have the greater biological impact. In spite of the progress made with respect to some pollutants, oxidants remained a nationwide problem and NO_2 levels continued to rise in most states. When it became obvious that many states could not meet all the standards, Congress passed the Clean Air Act Amendments of 1977, extending the date for meeting goals to 1982 and, in the case of O_3 and CO_2, to 1987.

With the change in political philosophy of the U.S. federal government in 1980, the future of the Clean Air Act has become precarious. Opponents of the legislation claim that the Act stifles industrial growth, constrains productivity, and bars the development of new energy sources by banning either new construction in polluted areas or expansion in areas that already have clean air (49).

Some of the new policymakers have released the following forecast for energy supply and demand over the next two decades: soil imports will hold steady until the mid-1980s and decline thereafter, coal consumption will climb from 3.5 million quads (quadrillion British thermal units) in 1980 to 7.5 million quads by 2000, and nuclear power will expand by a factor of four by the year 2000 (38).

In view of the uncertainty of such policy changes, we canvassed some of the leading atmospheric chemists for their predictions about the future quality of ambient air. As a group, these scientists foresee little chance for a decrease in the ozone level, inasmuch as the current strategy for control through the reduction of hydrocarbons has not been effective; on the other hand, they do not predict an increase in ozone above the present levels, relying instead on improved control technology to offset increased emissions. The experts disagree however about the future trend for NO_x. With hindsight, there is little agreement even about past trends for NO_x, as exemplified in the 1980 report of the Council for Environmental Quality (13), which concluded that NO_x data for 1974 and 1978 could be interpreted as an increase, decrease, or no change depending on which particular statistics were used.

Atmospheric chemists are extremely concerned about the acid precipitation problem. Even though gaseous SO_2 concentrations have been reduced in ambient air, to a large extent by the use of cleaner coal, particulate sulfate levels appear to be increasing. SO_2 and NO_x from tall stacks enter the upper atmosphere, and the resulting aerosols combine with water vapor in the air to form sulfuric or nitric acid. Unfortunately, at this time not even the chemists understand how SO_2 emissions relate quantitatively to the production of SO_4. Since 1970 the construction of tall stacks has accelerated; an estimated 429 stacks taller than 61 meters were built in the ensuing 10 years. By exporting pollution elsewhere, tall stacks reduce peak values at the source but increase pollution downstream.

Some of the chemists responding to our inquiry have warned us of an increase in CO_2, which will be a concern not in 10, but more likely in 50–100 years. A century ago the concentration of CO_2 in the air was about 280–300 ppm. It now ranges from 335 to 340 ppm and is expected to rise to at least 600 ppm in the next century. Although most scientists concede that CO_2 is increasing, they disagree on whether air temperatures are also increasing. In a recent report (15), National Aeronautics and Space Administration scientists claimed that CO_2 added to the atmosphere since the Industrial Revolution has already warmed the world climate. If fuel burning increases even at a slow rate, the study predicts a global temperature rise of 5 C in the next century, but if fuel burning rises rapidly, a rise of 6–9 C is anticipated. The opposition challenges some of the assumptions made, e.g., population growth rate, energy-consuming trends, new developments in solar energy, trends in energy conservation, and lack of knowledge regarding the extent to which oceans might remove CO_2 from the air.

A more immediate concern than the CO_2 issue appears to be the prediction that chlorofluorocarbons rising into the stratosphere are contributing to ozone destruction and thereby increasing ultraviolet radiation. In 1976 an O_3 reduction in the stratosphere of 7.5% was predicted, but more recent research indicates that a 16% reduction may be more realistic (1).

Several atmospheric chemists made the observation that improved instrumentation is revealing the presence of gases and particulate matter that were previously undetectable (26). Some examples include radicals (OH, H_2O_2) that participate in the interactive chemistry of nitrogen-sulfur hydrocarbon complexes, trace metals found as airborne particulate matter in ambient air, and a variety of organic compounds released from natural or human sources.

DOCUMENTATION OF AMBIENT AIR POLLUTION EFFECTS ON PLANT HEALTH

Heggestad's prediction of 15 years ago, that the threat of air pollution to agricultural crops would persist, has unfortunately materialized. Acute foliar symptoms resulting from oxidant continue to occur on sensitive species growing in polluted areas of the United States, and, near point sources, other plant symptoms can still be traced to SO_2 and HF emissions and occasionally to C_2H_4,

Cl_2, HCl, NH_3, and H_2S. What has become more explicit since Heggestad's presentation in 1967 is that ozone, the most pervasive of all ambient air pollutants, not only causes visible foliar lesions and reduces the value of a leaf crop, but also reduces the yield of other plant parts such as seeds, fruits, and tubers that also constitute commercial products. Thus, the threat of air pollution to agriculture has become an unequivocal reality. It is these cases that we want to emphasize, cases in which ambient air pollution (mainly ozone) so affects plant health that productivity is reduced.

Evidence of crop loss has been reported from states along the East Coast of the United States, including Massachusetts (35), New York (34), New Jersey (5), and Maryland, Virginia, and North Carolina (20–22) and from the West Coast in the South Coast Air Basin (39,40,50–52) and the San Joaquin Valley of California (3). In the interior of the country, crop loss has been reported in Minnesota (27) and Ohio (37) as a result of oxidant pollution. In the eastern United States and in California, yield reductions as high as 50–60% have been measured for some economically important commercial crops. In the East, pod production of snapbeans has been reduced 5–40% (20,34,35), soybean yield, 20% (22), fresh weight of tomatoes, 33% (34), and weight of potato tubers, 20–50% (5,20). In California, cotton production (bolls, lint, and seed) was reduced between 5 and 29% (3), weight of tomatoes, 23–50% (40), seed set in sweet corn, 42% (52), yield of lemon and orange, 32–54% (51), and grape yield between 12 and 61% (50). In the Midwest, the yield of potato tubers was reduced by as much as 25% (37) and soybean by 24% (27). The precise reduction is dependent, of course, on oxidant levels of that particular year, environmental conditions, and cultivar grown.

In addition to the demonstrated effects of oxidant pollution on crop yield, significant changes in crop quality due to oxidant have also been reported. The most striking example is the decrease in sugar and increase in acidity in Zinfandel grapes exposed to the polluted air of Southern California (50). A decrease was also found in the dry matter content of potato tubers when oxidant concentrations were unusually high in New Jersey (5). A significant reduction in the size of tomatoes (6718 VF) was reported in California (40) and of potato (Norland) in New Jersey as a result of ambient air pollution (5).

These adverse effects on crop yield and quality in the field have generally been associated with ozone levels in ambient air $\geqslant 0.01$ ppm. In the episodes reported for the East Coast, the number of hours $\geqslant 0.01$ ppm during any growing season ranges from one in Beltsville, MD, (20) to 125 in Yonkers, NY (34). In California the number of hours in that same concentration category ranged from several hundred to several thousands, depending on the geographical area (40, 51). Whereas the maximum reported ozone concentration approached 0.02 ppm in the East, in California the peak was at 0.69 ppm. The evidence is unequivocal that the frequency and severity of oxidant episodes in California far exceed those in the East. Additionally, it is clear that other oxidants, PAN and NO_2, also occur in greater concentrations in California than on the East Coast (33). It is revealing that in Thompson's experiments, navel oranges exposed to the total photochemical complex in California sustained more damage than those exposed to ozone alone. It may well be that the losses attributed to ozone in California are actually due to a mixture of oxidants. In the absence of supporting plant data and because levels of PAN and NO_x are lower in the East than in California, the same inference cannot be drawn about pollution damage to crops on both coasts.

When we turn from the agronomic and horticultural crops to forest or urban trees, far fewer data are available to relate ambient air pollution to plant health. The dramatic decline in ponderosa pine in the San Bernardino Mountains of California can be cited, but unfortunately the proof may never be sufficient to establish air pollution as the sole stress responsible (41). A condition described as "chlorotic dwarf" of eastern white pine, which has been observed in the forests of Ohio since 1959, has been attributed to a mixture of ozone and sulfur dioxide (8). In the East, oxidant symptoms have been observed on foliage of individuals representing various woody species in urban (43) and in forest situations (14). However, there is only a suggestion that oxidant pollution might reduce tree growth, and it is based on linear growth measurements. In Maryland, growth of Platanus seedlings was reduced 25% in ambient air compared to growth in filtered air (46), but since the seedlings were only 10–20 cm tall during the experiment, it is difficult to ascertain the meaning of the test in relation to established trees.

In New Jersey, evidence supports the conclusion that the growth of hybrid poplar (a supposedly O_3-sensitive species based on visible symptoms) is not adversely affected by ambient air pollution during a single growing season. Of course, it is possible that after many successive years' exposures, growth decreases could become evident. On the other hand, for about 7 years we have observed two Austrian pine genotypes, one sensitive and the other resistant to ambient air pollution; there has been no obvious difference between them in linear growth, although the sensitive trees always display O_3 symptoms and the resistant ones do not. We are inclined to believe that trees in the East do not very readily suffer growth reductions, considering that artificial ozone fumigations of 0.01–0.03 ppm for 6 hr/day, 7 days per week for 20–22 weeks have reportedly been required to reduce growth in even a few coniferous and deciduous species (16). Such high dosages as these have never been recorded in the East, and if they ever are, scientists will not be in condition to take growth measurements on trees.

Despite the widespread occurrence of O_3 symptoms on various plant species throughout the United States, these cited examples of effects of ambient photochemical pollution on plant health constitute the only documented cases we are aware of in the United States. Intuitively, we view them even now as the proverbial tip of the iceberg, but at this time, no other effects have been unequivocably demonstrated for commercially grown crops in the field, despite the many diverse effects that have been suggested by experiments in the laboratory or greenhouse.

WITHER GOETH AIR POLLUTION RESEARCH?

For whatever meaning can be read into the statement, we take note that the documented instances of air pollution damage have been found mainly in journals with environmental leanings rather than in journals of plant pathology.

To identify the recent research interests of plant pathologists in air pollution, we surveyed the articles appearing in *Plant Disease Reporter* or PLANT DISEASE and PHYTOPATHOLOGY from 1970 to 1980. Of approximately 140 papers dealing with pollution effects, three quarters concerned aspects of O_3 toxicity. By a ratio of 2:1, greenhouse or laboratory tests outnumbered field tests, and herbaceous plants drew more interest than woody species. The aspects of air pollution most frequently dealt with were: 1) dose response assays for agricultural and horticultural plant species based on the appearance of visible lesions, 2) the effect of environmental and plant factors on visible air pollution symptoms, 3) interrelationships between biotic diseases and air pollution, and 4) identification of resistant or susceptible genotypes within various species.

Although these studies provided much useful information, there was often an unbridgeable gap between results and their application in the real world. Oxidant dosages were frequently far in excess of those actually occurring even in the more contaminated areas of the United States. Field-grown plants were shown not to respond to a pollution episode in a manner similar to that of greenhouse or chamber-grown plants. Environmental and plant factors were assumed, but not proved, to exert the same influence in the field as in the greenhouse.

In the 1980s we see signs that plant pathologists are shifting gears in preparation for more realistic studies. They are beginning to recognize that terms such as "photochemical pollution" and "acid rain" are disarmingly deceptive. Atmospheric chemists and meteorologists alike have had to acknowledge the baffling complexity of chemical reactions and transformations involved in each pollution phenomenon. The plant pathologist can do no less. Although simplicity is the goal of our quest, we should heed the advice of the philosopher, Whitehead, who warned against the error of thinking that the facts are simple. We want to reflect now on the latest developments within the realm of photochemical pollution and suggest some approaches that might provide a better understanding and perhaps a solution of some practical air pollution problems.

MULTIPLE POLLUTANT STUDIES

Plant pathologists are recognizing that photochemical smog is a complex mixture of O_3, PAN, NO_2, and a variety of hydrocarbons and are now subjecting plants to combinations of the three oxidants. More importantly, they are now using levels of pollutants that actually occur in polluted ambient air.

In order to simulate more precisely the field situation, plants should be exposed to pollutants sequentially, following the characteristic diurnal periodicity displayed by the pollutant in ambient air. For example, in an area of high photochemical activity, concentrations of NO_2 peak in the morning and again late in the afternoon, whereas, O_3 levels are low in the morning and late evening and peak in midafternoon. Response to a pollutant may vary with the time of day because plants too have various diurnal rhythms. Among the many known circadian plant rhythms in plants are stomatal movement, photosynthetic capacity, root growth, and ion uptake. All these physiological processes have been related in various studies to plant response to pollution. By making experiments in growth chambers with a constant environment, the influence of these processes may be unrealistically diminished.

Future studies will have to address not only O_3, PAN, and NO_2 but also specific hydrocarbons in the photochemical mixture as they affect plant response. Circumstantial evidence has caused the aldehydes per se to be suspect as phytotoxicants for more than 40 years (2). The presence of aldehydes in ambient air plays a dual role; it influences the formation of O_3 and PAN and possibly exerts a synergistic effect on plants in the presence of other pollutants. It is well known that the hydrocarbon ethylene can induce a number of plant responses, ranging from hormonal to toxic, depending on its concentration. The consequences of combinations of O_3 and ethylene on plant response are unknown.

Future studies will have to involve combinations of oxidant and sulfur dioxide, especially if high sulfur coal is allowed to be used, and also combinations of oxidant and particulate matter pollution or acid precipitation, depending on the field problem under observation. Many trace elements may be found in environments also contaminated by O_3 and SO_2 (29). Even now it is known that low levels of Cd (4) and salt (9) can predispose plants to O_3 injury and that SO_2 can intensify Cd injury (28). Although some plant responses may seem to indicate the presence of a dominant pollutant, possible interactions should not be overlooked.

As studies are made with multiple pollutants at more realistic doses, it will be necessary to reevaluate the effect of plant and environmental factors on plant response, again using values that are realistic. It may be abundantly clear that temperature, humidity, light intensity, photoperiod, and plant fertility influence the response of greenhouse-grown plants to ozone, but how do these factors determine plant response in the field? A virus, fungus, or bacterium may predispose or protect a plant in the laboratory, but does the same situation occur in the field?

DIAGNOSIS OF AIR POLLUTION INJURY

In the early days of air pollution research (about 1950), the diagnosis of field symptoms seemed rather simple; just as long as one had sufficient knowledge to exclude mimicking symptoms induced by insects, cultural practices, environmental factors, etc. In light of recent developments, perhaps it is time to reexamine some of the old notions about pollution symptoms per se. PAN phytotoxicity is a case in point.

In the 1950s in California, PAN injury was first described as an under-surface glazing of petunia leaves (36). The description was liberally quoted in the literature, and any such symptom observed elsewhere in the field was attributed to PAN. In the 1970s, laboratory tests from Pennsylvania characterized the most frequent PAN response not as a glazing, but as a bifacial necrosis (10); to further complicate matters, the classic glazing symptom was able to be duplicated with a mixture of O_3 and SO_2 (32). Obviously the diagnosis of PAN injury needs more supporting evidence than mere symptom appearance.

Again, in the early days, the upper surface stipple or fleck on middle-aged leaves was accepted as an O_3 symptom. Since the work with combinations of O_3 and SO_2, we know that a mixture of gases can also be responsible for the very same symptom. Another example suggesting a need for clarification is that a mixture of O_3 and PAN produces not only O_3 and PAN symptoms but also a third unrelated type of symptom.

OZONE SENSITIVITY RATINGS OF PLANT SPECIES

In many of the current lists in which the sensitivity/resistance of various species to O_3 are rated, the basis for the rating is the appearance of visible foliar symptoms induced by a given dosage of the pollutant on greenhouse-grown plants. A species that has a greater percentage of leaves or surface area affected by O_3 relative to another species is termed "sensitive." These greenhouse-based relationships may not hold in the field for at least two reasons: 1) field-grown plants are physiologically and morphologically different from greenhouse plants, and 2) oxidant pollution is episodic rather than continuous in nature and is frequently accompanied by other pollutants.

But, more important than this inconsistency in lesion response between greenhouse and field vegetation (31), we question the propriety of basing relative sensitivity on apparent severity of the disease when, in fact, loss of yield is the real "bottom line" in crop production. In a given environment, greenhouse or field, two parameters (severity of symptoms and yield loss) may not be correlated (18). A certain species or cultivar may tolerate (sensu Schaefer, 47) the formation of lesions on a given percentage of leaves better than another species can, or, equally possible, the visible lesions may not be a valid representation of the total injury sustained by the plant. Oxidant lesions are expressed mainly on the older to middle-aged leaves, but in fact, leaves of all ages may have suffered a decrease in photosynthesis and an increase in respiration. The unexpressed injury may adversely affect the yield even more than the visible injury affects the leaves.

Recently an attempt was made to define the relationship between foliar lesions and biomass reduction in the field in seven tobacco cultivars and their 21 F_1 hybrids from a diallel cross (42). The correlation between visual injury and growth depression was not significant. Inasmuch as the experiment was of relatively short duration, other tests should be made to evaluate the long-term response for many plant species.

MECHANISMS OF O_3 INJURY

Despite the ackowledged importance of O_3 as the number one pollutant in the United States in the last 35 years, little monetary or intellectual investment has been devoted to determining the mechanisms of O_3 injury. One might adopt the attitude that as long as plants can be bred for O_3-resistance in the absence of this knowledge, then such a study is a luxury. On the other hand, the traditional scientific spirit demands that it be pursued. Would that we could foresee such a trend!

As we consider the most recent review of the initial events in ozone phytotoxicity (19), we realize that few giant steps have been taken since the 1960s when it was reported that ozone increases both lipid peroxidation of membranes and osmotic fragility of red blood cells. Although some indirect evidence has resulted from studies implicating the membrane as the initial site of ozone injury, little direct experimental evidence supports it. Plant pathologists dealing with biotic diseases have been preoccupied for the last 40 years with changes in membrane permeability (53), and their attempts to provide evidence for an interaction with receptor sites or enzymes bound to the plasmalemma have, for the most part, also yielded negative results. Currently they are questioning whether the pattern of permeability changes observed with pathotoxins suggests an initial effect on cell walls rather than on membranes. The key question of whether permeability changes are primary causal events in pathogenesis or whether they are merely another physiological symptom of injury remains to be answered. Air pollution researchers may well pose the same question.

The role of free radicals in ozone phytotoxicity is another area that could profitably receive more attention in future studies. A recent report (30) from the

USDA Laboratory in Beltsville claimed a marked correlation between the plant toxicity caused by free radicals and that caused by ozone. The resistance of certain cultivars of watermelon and bean to O_3 was attributed to high concentrations of superoxide dismutase, which has the ability to scavenge free radicals. Protection bestowed on plants by an antioxidant (EDU, *N*-[2-(2-oxo-1-imidazolidinyl) ethyl]-*N*-phenylurea) was likewise attributed to the induction of superoxide dismutase. Other interesting hypotheses concerning the role of free radicals could be tested. Ethylene is produced in situ under various stress conditions, including air pollution. Biochemists report that the hydroxyl radical is responsible for the production of ethylene from a precursor and that any compound that acts as a scavenger of hydroxyl radicals will inhibit the production of ethylene. A free radical-scavenging role has been postulated for cytokinin action and even for the process of photorespiration.

MECHANISM OF O_3 RESISTANCE

Just as few new developments concern the mechanism of ozone toxicity, few relate to ozone resistance. In 1964, a review (44) concluded that O_3 resistance is probably quantitative rather than qualitative, inasmuch as even "resistant" cultivars can be injured when ozone concentrations are increased. At the time the suggestion was made that any of at least four mechanisms could confer resistance: naturally high sugar contents, high levels of natural antioxidants, increased suberin content of mesophyll cell walls, and the closing of stomata in response to O_3. Almost 20 years have passed, and hardly a footnote can be added concerning the nature of ozone resistance. Just as Walker's discovery in 1932 of the resistance of yellow versus white onions to smudge disease led pathologists astray for many years in their search for a single trait for disease resistance, so also the discovery of stomatal closure in resistant onion has possibly misled air pollution researchers (12).

Recently there have been signs of interest in the genetic aspects of resistance. The tools of classical genetics, principally inheritance in diallel crosses, have been employed to determine the number of genes involved in the ozone responses and their general and specific combining abilities. In a recent review Ellingboe (11) expressed the sentiment that a combination of not classical, but molecular, genetics and plant pathology will have to be utilized in the future to gain a more incisive understanding of host-parasite relations. Air pollution researchers may be pressed to develop similar talents to fully understand their problems.

ASSESSMENT OF ECONOMIC CROP LOSS DUE TO OZONE POLLUTION

Plant pathologists have long been aware of the need for accurate and reliable estimates of crop losses due to disease. Yet, despite the critical decisions dependent on the information on the farm, in research, in government, and in industry, only sparse knowledge is available about the economic impact of most plant diseases. To the plant pathologist falls the burden of providing accurate information on the biological response of plants to known doses of pollutants before an economic assessment can be made. As modelers from the Brookhaven National Laboratories stated in 1981, "uncertainty introduced by estimating biological damage is merely compounded by the additional uncertainty introduced by modeling economic effects."

Realizing that foliar damage may not be a valid indication of plant damage, air pollution researchers have begun to measure the yield of various crops under polluted conditions. The current trend is to use paired open-top chambers with a filtered or nonfiltered air supply and to introduce known doses of O_3. At the end of a growing season the yields are compared. Two major objections are obvious: 1) the crops cannot be grown and harvested under standard commercial practices, and 2) the crops "see" only a single pollutant, ozone, and not the complex mixture of photochemical smog.

The most often cited "field study" used in estimating crop loss is that by Oshima et al (39) in California with alfalfa. Five-gallon (19-L) cans, each containing a single plant, were sunk into the ground in an area encompassing an ambient O_3 gradient in the South Coast Air Basin. Seasonal yield and average seasonal defoliation indices were correlated with seasonal O_3 doses. Because of the scarcity of data, there has been a tendency for modelers to use these particular damage functions for all parts of the United States, despite the fact that researchers have for years observed differences between California pollution problems and those, for example, of the northeastern United States. Southern California not only has higher O_3, but also higher NO_2 and PAN concentrations than other parts of the country. One may ask whether the alfalfa plants are responding to O_3, to all the oxidants, or to the combination of pollution and a dry, hot climate. On the other hand, the Northeast is considered "acid rain" country. Does acid rain influence the response of alfalfa to ambient O_3?

There is a truism that states that one who does not read history must live history. Air pollution researchers could benefit by close attention to the article by James (24). Among the suggestions that can be found in his review are:

1. Identical experiments should be made for at least 3 years in all geographical areas where the crop is important.
2. The major cultivars should be tested under normal farming conditions.
3. Harvesting procedures should be those used by the farmer.
4. The use of multiple treatments to produce various amounts of disease is superior to the use of paired plots where loss is calculated as a percentage of yield in a healthy plant.
5. Experimental design, shape, and size of plots should allow for the necessary precision of yield estimates. (Standard tests of cereal pathologists can detect only 10% differences.)

James presents a reprise of the methods for generating epidemics of different characteristics by the use of chemicals, isogenic lines, or cultivars. Although the procedures are admittedly difficult and expensive, air pollution researchers may be forced to undertake more rigorous procedures than have been used previously.

CONTROL OF OZONE INJURY TO PLANTS

The control of air pollution injury to plants is based on three different options: 1) elimination of the air contaminant, 2) protection of the plant with chemicals, or 3) substitution of a resistant cultivar for a susceptible one. As noted earlier, the Clean Air Act has reduced some air pollution levels and additional amendments could reduce others in the future. Chemical protection against ozone phytotoxicity has not been used commercially, although some compounds such as benomyl and EDU have proved effective on an experimental basis. The substitution of resistant cultivars has been a feasible step in areas threatened by pollution. To date the process has not involved a definite breeding program, but has included the selection of plants that are apparently free from visual symptoms of pollution injury.

Some 25 years ago a geneticist, Ryder (45), made the observation that vegetable crops in the Salinas Valley were so threatened by photochemical smog that a breeding program would be required to produce resistant plants for future use. Personal communication with him in 1981 revealed that the dire prediction had not materialized, nor had any breeding program. We made an informal poll of breeders throughout the United States to learn if they anticipated breeding programs for pollution-resistant crops.

A fruit breeder in a USDA laboratory in California gave a negative reply. A vegetable breeder in Beltsville, MD, did not think that the economic problem resulting from air pollution would become so acute as to justify reorientation of resources from more serious production problems. He noted, however, that vegetable breeders during the past two decades had been considering resistance to air pollutants in developing their breeding and progeny field evaluation strategy. By recording air pollution injury on susceptible materials, he said, they were able to make progress in improving yields of potato, tomato, cucurbits, and spinach. He turned the table on the air pollution researchers by noting the relative lack of correlation between the response of artificially fumigated plants and that of plants exposed to air naturally in the field. A tobacco breeder in Maryland reported that his crop has been deliberately bred for weather fleck resistance (rather than O_3 resistance, because the selection was done under field conditions). A source of resistance was identified in an old cultivar, Wilson, and it was crossed with Md. 609, to produce Md. 827, which had a good level of resistance. Because 25% of the Maryland acreage is planted to the resistant cultivar, losses from weather fleck are less than those occurring nationwide. According to the tobacco breeder, some of the factors that determine which particular problem will be investigated are: 1) economic loss, 2) source of resistance, and 3) inheritance of the resistance, which determines the difficulty of the selection program. A forest geneticist from Ohio replied that to date there had been no systematic effort to breed for resistance in forest trees, nor did he see an increased effort in the direction over the next 20 years *unless* some drastic changes were made in the nation's air quality laws. He emphasized that it is important to study and understand the genetic mechanisms involved in tolerance/sensitivity of trees to air pollution, but only as that knowledge will help us to determine 1) the physiological impact of air pollution on plant and 2) impacts on the genetic structure of plant populations, both natural and cultivated. Such knowledge will permit us to better define safe, reasonable standards for allowable air pollution concentration. An appropriate footnote to this paragraph is that the economic restraints to breeding pollution-resistant crops could be dramatically removed in the next two decades if protoplast (48) and anther culture replace traditional breeding methods.

ACID PRECIPITATION

Although O_3 has been rated as the major plant pathogenic air pollutant for more than 30 years, on the basis of documented adverse effects, acid rain has, in the last 5 years, gradually assumed a position of great importance based solely on its potential effects. General consciousness of the acid rain issue in the United States probably dates back to 1975 when the First International Symposium on Acid Precipitation and the Forest Ecosystem was held in Columbus, OH. Invited scientists from Western Europe, where pH measurements of rainfall had been made for more than 10 years, shared their experiences with American scientists who much more recently had become involved in laboratory studies with "simulated" rain. With respect to plant health, the most crucial report was that forests in southern Sweden showed 2–7% decrease in growth between 1950 and 1965. In explaining this phenomenon, Jonsson and Sundberg (25) "found no good reason for attributing [this] reduction in growth to any cause other than acidification." Although the evidence ought to be regarded as only circumstantial, the claim continues to be quoted in many reviews. As a result of simulated rain studies here and abroad, a variety of effects were cited, among them: foliar lesions on certain species, poor germination of seeds, leaching of nutrients from foliage, erosion of cuticular wax, decreased uptake of nitrogen by endomycorrhizae of tree seedlings, inhibition of reproduction of root-knot nematodes, inhibition of bean rust and halo blight, and inhibition of nodulation and fixation of nitrogen by *Rhizobium* in bean seedlings.

Despite the litany of possible effects, when the NATO International Conference on Effects of Acid Precipitation concluded in Toronto, Canada, 3 years later (1978), Hutchinson (23) emphasized that there was little solid evidence for the occurrence of visible or even detectable danger to terrestrial ecosystems caused by the acid precipitation events of the past 20 years. At this writing, researchers have still not unequivocally identified any adverse effect of acid rainfall per se on plants in a field situation in the United States. On the contrary, the few published items more often point to increased nitrogen or sulfur nutrition in less than well-fed plants as a result of acid rain in the field.

The history of acid precipitation and its effect on plants runs contrary to that of O_3, PAN, HF, and some of the minor air pollutants. In the latter cases, symptoms of unknown etiology were observed on vegetation in situ, and they were deemed of sufficient economic importance to warrant continued investigation until a cause could be identified. Sometimes, accurate diagnosis required several years—nearly a decade for ozone and only a year for PAN. The acid precipitation problem, on the other hand, has not had a similar origin, but, mainly because of the deterioration of certain aquatic systems associated with a low pH, the inference is made that terrestrial ecosystems must also be slowly deteriorating. Researchers are therefore faced with a task rather like "looking for a needle in a haystack." In view of the little success achieved or likely to be achieved from this approach, we would suggest a return to the rationale used for pollution studies in years past: to whit, identify plant health problems whose causes are unknown and test the involvement of acid rain. For example, in the northeastern United States, where acid rain occurs, the decline or dieback of maple, oak, and ash has been significant, and the problems cannot be traced to a single cause, although a variety of stresses have been investigated. The peach replant problem is also wanting for a cause, as is canker on sweet gum.

We suggest that agricultural crops are so well buffered that acid rain per se is not a problem to their health. However acid rain may be responsible for anomalous effects that are occasionally observed in the field where a pesticide does not perform as expected, being either ineffective or phytotoxic. We recall that when captan was first tested in the greenhouse against powdery mildew of bean, a captan-kaolinite formulation was significantly more toxic than a captan-talc formulation because of increased acidity (6). Acid rainfall might be involved in such a condition. Acid rain may also be responsible for predisposing host plants to certain parasites, inasmuch as SO_2 is said to suppress blackspot disease (*Diplocarpon rosae*) on rose and powdery mildew (*Microsphaera alni*) on lilac (17).

EPILOGUE

We have expressed our abiding concern about oxidant pollution of ambient air, which has been a serious problem to agricultural crops for more than 30 years. As actual measurements of crop yield continue to be made in areas with elevated oxidant levels, the magnitude of the problem will become even more evident. In addition, if regulations are relaxed to the extent that increasing amounts of high sulfur coal are utilized in oxidant-polluted areas, the threat to agriculture will be exacerbated. We have speculated that acid precipitation might have an adverse effect on woody vegetation, especially over a long period of time, but that certain agricultural crops might only be predisposed to injury from acid rain under very special conditions. We can foresee a time, much in the future, when changes in ultraviolet radiation and CO_2 concentrations might create changes in plant growth and reproduction. But what may emerge as a rather unexpected threat to plant health may be the combustion products of solid waste. As the oceans are closed to the dumping of waste and landfills are exhausted, much solid waste may have to be burned, and the by-products will end up as air pollutants. These vexing problems of the future we leave to the plant pathologists of the future.

ACKNOWLEDGMENTS

We gratefully acknowledge the assistance of those who replied to our inquiries, including the following: A. Altshuller, J. Pitts, J. Lodge, T. Gradel, P. Lioy, D. Pack, M. First, H. Hovey, M. Eisenbud, B. Goldstein, O. C. Taylor, W. Hooker, R. Webb, K. Aycock, J. Stavely, D. Houston, E. Ryder, and R. Worrest.

LITERATURE CITED

1. Anonymous. NASA reports drop in stratospheric ozone. Chem. Eng. News, Aug. 24, 1981, pg. 4.
2. Brennan, E. 1981. Effects of aldehyde on vegetation. Pages 256–276 in: Formaldehyde and other Aldehydes. Nat. Acad. Press, Washington, DC.
3. Brewer, R. F., and Ferry, G. Effects of air pollution on cotton in the San Joaquin Valley. Calif. Agric. 28(6):6-7.
4. Clarke, B., and Brennan, E. 1980. Evidence of a cadmium and ozone interaction on *Populus tremuloides*. J. Arboric. 6:130-133.
5. Clarke, B., and Brennan, E. 1981. The impact of ambient oxidant on foliar symptoms and tuber yield and quality of field-grown potato cultivars in New Jersey. Phytopathology 71:867.
6. Daines, R. H., Lukens, R. J., Brennan, E., and Leone, I. 1957. Phytotoxicity of captan as influenced by formulation, environment, and plant factors. Phytopathology 47:567-572.
7. Darley, E. 1968. Symposium on trends in air pollution damage to plants. Phytopathology 58:1075-1113.
8. Dochinger, L., and Seliskar, C. 1970. Air pollution and the chlorotic dwarf disease of Eastern white pine. For. Sci. 16:46-55.
9. Dochinger, L. S., and Townsend, A. M. 1979. Effects of roadside deicer salts and ozone on red maple progenies. Environ. Pollut. 19:229-237.
10. Drummond, D. B. 1972. The effect of peroxyacetyl nitrate on petunia (*Petunia hydrida* Vilm.). Ph.D. thesis, Pennsylvania

State Univ., University Park. 70 pp.

11. Ellingboe, A. H. 1981. Changing concepts in host-pathogen genetics. Annu. Rev. Phytopathol. 19:125-143.
12. Engle, R. L., and Gabelman, W. H. 1966. Inheritance and mechanism of resistance to ozone damage in onion, *Allium cepa* L. Proc. Am. Soc. Hortic. Sci. 89:423-430.
13. Environmental Quality. 1980. The 11th Annual Report of the Council on Environmental Quality. U.S. Govt. Printing Office, Washington, DC. 497 pp.
14. Gerhold, H. D. 1977. Effect of air pollution on *Pinus strobus* L. and genetic resistance. A literature review. EPA-600/3-77-002. 45 pp.
15. Hansen, J., Johnson, D., Lacis, A., Lebedeff, S., Lee, P., Rind, D., and Russell, G. 1981. Climate impact of increasing atmospheric carbon dioxide. Science 213:957-966.
16. Harkov, R., and Brennan, E. 1982. The effect of acute ozone exposures on the growth of hybrid poplar. Plant Dis. 66:587-589.
17. Heagle, A. S. 1973. Interactions between air pollutants and plant parasites. Annu. Rev. Phytopathol. 11:365-388.
18. Heagle, A. S., Philbeck, R. B., and Knott, W. M. 1979. Thresholds for injury, growth, and yield loss caused by ozone on field corn hybrids. Phytopathology 69:21-26.
19. Heath, R. L. 1980. Initial events in injury to plants by air pollutants. Annu. Rev. Plant Physiol. 31:395-431.
20. Heggestad, H. E. 1973. Photochemical air pollution injury to potatoes in the Atlantic coastal states. Am. Potato J. 50:315-328.
21. Heggestad, H. E., Heagle, A. S., Bennett, J. H., and Koch, E. J. 1980. The effects of photochemical oxidants on the yield of snap beans. Atmos. Environ. 14:317-326.
22. Howell, R. K., Koch, E. J., Rose, L. P. 1979. Field assessment of air pollution-induced soybean yield losses. Agron. J. 71:285-288.
23. Hutchinson, T. C. 1978. Conclusions and Recommendations. Pages 617–627 in: NATO Conference Series 1: Effects of Acid Precipitation on Terrestrial Ecosystems. T. C. Hutchinson and M. Havas, eds. Plenum Press, New York. 654 pp.
24. James, W. C. 1974. Assessment of plant disease and losses. Annu. Rev. Phytopathol. 12:27-47.
25. Jonsson, B. 1976. Soil acidification by atmospheric pollution and forest growth. Pages 837–842 in: Proc. Int. Symp. Acid Precip. For. Ecosyst., 1st. U.S. For. Serv. Gen. Tech. Rep. NE-23. 1074 pp.
26. Katz, M. 1980. Advances in the analysis of air contaminants. J. Air Pollut. Control Assoc. 30:528-557.
27. Kohut, R. J., Krupa, S. V., and Russo, F. 1977. An open-top field chamber study to evaluate the effects of air pollutants on soybean yields. (Abstr.) Proc. Am. Phytopathol. Soc. 4:88.
28. Krause, G. M., and Kaiser, H. 1977. Plant response to heavy metals and sulfur dioxide. Environ. Pollut. 12:63-71.
29. Leone, I. A. 1979. Symposium on detection and measurement of primary and secondary airborne particulates and their effect on plants. Phytopathology 69:998-1011.
30. Leshem, Y. Y. 1981. Oxy free radicals and plant senescence. What's New Plant Physiol. 12(1):1-4.
31. Lewis, E., and Brennan, E. 1977. A disparity in the ozone response of bean plants grown in a greenhouse, growth chamber, or open-top chamber. J. Air Pollut. Control Assoc. 27(9):889-891.
32. Lewis, E., and Brennan, E. 1978. Ozone and sulfur dioxide mixtures cause a PAN-type injury to petunia. Phytopathology 68:1011-1014.
33. Lonneman, W. A., Bufalini, J. J., and Seila, R. 1976. PAN and oxidant measurements in ambient atmospheres. Environ. Sci. Technol. 10:374-380.
34. Maclean, D. G., and Schneider, R. E. 1976. Photochemical oxidants in Yonkers, New York. Effects on yield of bean and tomato. J. Environ. Qual. 5:75-78.
35. Manning, W. J., Feder, W. A., and Vardara, P. M. 1974. Suppression of oxidant injury by benomyl: Effects on yields of bean cultivars in the field. J. Environ. Qual. 3:1-3.
36. Middleton, J. T., Kendrick, J. B., and Schwalm, H. W. 1950. Injury to herbaceous plants by smog or air pollution. Plant Dis. Rep. 34:245-252.
37. Mosley, A. R., Rowe, R. C., and Weidensaul, T. C. 1978. Relationship of foliar ozone injury to maturity classification and yield of potatoes. Am. Potato J. 55:147-153.
38. Norman, C. 1981. Reagan energy plan reluctantly unveiled. Science 213:520-522.
39. Oshima, R. J., Poe, M. P., Braegelmann, P. K., Baldwin, D. W., and Van Way, V. 1976. Ozone dosage-crop loss function for alfalfa: A standardized method for assessing crop losses from air pollutants. J. Air Pollut. Control Assoc. 26:861-865.
40. Oshima, R. J., Braegelmann, P. K., Baldwin, D. W., Van Way, V., and Taylor, O. C. 1977. Reduction of tomato fruit size and yield by ozone. J. Am. Soc. Hortic. Sci. 102:289-293.
41. Parmeter, J. R., and Miller, P. R. 1968. Studies relating to the cause of decline and

death of Ponderosa pine in southern California. Plant Dis. Rep. 52:707-711.
42. Petolino, J. 1982. Genetic analysis of ozone-stress responses in tobacco (*Nicotiana tabacum* L.) Ph.D. thesis, University of Maryland, College Park. 122 pp.
43. Rhoads, A., Harkov, R., and Brennan, E. 1980. Trees and shrubs relatively insensitive to oxidant pollution in New Jersey and southeastern Pennsylvania. Plant Dis. 64:1106-1108.
44. Rich, S. 1964. Ozone damage to plants. Annu. Rev. Phytopathol. 2:253-266.
45. Ryder, E. J. 1973. Selecting and breeding plants for increased resistance to air pollutants. Pages 75–85 in: Air Pollution Damage to Vegetation. Am. Chem. Soc., Washington, DC. 137 pp.
46. Santamour, F. S. 1969. Air pollution studies on *Platanus* and American elm seedlings. Plant Dis. Rep. 53:482-484.
47. Schafer, J. F. 1971. Tolerance to plant disease. Annu. Rev. Phytopathol. 9:235-252.
48. Shepard, J. F. 1981. Protoplasts as sources of disease resistance in plants. Annu. Rev. Phytopathol. 19:145-166.
49. Smith, R. J. 1981. The fight over clean air begins. Science 211:1328-1330.
50. Thompson, C. R., Hensel, E., and Kats, G. 1969. Effects of photochemical air pollutants on Zinfandel grapes. HortScience 4:222-224.
51. Thompson, C. R., Kats, G., and Hensel, E. 1972. Effects of ambient levels of ozone on navel oranges. Environ. Sci. Technol. 6:1014-1016.
52. Thompson, C. R., Kats, G , and Cameron, J. W. 1976. Effects of ambient photochemical air pollutants on growth, yield, and ear characteristics of two sweet corn hybrids. J. Environ. Qual. 5:410-412.
53. Wheeler, H. 1978. Disease alterations in permeability and membranes. Pages 327–347 in: Plant Disease, An Advanced Treatise. Vol. 3. G. Horsfall and E. B. Cowling, eds. Academic Press, New York. 487 pp.

CHAPTER 17

MINERAL DEFICIENCIES AND EXCESSES

EMANUEL EPSTEIN and ANDRÉ LÄUCHLI

Plant pathologists have in the main directed their attention and energies to those causes of ill health in crop plants for which pathogenic organisms are responsible. In addition, however, there is an array of diseases or disorders caused by abiotic factors. Foremost among these are drought and deficiencies or excesses of mineral elements. It is with these latter two conditions—stressfully low and stressfully high concentrations of inorganic ions in the soil—that the present chapter deals. The interplay between mineral disorders and diseases caused by pathogens has been discussed by Huber (19); it will not be covered in the present chapter, which deals with mineral nutrition and metabolism per se.

In addition to carbon, hydrogen, and oxygen, higher plants require for their growth a known minimum of 13 "essential" or "nutrient" mineral elements. For convenience, we divide them somewhat arbitrarily into "macronutrients" and "micronutrients." The former are present in healthy plants in concentrations ranging from about 0.1 to several percent of the dry weight, whereas micronutrients are present in the range of roughly 0.1–100 parts per million. Table 1 lists these elements in order of increasing concentration in plant matter. It includes the nonminerals oxygen, carbon, and hydrogen.

MINERAL DISORDERS AFFECTING PLANTS

Plants suffer deficiency diseases if any one of these essential or nutrient elements is present in the root medium at concentrations so low, or in chemical forms so poorly available for absorption by roots, that the internal concentration of the element falls to a level that is inadequate to meet normal structural or functional requirements, as evidenced by comparison with control plants grown at nonlimiting supplies of the element. The severity of the overall effect of the deficiency may range from slight growth inhibition to the death of the plant.

In addition to the elements of Table 1, cobalt is required for nitrogen fixation by both free-living and symbiotic nitrogen-fixing microorganisms. Other elements, although perhaps not essential in the strict sense of the word, are in many situations beneficial. Silicon is one such element. It imparts mechanical strength to cereal stems and thereby tends to minimize lodging. Sodium is another; in at least some crops, it is able to substitute for some although by no means all of the functions of potassium.

We have no assurance that the present list of essential elements is complete. There are only two experimental means for establishing essentiality. The most common is to grow the plant in a nutrient solution purged of the element under study to the extent feasible by available techniques. Failure of the plant to complete its life cycle under these conditions, when plants adequately supplied with the element thrive, constitutes proof of essentiality. If plants grow well in such deficient solutions, essentiality has not been demonstrated but neither has nonessentiality been established. The available techniques of purifying the nutrient solution may have left enough of the element in it to sustain the plant, or the element may have been supplied through the seed, from container walls, from the air, and from still other sources of contamination.

The second type of proof of the essentiality of an element is the demonstration that the element is a constituent of an essential metabolite; magnesium in chlorophyll is an example. Discovery of an essential metabolite that contains, say, vanadium as part of the molecule would constitute demonstration of the essentiality of vanadium.

By bringing sophisticated techniques to bear on the problem, animal nutritionists have in recent years discovered a raft of new essential trace (micronutrient) elements (24). No similar endeavor has been made in plant nutrition, and no generally essential plant nutrient has been discovered since chlorine was added to the list in 1954 (11).

Deficiencies are one of two kinds of mineral disorders that crops (and wild plants as well) are subject to. An excessively high concentration of soluble soil mineral constituents represents the other pathogenic condition involving mineral

TABLE 1. Concentrations of nutrient elements in plant material at concentrations considered to be adequate[a]

Element	Atomic weight	Concentration in dry matter		Relative number of atoms with respect to molybdenum
		μmol/g	ppm or %	
			ppm	
Mo	95.95	0.001	0.1	1
Cu	63.54	0.10	6	100
Zn	65.38	0.30	20	300
Mn	54.94	1.0	50	1,000
Fe	55.85	2.0	100	2,000
B	10.82	2.0	20	2,000
Cl	35.46	3.0	100	3,000
			%	
S	32.07	30	0.1	30,000
P	30.98	60	0.2	60,000
Mg	24.32	80	0.2	80,000
Ca	40.08	125	0.5	125,000
K	39.10	250	1.0	250,000
N	14.01	1,000	1.5	1,000,000
O	16.00	30,000	45	30,000,000
C	12.01	40,000	45	40,000,000
H	1.01	60,000	6	60,000,000

[a]Reprinted, by permission, from Mineral metabolism, by E. Epstein. Pages 438–466 in: Plant Biochemistry, J. Bonner and J. E. Varner, eds. ©1965 by Academic Press, New York.

elements. There are two classes of such solutes that between them account for widespread pathological conditions of crops: heavy metal ions and salts.

Ions of several essential micronutrient heavy metals may be toxic if present at excessively high concentrations in the medium and so may ions of heavy metal elements not known to be nutrients. Manganese and copper are examples of potentially toxic micronutrients; aluminum is the most damaging among the heavy metal elements not known to be nutrients. Toxicity of heavy metals goes hand in hand with soil acidity, which greatly increases their solubility and hence, their chemical and biological reactivity. Actual concentrations of heavy metals leading to toxicity are not very high; damaging concentrations are on the order of some parts per million in the soil solution. Heavy metal toxic soils are often low in nutrient status or "fertility."

Salts, mainly sodium salts, represent the second class of solutes that can cause pathological conditions if present in the soil at high concentrations. Salinity represents a double-barreled assault on the plant: by depressing the water potential in the soil (i.e., decreasing the solute potential of the soil solution) it renders soil water less readily available to the plant, and the specific ions responsible for the salinity of the soil may be toxic. Sodium and chloride are the most common culprits.

Sodium may be harmful even if its concentration in the soil solution is not high. Sodium adsorbed on the soil cation exchange surfaces to the extent that it occupies 15% or more of the total cation exchange capacity may exert toxic effects on account of this "sodicity," even in the absence of "salinity" (high concentrations of salts in the soil solution). Soils may combine salinity and sodicity. Ayers and Westcot (1) have described these conditions and their significance to crops.

While potentially toxic concentrations of heavy metals are on the order of some parts per million in the soil solution, salinity involves much higher concentrations, on the order of 1,000 parts per million or more. It is only at such concentrations that the colligative properties of the soil solution are sufficiently altered to affect the availability of water to plants. Toxic effects of specific ions of salts are also rare at concentrations lower than that. (Boron, a nonmetallic micronutrient, is often present in saline soil solutions at concentrations of 1 part per million or more—enough to be toxic to many crops.)

To sum up, the plant ills involving mineral elements are problems of too little and too much—too little "fertility" and too much in the way of heavy metals or salts. The total area affected by these pathogenic conditions approaches 4 billion hectares, or roughly 30% of the land area of the world (13).

STRATEGIES OF DEALING WITH DISEASES AND DISORDERS CAUSED BY MINERAL DEFICIENCY AND EXCESS

The principal strategy of dealing with biogenic diseases in crops has been the application of substances for control of the pathological conditions. Since these diseases are caused by pathogenic organisms, the substances applied are designed to control and suppress these organisms. This sort of strategy has led to the development of fungicides, bactericides, nematicides, etc. An additional strategy that has long been pursued is breeding crops that are resistant to diseases and pests.

What is the situation regarding strategies dealing with mineral disorders? The chief strategy of correcting mineral nutrient disorders in crops, comparable to one of the approaches in dealing with diseases, has been the application of substances; fertilizers, soil amendments, and micronutrient sprays have been and still are the principal means of overcoming or avoiding nutrient deficiencies. The main innovation in the years ahead will be an approach similar to the second one mentioned above, that is, breeding crops that grow and thrive under conditions that are less than minerally luxurious. Such crops are usually termed nutrient-efficient plants.

Until recently there has been little economic pressure to breed nutrient-efficient crops. Both the production and the large-scale application of fertilizers have been cheap, at least in the highly developed countries of the Western world. This has not necessarily been true of the developing countries, however. Now that the oil crisis has pushed the cost of energy dramatically upward, the prices of fertilizers as well as the costs of their distribution and application are skyrocketing to the point where the farmers of the developing nations have even less chance to affort them on a large scale. Even the economically more advantaged countries can only absorb the rising costs of fertilizers by raising the price of food and fiber. Thus, there is now economic pressure to "shift gear" and give impetus to the additional strategy of breeding nutrient-efficient plants. Modern genetic technologies such as genetic engineering may aid by complementing the classical selection and breeding techniques.

What has been said about breeding for mineral deficiency conditions is equally applicable to crop disorders caused by mineral excesses in soils. Heavy-metal soils often have not been used for crop production because of their high toxicity to most crops. The cure of mineral disorders of this type by soil amendments or the like is often either impossible or economically unfeasible. Hence only the selection and breeding of crops tolerant to heavy metals and acid soils can make such soils agriculturally productive. Mineral excesses caused by buildup of salts in soils (saline soils) have traditionally been fought by amending soil, drainage, and other physical means of controlling soil salinity. Because salinity is a pressing problem, especially in irrigation agriculture, the "soil-and-water approach" needs to be complemented by selection and breeding of salt-resistant crops.

Selection and Breeding for Mineral Deficiency Conditions

It has become increasingly evident in recent years that there are qualitative and quantitative variations in the nutrient requirements of plants and that these variations are under genetic control (18,29). Hence it is possible to select and breed plants that are fit for mineral deficiency conditions. Obviously, in the course of evolution numerous genotypes of plants have evolved that are adapted to soils poor in nutrients. Thus, many populations of wild plants are tolerant of low environmental concentrations of essential elements. Such wild plants may also be termed nutrient-efficient.

Wild plants adapted to poor soils appear to control their growth at a low rate, in keeping with the low nutrient status of their natural environment. Increasing the nutrient supply may not enhance the growth rate—the additional nutrient absorption may lead to toxic concentrations in the plant. On the other hand, cultivated species have traditionally been bred for fertile soils and respond to such soils with high growth rates. Increased nutrient supply can stimulate growth even further, within limits. When a nutrient deficiency situation develops in the soil, cultivated plants usually continue to grow until the onset of nutrient deficiency in the plant limits growth and yield. Such a nutrient-inefficient cultivated plant produces top yields only with application of appropriate quantities and kinds of fertilizer. If we are to select and breed a nutrient-efficient crop genotype that is tolerant of mineral deficiency conditions in the soil, we need to select a plant that gives the highest ratio of growth and yield to applied fertilizer rather than the maximal yield attainable with ample fertilization. More detailed accounts of the concept of nutrient efficiency versus nutrient inefficiency in wild and crop plants are presented by Chapin (7) and by Läuchli and Bieleski (21).

What are the chief elements likely to be in low supply in many agricultural soils of the world? For macronutrients, these are mostly nitrogen, phosphorus, and sometimes potassium. Sulfur, magnesium, and calcium appear to be of more limited and local significance. Nitrogen is the most heavily applied fertilizer in most situations. The rapidly increasing production costs for nitrogen fertilizers make breeding for efficiency in nitrogen utilization highly important. There are two main options to pursue. One is to make symbiotic nitrogen fixation more efficient, and the second one is to improve the nitrogen nutrition of those crops that do not form nitrogen-fixing symbioses, such as the cereal grains.

The situation may be even worse with regard to phosphorus. Not only are the agricultural soils of entire countries very phosphorus-deficient (e.g., Australia, New Zealand), but the resources in high-grade rock phosphates may eventually become limiting in the production of phosphorus fertilizers. A detailed understanding of the processes governing the movement of phosphate from soil to plant roots is crucial (3) if we are to make progress in breeding phosphorus-efficient genotypes. The development of host-mycorrhizal systems (28) will play an increasingly important role in crop phosphorus nutrition, as well as in other processes discussed in this chapter such as micronutrient nutrition and resistance to excess mineral stresses.

Among the micronutrients, iron is the element most frequently found to be deficient, particularly in alkali soils where the availability of iron is extremely low because of the low solubility of iron compounds at high pH. Selection and breeding for iron efficiency appear particularly promising, as many instances of iron efficiency and iron inefficiency in crop genotypes have been described and physiologically investigated (6). In addition, the development of micronutrient-efficient crops appears important with regard to boron, copper, zinc, manganese, and molybdenum.

Some recent reviews on breeding crops for mineral efficiency are as follows: Gerloff (17) and Gabelman (16) for N, P, and K; Loneragan (23) for P; Clark (8) for Ca, Mg, and Mo; Brown (5) for Fe; Loneragan (22) for B, Cu, Mn, Zn, and Co. General overviews have been given by Vose (29) and Gerloff and Gabelman (18). The successful development of crop genotypes adapted to low-nutrient environments depends to a large extent on the availability and knowledge of morphological or physiological markers of tolerance of a specific deficiency condition. In turn, the recognition of such markers depends on knowledge of the underlying physiological processes that govern nutrient efficiency or inefficiency. Unfortunately, our understanding of these physiological processes is still scanty; an up-to-date review has been presented by Gerloff and Gabelman (18). Physiological processes that may govern efficiency of nutrient utilization of crop genotypes are as follows: 1) nutrient absorption by the roots (influx from the soil minus efflux into the soil); 2) nutrient transport across the root and into the xylem vessels; 3) nutrient distribution in the plant to sites of utilization; 4) nutrient utilization in metabolism and growth.

Gerloff and Gabelman (18) think that at least some of the adaptations to low-nutrient soils by wild plants would be useful in breeding programs for cultivated species. In particular, efficient mechanisms for nutrient absorption by roots from low-nutrient soils, high efficiency in metabolic reactions activated by nutrients, and capacity for storage of nutrients during periods of relative abundance may all be important targets in such breeding endeavors. Tissue analysis for nutrients (4) appears to be the method of choice for detecting the onset of deficiency.

Selection and Breeding for Conditions of Mineral Excess

Heavy Metals

A genetic approach to the problems of heavy metal toxicity must be tailored to the particular situation prevailing. Soils may contain excessively high concentrations of heavy metals as the result either of natural geological processes or of human activities. The naturally occurring soils with heavy metals at concentrations so high as to be inhibitory to crops have become so as a result of the effects of water. Excessive leaching of the bases from soils in the humid tropics and other high-rainfall regions renders the soils acid, which in turn causes the heavy metals, especially aluminum and manganese, to go into solution. Manganese may be toxic even under nonacid conditions (pH 6 or even greater) if its concentration in the soil is high and if as a result of flooding it becomes reduced to the divalent form in which it is readily absorbed by plants.

Acid conditions from human activities have been largely the result of mining. Mining operations, especially in the last century, were so extensive that effluents from them affected large areas of land, while public insistence on pollution control was as yet nonexistent or ineffectual. These soils contaminated by heavy metals often lay close to normal, fertile soils. The rapidity with which the mine wastelands came into being, together with their proximity to ordinary, fertile land, created a stage on which evolutionary processes could come into play. Certain wild species, especially of the genus *Agrostis*, were found to have evolved ecotypes or physiological races tolerant to heavy metal that colonized the mine wastes in as short a time as 5 years (20). The extensive research by plant geneticists and physiological ecologists on wild species adapted and nonadapted to heavy metal laid much of the foundation for this aspect of stress physiology and its genetic control. Recognition of the genetic variability of this trait in turn furnishes the scientific rationale for projects of selection and breeding for heavy metal resistance in crops.

Actual selection of crops for resistance to metal toxicity, however, began without benefit of such scientific understanding. The southernmost state of Brazil, Rio Grande do Sul, has acidic, high-aluminum soils. Early in the century, certain established wheat cultivars yielded 800–1,000 kg/ha—acceptable yields in those days, as no liming was done and hardly any fertilizer used. Introduced cultivars failed. The different performance of the Brazilian and the introduced cultivars suggested the existence, in the Brazilian ones, of some tolerance factor that the breeders could not then identify (9). Nevertheless, progress was made in breeding wheat resistant to "crestamento" (burning or firing) throughout the first half of the century. It was not until the early 1950s, however, that Araújo identified aluminum as the "possible cause of wheat 'crestamento' " (2).

The efforts of the Brazilian breeders to bestow aluminum tolerance on wheat have been very successful. Adapted cultivars can tolerate soils in which 30% of the cation exchange capacity is occupied by aluminum. In large areas of the main wheat-producing states of Brazil no lime needs to be applied, the high aluminum content of the soils notwithstanding (9). The Brazilian work on aluminum-resistant wheat is the oldest ongoing project of selecting and breeding for resistance to heavy metal toxicity and the one that has made the greatest impact on the economy of an entire region.

In the United States, Neenan (25) in 1960 seems to have been the first to recognize differences in the aluminum tolerance of wheat cultivars (27). By now, the need to select and breed for conditions of excess aluminum, manganese, and some other heavy metals is widely recognized. The proceedings of a 1976 workshop edited by Wright (31) contains numerous contributions documenting the fact that the genetic variability necessary for such breeding exists in many crops and describing methods for screening for this type of tolerance. Physiological aspects of heavy metal toxicity have been described by Foy et al (15). A recent symposium volume (30) brings together a miscellany of papers on heavy metal stress in plants.

The need for pursuing the genetic approach to the metal toxicity problem will become more pressing with time. Liming can alleviate heavy metal toxicities, but in large areas of the world lime is not available and the importation of it is not economically feasible; yet the number of mouths to be fed increases inexorably. In those areas, the development of lines of crops tolerant to heavy metal is the only feasible alternative for increasing yields of existing crops and for the introduction of new ones now excluded because of heavy metal toxicity. Thus a genetic approach to the problems posed by this kind of mineral pathology, heavy metal toxicity, will loom large in the years ahead.

Salt

Salinity and sodicity affect huge areas in the arid and semiarid regions of the world where agriculture depends on irrigation. The soils of these regions not being heavily leached by rain, natural processes of weathering and biogeochemical cycling as well as irrigation itself cause a buildup of salt in the soil. All irrigation water, and especially the water available in dry regions, carries some burden of dissolved salt. The water evaporates from the soil and transpires from the plants while the salt remains. Inevitably, the salinity of the soil increases, unless excess salt is carried away to sinks through reclamation and drainage projects. But even with such measures, some level of salinity is an almost unavoidable feature of irrigation agriculture in dry, hot regions. Furthermore, the physical or engineering approach to salinity represents a huge cost in terms of dollars, power, water, and often, environmental impact.

In the western United States, large tracts of land will be disturbed in the recovery of coal, oil, tar, and gas. In the reconstitution and revegetation of this disturbed land, salinity will often be among the problems encountered (32).

All these considerations have given an impetus to a new approach to the problems posed by salinity and sodicity: the approach of breeding salt-tolerant crops (14). The biological rationale for such an approach is ample. Certain wild plants, the halophytes or salt plants, inhabit salt marshes, saline soils, and other habitats too saline for present-day crops. Until two decades ago, however, halophytes were of interest almost exclusively to ecologists, who studied their habitats, distribution, phylogenetic affinities, and physiological adaptations to salt stress.

In 1963, Epstein (10) drew attention to genetic aspects of biological membrane transport of mineral ions and their metabolism, with special reference to salt tolerance. He commented on the competence of halophytic plants in coping with salt and on the specialized, genetically controlled mechanisms of membrane transport of ions that they must possess. He hoped that studies of the genetic control of these processes in plants would furnish basic information helpful in breeding salt-tolerant crops.

Although wild halophytes are the salt-tolerant plants par excellence, much variability in this regard exists even in many crop species (12). We therefore have the options of selecting for this trait within species and of introducing salt tolerance into crop species from wild, salt-tolerant relatives. These conventional methods of selecting and breeding can be supplemented by such newer ones as somatic hybridization, cell and tissue culture, and recombinant DNA techniques (14,26).

The development of a genetic approach to the salinity problem has recently been traced by Epstein et al (14). There is now a considerable and growing effort along this line in many countries. It may safely be expected that in the future, the pathology of excess salt will be dealt with by a double-pronged attack: reclamation and drainage to reduce the stressful level of salinity or sodicity, as has long been done, and genetic means to develop new crops tolerant of these stresses.

RESEARCH

All pathological conditions in crops represent departures from the normal range that are deleterious to the growth, development, and productivity of the plants. Mineral deficiencies and excesses cause such departures by a plethora of processes. Deficiencies may result in the slow-down of protein synthesis, in inadequate activation of enzymes resulting in suboptimal supplies of essential metabolites or the buildup of toxic concentrations of intermediates, in derangements of membranes and other organelles, and in hormonal imbalances, to name but a few. Excessive concentrations of inorganic ions may displace functional ions from their active sites, cause local precipitation and inactivation of nutrients, compete with other ions in essential transport mechanisms, and denature proteins, among many possibilities. Whatever the primary lesion, it will in turn have catenary effects involving ever more processes throughout the entire metabolic machinery of the plant. Our understanding of these mineral pathologies is inadequate. Comparative physiological and biochemical studies with genotypes of a given species differing in respect to their toleration of or susceptibility to a certain mineral stress condition can be expected to advance our understanding of the underlying mechanisms. Such knowledge, in turn, will be useful in coping with these disorders by means ranging from cultural practices to the development of resistant genotypes.

OUTLOOK

This chapter carries a dual message. First, no less than pathogenic organisms, deficiencies and excesses of mineral elements are responsible for pathological conditions of crops over huge areas of the world. Far more collaboration than has been customary is called for between plant pathologists and plant physiologists specializing in mineral nutrition and metabolism. Second, the nature and extent of the soil factors making for mineral pathologies impose severe constraints on our ability to cure them exclusively by the application of fertilizers, amendments, and excess water. The difficulty is exacerbated by the ever mounting dollar and energy costs of these inputs. Therefore, a new approach is needed: the genetic approach of selecting and breeding crop genotypes tolerant of soils that impose mineral stresses on the plants. We are confident that on the occasion of the centenary of the American Phytopathological Society note will be taken of the enlargement of the scope of plant pathology that we here envision: the sharpened awareness of problems of pathological mineral metabolism and the potential of a genetic approach to them and the corresponding collaborative efforts with plant nutritionists and plant breeders.

ACKNOWLEDGMENT

This chapter is based in part on a report (13) prepared by E. E. at the request and with the support of the Office of Technology Assessment, United States Congress.

LITERATURE CITED

1. Ayers, R. S., and Westcot, D. W. 1976. Water Quality for Agriculture. FAO/UN, Rome. 97 pp.
2. Beckman, I. 1977. Cultivation and breeding of wheat (*Triticum vulgare* Vill.) in the south of Brazil. Pages 409–416 in: Plant Adaptation to Mineral Stress in Problem Soils. M. J. Wright, ed. Cornell Univ. Agric. Exp. Stn., Ithaca, NY. 420 pp.
3. Bieleski, R. L. 1976. Passage of phosphate from soil to plant. Pages 125–129 in: Reviews in Rural Science III. G. J. Blair, ed. Univ. New England, Armidale, Australia.
4. Bouma, D. 1983. Diagnosis of mineral deficiencies using plant tests. In: Encyclopedia of Plant Physiology. New Series, Vol. 15. A. Läuchli and R. L. Bieleski, eds. Springer-Verlag, Berlin. In press.
5. Brown, J. C. 1977. Screening plants for iron efficiency. Pages 355–357 in: Plant Adaptation to Mineral Stress in Problem Soils. M. J. Wright, ed. Cornell Univ. Agric. Exp. Stn., Ithaca, NY. 420 pp.
6. Brown, J. C. 1978. Mechanism of iron uptake by plants. Plant Cell Environ. 1:249-257.
7. Chapin, F. S., III. 1980. The mineral nutrition of wild plants. Annu. Rev. Ecol. Syst. 11:233-260.
8. Clark, R. B. 1977. Plant efficiencies in the use of calcium, magnesium, and molybdenum. Pages 175–191 in: Plant Adaptation to Mineral Stress in Problem Soils. M. J. Wright, ed. Cornell Univ. Agric. Exp. Stn., Ithaca, NY. 420 pp.
9. Da Silva, A. R. 1977. Application of the genetic approach to wheat culture in Brazil. Pages 223–231 in: Plant Adaptation to Mineral Stress in Problem Soils. M. J. Wright, ed. Cornell Univ. Agric. Exp. Stn., Ithaca, NY. 420 pp.
10. Epstein, E. 1963. Selective ion transport in plants and its genetic control. Pages 284–298 in: Desalination Research Conference. Natl. Acad. Sci.-Natl. Res. Council Publ. 942. NAS-NRC, Washington, DC. 451 pp.
11. Epstein, E. 1972. Mineral Nutrition of Plants: Principles and Perspectives. John Wiley and Sons, New York. 412 pp.
12. Epstein, E. 1978. Crop Production in Arid and Semi-Arid Regions Using Saline Water. A Report for the National Science Foundation—Applied Science and Research Applications. Dept. Land, Air and Water Resources, Univ. Calif., Davis. 107 pp.
13. Epstein, E. 1980. Impact of Applied Genetics on Agriculturally Important Plants: Mineral Metabolism. A Report to the Office of Technology Assessment, U.S. Congress. Dept. Land, Air and Water Resources, Univ. Calif., Davis. 42 pp.

14. Epstein, E., Norlyn, J. D., Rush, D. W., Kingsbury, R. W., Kelley, D. B., Cunningham, G. A., and Wrona, A. F. 1980. Saline culture of crops: A genetic approach. Science 210:399-404.
15. Foy, C. D., Chaney, R. L., and White, M. C. 1978. The physiology of metal toxicity in plants. Annu. Rev. Plant Physiol. 29:511-566.
16. Gabelman, W. H. 1977. Genetic potentials in nitrogen, phosphorus, and potassium efficiency. Pages 205–212 in: Plant Adaptation to Mineral Stress in Problem Soils. M. J. Wright, ed. Cornell Univ. Agric. Exp. Stn., Ithaca, NY. 420 pp.
17. Gerloff, G. C. 1977. Plant efficiencies in the use of nitrogen, phosphorus, and potassium. Pages 161–173 in: Plant Adaptation to Mineral Stress in Problem Soils. M. J. Wright, ed. Cornell Univ. Agric. Exp. Stn., Ithaca, NY. 420 pp.
18. Gerloff, G. C., and Gabelman, W. H. 1983. Genetic basis of inorganic plant nutrition. In: Encyclopedia of Plant Physiology. New Series, Vol. 15. A. Läuchli and R. L. Bieleski, eds. Springer-Verlag, Berlin. In press.
19. Huber, D. M. 1978. Disturbed mineral nutrition. Pages 163–181 in: Plant Disease, An Advanced Treatise. Vol. 3. How Plants Suffer from Disease. J. G. Horsfall and E. B. Cowling, eds. Academic Press, New York. 487 pp.
20. Humphreys, M. O., and Bradshaw, A. D. 1977. Genetic potentials for solving problems of soil mineral stress: Heavy metal toxicities. Pages 95–105 in: Plant Adaptation to Mineral Stress in Problem Soils. M. J. Wright, ed. Cornell Univ. Agric. Exp. Stn., Ithaca, NY. 420 pp.
21. Läuchli, A., and Bieleski, R. L. 1983. Synthesis and outlook. In: Encyclopedia of Plant Physiology. New Series, Vol. 15. A. Läuchli and R. L. Bieleski, eds. Springer-Verlag, Berlin. In press.
22. Loneragan, J. F. 1977. Plant efficiencies in the use of B, Co, Cu, Mn, and Zn. Pages 193–203 in: Plant Adaptation to Mineral Stress in Problem Soils. M. J. Wright, ed. Cornell Univ. Agric. Exp. Stn., Ithaca, NY. 420 pp.
23. Loneragan, J. F. 1978. The physiology of plant tolerance to low phosphorus availability. Pages 329–343 in: Crop Tolerance to Suboptimal Land Conditions. G. A. Jung, ed. Am. Soc. Agron., Madison, WI. 343 pp.
24. Mertz, W. 1981. The essential trace elements. Science 213:1332-1338.
25. Neenan, M. 1960. The effects of soil acidity on the growth of cereals with particular reference to the differential reaction of varieties thereto. Plant Soil 12:324-338.
26. Rains, D. W. 1981. Salt tolerance—New developments. Pages 431–456 in: Advances in Food Producing Systems for Arid and Semiarid Lands. Part A. J. Manassah, ed. Academic Press, New York. 676 pp.
27. Reid, D. A. 1977. Screening barley for aluminum tolerance. Pages 269–275 in: Plant Adaptation to Mineral Stress in Problem Soils. M. J. Wright, ed. Cornell Univ. Agric. Exp. Stn., Ithaca, NY. 420 pp.
28. Smith, S. S. E. 1980. Mycorrhizas of autotrophic higher plants. Biol. Rev. 55:475-510.
29. Vose, P. B. 1983. Effects of genetic factors on nutritional requirement of plants. In: Contemporary Bases for Crop Breeding. P. B. Vose and S. G. Blixt, eds. Pergamon Press, Oxford. In press.
30. Wallace, A., and Berry, W. L., eds. 1981. Trace Element Stress in Plants. J. Plant Nutr. Special Symposium Issue 3:1-741.
31. Wright, M. J., ed. 1977. Plant Adaptation to Mineral Stress in Problem Soils. Cornell Univ. Agric. Exp. Stn., Ithaca, NY. 420 pp.
32. Wright, R. A., ed. 1978. The Reclamation of Disturbed Arid Lands. Univ. N.M. Press, Albuquerque. 196 pp.

CHAPTER 18

PLANT HEALTH: A VIEW FROM ABOVE

RAY D. JACKSON

If we accept the premise that the world food supply is limited and that constraints on water and energy resources will restrict the supply even more, then we should examine every avenue for maintaining and improving plant health. The first phase of any such program is to evaluate plant health on a field basis and identify problems such as insufficiencies of water and nutrients and the occurrence of plant diseases and insect infestations.

At the present time the majority of plant health surveys are made from a moving vehicle, supplemented by a few walking excursions into portions of fields. The analysis consists of mental integration of what was seen, coupled with experience gained through the years. This time-honored means of evaluating plant health locates problems only after signs are visually apparent and yield-reducing damage is done. Obviously, if a method were at hand to rapidly survey large regions in detail, at frequent intervals, evaluations of plant health would be readily available and the best management practices could be applied that would maximize food production with a minimum input of water and energy.

If current research challenges are met successfully, a farm manager will have such a survey system constantly available. Using a small but powerful computer, the manager could interrogate sensors held by a satellite whose design will be radically different from the satellites of the 1980s. Images of each field of the farm would appear on a television screen. The manager would check each field, using various combinations of wavelengths of the energy reflected and emitted from the soils and plants. Certain combinations of wavelengths show water status, while others isolate areas of plant diseases, insect infestations, and nutrient deficiencies. The entire farm could be surveyed and problem areas identified in minutes. The development of such a survey system is technically feasible. A major research effort will be necessary to find the specific wavelengths and combination of wavelength intervals that will allow the identification and quantification of the various plant health features, with a minimum of interference from the many complicating factors.

Some readers, while scanning the preceding paragraphs, will feel that much of this has already been done. They know that aerial color infrared photography has proved useful for monitoring plant health, especially by plant pathologists and entomologists. Furthermore, the Landsat satellites have been operating for a decade. They wonder what is so challenging about doing more of the same. A short response to these doubts is that color infrared photography covers only a small portion of the electromagnetic spectrum that is useful for remote sensing, and the film must be retrieved, processed, and interpreted. Multispectral scanners obtain data in the photographic as well as in many other wavelength regions; the data are digitized and can be transmitted over great distances, providing vastly more information than is obtainable from the single band film. The earth resources satellites (Landsats) were heralded with much publicity and promise. Indeed, they have produced exceptional results. However, for agriculture, the benefits have been largely confined to market information through improved estimates of world grain yields. Information that could be used in day-to-day management decisions cannot be provided with this satellite system because data are obtained only once every 18 days (9 days if two satellites are operating), weather permitting, and availability of data is not immediate. Assuming that information could be obtained concerning a plant health feature, data would not be available in time to institute practices that could alleviate the situation.

The research challenges are many and need to be attacked by both biologically and physically oriented scientists. The following sections present some background information, some examples of the challenges, and a projection of the payoff.

REMOTE SENSING

Remote sensing usually refers to the study of objects from great distances. In this discourse, we will not be concerned with great distances, but will use the term "remote sensing" to mean the measurement of radiation reflected or emitted from an object. We consider measurements made with radiometers hand-held at a distance from a few centimeters to several meters from a target to be remotely sensed measurements, analogous to those made from space platforms.

The most frequently used sensor for assessing plant health is the human eye. Results, however, vary among individuals and no hard copy image is produced. Conventional photography produces a hard copy, but it is limited to the visible part of the electromagnetic spectrum, with the exception of color infrared film, whose sensitivity extends to about 0.9 μm. Contrary to a popular misconception, color infrared film is not affected by the temperature of the objects photographed. However, the emulsions used with this film are sensitive to ambient temperature, making the results dependent on film storage and developing conditions.

Energy reflected or emitted from objects can be detected in many parts of the electromagnetic spectrum. The three major parts currently used by remote sensing instruments are the reflected solar, thermal infrared, and microwave. The reflected solar region extends from 0.4 μm to about 3 μm and includes the visible (0.4–0.7 μm) and the near infrared (0.7 to about 3 μm).

The thermal infrared region extends from about 3 to 20 μm. In contrast to the reflected solar region, thermal infrared radiation is emitted and can, therefore, be related to the temperature of the target through the Stephan-Boltzmann blackbody law

$$R = \epsilon \sigma T^4 \quad (1)$$

where R is the emitted radiation, ϵ the emissivity of the surface material, σ the Stephan-Boltzmann constant, and T the temperature (°K) of the object. The demarcation between the thermal infrared and reflected solar is not sharp. Some energy is emitted below 3 μm and some is reflected above 3 μm. Water vapor absorbs thermal infrared radiation, causing some attenuation of this radiation by the atmosphere.

The microwave region extends from about 1 mm to nearly 1 m. Passive microwave systems detect energy naturally emitted from surfaces. Active microwave systems (radar) transmit energy in a particular waveband that is reflected from a surface, with the return beam measured by the instrument. Surface features are related to differences between the transmitted and the return energy levels. Since microwave transmissions are not affected by clouds or atmospheric haze, these systems have the distinct advantage of yielding surface information regardless of weather.

During the 1960s, airborne scanners were used to obtain images of farmlands in the visible, near infrared, and the thermal infrared regions. The major contribution of these flights was that they demonstrated the feasibility of obtaining, and the unique contributions to be expected from, remotely sensed data as an agricultural management tool. The launch of Landsat-1 (in 1972) provided impetus for the comparison of ground-collected data with remotely sensed data. A small instrument, called the "Landsat ground truth radiometer" (because it used similar wavebands as the multispectral scanner carried by the satellite) was developed to aid in the rapid collection of ground data (the term "ground truth" will be avoided here because it erroneously implies no error in the ground measurement). About the same time, portable infrared radiometers for surface temperature measurements became available. These instruments, hand-held or boom mounted, enabled rapid and repeated data to be collected over many small, differently treated plots—experiments impossible to conduct with aircraft or satellite-mounted instruments. During the past several years, these small radiometers have been used extensively by agricultural researchers. Several projects are underway whose goals are to develop the relationships between plant health parameters and remotely sensed data that will be necessary for the plant health monitoring system envisioned in the introduction to become a reality.

Experiments in agricultural remote sensing began from air and space platforms but have returned to the ground for the detailed research that must be done to develop the pertinent relationships. In the following section, a few examples of assessing plant health by ground-based, remote means are discussed. Obviously, we must return to space for the ultimate application of this research, but that is part of the challenge we face.

CURRENT RESEARCH IN PLANT HEALTH ASSESSMENT

The word "stress" has become a catchall term to signify a detrimental effect on plant growth. It is without precise physiological meaning but is useful for communication among scientific disciplines because of its implication of adversity. The term will be used here without further definition.

Plant growth is in concert with the soil and aerial environment. As with humans, the temperature of a plant is a measure of health when related to the environment. From energy balance considerations, it can be shown that plant temperatures are related to environmental parameters by the relation (7)

$$T_f - T_A = \frac{R_n r_a}{\rho c_p} \cdot \frac{\gamma(1 + r_f/r_a)}{\Delta + \gamma(1 + r_f/r_a)} - \frac{VPD}{\Delta + \gamma(1 + r_f/r_a)} \tag{2}$$

where T_f is the plant foliage temperature, T_A is the air temperature, R_n is the net radiation, r_a is the aerodynamic resistance to water vapor transfer from the plant surface to the atmosphere, r_f is the plant (or canopy) resistance to water flow from the root-soil interface to the leaves, ρc_p is the volumetric heat capacity of the plant material (essentially a constant), VPD is the vapor pressure deficit of the air, γ is the psychrometric constant, and Δ is the slope of the saturated vapor pressure-temperature relation. The air temperature and vapor pressure deficit are to be measured at the same height (usually 1.5–2 m above the surface). When water is not limiting, the foliage temperature decreases linearly with increasing VPD, at a constant air temperature. Foliage temperatures from 10 to 12 C below air temperature are not unusual (3). Equation 2 could be incorporated into models that predict disease and insect development in plant canopies, models that currently assume that air temperature is representative of the environmental temperature of the predators. Improved predictions of predator activity should result. Jackson (4) gives a detailed discussion concerning the development of equation 2 and a review of plant temperature research.

Plants exert control over their temperature by adjusting stomates to regulate the amount of water that evaporates from leaves. A plant growing in a soil in which water is not limiting will have a low value of r_f (the stomates will be open). When soil water is limiting, r_f increases and the plant temperature increases. The difference in temperature between water-limited and water-adequate conditions, for the same aerial environmental conditions, is an index of plant health. Any factor that restricts the movement of water to the leaves will cause an increase in temperature. Limited soil water is the most ubiquitous cause, but diseases, insects, and nutrient deficiencies are also causative factors.

Water Stress

Idso et al (3) and Jackson et al (7) have shown that plant water stress can be quantified. They measured foliage temperatures with infrared thermometers held at such an angle that the field of view of the instrument was essentially filled with foliage (minimum soil background). Wet and dry bulb air temperatures were taken (giving T_A and VPD), and net radiation (R_n) was measured or estimated from incoming solar radiation. With these data, the ratio r_f/r_a in equation 2 was evaluated. A crop water stress index (CWSI) was calculated graphically and from the relation

$$CWSI = \frac{\gamma(1 + r_f/r_a) - \gamma^*}{\Delta + \gamma(1 + r_f/r_a)} \tag{3}$$

where $\gamma^* = \gamma(1 + r_{fp}/r_a)$, with r_{fp} the minimum value of the foliage resistance obtained when the plants have ample available water. The CWSI ranges from 0 (no stress) to 1 (no transpiration) and provides a quantitative measure of stress.

Nutrient Stress

Laboratory studies of nutrient stress effects showed that mineral deficiences increased the reflectance of radiation in the visible wavelength region, whereas effects on near and middle infrared reflectance varied according to the specific mineral deficiency (1). Field measurements of corn canopies that received four nitrogen treatment levels showed that visible red reflectance increased and the near infrared reflectance decreased with decreasing nitrogen (14). The ratio of near infrared to the visible red radiances was related directly to the amount of nitrogen applied. Similar results have been reported for nitrogen-deficient sugarcane (6).

Disease and Insect Stresses

Much research has been done using color and color-infrared photography to detect plant diseases and insect damage. However, little has been done using scanners and radiometers, although the potential for success is great. Pinter et al (11) used a hand-held infrared thermometer to measure temperatures of sugar beet leaves in a field infested with *Pythium aphanidermatum*. Diseased beets were found to be 3–5 C warmer than adjacent noninfected plants. Plant temperatures provided a previsual signal of infection. Results with cotton infected with *Phymatotrichum omnivorum* were similar. Diseased plants were significantly warmer than healthy plants. However, with cotton, diseased plants wilted sooner than healthy plants, somewhat negating the previsual detection advantage of the remotely sensed temperatures.

The production of aflatoxin within cotton canopies is related to canopy temperature and moisture conditions. Canopy temperatures can be monitored with infrared thermometry. Microwave techniques can be used to estimate surface ($\simeq$ 5 cm) soil moisture beneath plant canopies (12). Remote sensing should be able to delineate areas where the potential for aflatoxin is high.

Insects and soilborne pathogens may invade plant roots or stems in a manner that disturbs the conducting tissue, resulting in a reduction of transpiration. This effect should result in a plant temperature increase and be detectable with thermal techniques. Leaf damage would be detectable by sensors that measure reflected radiation. Radar systems have not been used to monitor plant health but have proved useful for tracking insect migrations to and from agricultural fields (9).

THE CHALLENGES

In the previous section some successful plant health monitoring developments were discussed briefly, but the various factors that complicate the measurements and their interpretation were purposefully ignored. In this section, some of these factors are examined because they must be accommodated in the overall scheme if a viable monitoring system is to be developed.

The Canopy

The reflection and emission of electromagnetic radiation from plant canopies is influenced by the geometry of the canopy. Since plant stress can cause geometry changes (8), the interaction of radiation with the canopy geometry must be well understood if the several stresses are to be differentiated.

Energy from the sun and sky that strikes a field is either reflected, absorbed, or transmitted. The absorbed energy heats the soil and plant material, but some energy is dissipated through evaporation of water and some is returned to space by emission at longer wavelengths (about 3–20 μm). This emitted energy can be detected and related to the surface temperature (equation 1).

For a light-colored dry soil, about 25–35% of the energy is reflected and the remainder absorbed. If the same soil were wet, the amount reflected would be about half of that for dry soil. This holds for both the visible and the near-infrared regions. Green plants absorb much of the visible light, with only a small amount being reflected or transmitted. However, in the near infrared region, plants reflect and transmit most of the energy, with little being absorbed. It is these fundamental differences in reflectance properties that make possible the detection and quantification of green biomass by remote means.

Early in the growing season, before full cover is achieved, a remote sensing instrument receives a composite of energy emanating from both soils and plants. The interpretation of the composite scene presents a challenge. For example, if plant temperatures (from which a stress index can be calculated) are desired, they must be extracted from the plant-soil composite temperature. There is no known solution to the above example as stated. However, if additional information is known, say, the amount of plant cover, plant height, row direction, sensor view angle, and other pertinent factors, it may be possible to calculate the plant temperatures (5).

Another challenge arises when one attempts to determine plant cover, plant height, etc., from remotely sensed measurements. It is well known that several combinations of visible and near infrared bands are correlated with green biomass and the fraction of plant cover. However, these relationships are dependent upon the sun's elevation and azimuth angles. In other words, the result obtained will depend upon the time of day when the data were obtained. The composite soil-plant scene would record energy reflected from sunlit and shaded soil and plant materials. Reflectance from the shaded parts in the visible region is about 10% of that from sunlit parts since the energy striking the shaded areas is diffuse skylight. At low sun elevation with the soil shaded, the sunlit plant parts contribute most of the energy to the composite. At high sun elevations, with some soil sunlit, more energy would be reflected from the sunlit soil and the soil would contribute much more to the composite reflectance than at the low sun elevation. This situation is further complicated by the fact that much more near-infrared than visible energy is transmitted through plant leaves. About eight layers of leaves are necessary to absorb or reflect all the near-infrared radiation (2,8). Thus for many crops such as wheat, the near-infrared contribution to the composite scene at different sun angles could be different from that for visible light.

At this point the complexity of the situation can lead to the conclusion that it is impossible to resolve the problem. On the contrary, it is this very complexity that will provide us with the information that is needed concerning the canopy. Measurements made at different known sun angles can provide data that can be related to plant height, the amount of soil that is visible, and other factors. If it is successful, this research will allow the canopy geometry to be specified by remote means.

Once the canopy geometry can be assessed adequately, then it should be possible to determine small changes in leaf color caused by nutrient deficiencies. Insect and disease damage should be identifiable from the other forms of stress. To meet these challenges, it is essential that new combinations of wavelength intervals be examined. The results of these combinations must be related quantitatively to the various agronomic, pathological, and entomological parameters. Two major challenges remain: how to account for atmospheric effects on the radiation that is reflected or emitted from the soils and plants as it travels to a space sensor, and the design and construction of the space platform and sensor system.

The Atmosphere

Radiant energy in the visible and the near-infrared regions received by a remote sensor above the fields of interest is composed of energy that was reflected from the soil and plant surfaces and energy that was scattered by air molecules and particulate matter in the intervening atmosphere. Scattering due to the air molecules is called Rayleigh scattering, and scattering due to particulate matter is called Mie scattering. Rayleigh scattering, which is always present, is related to the inverse of the fourth power of the wavelength. Thus, the shorter the wavelength, the more scattering (more interference with remotely sensed data). Mie scattering is always present to some degree, since no atmosphere is completely clear, and increases as the amounts of pollutants increase. The two modes of scattering make rather complex the theoretical modeling that attempts to account for this factor. This has been an active research topic for several years but remains a challenge.

At the longer wavelengths, molecular and particulate scattering are insignificant but water vapor becomes a problem. Emitted radiation in the thermal infrared region is absorbed by water vapor. This effect is at a minimum, but not negligible, in the 8–14-μm interval, within which most infrared radiometers operate. Absorption effects can be accounted for with sophisticated computer programs if a measure of atmospheric humidity is known.

The challenge is to develop a method that accounts for both scattering and absorption effects from remotely sensed data only, so that "clean" data can be obtained directly from the sensor system. This may be possible by using certain combinations of wavelength intervals. At the reflected and thermal wavelengths, it is not possible to obtain data if clouds intervene. Cloud interference is one reason why a continuously operating sensor system is necessary. Data could then be obtained during clear periods whenever they occur.

The atmosphere has a negligible effect on passive and active (radar) microwave systems because of the relatively long wavelengths at which these systems operate. Weather independence is a major reason for their development. Most agriculturally oriented research with microwaves has been to evaluate soil moisture (12). Depending on the wavelength, vegetation may or may not affect the resulting signal. Some current research is directed toward vegetation measurement with radar systems. Their potential usefulness in plant health monitoring warrants the continuance of these efforts.

The Satellite

A satellite system for plant health monitoring must provide data at least hourly if not continuously. The data must be of high resolution and available in near real time ($<$ 30 min). Current satellites cannot meet these requirements. A challenge to our engineering colleagues is to develop a satellite system specifically designed for plant health monitoring.

Youngblood et al[1] proposed that a lightweight, propeller driven, electric airplane, with motive power from surface-mounted solar cells or microwave receivers, could be launched as a high-altitude, powered platform. This "forever airplane" or "poor man's satellite" would require only a few horsepower to run and would use aerodynamic principles learned from the Gossamer Albatross, the airplane that was pedaled across the English Channel by a man (10). The craft would have a wingspan of 60–70 m and weigh as little as 550 kg, including 220 kg of equipment. A small electric motor would slowly turn a 4–6 m propeller. Placed at an altitude of 30–35 km, the craft would not be influenced by weather and would encounter a relatively small amount of atmospheric drag. It could be programmed to be geostationary or to repetitively patrol a specific area.

The sensor system would receive energy in the visible, near infrared, thermal infrared, and microwave regions. Exact specifications await results from the many ground-based experiments that will be necessary to relate remotely sensed data to agronomic, pathological, and entomological factors.

THE PAYOFF

Once developed, the cost of a high-altitude, powered platform should be small compared to that of an orbiting satellite. One could be located over each major agricultural area to provide continuous coverage for that area. Data would be continuously telemetered to a ground station for storage. The satellite could also be interrogated with an advanced home computer system (home computers capable of processing satellite data will soon be marketed [13]). Farmers and farm consultants could readily process data and make management decisions based on information contained in images of fields shown on a televisionlike monitor or on a hard copy printout. Algorithms relating the various soil, plant, disease, and insect factors to remotely sensed parameters would be part of the computer software. At the press of a key, an image highlighting disease in a particular field would appear. The fraction of the field affected would be automatically calculated and the category of disease identified. Pressing another key would examine the area for water stress, etc. In addition to plant health surveys, the computer could do economic analyses. For example, if the survey of a field indicated that nitrogen was limiting plant growth, a yield projection could be made from present and past data to see if adding the nitrogen would improve yields by an economic amount.

Achieving this payoff will require persistence and ingenuity. In the words of John Gardner, "We face a series of great opportunities brilliantly disguised as insoluble problems."

[1]J. W. Youngblood, W. L. Darnell, R. W. Johnson, and R. C. Harriss. 1979. Airborne spacecraft—A remotely powered, high-altitude RPV for environmental applications. Presented at the Electronics and Aerospace Systems Conference, Arlington, VA. p. 6.

LITERATURE CITED

1. Al-Abbas, A. H., Barr, R., Hall, J. D., Crane, F. L., and Baumgardner, M. F. 1974. Spectra of normal and nutrient-deficient maize leaves. Agron. J. 66:16-20.
2. Allen, W. A., and Richardson, A. J. 1968. Interaction of light with a plant canopy. J. Opt. Soc. Am. 58:1023-1031.
3. Idso, S. B., Jackson, R. D., Pinter, P. J., Jr., Reginato, R. J., and Hatfield, J. L. 1981. Normalizing the stress degree day for environmental variability. Agric. Meteorol. 24:45-55.

4. Jackson, R. D. 1982. Canopy temperature and crop water stress. Pages 43-85 in: Advances in Irrigation. D. Hillel, ed. Academic Press, New York.
5. Jackson, R. D., Idso, S. B., Reginato, R. J., and Pinter, P. J., Jr. 1981. Canopy temperature as a crop water stress indicator. Water Resour. Res. 17:1133-1138.
6. Jackson, R. D., Jones, C. A., Uehara, G., and Santo, L. T. 1981. Remote detection of nutrient and water deficiencies in sugarcane under variable cloudiness. Remote Sensing Environ. 11:327-331.
7. Jackson, R. D., Reginato, R. J., Pinter, P. J., Jr., and Idso, S. B. 1979. Plant canopy information extraction from composite scene reflectance of row crops. Appl. Opt. 18:3775-3782.
8. Knipling, E. B. 1970. Physical and physiological basis for the reflectance of visible and near-infrared radiation from vegetation. Remote Sensing Environ. 1:155-159.
9. Lingren, P. D., and Wolf, W. W. 1982. Nocturnal activity of the tobacco budworm and other insects. Pages 211-228 in: Biometeorology in Integrated Pest Management. J. L. Hatfield and I. J. Thomason, eds. Academic Press, New York.
10. Petit, C. 1980. Beyond Gossamer. Science 80 1:93-94.
11. Pinter, P. J., Jr., Stanghellini, M. E., Reginato, R. J., Idso, S. B., Jenkins, A. D., and Jackson, R. D. 1979. Remote detection of biological stresses in plants with infrared thermometry. Science 205:585-587.
12. Schmugge, T. J. 1978. Remote sensing of surface soil moisture. J. Appl. Meteorol. 17:1549-1557.
13. Sloan, D. 1981. 32-bit minis process data from remote sensing satellites to evaluate natural resources. Computer 14:120,121.
14. Walburg, G., Bauer, M. E., and Daughtry, C. S. T. 1981. Effects of nitrogen nutrition on the growth, yield and reflectance characteristics of corn canopies. Rep. SR-P1-04044, LARS, Purdue Univ. W. Lafayette, IN. p. 19.

CHAPTER 19

QUANTIFYING THE EFFECT OF THE PHYSICAL ENVIRONMENT

PAUL E. WAGGONER

We know that the environment affects disease. But how much? By how many percent is disease in the field increased per degree of warmth, millimeter of rain, or kilometer of wind?

In the preceding chapters, others have demonstrated, factor by factor, the effect each can have on spores, bacteria or lesions. Now it is time to ask how we can measure the effect of environmental factors when together they explode or extinguish real epidemics.

Two strategies are available. The direct strategy is measuring disease and environment as Nature rolls the dice in a field and then deciphering the quantitative effect of each factor. The indirect strategy is measuring how the components of disease, say, germination of a spore or its takeoff toward a new host, are affected by each factor in a pot or petri dish and then assembling them to arrive at the quantitative effect of the factors in the natural environment. I shall start with the direct and conclude with the indirect.

THE FIELD

The Rubber Ruler

Knowledge of an object is generally thought to be uncertain until the object can be measured according to a numbered scale. It is understood that the scale, the ruler, should be rigid and the numbers equidistant, neither compressed at the ends nor stretched in the middle. Certainly we cannot expect to decipher the effect of environment upon disease if we measure disease with a rubber ruler.

A first step toward a rigid ruler is simply saying that the environment will be reflected, not in the addition of pathogens n but rather in their multiplication rate r′.

$$dn/dt = r'n \qquad (1)$$

We then turn practical and say that, although the number of lesions may be equal to the number of effective propagules that caused them, counting the number N of infected hosts is the practical thing to do. N and n will be about the same when they are small numbers, but when n becomes large, N will lag because of multiple infections. Thus we go on to say that the propagules will land at random, and the increase dN/dn in infested hosts per propagule will be

$$dN/dn = (K - N)/K , \tag{2}$$

where K is the total of healthy and diseased hosts.

Since (dN/dn) (dn/dt) = dN/dt, we can combine equations 1 and 2 to make

$$dN/dt = r'n\,(K - N)/K . \tag{3}$$

If we say further that the number N of infected plants is proportional to the number of lesions, then

$$r'n = rN \tag{4}$$

and hence

$$dN/dt = rN\,(K - N)/K . \tag{5}$$

This conception produced rulers for the measurement of disease and its relation to the environment. Equation 2 produces the multiple infection transformation (10) to transform the easily observed fraction $N/K = x$ of plants diseased into the number n of propagules and subsequent lesions controlled by environment and scattered at random to produce x. Further equation 5 produces Verhulsts's 19th-century logistic curve (12,23), which transforms the logits or $\log_e [x/(1 - x)]$ at two times into a multiplication rate r controlled by environment and unaffected by the supply K−N of hosts yet to be attacked.

For decades all seemed well with these rulers. For example, to decipher the effects of amount of rain, duration of wetness, and quantity of inoculum upon mildew produced by an exposure of healthy hops, Royle (21) measured disease by the multiple infection transformation. And Eversmeyer and Burleigh (9) transformed wheat leaf rust data into logits before relating it to environment.

As happens in either royal or scientific succession, these rulers have recently been challenged. In the beginning, Gregory (10) discovered that the propagules and thus infection would not always be randomly distributed; contagious disease is, after all, contagious—not random. Now we see that many diseases are better fit by a "contagious" distribution than by the random distribution formerly employed. Essentially, in these many diseases, more hosts are healthy and more have very many lesions than in a random distribution. This "over dispersion" with the variance greater than the mean can be caused by mixtures of random distributions, variations in either environment or susceptibility in the population, increased chance of a second lesion when a host has one, and a secondary multiplication and pruning of diseased tissue.

A standard "contagious" distribution is the negative binomial. The degree of over-dispersion in the negative binomial is indicated by the parameter k, which varies from infinity when the distribution is random to a small, positive value when over-dispersion is great.

A new multiple infection transformation is required for over-dispersion,

$$n = Kk\,[(1 - x)^{-1/k} - 1]\,. \tag{6}$$

This approaches the former transformation

$$n = -K \log_e (1 - x) \tag{7}$$

when k is large.

A modification of equation 5 for the increase dN/dt in "contagious" disease can also be made to provide rates for quantification of the effect of the environment. When propagules are produced in proportion to the number of infections or lesions—not to the number of infected hosts as in equation 4, equation 5 is replaced by

$$dN/dt = r\,Kk\,[(1 - x)^{-1/k} - 1](1 - x)^{(1 + 1/k)}\,. \tag{8}$$

Although the new equation seems forbidding, a simple outcome is discovered for k = 1, which is a reasonable value for several diseases. When k = 1, equation 8 simplifies to equation 5, the graph of logits vs. time is linear, and the rate r can be estimated from the linear change in logits.

In general terms, however, we can say that contagion or over-dispersion will cause a disease to increase more slowly than if its propagules were randomly distributed because there will be more wastage on multiple infections. Evidence of over-dispersion collected by several workers is cited by Waggoner and Rich (27), who provided more evidence and presented, in more detail than above, the consequences of the negative binomial distribution.

Having brought up the subject of equations that describe the increase in disease, I must mention some further alternatives. Jowett et al (13) derived the standard ones. One of these, the Gompertz, has its maximum increase dN/dt in severity at a severity of K/e instead of K/2 as in the logistic; and Berger (4) showed that this skewing of the rates fits the seasonal increase of nine diseases. Campbell et al (7) fit a cumulative frequency distribution with variable skewness to the seasonal increase of hypocotyl rot of snapbean. With such a variety of growth or empirical curves to fit the seasonal course, we must remember our purpose and make a choice.

Our goal was a ruler or scale whose numbers were unaffected by the multiplicative and contagious nature of the disease and by the changing number of available hosts, and, hence, could be related to the environment in a simple way. This being the goal here, equations 7 or 6 are logical for transforming severity to infections, and equations 5 or 8 are logical for transforming successive severities into a rate of change.

Regression

Having chosen scales or rulers and having measured disease and environment, one will then look to regression as the standard means of estimating quantitatively the relation between a physical factor of the environment and consequent disease. Our discussion will be more illuminating if we deal with a specific example, and I have chosen Royle's (22) observations of hop downy mildew. He observed the duration of wet leaves (RWD, hours), amount of rain (RA, millimeters), and number of spores in the air (AS, per cubic meter) during the 2 days that plants were exposed in a hop garden. He observed disease severity after the plants were incubated.

Since we are dealing with the success of a 2-day inoculation and not with an increase in disease severity, it is sensible to estimate the n infections per 100 plants by the standard multiple infection transformation of Royle's observations of severity and attempt to relate n—not the logarithm of n nor the logit of severity—to spores in the air and to moisture. The estimate from Royle's 27 observations in 1969 is

$$n = -3.29 + 1.05\ \mathrm{RWD} + 0.41\ \mathrm{RA} - 0.017\ \mathrm{AS}\ . \tag{9}$$

The first question to be asked about a regression equation is "Are the coefficients reasonable?" In controlled experiments, 2 hr of leaf wetness were required for any mildew, and 4 hr for much infection (21). Further, if n is related to RWD *alone* in the 1969 observations, the increase is 1–2 lesions per hour of moisture. Thus in equation 9, 1.05 lesions per hour of wetness is reasonable.

The increase of 0.41 lesions per millimeter of rain cannot be related to any controlled experiments. If n is related to RA *alone* in the 1969 observations, the increase is 4 or 5 lesions per millimeter of rain, much greater than the 0.41 in the regression equation 9.

Finally, we arrive at the end of equation 9 and the surprising estimate that additional spores AS actually decrease the lesions n! Although the graph of n versus AS *alone* shows no great correlation, it does show an increase of about one lesion per eight spores AS.

Why is the quantitative effect of rainfall RA so little and the effect of spores AS actually negative in the multiple linear regression 9?

The least-squares regression coefficients are adjusted for other variables in the regression. This is most easily seen in an equation for two independent factors. Consider

$$n = -2.41 + 1.17\ \mathrm{RWD} - 0.034\ \mathrm{AS}\ , \tag{10}$$

estimated from the same 27 observations of 1969. The −0.034 lesions per spores AS is affected by the slopes b_{nS} of lesions on spores, b_{nD} of lesions on duration, and b_{DS} of duration on spores. It is also affected by the correlation r_{SD} between spores and duration. The coefficient for AS in the two-factor equation can be calculated from:

$$[b_{nS} - b_{nD} \cdot b_{DS}]/[1 - r_{SD}^2]\ . \tag{11}$$

For the specific example, the effect of AS is

$$\begin{aligned}&[0.14\ \mathrm{n/spore} - (1.14\mathrm{n/hour})\ (0.15\ \mathrm{spore/hour})]/[1 - 0.38^2]\\ &\quad = -0.034\ \mathrm{n/spore}\ .\end{aligned} \tag{12}$$

When so many influences affect the coefficient in the multiple regression, it is small wonder that the coefficient or quantitative effect is sometimes surprising.

Hardened by this exercise, the reader may not be surprised to learn that the estimated effect of AS, and of RA too, change if we estimate the effects of just these two factors from the same 27 observations of hop downy mildew by Royle:

$$n = 0.67 + 1.96\ RA + 0.14\ AS\ . \tag{13}$$

The effect of AS now seems positive, and also its absolute magnitude is 5–10 times greater than in the preceding equations for n. Further, the estimated effect of RA has increased from 0.41 to 1.96 lesions per millimeter. The effects of RA and AS in this equation are close to their effects estimated by relating each alone to n; this is so because RA and AS were independent of each other in the sample.

Having seen the quantitative effect of the environmental factors vary according to which factors are considered, the reader may reasonably ask whether they vary between occasions. Fortunately, Royle repeated his observations in 1970, and in that year of fewer mildew lesions

$$n = -0.30 + 0.28\ RWD - 0.42\ RA + 0.023\ AS\ . \tag{14}$$

After seeing the effect of RWD decrease to a quarter and the effect of RA and AS actually reverse direction between the three-factor regressions for 1969 and 1970, equations 9 and 14, the reader has a right to be disillusioned.

All is not lost, of course, but before leaving regression for alternative means of quantifying the effect of the environment, two more questions should be asked. Are equations 9 and 14 plausible and usable? In a general way, it seems plausible to relate downy mildew to water and inoculum. Some experts might suggest, however, that temperature should be added.

Usability has to be appraised in terms of purpose. Certainly equations 9 and 14, one for 1969 and one for 1970, summarize the observations, including the correlations among the so-called independent factors RWD, RA, and AS. It is even conceivable that one of the equations could be used for prediction, as long as the weather and spore concentration—and the correlations among them—resembled 1969 or 1970.

Using the equations for control, however, does not seem possible. If, for example, the leaves were dried quickly without changing rainfall or spore concentrations, the slopes b_{DS} and b_{DA} as well as the correlations would be changed, and the coefficients in the equations as well as the quantity RWD would be changed. Further, the coefficient or effect of RWD may be 1.05 in 1969 and 0.28 in 1970 because it is correlated with a factor (say, leaf temperature) that is not in the equations, and changing RWD without changing the unspecified and correlated factor might change the coefficients and the predicted control.

A corollary complicates the estimation of effects from synthetic weather data. That is, the synthetic data must include a consideration of correlations between, say, rain and cold as well as the separate frequency distributions of precipitation and temperature. Put another way, the quantitative effect of rain on a disease in a future, changed climate will depend upon any changed relation between rain and temperature as well as upon the altered rain. So much for the tribulations of quantifying the effect of the environment by regressing disease on observations of the environment.

Law of the Minimum

After suffering the tribulations of regression, an experimenter may decide to cross-examine Nature rather than observe her. That is, one may decide to experiment rather than observe. Let us examine experimentation here, not for its own sake, but rather for what we can learn of our present subject of relating disease to uncontrolled environment outdoors.

Likely the experimenter would test the effect of, say, moisture by varying moisture while controlling leaf age, inoculum, temperature and light at constant and optimum values. That is, we would consider all but moisture as interference and strive to remove any limitations by the interfering factors to assure that moisture was the only limiting factor. This brings to mind the Law of Limiting Factors as Blackman (5) called it in 1905, or the Law of the Minimum as Liebig called it in the first half of the 19th century.

The Law states that when several factors affecting an outcome are present in abundance and one factor is deficient, increasing the deficient factor will likely change the outcome greatly whereas increasing the abundant ones will change the outcome little. The Law certainly does not apply perfectly (17), and in plant pathology, specifically, Rotem (18) has shown how an abundance of one factor, say inoculum, can "compensate" for an unfavorable temperature in the sense of causing a given severity of disease. Nevertheless, experimenting with one factor when others are controlled at their optimum does reveal most clearly the effect of the first, and we can ask whether in analyzing observations of uncontrolled environments there is a way of discerning and quantifying the effect of a factor when it is certainly limiting and the other factors are near their optima.

For the example of hop mildew, the Law of the Minimum can be written simply

$$n = \mathrm{AMIN}\,(b_D RWD,\ b_A RA,\ b_S AS)\ ,$$

which states that n is determined by AMIN, the least or minimum, of three simple proportionalities. To estimate the effects b_D, b_A, and b_S of duration of wetness, the amount of rain, and of spores in the air, one might perform three experiments, with wetness limiting in the first, amount of rain limiting in the second, and spores limiting in the third. Alternatively, the observations already made by Royle could be separated into three "experiments" according to which factor was limiting. In the first, duration RWD would be brief and limiting while millimeters of rain RA was large and spores AS were abundant; the effect b_D of RWD would be estimated from these observations alone. The b_A and b_S would be estimated from other observations where amount of rain or spores were limiting. A means has been proposed for separating the observations into those limited by each factor and of estimating the effects b (26).

When the regression, equation 14, and Law, equation 15, were fit to the data, approximately the same portion of the variability was explained by each (26). Thus, choosing between regression and Law was left to other considerations, such as the consistency and magnitude of the estimated effects of environment on lesions.

The effects of lesions per hour of wetness, lesions per millimeter of rain, and lesions per spores per cubic millimeter of air estimated for the two equations are given in Table 1. The estimates for the Law are larger and more consistent than the estimates for regression. Because of the similarity of estimating the parameters of the Law to experimenting with limitations of interfering factors minimized, the effects in the Law are nearer those observed in controlled experiments than are those estimated by regression. Having demonstrated the necessity of examining the estimated effects, I turn to another criterion of successful quantification.

The Extraordinary

Although arithmetic says that 10 times 10 equals 100, both the continuity of the host cultivar and the welfare of humans will be affected differently by 10 years of an ordinary 10% disease and one year of extraordinary 100%. Therefore, quantifying the way environment causes a few extraordinary outbreaks may be more important than minimizing the sum of squares of deviations from the disease of every year. I can report my failure to combine RWD, RA, and AS into a function that consistently discriminated the 24 cases with no disease from the 27 cases with some disease among Royle's 51 observations of 1969 and 1970. With somewhat more pride, I can report the outcome of predicting the 13 most severe cases of disease among the 51; in an improbable 11 of 13 cases the Law predicted disease more closely than did regression, but the mean squares of deviations from observations were not significantly different between regression and Law.

Likely we shall have to calculate less and think more to quantify the effect of environment on epidemics that rank with those examined by Klinkowski (14). New races and new cultivars figure in many great epidemics. Because the same physical environment that causes a 10-fold multiplication could increase disease from either 1/100 to 1/10% or from about 5 to 40%, the earliness of the season and the initial or primary inoculum also figure in great epidemics. In any event, pathologists have scarcely begun to think about quantitative explanations of extraordinary epidemics.

Physical Reasoning

Accustomed to mutable and complex organisms, pathologists are accustomed to considering the end product of their reasoning as "speculation," useful only in inspiring empirical tests. When dealing with the immutable and less complex physical processes of the environment, however, one may be able to deduce a quantitative effect rather than merely to speculate. Since it may be impractical to experiment with the environment of the whole outdoors, this exploitation of physical reasoning may be a godsend. A couple of examples will illustrate.

Long-distance dissemination on the wind is almost impossible to experiment with. Further, observing spores from aircraft is bound to be intermittent at best, and trapping them on the ground is plagued first by the rarity of spores from afar and then by clouds of other spores from nearby. It is fortunate, therefore, that

TABLE 1. Effects of three environmental factors on hop downy mildew[a]

	Change in number of lesions[b] attributable to					
	One hour of wetness		One millimeter of rain		Spores per mm^3 of air	
	1969	1970	1969	1970	1969	1970
Regression[c]	1.05	0.28	0.41	−0.42	−0.02	0.02
Law[d]	1.34	0.61	4.57	3.71	7.26	1.72

[a] From observations by Royle (22).
[b] Per 100 plants.
[c] Equation 14.
[d] Equation 15.

one can deduce whether a parcel of air, which visited the site just before an epidemic, had earlier visited a likely source of propagules. Calculation of the trajectory of the air can replace mere conjecture, on the one hand, and impractical experimentation and observation, on the other (11).

Nearby dissemination can also require physical reasoning rather than experiments or direct observation of spores. Aylor and Taylor (2) wished to determine the quantitative effect on the blue mold disease of ventilation in a tobacco shade tent. Would raising the wall of the tent slow the increase of blue mold by quicker drying of spores? Or would it accelerate the increase by blowing more spores from the margin to the center of the shaded field? Because natural occurrences of blue mold are rare in Connecticut and because inoculation is unconscionable among farmer's fields, the questions were answered by physical reasoning.

Drying of spores fallen on leaves was calculated by well-known theory from radiation, temperature, humidity and the ventilation with the wall up and down. The fate of spores airborne at the wall was calculated by another well-known theory; specifically, the number settling on soil, deposited on leaves and escaping through the roof of the tent were calculated from the deposition velocity of spores, the dimensions of the plants and the horizontal wind and vertical diffusivity with the wall up and down. As it turns out, raising the wall would scarcely change drying, while more spores would be deposited on the tobacco and blown through the roof to other fields. The important point is, however, that where experimentation and direct observation of the pathogen were impractical, the quantitative effect of the environment was reasoned from established physical theory.

Plots

Quantifiers of the effect of environment are justified in crying when they read "plots" and think of the ease with which other pathologists test fungicides or resistant cultivars. Nevertheless, we have opportunities for practical modification of the environment of a plot and the testing of its effect.

The physical environment of the roots can be modified locally. Irrigation and, with somewhat more difficulty, drainage or shelters from rain can be used to vary the water potential. Mulches can be used to accelerate or retard the seasonal or daily change in soil temperature as well as decrease the evaporation of water.

Despairing about plots for testing environments of the shoots is also unnecessary. As mentioned before, a shade tent changes evaporation and ventilation as well as radiation. Reflectors concentrate sunlight on a few plants. And Rotem et al (20) have shown how to use sprinkling to test the effect of leaf wetness.

THE POT AND THE DISH

Scientists, as contrasted to practitioners, like to say their goal is "know-why" rather than "know-how." By knowing why, they often mean explaining the operation of a complex organizational level, like a crop, by investigating the operation of less complex levels like single plants, single leaves, or single cells. In physics this might mean explaining the observed changes in pressure in a piston by investigating molecular movement. In pathology it might mean explaining epidemics by investigating molecular biology. Can we make an indirect attack on quantifying the effect of the environment? Can we compile laws about simple processes and then synthesize the quantitative effects of environment upon the complex process of a developing epidemic?

It isn't easy. Passioura (16) used dots on paper to make the point. Through a magnifying glass, one can measure the dots, a few at a time, compiling their sizes and even the correlation between sizes of adjacent dots. It is only, however, when we lay down the glass and leave the simple level of single dots to study the more complex organizational level of the entire array that the halftone picture of a man is seen. Needless to say, the picture would be doubly difficult to see in a view of the carbon atoms in the paper. Thus a nice warehouse of know-how about a simple level alone is not quite sufficient to quantify a complex level.

Unfortunately even our warehouse of knowledge of simple levels isn't nicely stocked for our task of reasoning upward from level to level. Passioura found in plant physiology that published research was concentrated on the top level of whole communities of plants and on the bottom level of molecules with little in the middle and connecting levels of organs, tissues, and cells.

But haven't we controlled epidemics with compounds discovered in experiments using microscope slides? Despite the obstacles that I have just described, scientists who have sabotaged pathogens by designing fungicides from the biochemistry of fungal metabolism may be sanguine about leapfrogging from test tube to field. Stopping a factory by throwing a wrench or an old wooden shoe into a particular gearbox is easier, however, than quantifying, from experiments with parts, how temperature changes the supply of products waiting to be shipped from the dock at the end of the general factory. Remembering that our goal concerns the running of the general epidemic rather than the sabotaging of a particular cell, we are carried back to Francis Bacon's *Novum Organum*:

> Then only, may we hope well of the sciences, when in a just scale of ascent and by successive steps not interrupted or broken, we rise from particulars to lesser axioms; and then to middle axioms, one above the other; and last of all to the most general.

The point is: estimating the quantities at the epidemic level is difficult enough from the adjacent level of a plant in a pot or a spore in a petri dish. Hence this section is entitled "The pot and the dish" rather than "The molecule and test tube."

Throughout the 75 years of our Society we have measured in pots and dishes the effect of environment on plants and spores. In the first volume of our journal one sees the title "Relation of temperature to spore germination and infection with *Cystopus*" (15).

If we are asked the three goals of the pot and the dish, we must answer, "Rates, rates and rates." A calculation of the outcome in the field will be an integration for weeks or seasons of the rates we estimated in controlled experiments. The question is "How fast?", and Aust et al (1), for example, prepared for integration by observing how fast a mildew fungus completed its incubation, not how many infections were seen after a couple of weeks.

To put some specific meat on these general bones, consider potato late blight. The quantity of blight should be limited and explainable in terms of the environment of the potato field and the survival of sporangia of *Phytophthora infestans* in controlled temperatures and humidities. When Crosier (8) exposed sporangia, presumably on glass, to 50% humidity at 20 C, they were all killed in 4 hr. His observations of spores on glass slides were born out by experience in several countries—until arid Israel had blight epidemics.

It was not enough, of course, to pile up observations of Israeli epidemics and even to correlate them with the weather. Scientists again sought to quantify the relation of the epidemic level by experiments at the explanatory level of controlled experiments. Rotem and Cohen (19) found 24 hr at 30 C and 50% humidity were required to kill the *attached* sporangia *on leaves*. Further, after mechanical traps showed spore flight had dwindled in the late afternoon, a few spores were deposited on exposed plants during the evening and night when conditions suited infection (3).

Thus Crosier's experiments before the fact failed to quantify the effect of environment on a limiting stage, the survival of propagules, but after the fact, observations of spores in Israeli fields allowed rationalization of the failure.

It was possible, of course, that hit-or-miss attempts to quantify the effect of epidemics from experiments on one stage of the pathogen were bound to fail, whereas a complete exposition of the pathogenic system would succeed. Hence, beginning 14 years ago, the entire life cycles of pathogens were converted into flow charts for computer programs that Bourke (6) called "full-blooded models." It seemed then, and still seems, that if we got the quantities or rates for each stage right, the whole assembly should calculate the quantitative effect of a factor on an epidemic.

Why have those dreams not yet come true? First, there has been more programming than testing. Then, some tests have been unsuccessful. The first simulator was of potato late blight (24), and it failed to explain the Israeli epidemics (25). The failure, however, was likely because of faulty rates, which can be corrected in light of new observations (3).

The greatest obstacles to simulating the whole epidemic from the components, however, will not be so easily removed. One obstacle is designing and interpreting controlled experiments for the subsequent calculation of rates in unsteady environments. The other is how to calculate the chancy movement of propagules to a new host because our goal is calculating spread across a field (28), not multiplication in the stirred contents of a basin. These are the critical matters for synthesizing the effect of environment upon an epidemic by controlled experiments upon its components.

MORAL

If a fable is well told, the moral can be inferred. Taking no chances, however, I write the morals, exemplifying them with water.

Quantifying the effect of an hour of wetness upon consequent epidemics requires our devising a logical and proven ruler or scale of disease that is not stretched by the amount of inoculum or the level of disease. It also requires a means of distilling from our observations a change in disease per hour of wetness that is stable from place to place, time to time, and even climate to climate.

We need put our minds to the effect of an hour of wetness on the forest fires of great epidemics as well as on the camp fires of ordinary epidemics.

When experiments are impractical, deductions from physical theories are godsends not to be confused with speculations.

Reasoning the effect of an hour of moisture in the field is impractical from the simple and distant level of, say, chemical reactions, but it is conceivable from the adjacent level of whole plants, organs, and spores. Even then, however, we are faced with difficulties of integrating rates in unsteady weather and calculating chancy trips of pathogens to new hosts.

LITERATURE CITED

1. Aust, H.-J., Hau, B., and Mogk, M. 1978. Effect of temperature and spore load on the incubation period of barley powdery mildew. Z. Pflanzenkr. 85:581-585.
2. Aylor, D. E., and Taylor, G. S. 1982. Aerial dispersal and drying of *Peronospora tabacina* conidia in tobacco shade tents. Proc. Natl. Acad. Sci. U.S.A. 79:697-700.
3. Bashi, E., Ben-Yoseph, Y., and Rotem, J. 1982. Inoculum potential of *Phytophthora infestans* and development of potato late blight epidemics. Phytopathology 72:1043-1047.
4. Berger, R. D. 1981. Comparison of the Gompertz and logistic equations to describe plant disease progress. Phytopathology 71:716-719.
5. Blackman, F. F. 1905. Optima and limiting factors. Ann. Bot. 19:281-295.
6. Bourke, P. M. A. 1970. Use of weather information in the prediction of plant epiphytotics. Ann. Rev. Phytopathol. 8:345-370.
7. Campbell, C. L., Pennypacker, S. P., and Madden, L. V. 1980. Progression dynamics of hypocotyl rot of snapbean. Phytopathology 70:487-494.
8. Crosier, W. 1934. Studies in the Biology of *Phytophthora infestans*. N.Y. (Cornell) Agric. Exp. Stn. Mem. 155. 40 pp.
9. Eversmeyer, M. G., and Burleigh, J. R. 1970. A method of predicting epidemic development of wheat leaf rust. Phytopathology 60:805-811.
10. Gregory, P. H. 1948. The multiple infection transformation. Ann. Appl. Biol. 35:412-417.
11. Hirst, J. M., Štedman, O. J., and Hogg, W. H. 1967. Long-distance spore transport: Methods of measurement, vertical spore profiles and the detection of immigrant spores. J. Gen. Microbiol. 48:329-355.
12. Hutchinson, G. E. 1978. Introduction to Population Ecology. Yale Univ. Press, New Haven. 260 pp.
13. Jowett, D., Browning, J. A., and Haning, B. C. 1974. Non-linear disease progress curves. Pages 115–136 in: Epidemics of Plant Diseases. Mathematical Analysis and Modeling. J. Kranz, ed. Springer-Verlag, Berlin.
14. Klinkowski, M. 1970. Catastrophic plant diseases. Ann. Rev. Phytopathol. 8:37-60.
15. Melhus, J. E. 1911. Relation of temperature to spore germination and infection with *Cystopus*. (Abstr.) Phytopathology 1:69.
16. Passioura, J. B. 1979. Accountability, philosophy and plant pathology. Search 10:347-350.
17. Rabinowitch, E. I. 1951. Photosynthesis, Vol. II, part 1. Interscience Publ., Inc., New York. 1,208 pp.
18. Rotem, J. 1978. Climatic and weather influences on epidemics. Pages 317–338 in: Plant Disease, An Advanced Treatise. J. G. Horsfall and E. B. Cowling, eds. Academic Press, Inc., New York. 436 pp.
19. Rotem, J., and Cohen, Y. 1974. Epidemiological patterns of *Phytophthora infestans* under semi-arid conditions. Phytopathology 64:711-714.
20. Rotem, J., Palti, J., and Lomas, J. 1970. Effects of sprinkler irrigation at various times of the day on development of potato late blight. Phytopathology 60:839-843.
21. Royle, D. J. 1970. Infection periods in relation to the natural development of hop downy mildew (*Pseudoperonospora humuli*). Ann. Appl. Biol. 66:281-291.
22. Royle, D. J. 1973. Quantitative relationships between infection by the hop downy mildew pathogen, *Pseudoperonospora humuli*, and weather and inoculum factors. Ann. Appl. Biol. 73:19-30.
23. Vanderplank, J. E. 1963. Plant Diseases: Epidemics and Control. Academic Press, New York. 349 pp.
24. Waggoner, P. E. 1968. Weather and the rise and fall of fungi. Pages 45–66 in: Biometeorology. W. P. Lowry, ed. Oregon State University Press, Corvallis. 171 pp.
25. Waggoner, P. E. 1976. Predictive modeling in disease management. Pages 176–186 in: Modeling for Pest Management. R. L. Tummala, D. L. Haynes, and B. A. Croft, eds. Michigan State Univ., East Lansing. 247 pp.
26. Waggoner, P. E., Norvell, W. A., and Royle, D. J. 1980. The Law of the Minimum and the relation between pathogen, weather, and disease. Phytopathology 70:59-64.
27. Waggoner, P. E., and Rich, S. 1981. Lesion distribution, multiple infection, and the logistic increase of plant disease. Proc. Natl. Acad. Sci. U.S.A. 78:3292-3295.
28. Zadoks, J. C., and Kampmeijer, P. 1977. The role of crop populations and their deployment, illustrated by a means of a simulator, EPIMUL76. Pages 164–190 in: Genetic Basis of Epidemics in Agriculture. P. R. Day, ed. N.Y. Acad. Sci., New York. 400 pp.

CHAPTER 20

PREDICTING THE ENVIRONMENT

ROBERT C. SEEM and JOSEPH M. RUSSO

The reader who has followed the logical sequence of the preceeding chapters in this section should have little doubt that environmental conditions have a primary influence on plant health and disease. It is not our purpose to reinforce this premise. In fact, it would be helpful if we could set aside our predominant concern for environmental effects on plants and their pests, and focus closely on the environment and the attempts to predict it. It is important to remember that our imperfect attempts to predict pathogen behavior in the environment (30; Chapters 14 and 32 in this volume) are often due to our limited ability to forecast the environment. Whereas plant pathologists do not forecast the weather, it is imperative that they understand and appreciate the science of weather and, more importantly, the extent and limitations of weather forecasting. We will attempt to describe some of the techniques of weather monitoring and forecasting, particularly as they relate to plant health. We will also present some case studies in which weather forecasting and crop management have been linked. We will conclude with a discussion of the sort of weather forecasting and associated services that we might expect in the future and the impact of these services (or lack of them) on improving plant health. First, however, we must present some fundamental concepts relating to plant disease, weather, time, and space.

THE TIME AND SPACE CONTINUUM

Plant disease and the environment are dynamic processes in both time and space. Figure 1 represents this concept on a geometric time scale of 1 second to about 200 years, both past and future, and a geometric space scale from 1 mm to 10,000 km. Superimposed on these scales are our present technologies of environmental monitoring of past events and forecasting of future weather events. The figure is divided on the time scale only to show the difference between monitoring and forecasting. The two are actually inseparable because without monitoring, forecasting is impossible. We have also identified various plant disease events and forecasts that will provide some perspective for plant pathologists. There is a general linear association between the two scales. The larger or more encompassing an event, the more time it takes for that event to occur, and conversely, the more rapid an event, the smaller is its area of spatial influence. To illustrate this in plant pathological terms, spore transport over many hundreds of kilometers takes several days (Fig. 1A) compared to the process of spore release that takes a second or less to occur and has immediate influence on only the adjacent few millimeters (Fig. 1C). Although the figure is only a "best guess" of the current situation, it is apparent that agricultural, and especially plant pathological, endeavors are on smaller time and space scales than are the general thrusts of meteorological endeavors. Typically, plant disease forecasts are made on a relatively short time scale (several hours to several days) and on a relatively small spatial scale (single field or farm, up to several kilometers). This happens to be the area in Fig. 1 least covered by present environmental monitoring or forecast systems. Herein lies the challenge for both plant pathologists and meteorologists.

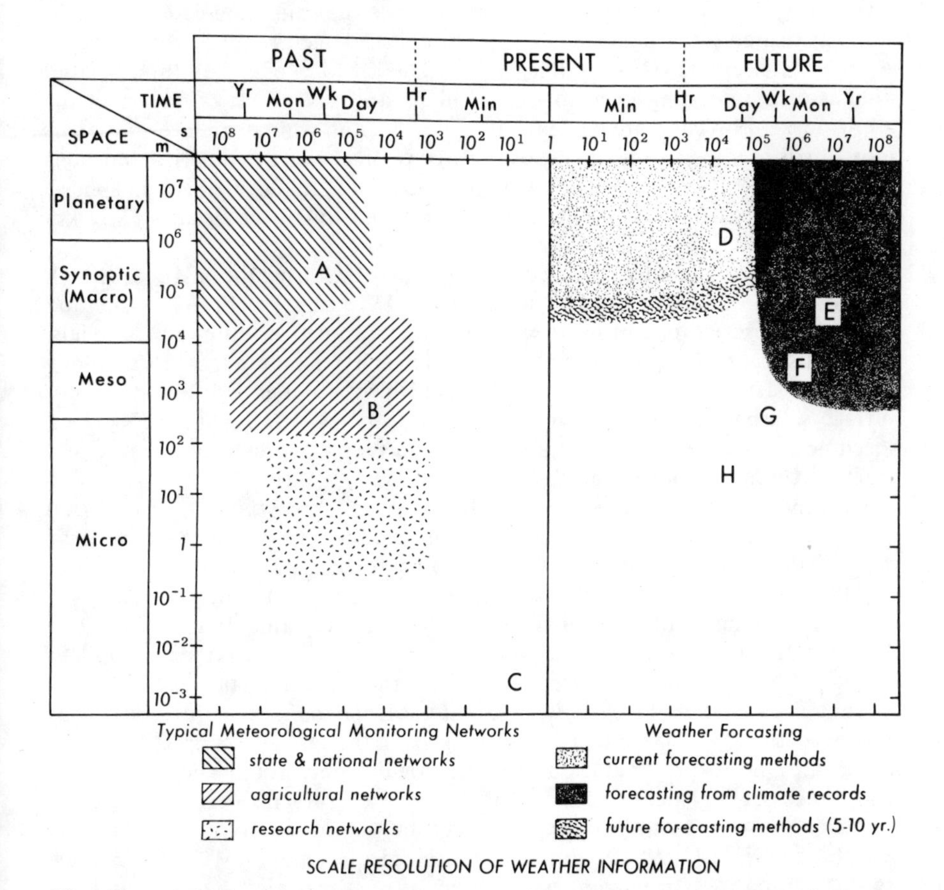

Fig. 1. The scales of dynamic change of environment and plant disease: past, present, and future. The technological limits of current meteorological monitoring networks and forecasting systems are presented in the shaded areas. Letters label important events concerning plant health: A, Spore dispersal in fronts and large storms; B, fungal infection processes; C, spore release mechanisms; D, synoptic potato late blight forecasts (3); E, Stewart's wilt forecast (4); F, stripe rust of wheat (19) and barley yellow dwarf (35) forecasts; G, potato late blight forecast (17); and H, turf Pythium blight forecast (23).

WEATHER PREDICTION

The synoptic, or weather map, scale is the predominant spatial scale used in weather forecasting (2). Vast amounts of meteorological data are plotted on a map, and analysis of the data reveal characteristic patterns that delineate the different weather systems. Measurements of movement and development of a given pattern at fixed time intervals coupled with the understanding of atmospheric dynamics result in a forecast of the extrapolated position and behavior of a weather system. Synoptic forecasting has been incorporated into plant disease predictions for relatively large-scale events (3,19,35).

Numerical weather prediction came into being when mathematical equations were used to describe the atmosphere's behavior. By the mid-1950s, operational computer models began providing quantitative estimates of large-scale flow patterns and their movement (31). These early, simple numerical models have given way, over the years, to primitive equation models. The most sophististicated form of a primitive equation model in current use is the Limited Area Fine Mesh (LFM) model (11,22). The LFM model has three spatial dimensions, consisting of a horizontal plane and seven vertical layers of the atmosphere. Forecasts are made for temperature and wind at all levels, whereas moisture is predicted only on the lower three vertical zones. Numerical output is presented on a polar stereographic map projection in the form of a hemispheric grid with actual forecast points 2.50 latitude and longitude or about 127 km apart.

The accuracy of numerical forecasts has improved during the past decade with the addition of model output statistics (MOS) (12,27). The MOS method correlates observations of local weather with output from numerical models. This statistical technique counteracts model bias by including local climatology in the forecast product. MOS statistics provide probabilities of different types of surface weather such as measurable precipitation, type of precipitation, wind speed and direction, maximum and minimum temperatures, dew point, cloud amount, ceiling visibility, and thunderstorms (16).

Staff meteorologists in regional weather service forecasting offices (WSFO) interpret numerical model predictions of weather events as part of their preparation of local forecasts. These tailored forecasts are disseminated in a timely manner to the public via dedicated telephone lines to teletype machines. The most common of these lines is the National Oceanic and Atmospheric Administration (NOAA) Weather Wire Service (20,21), which is the usual link to municipalities, industries, newspapers, radio, and television stations.

Numerous studies have shown that numerical guidance from models has improved the skill of weather forecasting over the last 20 years (5,28,31,33). There are, however, limitations to numerical forecasting. The two major drawbacks are logistical and theoretical. Logistic problems occur because the models require so many resources that can be provided only by the National Weather Service's (NWS) National Meteorological Center (NMC). The models must be initialized to current weather conditions provided by thousands of surface, atmospheric, and satellite observations. These data must be transmitted to the NMC and processed in a timely fashion so that the computer can run the models within an 8-hr operational deadline. The process is repeated every 12 hr and each run produces 12-, 24-, 36-, and 48-hr forecasts. The major component of the logistical problem is "computer power." The current model design cannot be improved with additional detail until the NMC acquires computers that can execute programs faster than their present machines.

The theoretical drawback of numerical forecasting remains the physical principles on which models are based. The fundamental principles of conservation of momentum, mass, and energy are mathematically formulated as three prognostic equations (the two horizontal components of the momentum equation and the thermodynamic energy equation) and three diagnostic equations (the continuity equation, the hydrostatic approximation, and the equation of state) (13). These equations represent a general, dynamic description of fluid motion in the atmosphere. They would become unmanageably complex if terms for local effects such as surface heating and friction were included in them.

LOCAL FORECASTS

The question often arises as to the utility of synoptic and numerical forecasting for local conditions, especially where weather information is needed at the farm level. Before we answer that question directly, let us consider some of the special problems of local or farm-level forecasts.

Accuracy and skill are the foremost measures of forecasting success. Meteorologists test their own as well as their model's ability to accurately and skillfully predict weather by various scoring rules. Forecasts are usually scored for a given period and area. For example, Charba and Klein (5) evaluated the probability of precipitation (POP) forecasts for the "tomorrow" or next-day period produced by 50 WSFOs. The evaluation consisted of a regression analysis of skill scores for POP forecasts. They found that skill scores had improved 7.4% over the last 10 years and attributed most of this improvement to the introduction of the LFM model and MOS forecasts in the early 1970s as well as to the experience gained by meteorologists using these guidance products.

The accuracy of regional forecasts will vary from region to region and season to season according to the large-scale weather systems experienced and the local environmental setting. Small-scale variations in weather variables due to local factors, such as topography, set a limit to the accuracy and skill of forecasts. Gedzelman (9) found that small-scale variations account for a large percentage of current forecasting errors and believes the greatest improvement left to be made on short-term weather forecasting accuracy will result from increased attention to, and understanding of, the micro- and mesoscale phenomena.

Currently available forecasts have not satisfied the desires and needs of farmers for weather information. Recent surveys indicate that growers demand greater accuracy and more localized forecasts (6,26). Plant pathologists experience the same demands when attempting to develop disease prediction systems. Are these requests too idealistic? What is realistically required to achieve these goals?

MONITORING NETWORKS AND RESOURCE CENTERS

Weather monitoring networks and resource information centers can significantly improve weather forecasts. Monitoring networks designed for agriculture can provide a detailed picture of current weather conditions within a particular region. Many plant disease forecasts must be initialized to current conditions before prediction, and a high-resolution weather monitoring system can provide accurate initial values. Equally important is the ability of a monitoring network to detect and quantify biases in regional forecasts when they are used at the local level. Once detected, the bias can be suitably corrected in local forecasts.

Resource centers that provide weather forecasts tailored for the agricultural community are best suited for localized forecasting. A discussion of specific examples appears in the next section; however, the most typical of these centers are the four Environmental Studies Service Centers (ESSCs) located in the midwestern and southern United States. Jointly supported by the Departments of Commerce and Agriculture, the centers provide detailed information usable by the agricultural industry. Local forecasts are provided by staff meteorologists, using both numerical and synoptic techniques. In other cases, states and universities have supplemented their pest management programs with meteorologists whose responsibilities include local weather forecasts suitable for implementing pest management strategies. An adequate monitoring network, access to regional forecasts, and a trained individual to synthesize and interpret the information can result in weather forecasts that are superior to those routinely provided by the NWS.

Even when local forecasts can be accurately and routinely made for the agricultural community, we are still faced with the complex problem of disseminating this timely information to the user. Weather information delivery is no less important than the actual preparation of a forecast, and it has received considerable written (6,26) and verbal[1] attention. No concensus has been achieved, although it appears that public radio and television are currently favored for general forecasts, whereas highly specialized information is disseminated most frequently by code-a-phone and interactive, computer-based information systems. Several different dissemination mechanisms will be discussed in the following section.

AGRICULTURAL WEATHER FORECASTING: SOME CASE STUDIES

We would now like to present some specific examples of integrating weather forecasts into crop production practices. This is by no means an inclusive list; however, it does represent different approaches to solving the same problem. Some of these systems have gained success and notoriety; others are still experimental.

Weather Forecasts for Agriculture in New York

The New York system for disseminating weather forecasts to the agricultural community was chosen as an example because it is simple and we have been associated with its recent development.

Weather forecasting for the agricultural industry in New York has been haphazard over the years. At best, county agents could persuade a meteorologist in one of the state's two WSFOs to participate in an early morning telephone conference call to discuss detailed weather conditions and such things as the

[1]Meeting on Methods of Improving the Dissemination of Weather Information to Agriculture and Forest Producers; Agric. Res. Inst.; September 28–29, 1978; University of Maryland. Workshop on Applications of Weather Data to Agriculture and Forest Production; Am. Meteorol. Soc.; March 26–27, 1981; Anaheim, CA.

likelihood and extent of apple scab infection periods and pesticide spray conditions. This information would be broadcast to growers on morning radio programs. At worst, growers could read the abbreviated forecast hidden somewhere in the middle of their morning newspaper.

The development of the interactive system for computer-aided management of pests (SCAMP) (29) made it possible to rapidly disseminate detailed weather information to county agents, who in turn could relay forecasts and advisories to growers. An agricultural meteorologist helped the SCAMP staff develop a program for providing NOAA weather wire forecasts to the agricultural community in New York State. After initially using manpower to enter NOAA forecasts into the computer at early morning hours, a black box was built that interfaced the weather wire directly to the computer. Further programming made it possible for the computer to scan all state, national, and international weather data coming in on the wire, select only the New York zone forecasts, and place these forecasts in their proper SCAMP file. In the process, we had to impress upon the NWS the need to consistently precede each zonal forecast with a computer-interpretable identification. They were not totally convinced and at times inexplicably changed some of the identifiers, which in turn made it necessary for SCAMP personnel to search for the lost file among such things as temperature reports for Melbourne and Djakarta. Nevertheless, the system does provide consistent forecasts in detail greater than is normally available to the county agent and grower.

The system has added utility, since forecast temperatures can be automatically transferred to models requiring temperature as an input. Three-to five-day temperature forecasts are characterized by phrases such as "below normal" or "much above normal." We translate these phrases to one standard deviation above or below the mean for "above" or "below" normal and two standard deviations above or below the mean for "much above" or "much below" normal. The standard deviations are calculated from climatological records of average weekly temperatures for reporting sites throughout the state.

With some additional programming, we will be able to retrieve from the weather wire state and national weather summaries as well as radar summaries of precipitation within the state. The system, once established, is inexpensive and does not require the continued special skills of an agricultural meteorologist. Of course, without an on-site meteorologist, we must accept less accurate and less localized forecasts.

NOWCAST

NOWCAST is a project designed to improve the dissemination of agricultural weather information (26). Developed by Colorado State University in conjunction with the National Aeronautic and Space Administration (NASA), the project provides detailed current and short-range weather information to the agricultural community via hourly broadcasts over educational television. Specially trained weather interpreters utilize satellite cloud imagery, radar, and upper air and surface weather data to provide information relevant to farmers. The initial project NOWCAST was done in Mississippi where the only pest-related information was summaries of conditions affecting pesticide spray applications.

Forest Weather Interpretations System

The Forest Weather Interpretations System is a cooperative project between the U.S. Forest Service, the NWS, the University of Georgia Office of Computing Activities, and other forest management agencies (24). The system was developed to deliver timely, interpreted site-specific weather information to forest managers; it has been tested in twelve southern states since the spring of 1978. Weather information includes observational and other numeric data from the Federal Aviations Administration 604 broadcast data line, which is automatically retransmitted to a computer at the University of Georgia, Athens. Personnel in participating WSFOs enter forecast data into the computer using a dial-up telephone line. The collected data are processed for error and reformatted before being made available to users in an interactive computer system.

The Forest Weather Interpretations System was designed to highlight weather events that are important to the operational management concerns of foresters. User-decision models as simple as matching weather conditions with their expected impact on forestry operations and as sophisticated as smoke dispersion and fire behavior are available on the system. Further development of models will require more research on the relationships between weather variables and biological processes that govern management decisions.

Satellite Frost Forecast System

The Satellite Frost Forecast System (SFFS) was developed by the Institute of Food and Agricultural Sciences of the University of Florida with support from NASA (18). It provides detailed forecasts of temperatures below freezing in Florida during critical periods of the cropping season.

The SFFS computer at the University of Florida, Gainesville, acquires temperature data sensed by a GOES (geostationary operational environmental satellite) from NOAA facilities in Suitland, MD, through an automated telephone link. The satellite data are partitioned into color bands and displayed as a thermal map on a color video terminal.

Remote automated data acquisition stations in Florida record air temperature, soil temperature, wind speed, and net radiation. The SFFS computer automatically interrogates the remote stations via telephone and uses the data as input into models. One model solves the surface energy budget equation and predicts temperature changes at each of the weather station sites. These predicted temperatures are converted to a thermal map with the aid of temperature patterns from past frosts and freezes. The predicted thermal maps can be used to program future satellite views and identify areas that may experience frost conditions.

Environmental Study Service Center

The ESSCs are part of the Agricultural Weather Service branch of the NWS. These centers combine agricultural meteorological resources of several states to provide more services efficiently. ESSCs are located in Auburn, AL; Stoneville, MS; College Station, TX; and West Lafayette, IN.

As an example, services provided to the state of Ohio by the West Lafayette (Purdue University) ESSC are listed below.

1. Agricultural forecasts. Content includes regular forecast products plus percent sunshine, rainfall amounts, humidity range, dry condition, dew intensity-duration, frost and freezing temperatures, soil temperatures, and livestock weather safety index.

2. Special agriculture weather bulletins. Bulletins are issued when emergent weather conditions such as floods, blizzards, freeze, heat wave, etc., are a serious threat to agriculture.

3. Agriculture advisories. Content includes an interpretation of weather effects on current crop situations, field operations, livestock conditions, weather-disease effects, and weather-insect relationships.

4. Status reports of weather variables. Real-time status reports include soil moisture, soil temperatures, growing degree days, heat units for insect evaluation, crop maturity, drying for crops, precipitation-evaporation summaries, temperature averages-departures, disease criteria checks, and climatology.

5. Agriculture weather observation stations. Agricultural weather observation stations are established and some existing stations with microcomputer systems are automated.

6. Research programs. Studies of weather are related to crops, operations, and livestock. Data bases are used to establish climatic risk for agricultural decisions.

7. Weekly weather and crop summaries. Centers will assist in providing data summaries.

8. Data requests. Data are available from a network maintained by the ESSC for agricultural users.

ESSC personnel coordinate their services with WSFO, extension, and research specialists. Information releases are disseminated through NOAA, the NWS weather wire, NWS VHF radio, newsletters, publications, computer files, and existing extension networks and/or other news services.

Project Green Thumb

Project Green Thumb is a joint agricultural weather program of the NWS and the Cooperative Extension Service (34). The program is divided into four functions: 1) data gathering by observers in rural areas, 2) preparation of localized weather and agricultural information, 3) dissemination of this information, and 4) education of the farm community as to how best to use the information.

The "Green Thumb Box" is the means by which a user is linked to the Green Thumb system. It is an electronic device that is connected to a standard television and phone line and allows a user to communicate with the project's computer. Several "pages" of information received from the computer can be saved and perused after ending the telephone call. Information includes weather, agricultural advice, markets, world climate, local meetings, and recent extension publications.

A Plant Disease Forecast System

There exist many other state supported projects or services, such as the Agricultural Weather Program of North Carolina (14). We note, however, that virtually none of these programs provides specific information for plant disease forecasts. We have mentioned some of the synoptic approaches to plant disease forecasting, but considering the recent technological improvements in weather forecasting, plant pathologists have failed to keep pace. This blanket condemnation is probably not justified, especially when we look back at Fig. 1; however, plant pathologists must make efforts to widen their scales of purview just as meteorologists must narrow theirs. Is this going to be a monumental task? Reconsider Waggoner's exhortations in Chapter 19. Also, consider the following example of an innovative approach to disease forecasting. M. A. Sall and colleagues (*personal communication*) have associated the MOS numerical predictions of probability of precipitation (POP) to the likelihood of *Phomopsis* cane and leaf spot disease of grapevines in California. Long-term analysis of *Phomopsis* indicated the probability of infection occurring when rain fell, and the POPs provided the probability of rainfall. The combination of probabilities resulted in calculated expectancy of disease incidence. The method is elegantly simple and quite usable in the decision process of disease management.

Hopefully the future will bring similar direct responses to the challenge of merging weather and plant disease forecasting technologies.

THE FUTURE—THE GOOD AND THE BAD

Let us now consider the future prospects of weather forecasting as it relates to plant health. We have already mentioned limitations of numerical forecasting models that will restrain future development. New experimental models may provide some improvement of forecast resolution. One such model, the Nested Grid Model (25), has a grid size smaller than that of the LFM. The Nested Grid Model starts out on a hemispheric scale and then collapses to a second grid and then a third grid much like a microscope switching lenses to more power and providing a narrower but more detailed field of view. The weather events occurring within a smaller grid are calculated from boundary conditions provided by the larger grid. This model has had some success in forecasting localized severe storm activity. Whether this model will provide better local forecasts for agriculture depends on its success in predicting small-scale weather variations from synoptic-scale flow patterns.

One recent development in the NWS that will have a definite impact on weather forecasting is a program called Automated Field Operations and Services (AFOS). AFOS is a large-scale effort toward nationwide automation of field operations and services (15). Automation will reduce the time it takes weather products to be developed and delivered to users. Professional level field personnel will be relieved of time-consuming routine tasks and, therefore, will spend more time on their forecasting and warning responsibilities. The AFOS program is designed to aid forecasters in their daily activities through the use of minicomputers, video displays, and rapid communications. In essence, it will computerize much of the forecasting operations of 200 stations throughout the United States. AFOS has already been installed in many locations and is currently undergoing a verification period at those sites.

The MOS program is undergoing revision to allow incorporation into AFOS. (The resulting acronym is LAMP—Local AFOS MOS Program.) Forecasts will be derived from a combination of predictors that have been tailored to local conditions and climate. LAMP will be able to provide improved forecasts of dew, precipitation, soil temperature, and maximum and minimum air temperature out to 5 days.

One continuous problem area has been precipitation forecasts, especially during the summer when prediction accuracy worsens due to convective rain (i.e., thunder storms). Some of the numerical models are being enhanced to specifically improve convective storm forecasts by linking MOS data to digitized radar. The newer radar systems, especially those that are computer-based, provide a high degree of resolution for the limited areas they scan. It should be possible to use digitized radar as a means of determining precisely where precipitation events are occurring as well as where they are likely to occur within several hours.

Similar methods are being used with imagery data from satellites. Sickler and Thompson (32) have demonstrated that the infrared thermal images produced by the GOES-1 can be used to provide accurate estimates of rainfall areas and amounts. The satellite data can be processed rapidly enough to provide flash flood warnings. Therefore, it is reasonable to assume that areas of precipitation could be easily mapped and linked to disease models that require precipitation as an input.

Improved technology in satellite sensing should allow accurate assessment of ground-level measurements. The Agriculture and Resources Inventory Surveys through Aerospace Remote Sensing (AgRISTARS) program was initiated within the last 5 years to determine the feasibility of integrating aerospace remote sensing technology into existing or future government data acquisition systems (1). The program is a cooperative effort of the U.S. Department of Agriculture, NASA, and NOAA. AgRISTARS consists of a number of projects that use remotely sensed temperatures, soil moisture, and snow depth as well as other measurements such as vegetation cover.

The detection of surface temperatures and moisture will be extremely important in the operation of plant disease forecast systems. Plant disease forecasting by remotely sensed data has the potential for truly worldwide utilization. Assuming that adequate information delivery systems are available to disseminate the results, generalized plant disease forecast models should operate equally well on satellite data from Chile as from California.

Back on the ground, weather monitoring networks will become more automated and hopefully more accurate and reliable. Until the incorporation of high-technology radar and satellites, local weather monitoring networks will become increasingly important.

The networks will not only provide greater resolution and allow for bias correction of forecasts, but they will also provide the necessary initialization of disease forecast models. We see a continuing role for the volunteer observer because of their low cost and dependability (assuming they are well trained and dedicated). Certainly automated instrumentation will augment and in some cases replace the observer; yet if high-resolution forecasts of weather and disease are to be achieved, a good observer network will likely be a major component of the system.

While there are bright prospects for predicting the environment, there are also some threatening clouds looming over us. Weather services in this country are provided by a single federal agency, the NOAA. Realistically, NOAA's basic efforts have been to improve large-scale weather forecasts and severe local storm warnings, to make this information available to the public, and to publish climatological data (6). Other agencies and organizations must take the responsibility to apply NOAA's weather forecasts and climatological data to specific needs. The only individuals regularly receiving special attention from NOAA are airplane pilots, because the safety and efficiency of every flight depends on weather. Notwithstanding the long-time support of agricultural weather programs, reductions in NOAA's funding will have an immediate and substantial effect on agricultural programs.

We are already seeing programs disappearing due to funding curtailment. NOWCAST projects, the SFFS, and even the ESSC program have recently been jeopardized because of spending cuts at the federal level. Another government venture, the National Climate Program (20,21), which is peripherally important to the topic at hand, has never been funded at its proposed level. Indeed, many of the regular services of the NWS will be phased out over the next several years (7).

The NWS itself has recently come under the scrutiny of the General Accounting Office (GAO) (10). Development of the AFOS system of nationwide automation has been hindered by poor project management and severe technical problems. The GAO claims that AFOS is at least 5 years behind schedule, has cost about 22 million dollars more than its original budget, and cannot meet its original requirements. It recommends a complete reevaluation of AFOS before its implementation nationwide. Any interruption or hampering of NWS forecasting and warning services due to the development of AFOS would be detrimental to agricultural users.

CONCLUSIONS

The message to agriculture is clear. Specific programs for applying weather information to agriculture are not of national priority and therefore will be relegated to states and universities or to the private sector. It has been calculated (8) that the annual cost of requiring and processing weather data by all civilian government agencies is about $2.29 per U.S. citizen. The cost of disseminating this information, which is done by nongovernment agencies, is $7.55 per citizen. From this we can project a substantial increase in the cost of procuring and disseminating weather information if not provided by public agencies. It is likely that low-demand information, such as that which would be useful for disease forecasts in most crops but not of importance to general agriculture, might be impossible or too costly to obtain. If weather information is available, the improved resolution of standard forecasts is not expected to be very great (Fig. 1). That leaves two options: 1) plant disease forecasts that utilized weather forecasts must be generalized to accommodate larger-scale weather predictions; or 2) disease forecasting, in general, will not be based on weather forecasts but must rely entirely on very localized monitoring of weather conditions. The first option has usually been avoided by plant pathologists, who cannot reconcile the loss of disease forecast precision with the benefits of using large-scale forecasts. The second option seems more realistic, especially in light of some of the recent developments in disease prediction using microprocessors (see Chapter 32).

In closing, let us hope that not all interest is lost in predicting plant disease from weather forecasts. We should accept our current problems as future challenges. Plant pathologists and agricultural meteorologists must work together if plant disease and environment prediction are to be successfully integrated.

A POSTSCRIPT

While discussing material in this chapter and perusing the literature, we came upon a noteworthy development in meteorology that can have a direct analogy in plant pathology. Meteorologists have developed, and in fact have been educated to utilize, computer models to augment their forecasting skills. These aids are neatly guised as "automated guidance of operational forecasts." Snellman (33) has observed that while the human-machine mix has allowed significant improvement of forecasts over recent years, it appears that some forecasters are now operating more as communicators and less as meteorologists. Forecasting skill scores leveled off and with some persons started to decline in recent years. In other words, some forecasters take their computer models at face value and do not attempt to augment the forecast by utilizing their skills as professional meteorologists. The analogy to plant pathology is that forecasts of plant health will increasingly rely on computer models, some of which may be exceedingly complex. At present, the analogy stops here, but we pose some questions: Will the new generation of plant pathologists that accompany these models be just communicators of machine prognoses? Will this "cancer," as Snellman calls it, erode the ability of plant pathologists to interpret events with biological, economic, and social meaning? We certainly hope not and, as both developers of models and educators in the profession, it is our duty to create the optimal human-machine mix. Let us hope we can learn from this example.

LITERATURE CITED

1. Anonymous. 1981. AgRISTARS Annual Report—Fiscal Year 1980. AP-JO-04111. Natl. Aeronaut. Space Adm., Greenbelt, MD. 62 pp.
2. Anthes, R. A., Ponosky, H. A., Cahir, J. J., and Rango, A. 1978. The Atmosphere. Charles E. Merrill Publishing Co., Columbus, OH. 442 pp.
3. Bourke, P. M. A. 1970. Use of weather information in the prediction of plant disease epiphytotics. Annu. Rev. Phytopathol. 8:345-370.
4. Castor, L. L., Ayers, J. E., McNab, A. A., and Krause, R. A. 1975. Computerized forecasting system for Stewart's bacterial disease of corn. Plant Dis. Rep. 59:533-536.
5. Charba, J. P., and Klein, W. H. 1980. Skill in precipitation forecasting in the National Weather Service. Bull. Am. Meteorol. Soc. 61:1546-1555.
6. Committee on Weather Information Systems. 1980. Weather-information systems for on-farm decision making. Board of Agriculture and Renewable Resources, Commission on Natural Resources, Natl. Res. Counc., Natl. Acad. Sci., Washington, DC. 98 pp.
7. Cressman, G. P. 1979. Proposed external user policies, National Weather Service weather and river information. Natl. Weather Dig. 4:7-10.
8. Federal Coordinator for Meteorological Services and Supporting Research. 1981. The federal plan for meteorological services and supporting research. FCMP1-1981. Natl. Oceanic Atmos. Adm., Rockville, MD. 135 pp.
9. Gedzelman, S. D. 1981. Limits on forecast accuracy due to small-scale weather variations. Bull. Am. Meteorol. Soc. 62:1570-1576.
10. General Accounting Office. 1981. Problems Plague National Weather Service ADP System. Report to the Congress of the United States. CED-82-6. U.S. GOA, Washington, DC. 122 pp.
11. Gerrity, J. P., Jr. 1977. The LFM Model, 1976: A documentation. Tech. Mem. NWS NMC 60. Natl. Oceanic Atmos. Adm., Rockville, MD. 68 pp.
12. Glahn, H. R., and Lowry, D. A. 1972. The use of model output statistics (MOS) in objective weather forecasting. J. Appl. Meteorol. 11:1203-1211.
13. Holton, J. R. 1972. An Introduction to Dynamic Meteorology. Academic Press, New York. 319 pp.
14. Johnson, G. L., and Perry, K. B. 1981. A New Agricultural Meteorology Program for the State of North Carolina. Pages 202–203 in: Conf. Agric. and For. Meteorol. 15th. Am. Meteor. Soc., Boston. 210 pp.

15. Klein, W. H. 1976. The AFOS program and future forecast applications. Mon. Weather Rev. 104:1494-1504.
16. Klein, W. H. 1982. Statistical weather forecasting. Bull. Am. Meteorol. Soc. 63:170-177.
17. Krause, R. A., Massie, L. B., and Hyre, R. A. 1975. BLITECAST: A computerized forecast of potato late blight. Plant Dis. Rep. 63:21-25.
18. Martsolf, J. D. 1981. Satellite frost forecast system. HortScience 16:586.
19. Nagarajan, S., Singh, H., Joshi, L. M., and Saari, E. E. 1976. Meteorological conditions associated with long-distance dissemination and deposition of *Puccinia graminis tritici* uredospores in India. Phytopathology 66:198-203.
20. National Climate Program Office. 1980. National climate program five-year plan. S/T 79-153. Natl. Oceanic Atmos. Adm., Rockville, MD. 102 pp.
21. National Climate Program Office. 1981. NOAA Climate Program Plan. S/T 81-48. Natl. Oceanic Atmos. Adm., Rockville, MD. 49 pp.
22. Newell, J. E., and Deaven, D. G. 1981. The LFM-II model, 1980. Tech. Mem. NWS NMC 66. Natl. Oceanic Atmos. Adm., Rockville, MD. 20 pp.
23. Nutter, F. W., Jr., and Cole, H., Jr. 1980. Preliminary results concerning the feasibility of developing a forecasting program for chemical control of Pythium blight. (Abstr.) Phytopathology 70:467.
24. Paul, J. T. 1981. A real-time weather system for forestry. Bull. Am. Meteorol. Soc. 62:1466-1472.
25. Phillips, N. A. 1979. The nested grid model. Tech. Rep. NWS22. Natl. Oceanic Atmos. Adm., Rockville, MD. 80 pp.
26. Priddy, K. T., and Marlatt, W. E. 1978. The potential benefit of improving the dissemination of agricultural weather information to the Mississippi cotton farmer. Final Rep., Grant No. NSG 5073. Earth Resources Dept., Colorado State University, Ft. Collins 80 pp.
27. Rieck, R. E., Sadowski, A. F., and Harrell, J. J. 1976. National Weather Service Forecasting Handbook No. 1. Facsimile Products. Natl. Weather Serv., Natl. Oceanic Atmos. Adm., Washington, DC. 294 pp.
28. Sanders, F. 1979. Trends in skill of daily forecasts of temperature and precipitation, 1966–78. Bull. Am. Meteorol. Soc. 60:763-769.
29. Sarette, M., Tette, J. P., and Barnard, J. 1981. SCAMP—A computer-based information delivery system for cooperative extension. N.Y. Food and Life Sciences Bull. 90. N.Y. State Agric. Exp. Stn., Geneva. 8 pp.
30. Shrum, R. D. 1978. Forecasting of epidemics. Pages 223–239 in: Plant Disease, An Advanced Treatise. Vol. 2. J. G. Horsfall and E. B. Cowling, eds. Academic Press, New York. 436 pp.
31. Shuman, F. G. 1978. Numerical weather prediction. Bull. Am. Meteorol. Soc. 59:5-17.
32. Sickler, G. L., and Thompson, A. H. 1979. Convective rainfall estimation from digital GOES-1 infrared data. Ref. Publ. 1034. Natl. Aeron. Space Adm., Greenbelt, MD. 90 pp.
33. Snellman, L. W. 1977. Operational forecasting using automated guidance. Bull. Am. Meteorol. Soc. 58:1036-1044.
34. Suchman, D. 1980. Project green thumb: A demonstration in Wisconsin. Bull. Am. Meteorol. Soc. 61:314-320.
35. Wallin, J. R., Peters, D., and Johnson, L. C. 1967. Low-level jet winds, early cereal aphid and barley yellow dwarf detection in Iowa. Plant Dis. Rep. 51:527-530.

CHAPTER 21

VIRUSES AND VIROIDS

MILTON ZAITLIN

When I accepted the challenge to highlight the emerging problems in the field of plant viruses and viroids, I did so with a hesitation based on an appreciation of the limitations imposed on such a prognostication by an ever-changing technology. From the time of the original disclosures that the tobacco mosaic disease was caused by what we now know to be a virus, the plant virus field has advanced beyond the descriptive stage in lock step with the development of procedures and techniques that have allowed new knowledge to be uncovered. Early pathologists were able to describe diseases and to study their host ranges, modes of transmission, and the effects of the environment on disease development and symptomatology. The only properties of the agents themselves that could be determined were those involving procedures in which stability of the infectious agents and crude plant sap or extracts could be determined; i.e., how long could infectivity be retained in that extract, how stable were the agents to heating or pH changes, etc. In the 1930s three discoveries permitted a major advance: methods were developed to purify a few viruses; virologists learned how to quantify them by biological and chemical means; and it was shown that viruses were immunogenic. These techniques enabled virologists to establish relationships among viruses and, particularly in the case of viruses of potato, to aid in diagnosis.

With the development of the electron microscope and the discovery in the 1950s that the nucleic acid of the tobacco mosaic virus was infectious, the era of modern plant virology began. The composition and structures of many viruses have since been elucidated, their coat protein sequences determined, and the number, size, and composition of the nucleic acids comprising their genomes become known. The complete sequences of the DNA of several strains of cauliflower mosaic virus and four viroids are known, and sequences of several RNA plant viruses are rapidly being generated. All of these discoveries were dependent upon the development of appropriate technology, which itself is in a constant state of change. When I was a graduate student (1950–54), for instance, I used a boundary electrophoresis apparatus for my thesis work and employed the Oudin test for my serological studies. None of my current students had heard of either technique; such procedures have passed from the scene. In recent years, we have seen a remarkable increase in technology that has permitted virus detection and analysis with a sensitivity and precision not even dreamed of a few years ago. For example, nucleic acid hybridization techniques allow the detection of picogram (10^{-12}-g) amounts of specific nucleic acids, and the complete sequence of a small nucleic acid can be determined with a few micrograms of material.

The point of the above is to stress that any predictions made here are limited by the technology we have at the moment and that there are still problems that we cannot address effectively. To understand the replication of viruses and viroids and the induction of disease, we need to know how these agents interfere with the host's metabolism to perturb it to induce the sequence of events that result in what we term disease.

As an indication of how tenuous prognostication can be, I refer you to the Leeuwenhoek Lecture given in 1959 by the late F. C. Bawden (for whom I admit an admiration), entitled "Viruses: Retrospect and Prospect" (2). He correctly assessed that we would have to look for the answer to the specificity of viruses in the "... rhythm of different patterns built up from four nucleotides." But in suggesting patterns for future emphasis he thought that studies on the virus infection process would be important and that "... plant viruses [will be found to] associate with the genetic mechanisms of their hosts in the same intimate way that bacteriophages do." Both predictions were not fulfilled. Unfortunately, we are still ignorant of much of the detail of the viral infection process. For instance, we still are uncertain about how viruses actually enter cells during mechanical inoculation and into which type of cell they enter. Recently in my laboratory we found that mesophyll cells are important sites of viral entry, in contrast to the accepted dogma that only epidermal cells receive virus upon mechanical inoculation (24). The way in which viruses enter isolated leaf protoplasts is also controversial. Unfortunately few people now work in the mechanism-of-infection field, so unless its popularity increases, a long time may pass before we get answers to these interesting questions.

Improved technology continues to help reveal surprising new types of viruses that would otherwise be considered more conventional. Recently, Randles and colleagues (21) in Australia have discovered a new group of viruses that encapsidate both linear and circular forms of viroidlike RNA in addition to the more conventional virus-sized RNA molecules. Without the gel technology that enables one to identify small circular and linear molecules and without the molecular hybridization techniques to show that the linear and circular molecules are different forms of the same RNA, this virus group would have been considered just another conventional one, when in fact, it is unique.

Finally, even the current sophistication does not provide all of the answers to what might appear on the surface to be straight-forward questions. For example, the nature of the etiologic agent of the big vein disease of lettuce that is transmitted by zoospores of *Olpidium brassicae* and by grafting remains unknown. A virus has long been suspected, but almost 30 years of research has failed to reveal it. Could it be an agent analogous to viroids, or something unique?

WHAT DOES THE FUTURE HOLD FOR PLANT VIROLOGY?

First and foremost, the aim of most plant virologists is to learn about viruses and their hosts with the ultimate goal of controlling disease. It is unrealistic to assume that within the foreseeable future control will be achieved by other than current practices—principally through breeding plant sources of immunity and resistance. In addition, an expanded knowledge of virus vectors, weed hosts, mode of transmission, etc., is required, and that information will continue to provide important weapons in the control arsenal. As will be discussed later, the new tissue culture technology may help to generate new sources of resistant material and to provide virus-free stocks of vegetatively propagated plants.

In addition, some new areas of research hold promise for exciting developments in plant virus and viroid research.

Plant Viruses and the Genetic Engineering of Plants

In the search for vectors to introduce foreign or modified genetic information into plants, one group of plant viruses, the caulimoviruses, has received considerable attention. Research to develop such a vector has attracted talented people oriented toward molecular biology. Application of that type of expertise has allowed a rapid advance in our knowledge of the molecular biology of one member of the group, cauliflower mosaic virus (CaMV). To date, the caulimoviruses constitute the only plant virus group in which the genome is known to be double-stranded DNA, and as such these are the most amenable to the recombinant DNA technology. Another group, the gemini viruses, containing single-stranded DNA could also be utilized as vectors but are just now attracting attention.

With recombinant DNA technology, the nucleotide sequence of the DNA of several strains of CaMV has been determined. Furthermore, the DNA has been cloned in a bacterial plasmid and, when excised from that plasmid, has been found to be infectious (13). It is known further that the DNA of the virus has six "open reading frames" (putative genes) and several intergenic regions. Such knowledge is essential if the virus is to be used to convey foreign genetic material into plants, because we must know where it is possible to insert new DNA without "killing" the virus. So far, one open reading frame (No. II) has been found to be nonessential, as some strains have deletions in the DNA of that region, making it a candidate area for the insertion of foreign DNA. There is another intergenic area in which very small pieces of DNA may be inserted (14) without inactivating the DNA. Unfortunately, the size of the inserted DNA that can be accommodated in open reading frame II is very small (about 250 base pairs), presumably because bigger inserts into the DNA make it too large to be packaged in coat protein to form virions (10). Unless means can be devised to overcome this limitation, the potential for CaMV as a vector for insertion of foreign DNA into plants is questionable. It will also be important to design the vector to ensure that the genes on the foreign DNA will be expressed in the host plant.

Another potential use of plant viral nucleic acid for amplification of nucleic acids has not been explored. Is it possible, for example, to ligate a messenger or other foreign RNA to one end (preferably the 5′) of a viral RNA, thereby allowing the virus to amplify this RNA? Based on the packaging limitations seen with CaMV DNA, it would seem essential that, of RNA viruses, only those with helical symmetry (i.e., rods) would be useful because they could possibly be elongated beyond their normal length without causing a packaging problem. However, it must be determined whether ligation of a foreign RNA to the 5′ end of the viral RNA will interfere with its normal replication.

Elucidation of the Molecular Basis for Symptomatology

We are beginning to come to grips with an understanding of how viruses and viroids perturb the host to cause disease. I recently discussed this question in some detail (27), so I will not belabor it here, but I think that it is important to stress this as an important area for future research. It is safe to say that we are completely ignorant as to how plant virus and viroid disease symptoms are elicited. Many studies have described symptoms in biochemical terms, but they tell what happens as a result of the disease, rather than describe its cause. It is axiomatic that the virus or viroid has to interact with the host's genetic apparatus to perturb it. Some recent work suggests a correlation between the content of pathogen double-stranded RNA and expression of symptoms in a plant (12,16), but many more confirmatory studies are needed to establish such a principle. One recent paper suggested that viroids might interfere with the normal messenger RNA splicing that occurs in eukaryotes (6). It is apparent that very slight changes in the sequence of a nucleic acid can result in drastic symptom modification. A severe and a mild strain of potato spindle tuber viroid differ from each other by only three of the 359 nucleotides of which they are composed (11). Thus, the symptom difference must be a result of the response of the plant to one, two, or three nucleotide exchanges at three possible sites in the viroid molecule. It is probable that similar subtleties exist among plant viruses, which might help to pinpoint a molecular basis for symptomatology. In my laboratory, for example, we have recently suggested that the difference in symptom behavior of a temperature-sensitive viral strain and its "parent" results from a very small change in the sequence of one viral-coded protein, which, of course, is a reflection of the coding sequences in the viral RNA (15).

How the viral or viroid nucleic acid interacts with its host and what systems are first perturbed are unknown, but the new technologies of cloning and nucleic acid hybridization should be helpful here. If, for example, a subtle interaction of viral nucleic acid and host genome can be detected by hybridization, clones of host DNA containing those interaction "hot spots" could be obtained and hopefully their genes could be resolved. This is a long term view because we know so little about the plant genes themselves, but in this way we should ultimately be able to identify the sites at which viruses and viroids first interact with the host.

Chemical Modification of Symptoms

It is well known that the symptoms expressed by a plant in response to a virus can be modified by the environment in which the plant is grown. Factors such as light, temperature, nutrition, water supply, and time of the year can influence disease severity, but they may or may not also influence the concentration of virus in the plant. Thus, it does appear that in some instances we can separate virus replication from symptomatology; i.e., perhaps we can suppress symptoms and not worry about the replication of the virus per se. In the 1950s and 1960s much attention was paid to the application of chemicals to plants to inhibit virus replication, with very little to show for it. Chemicals that suppress virus replication also seem to have deleterious effects on the host, so this line of research has essentially been abandoned. The two most promising chemicals that came out of these studies, 2-thiouracil and 8-azaguanine are both very phytotoxic. Recently it has been shown that 2-thiouracil works because it becomes incorporated into viral RNA and generates RNA that can no longer replicate itself (1). It is not inconceivable that a specific virus-inhibiting chemical can be found, but my prejudice is that it will be unlikely. It is essential to find a chemical that can inhibit specific virus replication steps but would have a minimal effect on the host, but most facets of the viral replication process depend on host systems. The one site that could be virus-specific is the replicase enzyme, which presumably synthesizes plant viral RNA. I say "presumably" because even though it is a good presumption that the virus contributes all or part of the enzyme to replicate its nucleic acid, evidence to support the presumption is scarce. In fact, there is a vocal school that postulates that the host is the contributor of the enzyme (8).

If we cannot inhibit the virus itself with chemicals, there is some evidence that chemical treatment could be used to alleviate virus symptoms. Tomlinson et al (25) showed that applications of carbendazim, a breakdown product of the fungicide benomyl, when applied as a soil drench caused a substantial reduction in the severity of symptoms caused by tobacco mosaic virus (TMV) in tobacco and beet western yellows virus in lettuce. Where TMV could be measured, there was little effect on viral replication. In fact, the virus content seemed to be slightly higher because the treated plants did not senesce as rapidly as the untreated ones. This chemical had no symptom-suppressing effect on tomato and cucumber plants inoculated with cucumber mosaic virus or lettuce inoculated with lettuce mosaic virus, so obviously it is not universally applicable. Nevertheless, it does demonstrate that, potentially, symptomatology and virus replication are separable, and it suggests that this line of research is worth pursuing.

Cross-Protection and the Control of Plant Viruses

Prior infection by a mild strain of a virus can lead to the protection of a virus-infected plant from disease caused by strains of that virus. Basically, if a virus is established in a plant, subsequent infection (and symptom production) of "related" viruses is suppressed, but "unrelated" ones can become established. Unfortunately, our understanding of the mechanism for the phenomenon is lacking. Cross-protection has also been used to confirm virus strain relationships; although the phenomenon can be demonstrated with viroids (19), its validity in determining viroid relationships has yet to be established. The phenomenon is used commercially to control tristeza in citrus (5) and TMV in greenhouse-grown tomatoes (7).

Utilization of cross-protection for virus control in the field seems to present certain risks beyond those experienced when the cross-protecting virus is used in a controlled or greenhouse environment. Viruses that may cause mild symptoms in one host can generate a serious disease in another, so that by introducing a substantial virus reservoir we may also increase the potential of increasing virus disease in surrounding crops. This caveat is of little consequence in the tristeza disease because the responsible virus has both a restricted host range and a specific vector relationship for its transmission, but it could be an important consideration when one uses mild strains of other viruses in the field. Knowledge of the response of other nearby plant species to the cross-protecting virus and of the vector relationships would be essential before any virus could be released in the field. Furthermore, there are known instances where mixed infections by two different viruses results in a synergistic enhancement of disease, giving further pause. In addition, the degree of biological stability of the mild strain must be determined. Could mutants with enhanced virulence be generated from that strain under environmental stress and thus exacerbate the disease?

It is considered that some viruses code for or generate "factors" that mitigate vector transmission. One attractive possibility would thus be to use cross-protecting strains that have lost the capacity to be transmitted by vectors, so that they could not be transmitted to nearby plant species. Naturally occurring strains of this type are known in a few viruses, and the generation of such mutants in others would certainly seem feasible.

Tissue Culture and the Control of Plant Viruses

With tissue culture techniques, significant advances in controlling plant viruses have already been seen. Production of virus-free stock of certain vegetative and fruit crops such as potatoes, strawberries, and raspberries is a common commercial practice, using propagules selected by apical or meristematic tip cultures, techniques that often can free diseased plants from viruses. In New York State, for example, all foundation stock disease-free potato "seed" is produced in this way. The techniques give the further advantage of allowing for rapid clonal propagation of desired stocks.

In recent years, regeneration of plants from isolated protoplasts has been achieved in a number of species, including crops such as alfalfa, asparagus, cassava, lettuce, and potato. One very surprising outcome of some of the studies was the phenotypic diversity observed in potato clones produced from protoplast regeneration by Shepard and his colleagues (23). It is not known yet whether this diversity is truly genetic or "epigenetic," but in a pragmatic sense, it presents the agricultural scientist with a spectrum of plants that vary in many characters, including differential responses to plant pathogens.

Tissue culture techniques may also be used to produce somatic embryos in a number of species, which can eventually result in regenerated plants. Such embryos may be produced directly from tissue explants or from callus in tissue culture. These techniques, plus the regeneration of plants from protoplasts, provide a wealth of material for the virologist to test for resistance or tolerance to virus. In addition, it may be possible to immunize the cells or embryos used to propagate improved plant cultivars with the mild virus strains and mutants suggested in the previous section, thus capitalizing on the cross-protection technique. There are potential problems that might mitigate against success, as it has been shown that viruses usually are lost from tissue cultures following repeated subculturing, and there is some evidence that culturing previously infected protoplasts can result in plants containing virus strains with enhanced virulence (22).

Protoplasts have been utilized to look for resistant cells in populations of leaf mesophyll cells. In some instances, not all of the cells are infected in leaves showing virus mosaic symptoms. Protoplast-derived plants from cells of those leaves have been examined to see if those that are derived from uninfected cells are resistant to reinoculation. So far the results have been disappointing, but more work along these lines is worthwhile. A further complication, when comparing resistance in protoplasts and resistance in the leaf, is reflected in the fact that protoplasts prepared from leaves of resistant plants can be fully susceptible to virus infection when inoculated in vitro (4). In these cases, it is probable that the resistance at the plant level reflects a defect in the cell-to-cell movement of the virus from sites of initial virus infection, rather than a restriction of virus replication per se (24).

Improvements in Diagnostic Techniques

In the last decade we have seen marked advances in the sensitivity of tests used for the identification and detection of virus- and viroid-infected plants. Biological assays still allow the most sensitive detection of viruses, but at times they present difficulties because of variations in the susceptibility of test hosts and because they are sometimes inconvenient and difficult to perform with viruses that are strictly vector dependent. New techniques, such as the enzyme-linked immunosorbent assay (ELISA) serological test because of its great sensitivity, have facilitated experimentation with certain viruses. For example, with the luteoviruses, which can only be transmitted by aphids and which replicate to only low levels in phloem tissue of their hosts, the ELISA test allows researchers to seek answers to questions that were formerly impossible, or at least very difficult to explore (17). Detection can be effective in tissue where the conventional aphid-feeding bioassay is less reliable. Furthermore, the test gives a more precise determination of virus concentration than does bioassay. Its great sensitivity enables virus quantification from a few grams of tissue or even individual leaves or from a few virus-carrying insects. A modification of the ELISA test using radiolabeled antisera was shown to facilitate detection of viruses in extracts at concentrations as low as 2 ng/ml (9). One disadvantage in the ELISA procedure is the marked specificity of the reaction, which reduces one's ability to detect strains and viruses related to the virus used to produce the antiserum.

The great sensitivity and selectivity of nucleic acid hybridization has been exploited in a test designed to detect potato spindle tuber viroid in potato plants (20). Nucleic acids contained in crude extracts of diseased tissues can be immobilized on special membranes and then hybridized with radio labeled probes of nucleic acids that are complementary to the viroid. This technique can be used to detect less than one nanogram of viroid. It should be a useful test with viroids because they engender no specific protein during their replication and hence are not detectable by conventional serological means, but the extra sensitivity is probably not necessary for viruses that can be detected by somewhat less sensitive (but technically simpler) serological tests.

RNA virus (and viroid) infection always results in the formation of double-stranded (ds) virus- and viroid-related RNAs. Gel electrophoretic procedures have been developed that can detect ds RNAs in plant extracts; the spectrum of ds RNAs from many viruses are characteristic and thus could possibly be used for diagnosis. Furthermore, detection of ds RNA in tissues of diseased plants in which the etiologic agent is unknown, but which is suspected to be virus in nature, can lend experimental support to a virus etiology for the disease (18).

Other Studies

The prospects for learning more about the details of viral replication are brightened by the introduction of many of the techniques described above. For instance, we are now capable of unraveling the manifold viral-related nucleic acid species generated in infected plants with complementary DNA plus in vitro labeling and cloning procedures, coupled with analysis by nucleic acid hybridization "blotting" techniques. Proteins also can be specifically identified in tissues by sensitive blotting procedures using specific antibodies, and it is even possible to get antibodies to proteins that one has not isolated or cannot isolate. The sequence of a known or putative viral protein can be deduced from the viral nucleic acid sequence. A small peptide of that protein can be synthesized in vitro and used as an immunogen; thus, antibodies specific for that protein can be produced (26). This technique should help us investigate the role of viral-specified proteins in the virus replication and infection processes.

There are a number of other questions that lack answers, and I am not as confident that we have the ability to get those answers soon. What, for instance, governs the host range of a virus, and what determines the specificity of the virus-vector relationship? Why are some viruses confined to certain tissues in a plant? These are difficult questions that will require the thoughtful concern of many people to provide individual approaches; unfortunately, plant virus workers are few and, with the funding problems now facing scientific research, I do not see an improvement in research support. Some areas within the plant virus discipline are particularly under-represented (vector workers, for instance), and progress in those areas will be less probable than in the more glamorous areas utilizing recombinant DNA and molecular biology technology.

CONCLUSIONS

To be sure, as indicated above, there are many unanswered questions in plant virology, and with our expanded capacity to investigate some of them we can foresee some answers. However, with the possible exception of advanced tissue culture techniques, I do not anticipate that most of the new research findings will have a significant impact on disease control. I do not expect, for instance, that chemicals will be employed to control virus diseases or that finding the viral replicase enzyme, or detailing the nature of viral nucleic acids, will help us design a control strategy. On the other hand, an expanded capability to detect, identify, and determine relationships between viruses will be useful in establishing guidelines for controlling, or at least living with, virus diseases. Simplified detection methods will help us identify plants that serve as viral reservoirs in the field, identify weed hosts, and point to methods of virus survival and to the mode of spread of diseases. Cross-protection should be explored more thoroughly as a potential means of disease control.

A word of caution for virus workers in the "basic" research area is in order, in my view. Once again, I turn to F. C. Bawden (3), who more than a decade ago counseled against placing an overemphasis on technology for its own sake.

> There is nothing easier than to put a virus through the current range of standard machines, some automatic or semiautomatic, that will purify it, photograph it, measure it, and analyze it, with a paper at the end containing the canonical measurements and pictures editors of journals readily accept, even though in essence it contains nothing new. It is much to ask someone to give up this easy approach to publication and tackle the more difficult problems in pathology, and I know it has been cynically said that "Scientific happiness is to have one experiment and continue doing it." Certainly, to add new knowledge to pathology will require more insight and ingenuity than to gain some information about the particles of

another virus by switching machines on and off, and seeking it may be more frustrating.

I see a tendency for such an attitude in some current plant virus studies. Cloning and sequencing procedures are becoming almost routine; hopefully we can avoid the temptation to clone and to sequence viral nucleic acids just because the technology exists. Rather we must use those techniques to generate answers to relevant questions. I hope I can resist the temptation myself.

LITERATURE CITED

1. Barta, A., Sum, I., and Foglein, F. J. 1981. 2-Thiouracil does not inhibit TMV replication in tobacco protoplasts. J. Gen. Virol. 56:219-222.
2. Bawden, F. C. 1959. The Leeuwenhoek Lecture, Viruses: Retrospect and prospect. Proc. R. Soc. B. 151:157-168.
3. Bawden, F. C. 1970. Musings of an erstwhile plant pathologist. Annu. Rev. Phytopathol. 8:1-12.
4. Beier, H., Bruening, G., Russell, M. L., and Tucker, C. L. 1979. Replication of cowpea mosaic virus in protoplasts isolated from immune lines of cowpeas. Virology 95:165-175.
5. Costa, A. S., and Müller, G. W. 1980. Tristeza control by cross protection: A U.S.-Brazil cooperative success. Plant Dis. 64:538-541.
6. Dickson, E. 1981. A model for the involvement of viroids in RNA splicing. Virology 115:216-221.
7. Fletcher, J. T. 1978. The use of avirulent virus strains to protect plants against the effects of virulent strains. Ann. Appl. Biol. 89:110-114.
8. Fraenkel-Conrat, H. 1979. RNA-dependent RNA polymerases of plants. Trends Biochem. Sci. 4:184-186.
9. Ghabrial, S. A., and Shepherd, R. J. 1980. A sensitive radioimmunosorbent assay for the detection of plant viruses. J. Gen. Virol. 48:311-317.
10. Gronenborn, B., Gardner, R. C., Schaefer, S., and Shepherd, R. J. 1981. Propagation of foreign DNA in plants using cauliflower mosaic virus as vector. Nature 294:773-776.
11. Gross, H. J., Liebl, V., Alberty, H., Krupp, G., Domdey, H., Ramm, K., and Sänger, H. L. 1981. A severe and a mild potato spindle tuber viroid isolate differ in three nucleotide exchanges only. Biosci. Rep. 1:235-241.
12. Habili, N., and Kaper, J. M. 1981. Cucumber mosaic virus-associated RNA 5. VII. Double-stranded form accumulation and disease attenuation in tobacco. Virology 112:250-261.
13. Howell, S. H., Walker, L. L., and Dudley, R. K. 1980. Cloned cauliflower mosaic virus DNA infects turnips (*Brassica rapa*). Science 208:1265-1267.
14. Howell, S. H., Walker, L. L.; and Walden, R. M. 1981. Rescue of in vitro generated mutants of cloned cauliflower mosaic virus genome in infected plants. Nature 293:483-486.
15. Leonard, D. A., and Zaitlin, M. 1982. A temperature-sensitive strain of tobacco mosaic virus defective in cell-to-cell movement generates an altered viral-coded protein. Virology 117:416-424.
16. Liang-Yi, K., Xi-Cai, Y., and Po, T. 1982. Double-stranded viral RNA content in tobacco leaves infected with virulent and avirulent isolates of tomato mosaic virus. Virology 118:324-328.
17. Lister, R. M., and Rochow, W. F. 1979. Detection of barley yellow dwarf virus by enzyme-linked immunosorbent assay. Phytopathology 69:649-654.
18. Morris, T. J., and Dodds, J. A. 1979. Isolation and analysis of double-stranded RNA from virus-infected plant and fungal tissue. Phytopathology 69:854-858.
19. Niblett, C. L., Dickson, E., Fernow, K. H., Horst, R. K., and Zaitlin, M. 1978. Cross protection among four viroids. Virology 91:198-203.
20. Owens, R. A., and Diener, T. O. 1981. Sensitive and rapid diagnosis of potato spindle tuber viroid disease by nucleic acid hybridization. Science 213:670-672.
21. Randles, J. W., Davies, C., Hatta, T., Gould, A. R., and Francki. R. I. B. 1981. Studies on encapsidated viroid-like RNA. I. Characterization of velvet tobacco mottle virus. Virology 108:111-122.
22. Shepard, J. F. 1975. Regeneration of plants from protoplasts of potato virus X-infected tobacco leaves. Virology 66:492-501.
23. Shepard, J. F. 1981. Protoplasts as sources of disease resistance in plants. Annu. Rev. Phytopathol. 19:145-166.
24. Sulzinski, M. A., and Zaitlin, M. 1982. Tobacco mosaic virus replication in resistant and susceptible plants: In some

resistant species virus is confined to a small number of initially infected cells. Virology 120:182-193.

25. Tomlinson, J. A., Faithfull, E. M., and Ward, C. M. 1976. Chemical suppression of the symptoms of two virus diseases. Ann. Appl. Biol. 84:31-41.
26. Wong, T. W., and Goldberg, A. R. 1981. Synthetic peptide fragment of *src* gene product inhibits the *src* protein kinase and crossreacts immunologically with avian *onc* kinases and cellular phosphoproteins. Proc. Nat. Acad. Sci. (USA) 78:7412-7416.
27. Zaitlin, M. 1979. How viruses and viroids induce disease. Pages 257–271 in: Plant Disease, An Advanced Treatise. Vol. IV. J. G. Horsfall and E. B. Cowling, eds. Academic Press, New York. 466 pp.

CHAPTER 22

BACTERIA[1]

A. K. VIDAVER

Plant pathologists who work with phytopathogenic bacteria face tremendous challenges and opportunities both in the near and distant future. Several remarkable discoveries and advances have been made in understanding the genetic and chemical basis of pathogenicity and plant damage, such as the sophisticated transfer of DNA from bacteria to plants by *Agrobacterium tumefaciens*, the role of plasmids in crown gall and olive knot disease, and the ice-nucleation phenomenon that can result in frost injury. Other notable advances have been made in the isolation of new classes of plant pathogenic bacteria, such as *Spiroplasma* spp., mycoplasmalike organisms, and rickettsialike organisms, in earlier or more specific diagnoses of some diseases, and in biological control. These advances show that the plant pathogenic bacteria offer new areas for biological research. When the clock is turned back to look at what was known about phytopathogenic bacteria and their interactions with plants at the 1958 Golden Jubilee of the American Phytopathological Society, these advances are all the more remarkable, given the limited number of people working in this area and the resources available. The near and distant future, as discussed here, holds more promise and surprises for us, and the challenges represent higher and more difficult peaks to climb in the unexplored terrain of phytobacteriology.

Plant health, to a plant pathologist, is the absence of disease and abnormality. Bacteria, a class of prokaryotes, are one type of microorganism that can affect plant health. Diseases caused by bacteria present special challenges in isolation, identification, and control. These challenges need to be met, especially with the increased demand for agricultural products. While the amount of crops currently lost to bacterial diseases cannot be determined accurately (7), there is a consensus that agriculture cannot continue to feed the world if these losses are not reduced. A basic understanding of this unique group of pathogens is needed; knowledge of bacterial genetics and physiology, host-parasite interactions, and ecology of

[1] Based in part on a talk presented at the 5th International Conference on Plant Pathogenic Bacteria, Cali, Colombia, August 1981.

plant pathogenic bacteria should lead to more rational control measures. These problems are some of the challenges we face in the coming decades.

ISOLATION OF FASTIDIOUS AND ANAEROBIC PROKARYOTES

The first step in definitive studies of bacterial diseases is the isolation of the causative agent. Some prokaryotes are not easily cultured, but much progress has been made in the last decade. Certainly more advances will be made in the culturing of fastidious rickettsialike and mycoplasmalike organisms. All pathogens in these two groups appear to be insectborne; perhaps some plant or insect component is essential for in vitro growth and reproduction.

There is accumulating evidence that some anaerobes, such as clostridia, may be plant pathogens (10,17). Anaerobic and microaerophilic animal pathogens are well known. Few attempts have been made to manipulate the gaseous environment in the isolation of phytopathogenic bacteria because it has been assumed that the majority of such bacteria are aerobic or facultatively anaerobic. Bacteria whose growth or pathogenicity is inhibited by oxygen may be a significant component of healthy and diseased plant microflora; they deserve more attention.

DIAGNOSIS AND IDENTIFICATION

The early detection of disease and the accurate diagnosis of a given disease are continuing challenges. Early diagnosis makes it easier to limit the spread of disease because physical containment (roguing, etc.) can be used with greatest effect. But our current procedures (6) can be described as primitive when compared with diagnosis of human and animal diseases. More sensitive and practical tests to detect incipient infections are required. If detectable levels of bacteria could be decreased to less than 10^3 colony-forming units per gram, fresh weight—currently the highest degree of sensitivity for detection by fluorescent antibody techniques (21)—control measures could be instituted much earlier. Such sensitivity of detection could be achieved by treating plant tissues in various ways to release whole bacteria or specific bacterial components. These materials might then be tested directly for the presence of bacteria or be concentrated before testing.

Specificity of identification is likely to be improved in several ways, as knowledge of the relationships between and among various classes of bacteria increases. The knowledge gained in taxonomy can lead to improved metabolic tests, serology, and analyses of proteins and nucleic acids by gel electrophoresis or chromatography. Identification and isolation of pathogenic determinants (cell components or compounds associated with the ability to cause disease) is likely to increase the specificity of antisera for diagnosis. As both investigator time and crops become more valuable, automation of identification tests will become desirable, if not necessary. Automated and rapid identification tests have been successful in medical microbiology. Plant pathologists working with large numbers of samples would find such methods particularly useful. In the future it may become common to have resident or regional laboratories, analogous to a hospital's diagnostic laboratory, that will service large farms, orchards, or plantations. Consulting services may fill this role; some have begun to provide diagnostic services for bacterial diseases in the United States. Such services generally have not provided laboratory or greenhouse tests, but this may change.

INTERACTIONS OF BACTERIA WITH PLANTS AND OTHER MICROBES

Bacterial interactions with plants generally are understood at only an elementary level. One of the greatest challenges for plant bacteriologists is to determine what occurs in the early stages of infection and to separate these occurrences into cause and effect. If the early stages of interaction between pathogen and host can be detected and understood, it may be possible to interfere with the early steps in an infection. Crown gall is the only disease in which attachment and early stages of infection are reasonably well understood (12,13). Such early interactions can be surprisingly rapid (9). In other diseases, bacteria can also begin the process of infection very quickly. We have found that bacteriocins applied to leaf surfaces 5–10 min after spray inoculation with *Pseudomonas syringae* pv. *syringae* and *P. syringae* pv. *glycinea* have little or no effect on lesion development in leaves, whereas spraying bacteriocins before inoculation markedly decreases the number of lesions (Vidaver and Thomas, *unpublished results*). Metabolic, ultrastructural, and physical methods of analyses have not been sensitive enough to detect such early perturbations in plants. Microchemical and microphysical methods of analysis in situ, as used in some animal studies, may provide insight into such early interactions.

Surface interactions between bacteria and plant cells are important in establishing infection (5,12,13,19). It would seem that more use could be made of tissue culture and protoplast systems for studying surface interactions. In these systems, both the chemical and physical properties of surfaces can be manipulated. Such studies may provide the insight needed to interfere with early infection processes.

Another kind of interaction of bacteria with plants is represented by the phenomenon of induced resistance. In this case, inoculation with generally avirulent or weakly pathogenic bacteria can elicit either localized or systemic resistance to pathogens inoculated later. There are now several such examples known, involving several different bacteria and plants, suggesting that this is a widespread phenomenon (5,19). The mechanisms of such induced resistance are complex and not likely to be well understood in the near future. However, mutants of both bacteria and plants are promising probes for determining the mechanism(s) involved in such induction. The challenge is to find and isolate bacterial components or analogues that can be produced cheaply for practical application of such resistance. Recombinant DNA technology may prove useful; if the "inducer" can be transferred to a noninducing bacterium, the genetic and biochemical basis of such induction will be easier to dissect and manipulate, and the "inducer" itself will be easier to produce in industrial quantities. Some crops, particularly perennials, are probably of sufficient value to warrant substantial investment in this area of research.

Yet another kind of interaction of bacteria with plants is constitutive resistance, in which the majority of cultivars of any crop are resistant to most microorganisms, including bacteria. In animals, many physical and biochemical factors contribute to constitutive or inherent resistance; many of these are known and are a continuing source of study, e.g., different blood fractions. In plants, the challenge is to identify such factors so that they can be analyzed for breeding purposes. Because they can be so readily manipulated and genetically marked and altered, phytopathogenic bacteria are particularly well suited for the complementary studies necessary to determine the generality or specificity of any effect detected in plants. If scientists knew how to control and transfer both constitutive and induced resistance, diseased plants would be a rarity.

Another type of interaction is that of bacteria with other microbes. The interactions of most interest to plant pathologists are those in which other microbes act synergistically to enhance infection and those in which infection is limited because of antagonism. The former type of interaction may be common but is rarely studied (11). Although it is difficult as yet to study such types of interactions, they may well be significant in the field, particularly in soil and water. Also, plant pathologists tend to study the predominant organism involved in so-called disease complexes, as in fungal stalk rots of corn and sorghum. In these diseases, the role of bacteria, if any, is not known.

Enhanced plant growth and yield by treatment of seed or transplants with certain bacteria (8) are postulated to be the result of an interaction between microorganisms, but it is not clear whether the benefits result from competition for nutrients or colonizing sites (receptors?) or whether antagonism of root pathogens, bacterial or not, occurs or some combination of interactions takes place. Understanding and controlling the physical, chemical, and environmental factors that promote root and aerial colonization of plants with beneficial bacteria is a formidable challenge. Quantitative and qualitative analyses of these factors should be possible, using bacterial mutants and isogenic cultivars. Disease suppression by biological control of pathogens, such as achieved with the control of the crown gall bacterium (14), is another important microbe:microbe interaction.

CONTROL

The ultimate aim of plant pathology is to control plant disease. It is striking that our only means of control of bacterial pathogens are, as yet, preventative. To my knowledge, we have no means to cure infected plants in the field; at best, they can undergo temporary remission of symptoms as long as treatment is continued. Are we limited to preventative control by plant structure and environment? Research into the uptake and distribution of potential bactericides into different plant organs and their effectiveness there will be increasingly important. It may well be that bacteria in plants are significantly different structurally and metabolically than in vitro, making the task more difficult. For example, the experimental systemic chemical techlofthalam is effective in plants against the rice pathogen *Xanthomonas oryzae*, but not in vitro (15). The compound may be altered in situ or else suitable conditions for in vitro assay have not been established.

The use of antibiotics and chemicals for the control of bacterial diseases of plants has had a long and checkered history and will not be discussed here. In brief, it is fair to say that in two situations, namely fireblight (a disease of *Malus* spp. caused by *Erwinia amylovora*) and certain mycoplasmalike diseases, selective antibiotic use has been successful in minimizing disease. Other compounds, e.g., copper derivatives, generally have given mixed results, being successfully used in some diseases and areas by some investigators. The future for nonmedical antibiotics as spray, seed, or rootstock treatment is not promising, given the regulatory and economic climate of the United States and other countries. The prospects for systemic bactericides are also not bright; to my knowledge none has been marketed in the United States or elsewhere. While there may be little interest in industry in systemic bactericides, this may be due to lack of success. The difficulty of such research was illustrated above. Yet this is an area that should be pursued and that may be productive in the future, as both bacterial and plant chemistry becomes better understood. The plant pathogens themselves may be sources of potent antibiotics, such as syringomycin, effective principally against a wide variety of fungi (20). The need is certainly there (1,3). New gaseous bacteriostatic or bactericidal agents should also be investigated for greenhouse crops.

An unexplored possibility for control is the potential use of selective metabolic inhibitors for disease in which temperature appears to be a critical factor. Plant pathologists are all familiar with bacterial diseases that are associated with certain temperature regimes. For example, *Pseudomonas syringae* pv. *phaseolicola* is considered a cool weather pathogen, whereas *Xanthomonas campestris* pv. *phaseoli* is considered a warm weather pathogen. Major metabolic changes may occur in these bacteria or in the plants at different temperatures. Temperature can markedly alter the ratio of the hexose monophosphate to the Entner-Doudoroff pathways for glucose catabolism in *P. fluorescens*, a close taxonomic relative of *P. syringae* (16). Thus it may be possible to use metabolic inhibitors as chemical control agents for bacteria, providing that they do not have adverse effects on plants. The need for new types of control agents has been pointed out many times (1,3); this need will intensify as agricultural products become more expensive.

Application of biological control, defined as the use of one biological entity to affect and minimize damage caused by another living organism, is still in its infancy (23). The effectiveness of *Agrobacterium radiobacter* strain 84 for control of most cases of crown gall in nursery stock has given impetus to the search for other bacteria effective in the biological control not only of bacteria, but also of fungi. Such controls will need to play a greater role in the future as currently available chemicals and antibiotics become ineffective, prohibited, or too expensive. For fastidious prokaryotes that have a resident phase in insects, it may be possible to have the insects feed on an appropriate antagonist. Indirect control of the pathogen by insect control should also be feasible, since most insects are subject to specific viral or bacterial diseases. The successes already achieved with insect control by polyhedrosis viruses and *Bacillus thuringiensis* serve as models. Also, in principle at least, with biological and genetic knowledge of effective antagonists, multiple genes for antagonism to pathogens affecting a single crop or even multiple crops might be incorporated into a suitable colonizing bacterium. Such genetic recombination is discussed later.

Postharvest diseases are currently given insufficient attention. Yet estimates by the Food and Agriculture Organization, United Nations (4), suggest that losses to such diseases are equal to or greater than losses to field diseases, particularly in developing countries. Several bacteria, notably the soft-rotting *Erwinia* and *Pseudomonas* species, are principal agents of such spoilage. Obviously their control could result in a quick gain in net usable yield.

Plant pathologists may well find that increasing the rate of plant debris degradation is a useful means of limiting pathogen populations. Plant debris is retained in low or minimum tillage practices, which are increasingly common because they conserve soil, moisture, and energy. Most, if not all, phytopathogenic bacteria can survive well in debris from infected plants (18) and with few exceptions survive poorly in soil, air, or water. Thus chemical or biological means to enhance breakdown of infested debris at controlled rates should prove useful for the control of several pathogens.

GENETIC ENGINEERING

One of the most exciting areas of basic and applied modern science is genetic engineering, or the in vitro construction of viable microorganisms with characters obtained from other microorganisms, plants, or animals. Bacteria are the most convenient microorganisms to use in these recombinant DNA studies. A program to develop the *Agrobacterium tumefaciens* tumor-inducing Ti plasmid as a vector or carrier to transfer desirable characters, such as production of storage proteins, between different plant species is underway in several laboratories. In addition, this area of research has enormous potential for enabling plant pathologists to understand host-parasite interactions at a molecular level. And, in principle at least, it should be possible to combine genes from growth-promoting bacteria to enhance plant growth and yield. There certainly are many technical difficulties ahead, but these are challenging areas of immense promise.

A specific example of genetic engineering is the potential to introduce genes determining disease resistance factors into biological control agents or into microorganisms for in vitro production. Identification of plant resistance genes would be difficult, but worthy of investigation. An *A. tumefaciens* plasmid or some other plasmid might be used to analyze such genes; e.g., random segments of plant DNA could be joined to a cloning plasmid that has one or more identifiable phenotypic markers and then transformed into a bacterium. If a cloned "resistance" gene produces a diffusible agent, it could be detected by transferring the plasmid-carrying bacterium to plates seeded with a pathogen of interest and looking for growth inhibition of the test pathogen. It may be critical to use low populations, since high populations of the pathogen could mask the inhibitory effect of a "resistance" gene. If such a screening method were successful, then the resistance gene(s) could be identified, or used for production of a novel control agent(s), analogous to antibiotic production. Resistance genes might also be transferred into a saprophyte that readily colonized the plant needing protection. Several investigators already are examining saprophytic bacteria that are good colonizers of various crop plants.

A more general test for transfer of resistance genes would require that the putative genes be transferred from the plasmid in the bacterium into the plant cell itself. There is the strong possibility of such a transfer to dicotyledonous plants, using the *Agrobacterium* plasmid system, but the transfer is not as likely for monocotyledonous plants, since nothing analogous to the *Agrobacterium*: dicotyledon system exists for monocotyledons. Gene transfer into this group of plants is thus all the more challenging. Transformation of suspension cell cultures or protoplasts remains to be explored; integration of DNA need not occur for expression of desired characteristics, if it can be maintained and replicated. Devising an appropriate assay system will be crucial. But if resistance genes can be transferred and assayed for, such procedures would revolutionize plant breeding.

These efforts taken together would supplement plant breeding efforts and enhance the useful lifetime of cultivars with desirable agronomic traits. The research proposed would be more difficult with plants carrying multiple genes for resistance, but that needs to be determined. For example, for purposes of recombinant DNA techniques, it may be feasible to link such genes, even if they are not linked in the plant genome.

The fact that most plants are resistant to bacterial and other infections means that inherent resistance is common. Using recombinant DNA techniques may lead to the discovery of the basis for this phenomenon. This would likely lead to the discovery of new biological principles, since nearly all current biological principles have been derived from animal and microbial studies.

USEFUL PATHOGENS

Plant pathologists are so accustomed to considering bacterial plant pathogens as undesirable that it is difficult to reconsider and ask whether their unique properties can be manipulated to advantage. The multiple uses of xanthan gum from *X. campestris* pv. *campestris* are reasonably well known; they range from oil drilling to salad dressings. More recently, extraction of commercially useful specific restriction endonucleases has occurred, and more can be expected. Other products, such as enzymes, toxins, bacteriocins, or bacteriophages might be useful in human medicine. Agrocin 84, the bacteriocin produced by *A. radiobacter* strain 84, is a substituted adenine nucleotide similar to the currently used cancer drug ara-C (22). If it could be produced in quantity, it might be tested for similar activity. The ice-nucleating property of *P. syringae* pv. *syringae* and some saprophytes is well established. If the material determining this activity could be isolated or synthesized cheaply, it might replace costly silver iodide for cloud-seeding purposes. The plant hormones produced by *A. tumefaciens, P. syringae* pv. *savastanoi* or *Corynebacterium fascians* might be exploited as plant growth regulators. Finally, we are all familiar with the plant-degrading enzymes of many phytopathogenic bacteria; these might be useful for biomass conversion. Such enzymes are now obtained principally from other microorganisms for industrial purposes. Plant pathogens may be a new and cheap source, and may provide novel enzymes. The possibilities discussed here are only a few; the reader undoubtedly could suggest other examples.

THE FUTURE

No one can accurately predict the future because of the complexity of today's world. However, some projections can be made with relative ease and certainty, and others cannot because of the lack of current knowledge and myriad interacting factors. It is probable that agricultural methods will change, as well as the crops and cultivars that are grown. It is likely that advances will be made in understanding the survival, spread, and disease potential of bacterial pathogens. Other areas are murky for the forecaster.

It has taken nearly 75 years to determine how closely related some dozens of generally host-specific xanthomonad plant pathogens are to each other (2). Recognition of such relationships has significance in developing control measures based on common properties, e.g., possible metabolic defects. The biological and taxonomic relationships among pathogens are sometimes obscured by current classifications; these relationships should be clarified.

Phytobacteriologists have traditionally concentrated on pathogens of economically important plants in temperate climates, but this emphasis may need reexamination. The bacteria of tropical plants, of forest trees, of weeds, and of fresh and salt water plants need more attention. Bacteria of tropical plants and perennial crops such as trees present special challenges because of the prolonged associations with the plant host. With increased cropping of both fresh and salt water plants, the prospect of problems like those of land plants are likely to increase. Pathogens of weeds deserve attention since biological control by plant pathogens should be technically feasible. With the increased costs of herbicides and some weeds becoming resistant to herbicides, such potential control appears attractive. Conversely, healthy weeds related to crop plants, e.g., johnsongrass (*Sorghum halepense*), may be sources of antibacterial products or resistance genes useful for genetic engineering.

Phytobacteriologists also would benefit from increased training in modern aspects of microbiology and plant sciences. These disciplines complement studies in traditional plant pathology and offer principles and techniques useful in studying bacterial plant pathogens and their interactions with plants. Without the incorporation of such principles and techniques into plant bacteriology, the pace of meeting the challenges outlined here will be much too slow.

And finally, both the private and public sector must be seen as vital to each other in advancing not only bacterial plant pathology, but also related sciences. Although success should always be sought in understanding and control of plant pathogenic bacteria, there must also be the freedom to fail. Without such freedom, at least for the academic community, agriculture and other sciences will be restricted to trying only what is safe. Science would suffer and be less able to meet the multiple challenges of the future.

The future is full of promise and challenge. At the centenary of the American Phytopathological Society, we will be able to look back at many advances in plant bacteriology and forward to the new challenges.

ACKNOWLEDGMENTS

I thank several of my colleagues for ideas and suggestions for this article and R. R. Carlson for editorial assistance of exceptional quality.

LITERATURE CITED

1. Beer, S. V., and Sherf, A. F. 1976. The dearth of and needs for new bactericides. Fungicide and Nematicide Test Results of 1975. C. W. Averre, ed. Am. Phytopathol. Soc., St. Paul, MN. 3:1-2.
2. Dye, D. W., Bradbury, J. F., Goto, M., Hayward, A. C., Lelliott, R. A., and Schroth, M. N. 1980. International standards for naming pathovars of phytopathogenic bacteria and a list of pathovar names and pathotype strains. Rev. Plant Pathol. 59:153-168.
3. Egli, T., and Sturm, E. 1980. Bacterial plant diseases and their control. Pages 345–388 in: Chemie der Pflanzenschutz und Schadlingsbekampfungsmittel, Vol. 6. Springer-Verlag, Berlin.
4. Food and Agriculture Organization. 1981. Food loss prevention in perishable crops. Agric. Serv. Bull. 43. FAO, United Nations, Rome.
5. Goodman, R. N. 1980. Defenses triggered by previous invaders: Bacteria. Pages 305–317 in: Plant Disease, An Advanced Treatise. Vol. 5. J. G. Horsfall and E. B. Cowling, eds. Academic Press, New York. 534 pp.
6. Grogan, R. G. 1981. The science and art of plant-disease diagnosis. Annu. Rev. Phytopathol. 19:333-351.
7. Kennedy, B. W., and Alcorn, S. M. 1980. Estimates of U.S. crop losses to procaryote plant pathogens. Plant Dis. 64:674-676.
8. Kloepper, J. W., Leong, J., Teintze, M., and Schroth, M. N. 1980. Enhanced plant growth by siderophores produced by plant growth-promoting rhizobacteria. Nature 286:885-886.

9. Lippincott, B. B., Whatley, M. H., and Lippincott, J. A. 1977. Tumor induction by *Agrobacterium* involves attachment of the bacterium to a site on the host plant cell wall. Plant Physiol. 59:388-390.
10. Lund, B. M., and Brocklehurst, T. F. 1978. Pectic enzymes of pigmented strains of *Clostridium*. J. Gen. Microbiol. 104:59-66.
11. Maino, A. L., Schroth, M. N., and Vitanza, V. B. 1974. Synergy between *Achromobacter* sp. and *Pseudomonas phaseolicola* resulting in increased disease. Phytopathology 64:277-283.
12. Matthysse, A. G., Holmes, K. V., and Gurlitz, R. H. G. 1981. Elaboration of cellulose fibrils by *Agrobacterium tumefaciens* during attachment to carrot cells. J. Bacteriol. 145:583-595.
13. Merlo, D. J. 1978. Crown gall—A unique disease. Pages 201–213 in: Plant Disease, An Advanced Treatise. Vol. 3. J. G. Horsfall and E. B. Cowling, eds. Academic Press, New York. 487 pp.
14. Moore, L. W., and Warren, G. 1979. *Agrobacterium radiobacter* strain 84 and biological control of crown gall. Annu. Rev. Phytopathol. 17:163-179.
15. Nakagami, K., and Tanakea, H. 1980. Effect of techlofthalam spray on *Xanthomonas oryzae*, the causal pathogen of rice bacterial leaf blight in the rice leaf. J. Pesticide Sci. 5:511-516.
16. Palumbo, S. A., and Witter, L. D. 1969. The influence of temperature on the pathways of glucose catabolism in *Pseudomonas fluorescens*. Can. J. Microbiol. 15:995-1000.
17. Perry, D. A., and Harrison, J. G. 1977. Pectolytic anaerobic bacteria cause symptoms of cavity spot in carrots. Nature 269:509.
18. Schuster, M. L., and Coyne, D. P. 1974. Survival mechanisms of phytopathogenic bacteria. Annu. Rev. Phytopathol. 12:199-221.
19. Sequeira, L. 1980. Defenses triggered by previous invaders: Recognition and compatibility phenomena. Pages 179–200 in: Plant Disease, An Advanced Treatise. Vol. 5. J. G. Horsfall and E. B. Cowling, eds. Academic Press, New York. 534 pp.
20. Sinden, S. L., DeVay, J. E., and Backman, P. A. 1971. Properties of syringomycin, a wide spectrum antibiotic and phytotoxin produced by *Pseudomonas syringae*, and its role in bacterial canker-disease of peach trees. Physiol. Plant Pathol. 1:199-213.
21. Slack, S. A., Kelman, A., and Perry, J. B. 1979. Comparison of three serodiagnostic assays for detection of *Corynebacterium sepedonicum*. Phytopathology 69:186-189.
22. Vidaver, A. K. 1983. Bacteriocins—The lure and the reality. Plant Dis. 67:471-475.
23. Vidaver, A. K. 1982. Biological control of plant pathogens with prokaryotes. Pages 387–397 in: Phytopathogenic Prokaryotes. G. H. Lacy and M. S. Mount, eds. Academic Press, New York.

CHAPTER 23

FUNGI

E. S. LUTTRELL

The central problem facing mycologists and plant pathologists is development of theory, the organization of knowledge of fungi and of diseases into systems that permit broad generalizations and provide a basis for reliable predictions. Meeting the challenge requires recognition of a commonality of purpose among mycologists and plant pathologists and a pooling of resources of the two sciences to attain common objectives. The relationship between mycology and etiology, with its emphasis on life cycles and developmental morphology of fungi, is obvious. The search for convenient morphological characters on which recognition can be based also is common both to identification of fungi and to diagnosis of diseases. Emphasis, therefore, will be placed on three areas: 1) physiology of fungi, among which parasitism in some form probably is more frequent than is generally appreciated, and pathology, used here in the restricted sense to refer to study of the morphological and physiological changes that occur in the host during the course of infection; 2) ecology of fungi and epidemiology of disease; and 3) classification of fungi (taxonomy) and classification of diseases (nosology).

Knowledge of fungi is organized in the taxonomy, the general system of classification. The most wanted predictions from the system are those on behavior of fungi. These are the predictions most difficult to make. Various forms of parasitism, for example, have developed repeatedly during the evolution of the fungi, and at least superficially similar forms of parasitism occur in distantly related groups (22). The question for mycologists is whether data on physiology and ecology can be adequately accommodated in the general system of classification or whether special systems based on physiology or ecology, which are immediate necessities, are also the ultimate answers. The question for pathologists is whether a classification of diseases based on pathology or epidemiology would prove more useful than the customary "diseases-caused-by" classification based on etiology.

PHYSIOLOGY OF FUNGI AND PATHOLOGY

Nutritionally, fungi may be divided into parasites, commensals, symbionts, and saprotrophs. Parasites may be subdivided into biotrophs, hemibiotrophs, and perthotrophs. Biotrophy is synonymous with ecologically obligate parasitism

(3). Biotroph (19) includes de Bary's (7) category "obligate parasites" and skims from his category "facultative saprophytes" most of the examples he gave since these (Ustilaginales, Clavicipitaceae, Taphrinales, Exobasidiales) were distinguished by their ability to grow in culture, an attribute that has little relevance to what occurs in nature. Ability of the fungus to grow in culture on defined media does, however, offer advantages in experimental work directed toward an understanding of biotrophy, and culturing on dead host tissue may suggest whether the restriction to biotrophy results from resistance of the substrate or from competition. Biotrophs require a continuing supply of nutrients from living host cells for sporulation and possibly may be characterized also by the ability to sporulate under this circumstance, an ability that may be lacking in fungi in which sporulation is induced by a shift in nutritional status. Survival outside of host tissues is by means of resting spores, sclerotia, or stromata containing reserves of food presumably obtained from living host cells. In addition to the ability to surmount the innate physical and chemical barriers of the host, biotrophs must be able to avoid activating the induced defense mechanisms that constitute the generalized response of plant tissues to invasion or injury. Knowledge of biotrophs and the pathology of the diseases they cause has been based primarily on studies of superficial or intercellular haustoriate parasites in the Peronosporales, Erysiphales, and Uredinales that cause local lesion diseases. Biotrophs occur also in the Chytridiales, Taphrinales, Clavicipitaceae, Phyllachoraceae, Meliolales, Hemisphaeriales, Exobasidiales, and Ustilaginales, as well as in isolated positions in many families of fungi. They may be readily culturable or nonculturable. Infections may be local or systemic. The mycelium may be superficial and haustoriate, subcuticular, intercellular and nonhaustoriate, intercellular and haustoriate, or intracellular. The assumption must be that basic principles can be elaborated and that information produced by host-parasite systems most favorable for experimental work can be extrapolated to similar systems. Nevertheless, it is evident that all aspects of biotrophy are not adequately represented by the rusts and mildews on which research has been concentrated and that concepts derived only from these fungi are too narrowly based.

Hemibiotrophs (22), like biotrophs, are capable of overcoming the innate resistance mechanisms of host plants and of establishing infections but sporulate and complete their development after death of the tissue they have colonized. Generally, they can be grown in culture, but growth may be poor, and sporulation may be difficult to induce. Hemibiotroph corresponds to de Bary's category "facultative saprophytes" with exclusion of the culturable biotrophs. The type example of hemibiotrophy is the black rot foliage disease of *Vitis rotundifolia* caused by *Guignardia bidwellii* (21). Infection occurs on immature leaves, and colonization proceeds without production of symptoms until the leaves reach maturity. At this point, invasion of the fungus is abruptly halted. The colonized areas die suddenly and appear as circular brown spots. Pycnidia develop in the dead tissue. As the leaves die in the fall, the fungus advances into the previously unoccupied tissue. During the winter, ascocarps develop over the entire surface of the fallen leaf. The ability of the fungus to colonize immature leaf tissue suggests that before maturity the induced response system of the leaf is not developed to the point at which the fungus is unable to avoid recognition. Many hemibiotrophs, however, can infect and colonize mature leaves, and many colonize only senescent leaves. Incubation period and lesion size apparently depend on the time during which the fungus can escape recognition or modify the host response. In the pepper spot disease of clovers (*Trifolium* spp.) caused by *Leptosphaerulina trifolii*, the response of the host is almost immediate, and the infection is restricted to tiny black flecks in the leaves. These, however, remain as latent infections. When the leaf, or any area of the leaf, dies from any cause, the fungus may produce mature ascocarps in the dead tissue within a week. Both *G. bidwellii* and *L. trifolii* represent an extreme in the colonization of dead tissue as assessed by the subsequent distribution of sporocarps outside of the original lesions. In some hemibiotrophs, sporocarps on dead overwintered leaves are restricted to the area of tissue originally colonized in the living leaf. Questions remain concerning the time of invasion of tissue surrounding the lesions, whether after death of the tissue or while it is moribund, and the factors that limit colonization in this phase. The usual difficulties in classification arise in separating biotrophs and hemibiotrophs. *Phytophthora infestans*, which Cooke (6) used as an example of a hemibiotroph, seems better classified as a biotroph since sporulation occurs on living tissue near the margins of the lesions. *Venturia inaequalis* is more difficult to classify because of the prolonged biotrophic phase in which it produces conidia on living host tissue. Consideration of its entire life cycle, which includes development of ascocarps on dead leaves, indicates that it may be more comfortably accommodated among the hemibiotrophs.

Emphasis on culturability of fungi has clouded thinking on parasitism. The hypersensitive reaction as defined in terms of host cell death was assumed to be a sufficient explanation for restriction of infections by "obligate parasites." This explanation could not apply to hemibiotrophs, and it is now evident that it does not apply to biotrophs either. Cell death is not an invariable feature of the induced response (30), and other elements, such as phytoalexin production, equally effective in their fungistatic effects against biotrophs and hemibiotrophs must be sought as primary factors. A parallel to reasoning on the hypersensitive reaction as a defense mechanism against biotrophs obtains in reasoning on cell death as a disease induction mechanism in hemibiotrophs. Because hemibiotrophs are capable of growth and sporulation on dead tissue, it is assumed that killing of tissue must be the primary feature of their parasitism. Toxins are responsible for many of the characteristic symptoms of diseases, and toxins and enzymes that alter host tissues coincident with or in advance of hyphal invasion (23,27) may function to varying degrees in colonization by hemibiotrophs, as enzymes do also in colonization by biotrophs that depend on penetration of host cell walls. The underlying factor in cell death associated with hemibiotrophs, however, may be the hypersensitive reaction of the host. Infections by both biotrophs and hemibiotrophs are limited by the induced response of the host, in which hypersensitivity serves at least as a marker of the site of the action. Biotrophy depends on avoidance of the induced response; hemibiotrophy, as most vividly illustrated by the extreme example of the *Trifolium-Leptosphaerulina* combination, represents an exploitation of this induced response.

Münch (24) distinguished perthotrophs from saprotrophs as fungi that derive their nutrients from parts of living plants that they themselves have killed. This group approximates most closely de Bary's category "facultative parasites," which included fungi such as *Rhizopus* that normally complete their development on dead organic matter but also are capable of attacking living tissue. Harter and Weimer (11) demonstrated that mycelium of *Rhizopus* growing in dead tissue of wounds in storage roots of sweet potato produces degradative enzymes that kill living tissue in advance of penetration. *Sclerotium rolfsii*, operating from a food base of organic debris or from sclerotia (26), through production of oxalic acid and enzymes (2) destroys the natural barriers and underlying tissues of seedlings and even of mature stems before penetration. On the basis of these models, perthotrophs may be defined as fungi that are capable of killing host tissue in advance of penetration although they are unable to penetrate and infect unaltered living tissue. Under this restricted definition most of Münch's perthotrophs would be classified as hemibiotrophs, and perthotrophs would represent a relatively small group of pathogenic fungi more closely related nutritionally to saprotrophs than to the majority of parasites. Münch, in fact, proposed a dichotomy separating fungi into parasitotrophs and necrotrophs; necrotrophs were further subdivided into perthotrophs and saprotrophs.

Nutritional relationships between perthotrophs and saprotrophs, which depend entirely on dead organic matter, are obviously close. The general concepts of biotrophy also apply to saprotrophs. Host-parasite relationships, host resistance, and host specificity are merely special cases of substrate-fungus relationships, substrate resistance, and substrate specificity. Colonization of dead plant substrates is similar to colonization of living tissue, as, for example, in the traversal of plant cell walls by narrow penetration pegs. Innate structural and chemical resistance factors are common to all substrates. In onion smudge, a classic example of a disease in which host resistance is based on innate chemical factors, resistance in fact depends on phenolics that are present in the dead outer scale leaves of the onion bulb (33). The predominance of relatively undigestable celluloses and lignins increases the resistance of dead plant substrates and leads to substrate specificity in saprotrophs that is determined by enzymatic capabilities as well as by degrees of tolerance for toxic compounds. An example of substrate specificity in saprotrophs is the restriction of *Coccomyces triangularis* to dead branches of a single species, *Quercus alba* (28). Investigations of substrate specificity, however, should proceed on a cautionary note. Webster (35) indicated that *Leptosphaeria acuta*, which produces ascocarps on dead herbaceous stems of *Urtica dioica*, can be consistently isolated from living symptomless stems of this plant. The phenology of a myriad of supposedly saprotrophic ascomycetes on dead leaves and stems is suspiciously like that of hemibiotrophs: they produce ascospores at the start of the growing season when the substrate is available only in the form of living tissue. Specificity may be determined in an uninvestigated parasitic phase. A number of "saprotrophic" fungi produce symptomless infections of living plants (20). Most surprising is the frequency with which species of *Xylaria* and *Hypoxylon* occur as symptomless parasites in leaves and the fact that specificity is lower in this parasitic phase than in the saprotrophic phase in which the ascocarps develop on woody substrates.

Commensals depend for nutrients on an intimate association with the host but differ from parasites and symbionts in that they have no effect on the host. In endobionts (20), the intimacy of the association and the nutritional dependence on the host are not in question, and the assumption would be that diversion of any amount of host energy would have some detrimental effect. Because of the difficulty of proving a negative, convincing proof to the contrary would depend on demonstrating a beneficial effect, in which case the choice would be between parasitism and symbiosis. A striking aspect of these presumed commensalistic associations is the high level of compatibility with the host achieved by endobionts that are primarily saprotrophic. For ectobionts the question of intimacy of association increases as the site moves from the phyllosphere to the rhizosphere. The first assumption would be that nutritional dependence is on substances normally exuding from the host. The burden of proof would rest on the assumption of parasitism and would require demonstration of some alteration in the host that would increase nutrient flow to the fungus. Host specificity is more difficult to explain, since the definition of commensalism excludes the induced defense mechanisms that are the primary concern of attempts to explain specificity of parasites at the host cultivar level. The obvious explanation of specificity in terms of nutritional adequacy of host exudates suggests that commensals are more discriminating than parasites in their nutritional requirements. Such a challenging assumption might stimulate more interest in specificity and host resistance at the species level, where induced defense mechanisms may be relatively less important than at the cultivar level.

For practical reasons, studies on symbiosis, in which the intimate association between host and fungus is of mutual benefit, have dealt primarily with mycorrhizae. A point of theoretical interest is the higher level of host compatibility and specificity exhibited by ectomycorrhizal fungi, which are culturable and are closely related taxonomically to saprotrophs, in comparison with the endomycorrhizal fungi, which have not been grown in culture. The orchidaceous and ericalean "mycorrhizal" fungi (12) are of even greater theoretical significance. The orchids almost certainly must be considered parasites or sedentary predators on the fungi that enter their tissues. Although experimental evidence is fragmentary and is derived almost entirely from studies on *Monotropa* in the Ericales, it seems almost as certain that all "saprophytic" vascular plants are parasitic on fungi (9). Studies on *Monotropa* also indicate a flow of nutrients from chlorophyllous plant to "mycorrhizal" fungus to achlorophyllous plant. It is this latter aspect, the transfer of energy from fungus to vascular plant, that most needs investigation. In all of these associations, including the true endomycorrhizae, digestion by the host of intracellular fungus arbuscules and pelotons is suggestive of phagotrophy, and study of these fungus structures may be of relevance in determining the nature of haustoria.

ECOLOGY AND EPIDEMIOLOGY

Ecology is the study of interacting populations of organisms in relation to their physical environment. The basic unit is the community. A community (biocoenosis) is a recognizable association of populations of different species that, through the integrated activities of its constituent species, is adapted to the exploitation of the physical environment of a particular geographical area at a particular period. To give explicit recognition to the interaction of the biotic components with the physical factors of the site, the community may be referred to as an ecosystem (biogeocoenosis). Ecosystem ecology often carries the unwarranted connotation of a holistic approach, which has among its premises the assumption that some understanding of the system may be arrived at through study of the system as a whole even though the functioning of many of its components may remain obscure. Parasitic fungi are peculiarly adapted to this approach because they are biologically labeled by their pathogenicity.

Regardless of the multiplicity of factors that may be involved, there is little question of the influence on the North American oak-chestnut ecosystem of the input of *Endothia parasitica.* In a reductionistic approach, subsystems or subcommunities may be studied with only tacit recognition of the ecosystem, or with explicit interest in their functioning as components of the ecosystem (5). The scope of investigation may be further narrowed to populations of a single species, as in population ecology and population genetics (17,25). Fungi also are peculiarly adapted to the reductionistic approach because, as microorganisms, they may be studied experimentally in controlled microcosms. Epidemiology is the study of disease in ecosystems. Epidemiological investigations may be reduced to consideration of the population ecology of a single parasitic fungus. The epidemiology of the late blight disease in potato communities, for example, is essentially a reflection of the population ecology of the causal fungus, *P. infestans.* In agricultural systems (agroecosystems), however, studies of subsystems must be explicitly related to the overall functioning of the system because agroecosystems are controlled systems with specified production goals. Management of these systems extends from control of the biotic components of dominant crop plants, competing vascular plants, insects, and symbiotic and parasitic microorganisms and their competitors to control of the physical components through such practices as irrigation and soil amendment, and the effects of these factors must be considered at all levels of the system.

Since fungi are heterotrophic, the composition of fungus communities is primarily dependent on the substrate. The potential diversity of the community is inversely related to the resistance of the substrate. The realized diversity is determined by competition. Competitions are influenced by the physical factors of the environment. Communities developing on identical substrates may differ in an aerial habitat and in a soil habitat, one important factor being the rapid and extreme fluctuations in wetting and drying in the aerial habitat. This habitat difference is especially significant in the survival of hemibiotrophs and leads to the disease control recommendation of burying crop residues by plowing. A further consequence of heterotrophy is that succession in fungi (32) can lead only to exhaustion of the substrate and extinction. However, on renewable substrates such as litter from annual leaf fall, there may be cyclical climax communities of fungi that are characteristic of the ecosystems in which they occur (36). Difficulties in controlling soilborne diseases and early recognition that such measures as fungicide applications directed specifically against the pathogen might operate indirectly through their influence on the competitive balance of soil organisms led to the preoccupation of epidemiologists with soil communities. These communities have also been the primary interest of fungal ecologists, despite the fact that ecological problems might be more readily approached through study of the relatively less complex aerial communities. The concern of epidemiologists with ectobiotic aerial communities derives from their influence on infection by parasitic fungi (34). Another functional aspect is recognized in studies of the influence of ectobiotic fungus communities on twigs of conifers on flow of nutrients through forest canopies, with consideration extended to the peer populations of algae, bacteria, and small invertebrates (5). The frequency with which "secondary fungi" are obtained in routine attempts to isolate culturable parasites should direct the attention of epidemiologists to the endobiotic communities that have excited the interest of fungal ecologists (20). Oddly, however, the best examples of competitive relationships in endobiotic fungus communities have come from studies of rust diseases (4). Studies such as these on rusts and on suppression of pathogens in soil (18,29) suggest that an understanding of the natural balances in fungus communities that may be generally effective in reducing levels of disease is of greater importance than the occasional discovery of more spectacular means of biological control of specific diseases. Although practical considerations have led epidemiologists to far-ranging investigations on fungus communities, it should be apparent that the primary responsibility for development of general theory with general applications rests with fungal ecologists.

TAXONOMY AND NOSOLOGY

In the absence of undergirding theory, such as the theory of evolution on which the taxonomy is based, classifications of diseases are entirely pragmatistic, with predictive value in disease control as the determinant. The criticism of an etiological classification as a classification of pathogens rather than diseases is valid so long as the taxonomy is used only as an outline for discussion of diseases. Although the taxonomy may serve as a preliminary guide, disease classification requires a grouping of specific diseases on the basis of their overall characteristics, including pathology, epidemiology, and applicable control measures as well as etiology. The test of the classificatory hypothesis is the goodness of fit between the resulting classification of diseases and the taxonomy. The classification of symbioses provides an example. Investigations on all aspects of mycorrhizal relationships, which have attracted attention from both mycologists and pathologists, have resulted in the recognition of two distinctive groups, the endo- and the ectomycorrhizae, and this division is reflected in the taxonomy of the fungal symbionts.

Alternative starting points for disease classification are available. In 1774 Fabricius (8) presented a classification based on pathology in which he recognized "genera" of diseases and treated individual diseases as "species" under the genera. There is a remarkable correspondence between Fabricius's genera and the discussion topics outlined by Horsfall and Dimond (13) nearly 200 years later: Rendering unproductive—Reproduction is affected; Wasting—The host is starved; Decaying—Tissue is disintegrated; Rendering misshapen—Growth is affected. The most notable modern example is the hierarchical classification advanced by Kommedahl and Windels (14), which groups root diseases into pathogen-dominant diseases and host-dominant diseases with tissue-specific pathogens vs tissue-nonspecific pathogens and macerative vs toxicogenic pathogens as major subcategories. This contribution goes beyond the delineation of categories to the classification of specific diseases and to generalizations on control based on the classification. Such attempts at classification may be most useful in drawing attention to present deficiencies in comparative pathology. The selection of fungus taxa given to illustrate the diversity in biotrophy, for example, was necessarily based in part on observations made by taxonomists on herbarium specimens. Such attempts point also to difficulties both in concept and in terminology in distinguishing between fungus and disease. Despite its title, the proposed nutritional classification of fungi might be considered a pathological classification since it is based on the interaction of fungus and host throughout the disease cycle. Kranz (16) has proposed a comparative epidemiology and a classification of diseases based on epidemiology. The progress curves of diseases on wild and cultivated plants he (15) analyzed in a series of earlier papers illustrate the possibilities and should be of interest to fungal ecologists as well as pathologists. The basic epidemiological division, however, is between airborne and soilborne diseases, and it is frustrating to discover the inevitability with which all approaches to disease classification converge on this one point.

A reciprocal flow of information occurs between nosology and taxonomy. From a few examples of the effectiveness of selective fungicides, pathologists extrapolate through the taxonomy to generalizations on the use of specific fungicides against major groups of fungi. Selective fungicides then may be used as taxonomic tools in the classification of both saprotrophic and parasitic fungi (10). In this connection it should be noted that studies on the nature and inheritance of fungicide tolerance have been made on convenient genetic systems provided by saprotrophic fungi (31). Taxonomy is the study of populations. Much of the information on fungus populations, and particularly on subspecific populations, is generated by plant pathology, and mycologists and pathologists have a joint responsibility for incorporating these data into the taxonomy. Mycologists also must share with pathologists in the opportunities for study of fungal populations, population shifts, and speciation (17,25) that are afforded by agroecosystems. Because agricultural systems are controlled, they are essentially experiments, laid out on a grand scale and freely available for scientific exploitation.

When mycological flights of fancy end, they return to the same point. The persistent need, perceived by all who work with fungi or the problems they cause, is for the definition of species, the organization of species into classifications however tentative, and the provision of some means of identification (1,28,37). This is the challenge least likely to be accepted.

LITERATURE CITED

1. Barr, M. E. 1978. The Diaporthales in North America. J. Cramer, Lehre, Germany. 232 pp.
2. Bateman, D. F., and Beer, S. V. 1965. Simultaneous production and synergistic action of oxalic acid and polygalacturonase during pathogenesis by *Sclerotium rolfsii*. Physiol. Plant Pathol. 55:204-211.
3. Brian, P. W. 1967. Obligate parasitism in fungi. Proc. R. Soc. (London) Ser. B 168:101-118.
4. Byler, J. W., Cobb, F. W., Jr., and Parmeter, J. R. 1972. Effects of secondary fungi on the epidemiology of western gall rust. Can. J. Bot. 50:1061-1066.
5. Carroll, G. C. 1979. Forest canopies: Complex and independent subsystems. Pages 87-107 in: Forests: Fresh Perspectives from Ecosystem Analysis. R. H. Waring, ed. Ore. State Univ. Press, Corvallis. 200 pp.
6. Cooke, R. 1977. The biology of symbiotic fungi. John Wiley & Sons, New York.
7. de Bary, A. 1887. Comparative morphology and biology of the fungi, mycetozoa, and bacteria. English translation of de Bary (1866), 2nd ed., 1884, by H. E. F. Garnsey and I. B. Balfour. Clarendon Press, Oxford. 525 pp.
8. Fabricius, J. C. 1774. Attempt at a Dissertation on the Diseases of Plants. English transl. by M. K. Ravn, 1926. Phytopathol. Classics No. 1. Am. Phytopathol. Soc., St. Paul, MN. 66 pp.
9. Furman, T. E., and Trappe, J. M. 1971. Phylogeny and ecology of mycotrophic achlorophyllous angiosperms. Q. Rev. Biol. 46:219-225.
10. Hall, R. 1979. Fungitoxicants and fungal taxonomy. Bot. Rev. 45:1-13.
11. Harter, L. L., and Weimer, J. L. 1921. Studies in the physiology of parasitism with special reference to the secretion of pectinase by *Rhizopus tritici*. J. Agric. Res. 21:609-625.
12. Hayman, D. S. 1978. Endomycorrhizae. Pages 401-442 in: Interactions Between

Non-Pathogenic Soil Microorganisms and Plants. Y. R. Dommergues and S. V. Krupa, eds. Elsevier Sci. Pub. Co., Amsterdam. 475 pp.
13. Horsfall, J. G., and Dimond, A. E., ed. 1959. Plant Pathology. Vol. 1, The Diseased Plant. Academic Press, New York. 674 pp.
14. Kommedahl, T., and Windels, C. E. 1979. Fungi: Pathogen or host dominance in disease. Pages 1-103 in: Ecology of Root Pathogens. S. V. Krupa and Y. R. Dommergues, eds. Elsevier Sci. Pub. Co., Amsterdam. 281 pp.
15. Kranz, J. 1968. Eine Analyse von annuellen Epidemien pilzlicher Parasiten. III. Über Korrelationen zwischen quantitativen Merkmalen von Befallskurven und Ahnlichkeiten von Epidemien. Phytopathol. Z. 61:205-217.
16. Kranz, J. 1978. Comparative anatomy of epidemics. Pages 33-62 in: Plant Disease. Vol. 2, How Disease Develops in Populations. J. G. Horsfall and E. B. Cowling, eds. Academic Press, New York. 436 pp.
17. Leonard, K. J. 1978. Polymorphism for lesion type, fungicide tolerance, and mating capacity in *Cochliobolus carbonum* isolates pathogenic to corn. Can. J. Bot. 56:1809-1815.
18. Lin, Y. S., and Cook, R. J. 1979. Suppression of *Fusarium roseum* 'Avenaceum' by soil microorganisms. Phytopathology 69:384-388.
19. Link, G. K. K. 1933. Etiological phytopathology. Phytopathology 23:843-862.
20. Luginbühl, M., and Müller, E. 1980. Endophytische Pilze in den oberirdischen Organen von 4 gemeinsam an gleichen Standorten wachsenden Pflanzen (*Buxus, Hedera, Ilex, Ruscus*). Sydowia, Ann. Mycol. Ser. II. 33:185-209.
21. Luttrell, E. S. 1946. Black rot of muscadine grapes. Phytopathology 36:905-924.
22. Luttrell, E. S. 1974. Parasitism of fungi on vascular plants. Mycologia 66:1-15.
23. Mercer, P. C., Wood, R. K. S., and Greenwood, A. D. 1975. Ultrastructure of the parasitism of *Phaseolus vulgaris* by *Colletotrichum lindemuthianum*. Physiol. Plant Pathol. 5:203-214.
24. Münch, E. 1929. Ueber einige Grundbegriffe der Phytopathologie. Z. Pflanzenkr. 39:276-286.
25. Puhalla, J. E. 1979. Classification of isolates of *Verticillium dahliae* based on heterokaryon incompatibility. Phytopathology 69:1186-1189.
26. Punja, Z. K., and Grogan, R. G. 1981. Mycelial growth and infection without a food base by eruptively germinating sclerotia of *Sclerotium rolfsii*. Phytopathology 71:1099-1103.
27. Scheffer, R. P., Nelson, R. R., and Ullstrup, A. J. 1967. Inheritance of toxin production and pathogenicity in *Cochliobolus carbonum* and *Cochliobolus victoriae*. Phytopathology 57:1288-1291.
28. Sherwood, M. A. 1980. Taxonomic studies in the *Phacidiales*: The genus *Coccomyces* (*Rhytismataceae*). Farlow Herb. Occasional Papers 15:1-120.
29. Shipton, P. J. 1977. Monoculture and soil borne plant pathogens. Annu. Rev. Phytopathol. 15:387-407.
30. Tani, T., Yamamoto, H., Onoe, T., and Naito, N. 1975. Initiation of resistance and host cell collapse in the hypersensitive reaction of oat leaves against *Puccinia coronata avenae*. Physiol. Plant Pathol. 7:231-242.
31. Van Tuyl, J. M. 1977. Genetics of fungal resistance to systemic fungicides. Meded. Landbouwhogesch. Wageningen 77(2):1-136.
32. Visser, S., and Parkinson, D. 1975. Fungal succession on aspen poplar leaf litter. Can. J. Bot. 53:1640-1651.
33. Walker, J. C., and Stahman, M. A. 1955. Chemical nature of disease resistance in plants. Ann. Rev. Plant Physiol. 6:351-366.
34. Warren, R. C. 1972. Interference by common leaf saprophytic fungi with the development of *Phoma betae* lesions on sugarbeet leaves. Ann. Appl. Biol. 72:137-144.
35. Webster, J. 1979. Dialogue on the ecology of anamorphs and teleomorphs. Page 625 in: The Whole Fungus, Vol. 2. B. Kendrick, ed. Nat. Museum Nat. Sci., Ottawa.
36. Wicklow, M. C., Bollen, W. B., and Denison, W. C. 1974. Comparison of soil microfungi in 40-year-old stands of pure alder, pure conifer and alder-conifer mixtures. Soil Biol. Biochem. 6:73-78.
37. Ziller, W. G. 1974. The tree rusts of western Canada. Pub. 1329. Can. For. Serv., Victoria, B.C. 272 pp.

CHAPTER 24

NEMATODES

M. V. McKENRY and H. FERRIS

Plant-parasitic nematodes can exert substantial stress on plant health (18). The intensity of their challenge varies with geographic region, cropping intensity and sequence, cultivar selection, soil conditions, and nematode community structure.

Nematologists, over the past four decades, have responded to this challenge. Numerous agricultural practices designed to reduce nematode numbers have been tested and refined. It is possible to mitigate almost any plant-parasitic nematode problem except those pertaining to established perennial crops. Quarantines, lengthy nonhost crop rotations, soil heating, eradicative soil fumigation, and resistant planting stocks represent examples of powerful tools for crop and nematode management. Nematologists have used these same tools to improve their recognition of nematode problems and their understanding of nematode biology. The major limitation to the practical utilization of these tools has been their cost; whether the cost is borne by the grower, consumer, or the rest of the society, these approaches are expensive. Each should be evaluated in terms of the expected benefit. The challenge to nematologists at the applied level has been in the development of cost-efficient nematode reduction practices. Frequently cost-effectiveness has been ignored in favor of degree of control. For example, a 10-year nonhost rotation is a relatively sure method of reducing nematode populations, especially when used in combination with soil fumigation or sanitation procedures. In practice, of course, in a cash economy with limited land resources, farmers cannot afford such luxuries. Nematologists have been developing regulatory, chemical, cultural, and other nematode reduction strategies, in combination with the current status of the understanding of nematode biology, to mitigate nematode problems. There are environmental and agricultural trade-offs to each management strategy. This basic challenge is not expected to change in the decades ahead. Nematode components of crop and pest management systems are studied in association with diverse food and fiber crops on a worldwide basis.

Since nematodes and the damage they cause are of relatively low visibility to growers, considerable effort is necessary to understand nematode biology and to bring about grower awareness of nematode diseases. This has been accomplished largely through demonstration field plots and nematode sampling results and extension bulletins made available to individual farmers and homeowners. It is incumbent upon phytonematologists to teach the agricultural and governmental sectors along with interested public to predict the nematode damage in a given field or situation. We have an obligation to raise, but not exaggerate, the importance of nematodes.

Nematology is a young discipline. Usually the nematologist involved with implementing practical control strategies also has responsibilities in some aspect of fundamental research. There are, however, chemists, parasitologists, biologists, geneticists, zoologists, gerontologists, ecologists, and computer scientists who utilize nematodes primarily as test organisms to answer fundamental questions (19). These scientists may have little direct interest in the health of plants. The evolution of this divergence is natural, but it demands that nematologists maintain active communication links to the diverse and expanding literature on nematodes.

This chapter represents an attempt to project challenges of both a basic and applied nature, primarily as they relate directly to plant health. The phylum Nematoda is considered to consist of about 50% marine forms, of which the biology is largely unknown, 25% that feed on fungi and bacteria, 15% parasitic or predaceous on animals, and 10% that parasitize higher plants. We elect to consider only the latter. These plant parasites are included in only two of the 12 orders of the phylum.

PREDICTING A NEMATODE PROBLEM

Predictability is an important component of plant nematology because prevention and avoidance provide the cornerstone of economic control methods. Plant-parasitic nematodes are present in soil or plant tissues. The obvious steps in predicting a potential nematode problem include: 1) separation of nematodes from soil or tissues, 2) identification of potentially damaging species, 3) determination of the population of each species, and 4) estimation of the potential damage to the specific crop by relying on previous field or greenhouse correlations.

Decades of nematological experiences have demonstrated that this approach sometimes has merit. However, failures are sufficiently frequent to shake confidence and reaffirm our lack of total understanding of the biology involved. There are some apparent shortcomings in the current technology. First, not all nematode life stages are readily separated from soil and plant material. Second, the efficiency of the extraction procedure varies with the nematode genus, the soil texture, and the extraction technician. Unfortunately, extraction efficiencies have not been quantified. In practice, one cannot select the best method of extraction until the nematodes to be extracted have been identified and the objectives specified. For example, the extraction method selected should maximize recovery of the nematode parasites of the anticipated next crop. Third, speciation of nematodes requires considerable experience, investment of equipment, and reference material. Fourth, the laborious technical jobs of sampling, extraction, and counting are a source of discrepancies, especially when comparing values from one technician to those from the next. Fifth, soils usually contain several potential plant parasites. Unfortunately, most of the pathogenicity data are derived from individual species on specific host cultivars.

Synergistic and interaction effects are largely unknown. Equally problematic is the vast amount of field data that have been based on use of a broad-spectrum biocide as a means of providing a nematode/no-nematode test situation, with the implicit assumption that yield differences reflect nematode influence.

To the grower or technical farm advisor, the need for a standardization of methodologies is apparent. However, the danger that standardization may retard progress and innovation should be realized and avoided. Applied personnel need to be able to interface the nematode status of a field with other field characteristics or pest problems (1). Except for a few crop or nematode situations, however, there is no routine method of accomplishing this goal. At least, we have not sufficiently explored the strategy. It is the continuing problem of validation of hypotheses.

In the area of soil sampling, we continue to need basic guidelines for identifying the number of subsamples needed for a given field situation. Additionally, there is a need to standardize or remove subjective error from the sampling and extraction procedure. This cannot be accomplished without fundamental investigations followed by confirmations from various laboratories.

THE NEMATODE-PLANT-SOIL INTERFACE

The Nematode Component

The area of nematode taxonomy received voluminous attention by early nematologists. A necessary requirement of biological research is an identifying label for morphological or physiological variants. The vast majority of this work has been based on morphological determinations under the light microscope, with very few companion studies of a biological nature. During the past decade use of the scanning and transmission electron microscopes has provided valuable support to the early studies. The challenge and frustration in taxonomy is that nematodes do not possess a great number of distinguishing morphological characteristics. Further, liberal translation of the species definition is necessary in pathogenic organisms. Also, behavioral characteristics that may enhance species determinations or relationships are difficult to assess with preserved specimens or slides. In some cases the identification of a nematode species cannot be completed until verified by a single taxonomist somewhere in the world. This makes quick decisions by a plant doctor relative to a regulatory or pest managment approach more difficult. The alternative solution in such cases is chemical control as insurance against nematode damage.

In the area of taxonomy there is a fundamental constraint that hinders application of practical information for several important plant parasites. For example, the major nematode genus detrimental to crop production worldwide is undoubtedly the root knot nematode, *Meloidogyne*. The inability of the applied nematologist to reliably, accurately, and readily identify the individual species of the genus is an incredible deterrent to the application of any pest control approach aside from the use of chemicals. Major efforts in this direction are being made through North Carolina State University in the International Meloidogyne Project (16). Perhaps too much reliance and hope is placed on the outcome of these studies by nematologists within the United States. Nematologists in each state should be identifying local populations of root knot.

They should be determining the host ranges by any means of identification. This would allow selection of resistant cultivars or the use of appropriate rotation schemes. Our work in California vineyards and orchards indicates that mixed populations are common in perennial crops; however, the selection of a proper intercropping or cover-cropping plant should certainly include consideration of the potential for root knot nematode population increase. Identification of mixed populations of root knot nematodes in fields used for annual crop production might indicate patterns of crop rotation that result in a balance of the population. Thus, rather than allowing one of the populations to become dominant and destructive to crops, we could use a stabilizing selection approach similar to that proposed by Jones et al (7) for management of biotype composition of the golden nematode in potato fields.

The taxonomic challenge is not limited to *Meloidogyne*. Consider *Xiphinema americanum* or *Ditylenchus dipsaci*. These two species complexes likely require a multifaceted approach including morphological, biochemical, and host range studies to clarify logical groupings and relationships.

Gaps are apparent in our knowledge of nematode dispersal, motility, overwintering, and survival. New information casts doubt on earlier notions that nematodes move only short distances. Recent studies (15) have indicated that in 3 days' time approximately 33% of a *Meloidogyne* population can move a vertical distance of 25 cm. Movement across such distances was somewhat unexpected, based on numerous field experiences and observations. Once again, many of those field experiences involved fumigated versus nonfumigated soils. For example, in row crops it has been a common practice to apply soil fumigants in bands and thus treat only the immediate vicinity of the seedling. The success of this practice would suggest that nematodes either do not move readily or that nematodes are most devastating in the early stages of root development. A recent paper described a nematode-inhibiting factor, apparently biological in nature, that can be present in soil following a fumigation (12). Other work indicates a minimum clay component requirement to allow adsorption of attractant factors from plant roots to accelerate nematode movement. There is a danger in extrapolating information from diverse testing conditions. Apparently, however, nematodes are capable of greater motility than generally recognized.

For plants grown in regions with a distinct winter period, there is considerable value in identifying the overwintering stages and their longevity. Active juvenile stages may deplete energy reserves and lower their inoculum potential. However, from an advisory standpoint, these are the forms most readily extracted from soil, and therefore perhaps an inappropriate measure of the potential threat to the next crop. Conceptually, in perennial crops even the adult female stage of root knot nematode may overwinter. This would accelerate the phenological calendar of the nematode community the following year. However, some rather obvious questions have been ignored due to technological difficulties.

The Host Component

Nematologists need to devote greater attention to the host plant. Nematologists commonly report the nematode population as nematodes per volume or weight of soil. These units are useful, but there would appear to be greater practical relevance to reporting the numbers of nematodes on a root basis or at least relating the amount of root per volume of soil. It is well documented that root knot feeding sites act as metabolic sinks for photosynthate (11). The stress potential of the nematode in perennial crops may be more appropriately related to the leaf area supporting these sinks than to unit of root in quantification of effects. Nematode sampling by the authors indicated that each grapevine in a specific established vineyard was associated with $2-10 \times 10^6$ plant-parasitic nematodes (3). Meanwhile, the grapcvine possessed seasonal flushes of root activity, but the total number of root tips that would be targets of the nematode were in the order of $3-6 \times 10^4$. With quantitative data such as these, better concepts of the value of the timing and requirementsof nematicidal agents can be developed.

Focusing from the macroscopic to the microscopic and biochemical levels, there are a series of unanswered questions relative to host resistance, tolerance, phytoalexins, and other defense mechanisms. Discussion in these areas is limited unless terminology is repeatedly defined. Enter into the discussion a geneticist or plant breeder, and a contrary viewpoint emerges. Enter also into the discussion a biochemist or physiologist with comments on inhibitory components such as alkaloids, glycosides, and phytoalexins while simultaneously mentioning nematode-attracting substances such as the amino acid proline, and new insights alter our understanding. It becomes obvious that there is a tremendous need for interdisciplinary study in this area (6). Further, greater attention should be given to one or two hosts in an effort to develop an in-depth but integrated understanding.

A number of scenarios can be developed to emphasize the need for more precise information on the specific mode of action of resistance. For example, the rootstock Nemaguard is the most important root knot nematode deterrent in the deciduous tree fruit industry of the United States. The mechanism of resistance is an inability of the nematode to develop a feeding site (9). The nematode enters the root but apparently exits after a brief period. Consider a young seedling with 10–50 root tips planted into field soil with 1×10^6 plant-parasitic nematodes per tree site. Substantial physical root damage can occur during the first year. This resistance mechanism will have greater value in a field maintained free from weed hosts of the nematode. Consider instead, a mechanism for resistance that involves a physical, biological, or chemical barrier at the root surface (4). The presence of a more attractive or more accessible trap root system as companion to the agronomic crop may serve as an advantage to the grower.

The Ecosystem

The soil is a combination of physical, chemical, and biological subsystems that constantly influence the outcome of the nematode-host interface. Perhaps the major challenge here for nematologists is that once again an interdisciplinary approach would be appropriate. Fewer than 20% of the soil-inhabiting nematode species are plant parasitic. Unfortunately advances in our understanding of nematode community structure and biological management are limited by the fact that most nematologists who are oriented toward plant parasites are unable to recognize any but those species. Among the remaining 80% of soil-inhabiting nematodes, there are various trophic levels including fungal feeders, bacterial feeders, protozoan feeders, insect parasites, annelid parasites, higher animal parasites, nematode predators, and others. It is highly likely that these organisms are contributing to nutrient cycling in the soil system. Thus, the nematode community structure and trophic level composition could be a strong indicator of soil fertility and productivity. Similarly, a large number of other soil-inhabiting organisms are predaceous, parasitic, or inhibitory to plant parasites. The management approaches used to reduce preplant and postplant populations of plant parasites may be equally detrimental to the biological antagonists of nematodes. Such a situation is not uncommon in pest biology. The need for an expanded approach to soil biology is apparent.

There is little application of ecological principles to the study of nematode community structure and particularly to the manipulations of nematode biological antagonists for management of plant-parasitic nematodes (13). Considerable conceptual and practical advances are being made in this area by scientists in other pest disciplines. For example, entomologists have developed and verified hypotheses for the probability of successful establishment of a biological antagonist into a system in which the niche is already occupied by other antagonist species and into systems in which there is a void at this trophic level. Further, the hypotheses are expanded to cover the case of multiparasite introductions and their probability for successful establishment relative to the number of incumbent species in that niche region in the system. Yet a further consideration in these hypotheses is the probability of management success relative to the number of antagonistic species successfully introduced. The hypotheses would indicate that the probability of a k level (carrying capacity) reduction for the target pests in the system increases with the number of antagonists either present or successfully introduced. The rationale is that introductions of several species of antagonists would generate a dampening of the typical predator-prey oscillations since the antagonists may be differentially influenced by the environment and have differential alternative food sources. Consideration of the basic biological and ecological principles encompassed by this rationale might provide a logical framework for the study of nematode management.

The whole area of biological antagonists has received considerable attention in recent years (17). It is apparent that in several nematode species, egg parasites can be of considerable importance in reducing populations. As with other soil organisms, it may take a number of years of continuous cropping of annual crops before the antagonists are increased to a sufficiently high population to reduce or manage the parasite population. With *Heterodera avenae* in England, there are considerable losses for the first 2 or 3 years of continuous cropping (8). However, after 5 or 6 years, the *H. avenae* populations are maintained at or below the economic threshold. Of course, this threshold level is somewhat higher than would be tolerable in warmer regions because of the increased number of generation times with higher temperatures. There may be specific crops or cropping sequences that enhance the establishment of nematode antagonists. Viable research goals in this area include the solution of establishment problems referred to above, determination of the effects of standard cultural practices on the establishment and survival of the antagonists, and considerations of soil disturbance, tillage, and pesticide effects.

Certain nematode species have long been identified as predators upon specific insect populations. In the last decade the epidemiological aspects of this predator have been studied in considerable detail (14). Most recently, nematode species such as *Neoaplectana carpocapsae* have been mass produced and applied as sprays to control practical insect problems of plants. There have been successes and failures in this area (10). Today there are several laboratories mass-producing these potential biological protectors of plant health.

THE NEMATODE-AGRICULTURE INTERFACE

Some Dangers of Specialization

During this century some predictable changes have occurred in the activities and interests of nematologists. Early workers in the field were generalists; they had responsibilities or interests in such areas of agriculture as plant pathology or entomology and often strayed into the study of nematodes from these areas. Their broad general interests are reflected in the writings and activities of such pioneers as Cobb, Chitwood, Christie, Godfrey, Steiner, Allen, and many others. Often, their interest in nematology was the result of a specific encounter with a field problem. Sometimes the interest was purely one of curiosity in biological phenomena.

The impact of various nematodes on food and fiber production was obvious for the few nematodes that cause distinctive symptoms in the host. However, the impact of many nematode species only became apparent through technological advances that resulted in nematode control and yield improvement. Management successes generated advances in several areas, including promotion of grower efficiency in nematode and crop management (extension activities), a stimulus to the agricultural pesticide industry, and an increase in the number of nematologists! The economics of these simultaneous and interdependent events have not been studied; the results of such a study might well be interesting. The expanded nematologist population researched many fundamental questions, some of which were only of long-term interest to applied agriculturalists.

Few of the nematologists of the 1980s have received fundamental training in agriculture. They have received training in a great number of diverse subjects, including soil science, ecology, parasitology, computer science, integrated pest management, biochemistry, and genetics. Many working in basic areas have only limited experience with the range of histopathological phenomena and symptoms associated with nematode parasitism of plants. Similarly, they may be unfamiliar with the range of nematode morphology, biology, and ecological phenomena exhibited by the phylum. We do not contend that it is important for nematologists to have worked with specific groups of nematodes or to be involved in practical field problems. Each individual is either assigned specific job responsibilities or weighs the working objectives and priorities according to personal interests and expertise. However, we do maintain that it is of paramount importance that nematologists maintain the ability to interact and communicate with each other and that such interaction be formalized through cross-linkages in objectives to promote progress toward common goals.

Further on the subject of communication, while nematologists have communicated well with each other over the past several decades, it is apparent that contemporary nematologists have not excelled in communicating with their clientele and, specifically, with production agriculture. This failure stems from both sides. Insects and mites can be seen crawling, flying, defecating, or chewing on and around plant parts. Stem and leaf fungal and bacterial pathogens or damage from them is usually readily visible. Likewise the competitive effects of weeds, and the damage of vertebrate pests can be readily observed. However, recognition and monitoring of nematode presence is much more difficult. The pathogenicity is more insidious and less apparent at the plant and farm level. In short, there is minimal visibility of nematodes as biotic constraints of plants. Hence, it is important that the scientists and individuals with expertise and understanding in this area provide their clientele with a realistic appraisal of the pest potential of this group of organisms in various cropping situations.

The pressures that direct the course of production agriculture are constantly varying in their relative impact. They include water supply, energy costs, economic and marketing strategies, requirements for increased productivity, subsidies, labor force availability and contracts, new equipment, new agricultural chemicals, new plant germ plasm, and new tillage and pest management philosophies. Consequently, practitioners of plant health must deal with food and fiber production in a state of dynamic flux. Their management recommendations must be customized to individual situations and incorporate consideration of the economic and environmental climate for that situation. At the same time, the management recommendations must be sufficiently robust to stand up under stochastic variations in the pressures upon the particular production system. The changing baseline for crop management challenges our understanding of the relationship between pest biology and plant growth. The complexity of the production systems has resulted in a disciplinary emphasis on various aspects. Thus, the role of the plant nematologists may be perceived as a requirement to provide a basic understanding of the biology and ecology of nematodes and, further, to provide the philosophy and tools for their management and control. Common to all of the disciplines that are constituent to agricultural production is an umbrella higher level of organization; perhaps a commodity research institute or an agricultural experiment station. Frequently, however, the individual specialist may lose sight of the required integration of information across disciplines to promote understanding of the behavior of the production system as a whole.

As with all the disciplines within agricultural production, phytonematology has both basic and applied aspects. The rationale behind pursuing the basic studies in plant nematology is to provide the background information and tools for practical applications. It is important that nematologists at all levels of research have among their objectives and priorities the interests and needs of their clientele and funding sources. This is somewhat easier to do at the advisory and grower interaction levels than in university or research institute laboratories. Ultimately, sources of funding stem from the tax-paying public or by taxes levied on commodities and passed on to the consumer. The basic researcher is doing a disservice to this clientele if no effort is made to ensure the extension of research progress or its integration into a higher level of organization to promote understanding of the biological systems. The responsibility for this integrational activity rests upon both the basic and the applied researcher. Too often the claim is made that applied researchers are paying no attention to the progress and results generated through basic research activities or that basic researchers are paying little or no attention to the problems encountered by applied nematologists in the field. The appropriate administrative linkages usually exist between and among the individuals involved, but the individuals must make an effort to communicate with each other in a common language. This communication would be fostered by a sharing of research objectives and a requirement for accountability in progress toward these objectives.

As nematologists respond and react to changes in agriculture, it is of paramount importance that interface be maintained with biological development in diverse areas of plant health.

Nematode Damage and Management

The impact of nematodes upon plant growth is dependent upon and influenced by various extrinsic environmental factors and management practices. These include soil texture, soil moisture regime, time of planting, cropping history, host susceptibility, geographic region, development degree days, and others. The prediction of a damaging level of nematodes has to be customized to the extrinsic factors associated with an individual field and further modified by the economics of production in that field. However, all too frequently applied nematologists have ignored the fund of information available on the ecology and biology of nematodes and have recommended the management or control of nematodes on a "one size fits all" basis. Field conditions associated with nematode presence and damage potential must receive greater attention by nematologists if agriculture is to be served adequately and efficiently. Concomitantly, nematode visibility must be enhanced if the plant damage caused by nematodes is to be reduced. The yield of a nematode-infected plant may be reduced from 10 to 50% with only such subtle symptoms as "stunting and chlorosis" associated with the presence of the nematode. A grower may plant in nematode-infested soil and may experience considerable yield reduction without ever fully realizing the extent of the loss. Wallace (20) summarized the situation:

> . . . our assessment of the importance of nematodes in agriculture relies mostly on faith, or on those situations where the crop is decimated. Perhaps the less dramatic losses that miss our notice, and which cover much larger areas, are in the long run the most important.

The development of resistant and tolerant cultivars is an expensive and long-term process. Numerous examples attest to the benefit of this form of nematode control; however numerous breeding programs continue without nematological input. In the enlightened atmosphere of agricultural research today, it seems absurd that plant breeders would work in a vacuum relative to nematode and other pest problems. By the same token, it is a failure on behalf of nematologists that they have not brought nematode problems more forcefully to the attention of plant breeders. Much emphasis in plant breeding approaches has been toward the selection for lack of symptoms or pest evidence. This frequently results in unacceptable cultural, yield, or quality properties. Field selection for vigorous growth and tolerance however, may result in more productive breeding programs. An example exists within the grape industry in California. Root knot-resistant rootstocks of grapevine confer a vigor to the scion that results in an increase in vegetative production of the vine at the expense of yield in many soils. On the other hand, these rootstocks are successful in sandy, water- and nutrient-deficient soils that are also conducive to the nematode. In sandy loam and loamy sand, conditions less favorable to the nematode, the cultural aspects of the rootstocks become limiting. The widely grown Thompson Seedless and other cultivars of mideastern origin have a considerable degree of tolerance to root knot nematodes compared with cultivars of European origin (4). This tolerance is sufficient to minimize yield losses due to root knot nematodes in all but extremely sandy locations. An innovative approach to nematode management in sandy loam and finer textured soils where root knot might be damaging to extremely susceptible cultivars would be to use Thompson Seedless as a rootstock. Similarly, crosses between Thompson Seedless and the immune or resistant rootstock may improve their cultural characteristics and incorporate another type of resistance.

Rotation schemes have provided the most common method of control of nematodes. Each crop plant may be attacked by one to several nematode species, yet on a regional level there are numerous examples of the avoidance of nematode damage through proper selection of rotation schemes. Literature available on this subject is sparse. There is a need for the cataloguing of levels of resistance and tolerance of various cultivars of agricultural crops. This will indicate the probability of avoiding damage from the current crop and any danger of generating a nematode problem for planned subsequent crops. Much productive work can be done in this area in small field plots and laboratory and greenhouse experiments. The availability of minicomputer systems and electronic data manipulation provides efficient means of storing and accessing such data bases.

Nematicides are available that prevent or delay crop damage and the increase of nematode populations. However, nematode increases can be expected within several months or years following application, depending upon the efficacy of the specific treatment and upon the climate for reinfestation for the field. A major challenge to the chemical control of nematodes is that the current small arsenal of nematicides is becoming smaller (5). The current nematicides are generally byproducts of the petroleum industry, and most have already outlived their patents. Replacement chemicals are not being developed because of the relatively small size of the nematicide market and the high cost of pesticide registration. Meanwhile, heightened public awareness over environmental concerns demands more information on a wide range of environmental effects. Ultimately, the consumer will pay the cost of the necessary research, which will become incorporated into the price of the nematicide.

We think that within the next few years, nematologists and other pest scientists will become more conscious of the need for cost efficiency and of the concept of optimization in pest management. This will involve applying an adequate level of management so that the difference between the crop value at that management level and the cost of the management is maximized. Of course, this all requires knowledge of the relationship between plant yield and growth and numbers of nematodes. It further requires knowledge of the relationships between the nematode and its environment and the plant and its environment and of the interaction of the two. Further knowledge is required on the efficacy of the management practices under different environmental conditions and at different levels of management, since agriculture deals with plantings, not individual plants. It involves considerable biological and environmental monitoring, which will have to be cost effective and reliable.

Integration and Consideration of the Whole System

Throughout, we have alluded to a major challenge to be met across all pest and plant disciplines contributing to agriculture productivity. That is the need to integrate the information and philosophies of the individual disciplines into a higher level of understanding of the behavior of the whole system. In recent years, systems approaches have been used for integration across levels of organization in agriculture and to aid in understanding the behavior of agricultural systems. The systems approach seeks to use basic information derived through reductionist disciplinary research in which the impact of one or only a few variables are studied. This phase is the natural progression of the pattern that the development of the science of nematology has taken. It conforms with the evolutionary pattern of all biological sciences with application in agriculture. After a period of necessary reductionist activities, there follows a period of integrationist activities. The integrators are the mediators of the progress of understanding the system. In this case, the systems analyst, or applied plant health specialist with training in systems analysis, is the integrator of the basic information developed in all pest and plant disciplines. The objective is to understand the outcome of perturbations of the system across all of the interacting components, to allow testing and understanding of alternative crop production and management systems. In this process, there is need for close interaction between the systems analyst and the applied biologist. Further, it is important that the pest and plant biologists have training and fundamental understanding of this systems approach to foster communication. Through this process, the computer becomes one of the many tools of agriculture. As with all other tools, it does not provide all the answers needed by the practitioner. However, the modeling and analytical activities of the systems analyst, using the computer as a tool, provide for the coupling of the information sources in solving complex problems in applied plant production.

In the understanding and modeling of complex agricultural systems, it is convenient to organize the available information into a series of interacting subsystems. Conveniently, these subsystems would represent the individual pest species of importance and a central plant subsystem mediating their interaction. By developing models of the growth of the plant with time and the development or increase of the pest populations with time and prevailing environmental conditions, pest interactions with the plant and consequently with each other can be predicted. First attempts at such modeling, linkage, and prediction do little more than reveal gaps in our knowledge and misconceptions in our understanding of the system, which promotes further appropriate experimentation. When better models are developed and validated, they will provide a vehicle through which management alternatives can be tested before field application. Besides such pest and plant growth models, other models that are important and that promote the cataloguing of available information are management decision and procedure models and economic analysis models.

These efforts of integration at the whole plant or crop production level follow certain prescribed steps:

1. The objectives and goals of the integrative effort are specified.
2. The limits of the production system under consideration are defined.
3. Key subsystems are identified (plant and key pests).
4. The extrinsic parameters influencing the plant and pests are identified (environmental conditions, soil parameters, cropping history).
5. Models are developed of the key subsystems at appropriate levels of complexity.
6. Coupling structures with the plant model are developed for each of the pest models.

7. The behavior of the model is verified. Problems are identified and data gaps are filled by further experimentation.

8. The validated model is used to test and study the simulated behavior of the system and the response to management.

9. Further developments of the model can include higher order interaction (biological antagonists of pests and pest community structures).

The equipment, technology, and implementation packages utilized and developed in the systems approach will add visibility to each component subsystem (in our case, the nematodes). The approach, of course, does not replace the need for basic field observations. Computers will not develop chemical, biological, or other agents of management, but their use will enhance development of such agents. We look to the implementation of the systems approach to aid the visualization and solution of specific challenges and also to allow analysis of various problems and interaction in the context of the behavior of the whole system. Integrated pest management and crop management projects are being developed nationwide as practical vehicles to fit pest control disciplines into the broader picture of crop production. Unfortunately, few nematologists have sufficient agricultural training to fit into these broad study groups (2). There is a need to expose more nematologists to field situations or agricultural training as well as to encourage their development of computational and analytical skills.

IN CONCLUSION

Many of the above considerations were related to the applied aspect of nematology. Supportive to any applied work, however, is the background basic research activity. We have attempted to stress the necessity of examining the biological and crop production system as a whole from a basic research and information gathering standpoint. In other words, we promote a systematic approach to basic research by cataloguing the current state of knowledge through a defined analytical approach and identifying areas in which useful information is lacking. Less efficient, we feel, would be to have basic researchers being at the demand of the perceived needs of applied researchers. However, we strongly support interaction and communication between the two groups, if, indeed, they are even separable into two groups.

In many regards, the problems of applied nematology have been perceived and dealt with in piecemeal fashion. Challenges have been answered with short-range solutions appropriate to the immediate interests of the agricultural clientele. The major roadblocks that limit our understanding and implementation of efficient nematode management seem to have changed little in recent decades. We still need systemic, nonphytotoxic, target-specific nematicides; cultivars with tolerance to nematode damage conferred through a broad genetic base to reduce the potential for virulent pest biotypes; efficient measurement of nematode populations; prediction of potential damage; and economic means of nematode management. All of these have been talked about and written about, but there have been few recent milestones of achievement.

Optimization of nematode and pest management will require better understanding of the links within the system as a whole and between the multiple subsystems that are components of it. It will require the intestinal fortitude to step outside of our scientific jargon and protective specialization to address the problems of applied nematology and applied crop production. Challenges we have suggested are basically similar to those facing applied sciences elsewhere. Conceptually, many of the specific challenges we suggest here are not unique to nematology. In fact, they are not even very recent priorities within nematology. In 1960, Steiner wrote an introductory chapter to a book edited by Sasser and Jenkins in which several of the priorities outlined in this text were listed, as well as several other areas of interest that are still applicable. Our colleagues in other pest and plant production disciplines are solving problems of a similar nature. Information interchange with scientists in these disciplines will speed advances within nematology. It is our hope that current integrationist activities will improve our ability to communicate adequately within the framework of production agriculture.

LITERATURE CITED

1. Anonymous. 1981. Grape Pest Management. Publ. 4105. Dep. Agric. Sci., Univ. Calif. Berkeley. 312 pp.
2. Bird, G. W. 1980. Nematology—Status and prospects: The role of nematology in integrated pest management. J. Nematol. 12:170-176.
3. Ferris, H., and McKenry, M. V. 1974. Seasonal fluctuations in the spatial distribution of nematode populations in a California vineyard. J. Nematol. 6:203-210.
4. Ferris, H., Schneider, S. M., and Stuth, M. C. 1982. Probability of penetration and injection by root knot nematode, *Meloidogyne arenaria* in grape cultivars. Am. J. Enol. Vitic. 33:31-35.
5. Gilpatrick, J. D. 1979. Contemporary Control of Plant Diseases with Chemicals: Present Status, Future Prospects, and Proposals for Action. Am. Phytopathol. Soc., St. Paul, MN. 170 pp.
6. Gommers, F. J. 1981. Biochemical interactions between nematodes and plants and their relevance to control. Helminthol. Abstr. Ser. B, Plant Nematol. 50:9-24.
7. Jones, F. G., Panott, D. M., and Ross, G. J. S. 1967. The population genetics of the potato cyst nematode, *Heterodera rostochiensis*: Mathematical models to simulate the effect of growing eel worm-resistant potatoes bred from *Solanum tuberosum* subspp. *andigena*. Ann. Appl. Biol. 60:151-171.
8. Kerry, B. 1980. Biocontrol: Fungal parasites of female cyst nematodes. J. Nematol. 12:253-259.
9. Kochba, J., and Samish, R. M. 1971. Effect of kinetin and l-naphthylacetic acid on root knot nematodes in resistant and susceptible peach rootstocks. J. Am. Soc. Hortic. Sci. 96:458-461.
10. Lindegren, J., Hoffman, D. F., Collier, S. S. and Fries, R. D. 1979. Propagation and storage of *Neoaplectana carpocapsae* Weiser using *Amyelois transitella* (Walker) adults. Agric. Res. (Western Region), USDA-SEA, Berkeley, CA.
11. McClure, M. A. 1977. *Meloidogyne incognita*: A metabolic sink. J. Nematol. 9:88-90.
12. McNamara, D. G. 1980. The survival of *Xiphinema diversicaudatum* in plant-free soil. Nematologica 26:170-181.
13. Norton, D. C. 1978. Ecology of Plant-Parasitic Nematodes. John Wiley & Sons, New York. 268 pp.
14. Poinar, G. O., Jr. 1975. Entomogenous Nematodes, a Manual and Host List of Insect Nematode Associations. E. J. Brill, Leiden, The Netherlands. 317 pp.
15. Prot, J. C. 1978. Vertical migration of four natural populations of *Meloidogyne*. Rev. Nematol. 1:109-112.
16. Sasser, J. N. 1980. Root knot nematodes: A global menace to crop production. Plant Dis. 64:36-41.
17. Sayre, R. M. 1980. Promising organisms for biocontrol of nematodes. Plant Dis. 64:527-532.
18. Society of Nematologists. 1971. Estimated crop losses due to plant parasitic nematodes in the United States. Spec. Publ. 1. J. Nematol., Suppl. to Vol. 3.
19. Van Gundy, S. D. 1980. Nematology situation in the United States and priority areas in research. Bull. OEPP 10:393-402.
20. Wallace, H. R. 1973. Nematode Ecology and Plant Disease. Edward Arnold, London. 223 pp.

CHAPTER 25

ARTHROPODS

DALE M. NORRIS

Many arthropods derive shelter and/or nutrition from plants, which thus are termed "hosts." By "taking shelter" or "taking nutrition," arthropods frequently challenge the health of such plants, i.e., they cause "injury" and "disease." For decades, concerns about the semantics of "disease" versus "injury" prevailed over interests in developing a more dynamic understanding of arthropod-associated pathogenesis in plants. However, in our modern efforts to understand plant health, we can no longer afford to linger in debates of such semantics but rather are compelled to develop the greatest possible dynamic insights into the subject. Toward this modern goal this chapter addresses arthropods as biotic environmental constraints on plant health. It focuses especially on some major means by which arthropods impair plant health, on current emphases in related research, and on future directions in such investigations regarding plant health.

PLANT-INHABITING ARTHROPODS

The major groups of arthropods involved in affecting plant health are classified in the Class Insecta, i.e., insects, and Order Acari, i.e., mites and ticks (Table 1). The sowbugs (pillbugs) in the Order Isopoda also include important pests of cultivated plants. Harvestmen or daddy longlegs in the Order Phalangida feed on juices and exudates from plants, and some millipedes attack living plants. *Oxidus gracilis* is a common millipede pest of plants grown in greenhouses. Finally, symphylans (Class Symphyla), e.g., *Scutigerella immaculata*, are occasional pests of plants in greenhouses.

HOW ARTHROPODS IMPAIR PLANT HEALTH

It seems safe to say that arthropods, especially insects, impair plant health by "every conceivable means." Some major activities of arthropods that lead to plant death are:

1. surface chewing (feeding) on any portion of a plant, (e.g., leaves, buds, stem, roots, or fruits);
2. sucking sap from cells in these plant parts;
3. boring, mining, or tunneling in these plant portions;
4. causing changes in chemicals (i.e., phytoallactins), which induce alterations (e.g., galls) in plants;

5. using portions of plants to construct shelters or as repositories, (e.g., for deposited eggs);
6. employing plants as a substrate on which to maintain mutualistic species, (e.g., ants tending aphids on plant leaves);
7. transmitting pathogenic biological agents into plant tissues; and
8. cross fertilizing plant-pathogenic agents, (e.g., certain rust fungi).

Applied zoologists historically have especially concerned themselves with categories 1–3,5, and 6. Death from these activities has been considered as simply from physical injuries; however, sucking of sap and boring, mining, or tunneling in tissues involve other dimensions of plant health impairment. Sap withdrawal is an integral parameter of many arthropod-induced phytoallactic states (category 4), and of arthropod transmissions of biological pathogenic agents (category 7). Boring, mining, and tunneling in plants especially create "courts of entry" or "inoculation sites" for pathogenic agents (category 7); thus, the historical disciplinary distinctions between insect-induced injuries and pathogen-induced diseases in plants should be reexamined in the contexts of modern biology. It is becoming increasingly clear that symbiotic multispecies (supraspecies) complexes of arthropods and other biological forms (e.g., microbial agents) are the more likely causal entities responsible for biotic environmental constraints on plant health. This view of symbiotic biological complexes of species as causes of plant death focuses important scientific attention on the ecological and nutritional interdependencies that exist among species.

Arthropod-transmitted pathogenic biological agents (category 7) have been recognized for nearly a century as major causes of impaired plant health. A specialized aspect of such transmissions involves the cross fertilization of plant pathogenic agents, e.g. certain rust fungi, by arthropods (category 8). Scientific investigations in this overall subject area of transmission have contributed greatly to the expanded considerations of complexes of species, rather than single species, as the more likely inducers of plant disease (25).

Arthropod production of phytoallactins that cause changes (e.g., galls) in plants (category 4) is another of the more important means of plant death, but the

TABLE 1. Arthropod pests of plants

Scientific classification[a]	Common name(s)
Subphylum—Chelicerata	
Class—Arachnida	Arachnids
Order—Phalangida	Harvestmen (daddy longlegs)
Order—Acari	Mites and ticks
Suborder—Tarsonemini	Mites
Suborder—Tetrapodili	Gall mites
Family—Eriophyidae	Eriophyid mites
Suborder—Prostigmata	Spider mites and harvest mites
Subphylum—Mandibulata	
Class—Crustacea	Crustaceans
Order—Isopoda	Sowbugs (pillbugs)
Class—Diplopoda	Millipedes
Order—Polydesmida	Millipedes
Class—Symphyla	Symphylans
Class—Insecta	Insects

[a] Source: Borror et al (1).

physiology and biochemistry of such causes of disease are very poorly understood.

When all the categories of plant death are considered, it becomes very clear that arthropods are major participants in the biotic environmental constraints on plant health.

CURRENT EMPHASES IN RESEARCH ON ARTHROPOD-ASSOCIATED PATHOGENESIS IN PLANTS

Major efforts in this subject area have been focused particularly on 1) arthropod, especially insect, transmission of viruses, mycoplasmalike organisms (MLOs), spiroplasmas, or rickettsialike organisms (RLOs); 2) coevolution between plants and arthropods; 3) plant resistance to arthropods; and 4) symbiotic complexes of species as causal entities. Exciting progress has occurred in each of these specialties of research.

Arthropod Transmissions

The current status of information in this very active and productive field has been detailed in several volumes (10,11,18,32). Most known arthropod vectors of viruses, MLOs, spiroplasmas, and RLOs are insects, and these are largely in the order Homoptera. Aphids, leafhoppers, and planthoppers include major groups of such vectors.

More than 180 species of aphids are now reported as vectors of at least one plant virus. Their roles in the transmission of other organisms, however, remain unclear. Advances in the understanding of the ecology of aphid transmissions of plant viruses have provided important insights into new strategies for suppressing such spread. One such important insight involves the fact that aphids especially follow a glucoside gradient to position their mouthparts in the phloem where transmission can occur. Thus, the processes of chemoreception in insects, and aphids in particular, are important to such transmissions. Major advances have occurred in our understanding of how organisms perceive chemicals in their environments, e.g., Norris (28), and such information has provided bases for developing practical techniques to disrupt aphid probing to the phloem. Such behavior alteration can prevent virus transmission. An important practical development that involves disrupting such sensory functions is the application of mineral and vegetable oils to plant surfaces. Such oils effectively interrupt noncirculative transmissions by interfering with the probing activities or impeding virus adsorption on the aphid's stylets. Another practical development is the use of insecticides, including systemics, to reduce the spread of circulatively transmitted viruses. Such pesticides can accomplish this reduction because aphids must feed for relatively long periods if they are to transmit circulative plant viruses, and some insecticide applications can kill the insect during this period. Our increased knowledge of transmission ecology also has contributed to the selection and breeding of plants that are not preferred by, or are resistant to, aphid vectors. More details of these developments will be discussed later.

There are about 130 species of leafhoppers, i.e. Cicadellidae, that are known to transmit more than 70 causal agents of plant disease. These agents include viruses, MLOs, spiroplasmas and RLOs. Through increased understanding of the ecology of transmissions, it is now known that although most leafhopper transmissions are circulative, a small number of viruses, e.g., rice tungro virus, are transmitted in noncirculative manners.

Applications of insecticides have proven effective in reducing the incidence of circulative transmissions by leafhoppers. The basis for this practical use is the same as that previously discussed for aphid vectors of circulative agents. Leafhoppers also usually have specific interactions with plants, e.g., they mostly feed in phloem during transmissions. As with aphids, leafhoppers apparently rely on chemical messengers to guide their mouthparts to the phloem of a plant. Therefore, there is considerable ongoing research to understand the chemoreception involved in feeding by leafhoppers.

These current emphases in the investigation of leafhopper vectors are generally true also for the third most important group of vectors, the planthoppers. This is true because these two kinds of Homoptera have very similar transmission ecology and vector behavior (17).

Research also has shown clearly that another group of Homoptera, whiteflies (Family: Aleyrodidae), are important vectors of plant viruses. Three species transmit viruses that cause mosaic, leaf curl, or yellowing symptoms. Transmissions are largely found in tropical areas of the world.

Coevolution Between Plants and Arthropods

Theories and facts in this research area have increased severalfold during the last decade. Useful summaries are found especially in Feir (4), Harborne (9), and Wallace and Mansell (34). Our present understanding of this subject indicates clearly the dynamic natures of interactions between plants and arthropods. However, in spite of such evidence, economic zoologists still tend to view such interrelationships in static terms. Such views frequently lead to the development of methods of arthropod suppression that, at best, are of short-term usefulness.

Current evidence from studies of coevolution further suggests that we commonly overreact to arthropod occurrences on plants that we value. Thus, we frequently consider arthropods as significant constraints on plant health when they are not. Examples of such situations include *Scolytus quadrispinosus* beetle feeding on *Carya* trees, *Neodiprion* larvae feeding on *Pinus banksiana* needles, and *Ostrinia nubilalis* larvae feeding on *Zea mays*. *S. quadrispinosus* beetles feed almost exclusively on the current year twig growth of *Carya* trees (7). The trees contain relatively deterrent and feeding-inhibitory chemicals in older twig tissues, which largely restrict the beetle's feeding to the current growth. Such tissues provide the beetle with its required nutrition, and the feeding does not significantly reduce the vigor of *Carya* trees. Instead, the beetle pruning of some current shoots stimulates lateral branching, which generally leads to a greater density of foliage. The increased foliage can improve the efficiency of *Carya* trees in photosynthesis. With *Pinus banksiana* trees, the loss of juvenile needles from current-year shoots will kill the trees; however, older needles can be completely removed and the trees still will maintain their vigor (13). Larvae of *Neodiprion* sawflies have coevolved with this tree so that they feed only on older needles and thus obtain required nutrients. The juvenile needles contain feeding-deterrent amounts of 13-keto-8(14)-podocarpen-18-oic acid, but older needles do not.

With *Z. mays*, young highly photosynthetic tissues contain concentrations of specific chemicals (e.g., benzoxazolinones) that deter the first-instar larval feeding of *O. nubilalis*. Older tissues of the corn plant do not contain such high concentrations, and such larvae feed significantly more on such parts of the plants (14). The larval feeding on older corn tissues does not usually threaten the plant, but larval destruction of young tissues can. Such insights from studies of coevolution strongly indicate the need for establishing sound ecological and economic thresholds of damage for each arthropod-plant interaction situation.

Plant Resistance to Arthropods

Although the development of our understanding of plant resistance to arthropods is several decades behind that for plant resistance to microbial agents, major advances in the former have occurred especially during the last two decades (19). Both the related proven usefulness of plant resistances to microbes and the now rapidly expanding array of such successes against arthropods indicate that these approaches have been, and will continue to be, among the more successful.

Results from much recent research on analyzing stress phenomena in plants have tended to indicate that plants have evolved strategies for coping with complex stressful environments. They apparently have not usually evolved strategies specifically directed at given biological constraints (e.g., arthropods) in such environments. This situation should mean that efforts to develop plant resistances to arthropods should benefit from progress that has already occurred in developing resistances to other entities (e.g., microbes) that cause stress in plants. The involved logic is that plants employ common principles in dealing with stresses caused by different factors. An example is that given phenolic chemicals in plants inhibit both microbial and certain arthropod constraints on the plant's health. Further investigations of unifying principles involved in plant resistances to environmental stresses therefore should be especially rewarding.

Effective plant resistance to arthropod-associated constraints on plant health in natural environments usually involves a complex of morphological, anatomical, physiological, and biochemical characteristics (29). For example, rice resistance to the stem borer, *Chilo suppressalis*, results from the interactions of leaf blades with a hairy upper surface, tight leaf sheath wrapping, small stems with a ridged surface, and thickened hypodermal layers (30). Cotton resistance to the boll weevil, *Anthonomus grandis*, involves frego bracts, red plant color, increased pubescence, and rapid fruit set (15). Soybean resistance to several agromyzid beanflies involves, at least, trichome density of the abaxial surface of the leaf, leaf area, leaf moisture content, and stem diameter (2). However,

TABLE 2. Plant resistance to arthropod vectors of plant-pathogenic biotic agents

Insects	Plant	Resistance Factor
Aphids	Potato	Insect-trapping hairs
Myzus persicae	Tobacco	Nicotine content
Eriosoma lanigerum	Apple	Sclerenchymatous tissue
Brevicoryne brassicae	Brussel sprouts	Open foliage
Myzus persicae	Potato	Age of leaves
Perkinsiella spp.	Sugarcane	Reduced duration of phloem ingestion and salivation
Nephotettix cincticeps	Rice	Reduced duration of phloem ingestion and salivation
Nilapervata lugens	Rice	Oxalic acid feeding deterrent

resistant cultivars reported in field conditions are attributed, at least largely, to one plant character. These reports seem to be more common when resistance is judged on one criterion (e.g., lack of transmission of a plant pathogenic agent by a vector arthropod). Some reported examples of such plant resistance are given in Table 2. Even though the judged resistance in these situations apparently is highly correlated with one character, several factors likely are significantly involved.

Research on plant resistance to arthropods appears to be one of the more promising lines of current investigations. The longer-term usefulness of specific findings will probably correlate well with the extent to which the multiple parameters of each resistance are elucidated.

Symbiotic Complexes of Species as Causal Entities

Some of the recent progress in this subject has been addressed elsewhere (25,31). Apparently the importance of symbiotic complexes of species in the biological constraint of plant health is currently being more widely recognized. Such realization is probably one of the more important developments that has occurred in the subject field of causation of biological constraints on plant health. After all, as Fawcett (3) stated, "Nature does not work with pure cultures." Thus, the quicker humans can move beyond using just pure-culture techniques to attempt to interpret causation, the better! Perhaps the most progress in this subject has recently occurred in regard to plant-pathogenic microbes that are ectosymbiotes of wood-inhabiting insects, especially ambrosia beetles (e.g., Scolytidae and Platypodidae) (5,6,22,23,25,26). These studies have established clearly that the insect and a coevolved complex of ectosymbiotic fungi, bacteria, and yeasts jointly constrain the health of the host plant. The interrelations between the insect and the microbes in the plant are obligatory and mutualistic (26).

During the last decade, leafhopper vectors also were shown to transmit complexes of plant viruses and MLOs (16). These findings made it necessary to reexamine which entities are actually biological constraints on plant health.

The current knowledge on symbiosis and plant health strongly suggests that continued research in this subject area is likely to bring a situation where most causal entities are described as complexes of species.

SOME FUTURE EMPHASES IN RESEARCH

Three areas of research that seem especially worthy of future emphases are 1) the energy exchanges involved in the chemical senses and chemical ecology, 2) symbiosis and the continuum of life, and 3) induced resistance in plants to biological constraints.

Energy Exchanges in the Chemical Senses, and Chemical Ecology

Although input from all sensory modalities obviously is important in determining arthropod behavior, research interest and recent advances in our knowledge of chemical ecology and the chemical senses seemingly have set the stage for major breakthroughs in our understandings of unifying principles in the realms of the chemical senses.

Hundreds of compounds are now proven messengers between arthropods and their environments, and the question, "Is there order (e.g., a code) to the exchange of energy that occurs between the sensory receptors of the arthropod and the messenger molecule, ion, or free radical?" deserves an answer. I have hypothesized that there is such a code. If it exists, deciphering it could be as important to understanding why an organism occurs in given ecological niches as revealing the molecular genetic code has been to comprehending dynamically the inherited potential of an individual.

"Are there scientifically based data to indicate already that energy exchanges in the chemical senses are ordered?" The answer is "yes" (29). Evidence indicates that each energy-transducing sensory receptor in arthropods probably functions only within a discrete range of electromotive or chemicomotive (e.g., protonmotive) forces (27–29). If this were true, then an individual only would receive a message about the status of its environment from ligands that could exchange or share energy within comparable ranges of electromotive or chemicomotive forces. A second fundamental fact is that each messenger is, at least, a "double agent." This means that each messenger contributes to the excitation of given energy-transducing macromolecules in certain arthropods and to the inhibition of transducers in other individuals. Thus, "poison" to one species or biotype is "pleasure" to another. As examples, the sesquiterpenoid, gossypol, which is found in the cotton plant, attracts (i.e., is kairomonic to) the cotton boll weevil (*Anthonomus grandis*), but repels (i.e., is allomonic to) *Heliothis* spp. larvae and *Epicauta* blister beetles (20). In *Gossypium*, the flavonoid quercetin stimulates feeding by *A. grandis*, but in *Quercus* it inhibits the smaller European elm bark beetle, *Scolytus multistriatus* (29).

In addition, it is now established that the order of relative effectiveness (e.g., repellency) among a class of phytochemical messengers (i.e., naphthoquinones or flavonoids) for arthropods can be highly correlated with the combined relative oxidation state and hydrogen-bonding capability of each ligand (24,27,28,29). The flavonoids are extremely common and important chemicals in plants, and among them the C_3 unit, which connects the two rings in such compounds, exists in a discrete range of oxidation states (Table 3). It is most interesting and apparently biologically significant that each species of plant contains a mixture of flavonoids that provide an unique array of redox states among their C_3 units. Thus, just considering flavonoids, species of plants apparently possess unique arrays of messengers and the physicochemical capabilities for interacting energetically in a highly ordered fashion with the unique sensory receptors of

TABLE 3. Oxidation states in the C_3 units of different types of flavonoids

Flavonoid	Structure of C_3 Unit[a]
Flavan-3-ols	A–CH_2–CHOH–CHOH–B
Hydrochalcones	A–CO–CH_2–CH_2–B
Chalcones	A–CO–CH=CH–B
Flavanones	A–CO–CH_2–CHOH–B
Leucoanthocyanidins	A–CHOH–CHOH–CHOH–B
Flavones	A–CO–CH_2–CO–B
Anthocyanidins	A–CH_2–CO–CO–B
Benzalcoumaranones	A–CO–CO–CH_2–B
Flavanonols	A–CO–CHOH–CHOH–B
Flavonols	A–CO–CO–CHOH–B

[a] A and B represent the corresponding rings of the flavonoid structure.

each species and biotype of arthropod. Ignoring other parameters, the aforementioned types of physicochemical characteristics of such environments would provide a basis for ordering the distributions and abundances of species or biotypes of arthropods within, or among, these habitats.

If future investigations can further describe arthropod-plant interactions in such highly ordered physicochemical terms, then such information should provide new bases for highly specific disruptions of such interrelations in practical situations.

Symbiosis and the Continuum of Life

An era of considering multiagent causation of biological constraints on plant health will unfold, and research on arthropod involvements in causation will increasingly focus on such an animal as just one identifiable species within "discernible continua of several biological entities" that interact symbiotically to affect plant health. The reality of such "continua" has been clearly demonstrated by extensive studies of scolytid beetles and the organisms that interact symbiotically with such insects in causing pathogenesis in plants. Such investigations have shown that such discernible causal continua (i.e., multispecies complexes) include the insect, mites, humans, nematodes, fungi, bacteria, yeasts, protozoa, etc. Thus, future pioneering research on the biological agents of plant death will study causal continua that consist of as many symbiotic entities (e.g., species) as can be discerned under the given experimental conditions. Studies of pure-culture causation should be considered strictly artifactual. This expanded complexity of experimentation on causation will be readily feasible through computer-programmed complexes of experimental environments.

Results from investigating arthropods as participating species in continua of different symbiotic entities that jointly cause plant death should provide economic zoologists with a significant number of new strategies for suppressing such arthropods. As an example, variously disrupting fungal, bacterial, yeast, protozoa, nematode, etc. symbiotes of such arthropods may effectively curtail them as biological constraints on plant health.

Induced Resistance in Plants to Biological Constraints

In recent years, induced resistance to most major biotic constraints on plant health, except humans, has been recognized (12). The advances in understanding such inductions against arthropod invaders are limited (21,29), but the evidence is exciting. Further interdisciplinary investigations in this field will probably provide the bases for some of the most important developments that have occurred in the realm of methods for the suppression of arthropod attacks on plants.

Research on the subject should provide quick and probably enduring returns on efforts. It is now apparent that many existing plant species exhibit previously unrecognized, induced (i.e., increased) resistance to agents after an initial attack. Thus, increasing our understandings of such "in place" resistances should prove immediately useful in improving strategies to suppress the arthropod-associated death of plants. Because most phytopathogenic arthropods are symbiotic with microbes, nematodes, protozoa, etc., existing inducible resistances in plants to such phytopathogenic symbiotes should significantly disrupt the interrelations between such symbiotes and the arthropod of concern. These indirect, but real benefits also should be quite quickly translatable into improved practical suppression methods.

More fundamental investigations into the basic physicochemical mechanisms of induced plant resistance to arthropods should provide dynamic bases for several future strategies for creating plants that possess a greater inducible resistance to such agents. One apparently major fundamental basis for such inducible plant resistance involves the release (e.g., hydrolysis) of aglycones from their bound state in glycosides, upon initial attack of the plant. It is well established that both the quality and quantity of aglycones significantly influence the suitability of plants to arthropods (24,29). A given aglycone is allomonic to some arthropods but kairomonic to others. As discussed previously in this chapter, the qualitative array of aglycones, e.g. flavonoids, differs among species and varieties of plants. Thus, selection, breeding, and other biotechniques to alter the aglycone compositions of plants should significantly change their capabilities for induced resistance to specific arthropods.

Because the state of oxidation and the hydrogen-bonding capability of an aglycone especially influence its effectiveness as an allomone to arthropods, the development of biotechniques that readily enable the precise regulation of these physicochemical parameters among major classes of aglycones in plants should be highly useful.

The induction in plants of inhibitors of insect proteinases is perhaps the most exciting discovery that has occurred in the realm of induced plant resistance to arthropods (8,33). Such inhibitors may disrupt basic metabolic functions, such as digestion, in arthropods. The inhibitors are peptides or phenolics, and at least some are induced systemically throughout the attacked plant, not just at the specific attack site. Although the understanding of these phenomena in plants is still sketchy, the possible practical ramifications from these more recently recognized capabilities in plants rather boggle the mind. In addition to biotic induction, heavy metals (e.g., mercury), sulfhydryl agents (e.g., *N*-ethyl maleimide), disulfide-reducing agents (e.g., reduced glutathione), ultraviolet light, exposure to radioactivity, and other parameters of environmental stress may induce such systemic proteinase inhibitors. Thus, future research is likely to develop a rather diverse arsenal of abiotic and biotic inducers of these inhibitors for practical use under a variety of plant and environmental situations. Further developments in this realm of induced systemic proteinase inhibitors may bring such resistance in plants close to the status of induced immunity in animals to certain pathogens.

SUMMARY

Our capabilities and potentialities for alleviating arthropod-induced constraints on plant health in the future appear to be very encouraging. Rather extraordinary advances in our understandings of several pertinent aspects of biology during the 1970s have especially established the foundation for this optimism. These areas of important progress in understanding include 1) biochemical parameters of plant-arthropod coevolution, 2) symbiotic complexes (i.e., discernible continua) of biological constraints on plant health, 3) physicochemical mechanisms of perceptions in the chemical senses, and 4) inducible plant resistance to biotic and abiotic constraints.

Ingenious combined utilization of these relatively new pools of knowledge, the proven and currently practiced means of arthropod suppression, and the diverse capabilities of other emerging biotechnology should create "brave new world" strategies for alleviating arthropod constraints on the health of our valued plants.

LITERATURE CITED

1. Borror, D. J., DeLong, D. M., and Triplehorn, C. A. 1976. An Introduction to the Study of Insects, 4th ed. Holt, Rinehart, Winston, New York. 852 pp.
2. Chiang, H. S., and Norris, D. M. 1982. Morphological and physiological parameters of soybean resistance to agromyzid beanflies. Environ. Entomol. 12:260-265.
3. Fawcett, H. S. 1931. The importance of investigations of the effects of known mixtures of organisms. Phytopathology 21:545-550.
4. Feir, D. 1977. Physiology and biochemistry of insect-host interactions. Pages 211–277 in: Proc. Int. Congr. Entomol., 15th. Entomol. Soc. Am., College Park, MD. 824 pp.
5. Francke-Grosmann, H. 1965. Symbiosen bei xylomycetophagen und phloemphagen Scolytoidea. Material Organismen 1, Beiheft 1, pp. 503-522.
6. Francke-Grosmann, H. 1967. Ectosymbiosis in wood-inhabiting insects. Pages 141–205 in: Symbiosis, Vol. 2. S. N. Henry, ed. Academic Press, New York.
7. Goeden, R. D., and Norris, D. M. 1963. Feeding characteristics of adult *Scolytus quadrispinosus* Say. Proc. North Central Branch, Entomol. Soc. Am. 18:64-65.
8. Green, T. R., and Ryan, C. A. 1972. Wound-induced proteinase inhibitor in plant leaves: A possible defense mechanism against insects. Science 175:776-777.
9. Harborne, J. B. 1978. Biochemical Aspects of Plant and Animal Coevolution. Academic Press, London. 618 pp.
10. Harris, K. F., and Maramorosch, K. 1977. Aphids as Virus Vectors. Academic Press, New York. 559 pp.
11. Harris, K. F., and Maramorosch, K. 1980. Vectors of Plant Pathogens. Academic Press, New York. 467 pp.
12. Horsfall, J. G., and Cowling, E. B. 1980. Plant Disease: An Advanced Treatise. Vol. 5. How Plants Defend Themselves. Academic Press, New York. 534 pp.
13. Ikeda, T. F., Matsumura, F., and Benjamin, D. M. 1977. Chemical basis for feeding adaptation of pine sawflies *Neodiprion rugifrons* and *Neodiprion swanei*. Science 197:497-499.
14. Klun, J. A., Tipton, C. L., and Brindley, T. A. 1967. 2,4-Dihydroxy-7-methoxy-1,4-benzoxazin-3-one (DIMBOA), an active agent in the resistance of maize to the European corn borer. J. Econ. Entomol. 60:1529-1533.
15. Lukefahr, M. J., Bottger, G. T., and Maxwell, F. G. 1966. Utilization of gossypol as a source of insect resistance. Pages 215–222 in: Proc. Annu. Cotton Dis. Council, Cotton Defoliation Physiology Improvement Conf. Nat. Cotton Council, Memphis, TN.
16. Maramorosch, K. 1979. How mycoplasmas and rickettsias induce plant disease. Pages 203–217 in: Plant Disease: An Advanced Treatise, Vol. 4. J. G. Horsfall and E. B. Cowling, eds. Academic Press, New York. 466 pp.
17. Maramorosch, K. 1980. Insects and plant pathogens. Pages 137–155 in: Breeding Plants Resistant to Insects. F. G. Maxwell and P. R. Jennings, eds. John Wiley & Sons, New York. 683 pp.
18. Maramorosch, K., and Harris, K. F. 1979. Leafhopper Vectors and Plant Disease Agents. Academic Press, New York. 654 pp.
19. Maxwell, F. G., and Jennings, P. R. 1980. Breeding Plants Resistant to Insects. John Wiley & Sons, New York. 683 pp.
20. Maxwell, F. G., Lefever, H. N., and Jenkins, J. N. 1965. Blister beetles on glandless cotton. J. Econ. Entomol. 58:792-793.
21. McIntyre, J. L. 1980. Defenses triggered by previous invaders: Nematodes and insects. Pages 333–343 in: Plant Disease: An Advanced Treatise, Vol. 5. J. G. Horsfall and E. B. Cowling, eds. Academic Press, New York.
22. Norris, D. M. 1965. The complex of fungi essential to *Xyleborus sharpi*. Holz Organismen 1:523.
23. Norris, D. M. 1975. Chemical interdependencies among *Xyleborus* spp. ambrosia beetles and their symbiotic microbes. Material und Organismen, Beiheft 3, pp. 479-488.
24. Norris, D. M. 1977. Role of repellents and deterrents in feeding of *Scolytus multistriatus*. Pages 215–230 in: Host Plant Resistance to Pests. P. A. Hedin, ed. Am. Chem. Soc., Washington, DC. 286 pp.
25. Norris, D. N. 1979. How insects induce

disease. Pages 239–255 in: Plant Disease: An Advanced Treatise, Vol. 4. J. G. Horsfall and E. B. Cowling, eds. Academic Press, New York. 466 pp.

26. Norris, D. M. 1979. The mutualistic fungi of Xyleborini beetles. Pages 53–63 in: Insect-Fungus Symbiosis. L. R. Batra, ed. Allanheld, Osmun & Co., Montclair, NJ. 276 pp.
27. Norris, D. M. 1979. Chemoreceptor proteins. Pages 59–77 in: Neurotoxicology of Insecticides and Pheromones. T. Narahashi, ed. Plenum Publishing Corp., New York.
28. Norris, D. M. 1981. Perception of Behavioral Chemicals. Elsevier/North-Holland Biomedical Press, Amsterdam. 328 pp.
29. Norris, D. M., and Kogan, M. 1980. Biochemical and morphological bases of resistance. Pages 23–62 in: Breeding Plants Resistant to Insects. F. G. Maxwell and P. R. Jennings, eds. John Wiley & Sons, New York. 683 pp.
30. Patanakamjorn, S., and Pathak, M. D. 1967. Varietal resistance of the Asiatic rice borer, *Chilo suppressalis* (Lepidoptera: Crambidae), and its association with various plant characteristics. Ann. Entomol. Soc. Am. 60:287-292.
31. Powell, N. T. 1979. Internal synergisms among organisms inducing disease. Pages 113–133 in: Plant Disease: An Advanced Treatise, Vol. 4. J. G. Horsfall and E. B. Cowling, eds. Academic Press, New York. 466 pp.
32. Russell, G. E. 1978. Plant Breeding for Pest and Disease Resistance. Butterworth Sci. Publ., London. 485 pp.
33. Ryan, C. A., and Green, T. R. 1974. Proteinase inhibitors in natural plant protection. Rec. Adv. Phytochem. 8:123-140.
34. Wallace, J. W., and Mansell, R. L. 1976. Biochemical Interaction Between Plants and Insects. Plenum Press, New York. 425 pp.

CHAPTER 26

WEEDS AND HIGHER-PLANT PARASITES

F. W. SLIFE

Weed science has made remarkable progress in the past 35 years and clearly has been a major factor in the steep rise of agricultural productivity during the 1945–1980 period.

Weeds, like other plant pests, were recognized early in our agricultural history as deterrents to production, but very few tools were available for their control. Rotations, fallowing, row cultivation, and hand pulling or hoeing were the main means available. Although selective weed control became available about 1900, it was used only sporadically until 1945, when 2,4-D was introduced. This herbicide was the very foundation of our modern weed control practices. It was adopted quickly and used widely for broadleaved weed control in monocotyledonous crops around the world. Early applications of 2,4-D were made with sprayers of very high gallonage because the concept of low-volume sprayers was not introduced until 1947. Another milestone in the development of weed control was *N-N*-diallyl-2-chloroacetamide (CDAA), introduced in 1956. This selective chloroacetamide herbicide controlled annual grass weeds when applied as a preemergence treatment. It was the first soil-applied herbicide to gain wide use on a variety of crops. Following 2,4-D and CDAA, major developments during the 1960s were the introduction of atrazine, trifluralin, and the thiocarbamate herbicides. Trifluralin introduced the preplant incorporation concept, and this method of using herbicides has remained prominent because of consistent performance.

During the 1970s, additional new chemicals were introduced, and herbicide combinations to give broad-spectrum control were used more widely. The benefits of combining herbicide treatments with cultural practices were recognized, and the introduction of integrated pest management programs brought more refinement to weed management programs (7). Crops grown on limited acreage—vegetables, ornamentals, citrus, etc.—have benefited greatly from the use of herbicides. Interregional programs in cooperation with the chemical industry have been successful in obtaining new clearances for minor crop use of many pesticides.

As we enter the 1980s, weed control in a wide variety of crops has never been better in our agricultural history. Herbicides will continue to be a vital part of future programs, simply because, used judiciously, they are more effective than any other present weed control alternatives.

CURRENT PROBLEMS

The intensive use of herbicides over the past 20 years, however, can be viewed with mixed blessings. On one hand, it has been economically sound and very successful; on the other hand, it has developed so rapidly that some undesirable side effects are beginning to appear.

Examples of these effects are herbicide-resistant weeds, weed spectrum changes, herbicide residues in soil, and accelerated degradation of herbicides in soil.

Herbicide-Resistant Weeds

The development of pest resistance with the repeated exposure of the same pesticide or class of pesticides has been recognized in plant pathology and entomology for many years. Herbicide-resistant weeds have developed more slowly, but they have now been identified in a number of areas (6). The most frequent example has been associated with the continuous use of atrazine for about 10 years. Resistance has developed in a rather wide variety of species, both dicotyledons and monocotyledons, indicating tremendous genetic variability in those plants we call weeds. Increased vigilance will be needed to identify the development of herbicide-resistant weeds early, before they become a major problem in a given area. Even more important is the development of herbicide management systems that use a wide variety of herbicides. By rotating the herbicide treatment, the onset of weed resistance could be delayed and probably eliminated altogether.

There is a strong possibility that weed resistance is developing because of our intensified cropping systems and their associated herbicide treatments. The considerable acreage of corn grown in a monoculture system in the corn belt could give rise to resistant weeds if the weed management system does not have some variation. In the same area, millions of acres are devoted to alternating corn and soybean crops. In all probability, atrazine is being used on corn and metribuzin is a common treatment on soybeans. Both are triazine herbicides with the same mode of action in sensitive plants. This cropping system with these particular herbicide treatments would be a prime candidate for the development of weed resistance.

Weed Spectrum Changes

Before the widespread use of herbicides, weed spectrum changes were gradual and influenced primarily by cropping systems (2). Apparently enough competition existed among species to prevent rapid changes in the weed spectrum. Herbicide use has been dramatic in shifting weed populations. The first example was the introduction and widespread use of 2,4-D. In a relatively few years, the predominant dicotyledonous weed species became a minor problem and the minor monocotyledonous weed species became dominant. Herbicide combinations chosen to fit the weed spectrum in a given area have helped correct the shift in species but have not completely eliminated the problem. Some weed species considered to be of minor importance have suddenly emerged as major problems. These species are not well controlled by the herbicide treatments and this, coupled with the lack of competition from other weeds, has allowed them to increase rapidly. The problem of weed spectrum changes is more acute in no-till agriculture because preplant herbicides cannot be used and the opportunity for cultivation after planting has generally been eliminated. Perennial weed species in particular have been troublesome in continuous no-till practices, and many of these are perennials of noncultivated areas. In certain areas, the rapid increase in perennials has prompted producers to return to the moldboard plow for 1 year after 3 years of no tillage.

In row crops there are a number of examples of annual or perennial weed species that increase rapidly because they are not well controlled by the present herbicide treatments. Velvetleaf (*Abutilon theophrasti*) has become a very prominent weed in the upper Midwest, although historically it has been a minor species. The nightshades (*Solanum* spp.), both annual and perennial, have greatly increased in intensity over a wide area in the United States over the past 5 years. Again, the lack of adequate control, the reduced competition from other weed species, and the reduction in row crop tillage seem to be the causative factors.

Herbicide Soil Residues

Occasionally some of the triazine and dinitro aniline herbicides do not completely degrade in one season and hence can affect sensitive crop growth the following year. Although very little crop loss has occurred, it has restricted some use of these valuable herbicides. The factors conducive to undesirable soil residues have not been clearly defined.

Accelerated Herbicide Degradation

Eptam is widely used in several crops. Specifically this thiocarbamate herbicide has been used intensively for control of weedy members of the sorghum family. It has been the leading treatment for the control of wild cane in corn fields in the corn belt. Recently it has been shown that annual use on the same fields has led to accelerated degradation of the herbicide. Weed scientists have known that this could occur since it was demonstrated with 2,4-D in laboratory studies in the late 1940s. Eptam appears to be the first example of this phenomenon in the field.

BASIC STUDIES

To solve these problems and others, more attention will be needed on basic weed biology, ecology, and physiology as they relate to herbicide use. Weed seed germination patterns and requirements for germination are not well understood, and yet most of our herbicides are soil applied. The growth and development of individual weed species has not been explored fully in terms of response to control measures at different stages of growth. These studies might lead to control measures that would reduce weed interference and prevent weed seed production and yet not eliminate the weed plant.

Multidisciplinary research is needed on the impact of weed control measures on the soil organism complex, from the standpoint of both beneficial and pathogenic organisms. Certainly the drastic reduction or elimination of the weed flora in an area must affect the soil organism complex. These soil organism changes could lead to reduced soil degradation of herbicides, or it could create conditions for more rapid degradation if part of the soil organism complex had disappeared.

Herbicide physiology studies have contributed to a more thorough understanding of plant metabolic processes. Research in this area has increased in the past few years, and hopefully it will continue to expand. Discovering the exact mode of action of herbicides could lead to help in the design of new and more efficient compounds and in the more efficient use of present herbicides.

In addition to increased emphasis on weed biology, ecology, and physiology, much more emphasis is needed on the development of weed control methods other than by soil-applied herbicides. Our very high dependence on herbicides creates a risk, in that there is no assurance that the chemical industry will continue its investment in this area.

BIOLOGICAL CONTROL

Biological control historically has largely been limited to the control of perennial weeds in noncultivated areas (1). Success in this endeavor has been high in relation to the effort put forth. Insect predators have been the most successful in the past, but more recently research has intensified on the use of plant pathogens, particularly fungi (3). Plant pathologists and others have well documented the occurrence of pathogenic fungi, viruses, and nematodes on weeds as well as on crops, but the endemic populations only occasionally are high enough to have a material effect on weed growth and resulting weed seed production. The mass culture of particularly virulent strains of fungi and their application in large doses to relatively young weed plants has been highly successful. Two indigenous special forms of *Colletotrichum gloeosporioides* f. sp. *aeschynomene* (CGA) and f. sp. *jussiaea* (CGJ) have been outstanding for the control of northern jointvetch in rice and soybean and for water primrose in rice. Dry formulations of the pathogens, prepared by the UpJohn Co., are equivalent to liquid preparations. Registration by the Environmental Protection Agency was granted for CGA in 1982. *Phytophthora palmivora* received registration in 1981 for the control of the milkweed vine (*Morrenia odorata*). This very serious weed in Florida citrus has not been well controlled by other means. Abbott Laboratories is producing the material commercially. Walker (11) has recently reported an isolate of *Fusarium lateritium* that is pathogenic to spurred anoda (*Anoda cristata*), prickly sida (*Sida spinosa*), and velvetleaf (*Abutilon theophrasti*). Although okra and hollyhock are also affected, corn, cotton, soybean and 18 other representative crop and weed species in eight families are resistant. The development of this pathogen could have a substantial effect because of the widespread occurrence of the sensitive weeds.

Research on the mycoherbicides is intense, particularly in the southern states, and this, combined with increased surveillance in other countries for potential biological control agents, indicates that biological weed control will be an important strategy in the near future.

Allelopathy is a potential weed control method (5). Some strains of some crops produce compounds that can interfere with weed seed germination and seedling growth. Breeding programs designed to develop this trait could supplement weed control programs.

POSTEMERGENCE HERBICIDES

Another possible alternative to our present chemical weed control programs is the development of a wider array of postemergence herbicides. At the present time, many of the herbicide treatments are applied to the soil before or immediately after crop planting. The treatments used are based on the history of the weed problem on a particular field. Because these treatments have been successful, the weed potential in some fields is now much lower. If appropriate postemergence herbicides were available, the decision of whether to use herbicide treatment could be delayed until the weed problem was apparent. Postemergence treatments would also allow the use of threshold concepts of weed control to be utilized more fully.

FUTURE PROBLEMS

There are other constraints to weed control that will have to be solved in the next few decades. They are: tillage changes, new weeds, and the impact of government regulations.

Tillage Changes

Interest in soil conservation is widespread and is expressed by both the producer and the general public. No-till farming, now practiced successfully on more than 1 million hectares, continues to expand (10). In addition to its use on sloping land with high erosion potential, its use on some of the more level land is also increasing. Significant progress has been made in small grain regions of the western states in terms of reducing wind erosion during the fallow season. Herbicides are used instead of repeated tillage, and under this system soil moisture conservation is higher and wind erosion much less.

Some form of reduced tillage is now being practiced in almost all areas of the United States (10). The reduced tillage systems are variable, but the major objective is to leave more crop residue on the surface for erosion control and to reduce the energy costs associated with tillage. There has been a dramatic increase in the use of the chisel plow as a replacement for the moldboard plow.

It seems obvious that some form of reduced tillage will be practiced widely, providing it is profitable for the producer to do so. The challenge of controlling agricultural pests is paramount, and, in fact, pest control will likely determine the types of reduced tillage programs and extent to which they will be used.

During the 1970s, the major type of weed management system developed for many crops was a soil-applied treatment applied either as a preemergence or preplant (incorporated) treatment. Cultivation or postemergence herbicides were used to supplement the primary control practice. During the last half dozen years, the preplant treatments have increased rapidly and have become the dominant treatment system. The weed management systems used by producers have evolved under a soil management system of clean tillage with little or no surface residue.

It is apparent, then, that reduced tillage that leaves crop residue on the surface for erosion control is in conflict with many of the weed practices presently used. Incorporating herbicides also incorporates the crop residue, and the surface residue intercepts some of the herbicide applied as a preemergence treatment. Incorporated treatments are perhaps in the greatest conflict since some of the crop residue is lost at the time when surface erosion potential is at its highest peak. In the early 1960s producers changed their tillage practices in order to utilize preplant herbicides; in the 1980s weed scientists will have to develop new weed control methods that are compatible with the reduced tillage concept.

New Weeds

Most of the plants we call weeds were brought to North America unintentionally by the early settlers of this continent. Little attention was given to the importation of seeds and plant material in terms of potential weed problems. In spite of the fact that through the years more regulations have been imposed on imported seeds and plants, we continue to bring in new weed species. The following are a few examples of imported weed species.

In the 1830s, johnsongrass (*Sorghum halepense*) was introduced as a potential forage crop and in ensuing years has become a very major weed problem in the southern half of the United States. Prickly pear cactus (*Opuntia* sp.) was introduced as an ornanental and subsequently has become a major range weed problem in the Southwest. Giant foxtail (*Setaria faberi*) was identified in 1932 in millet seed imported from China. It apparently slowly migrated to the Midwest and now is the major annual weed species found in the corn belt. This species serves as a good example of how difficult it is to predict the potential adaptability of new or introduced species. Giant foxtail is not reported to be a serious weed problem in any other country in the world, and yet it apparently has found its most desirable niche in our corn belt. Witchweed (*Striga asiatica*) was first recognized in 1956, and much more recently milkweed vine (*Morrenia odorata*) was introduced into the citrus area of Florida.

One of the latest new weed problems is wild proso millet (*Panicum milliaceum*). It was recognized as a weed problem less than 10 years ago and since that time has infested several million acres in lower Canada, Wisconsin, and Minnesota. It is not known whether this is an introduced species or whether it originated in North America. Since the plants appear to be relatively uniform, with little or no variation, it may be an indigenous species. This annual species is a prolific seed producer, and control is difficult because it continues to germinate throughout the growing season. Only time will determine the adaptable range of this species. Scattered infestations were found in northern Illinois and Iowa in 1980 and 1981.

New weed problems will continue to plague our agriculture in the decades to come. Hopefully we will be able to slow down or eliminate the introduction of imported species, but our present species will continue to evolve and present new problems.

Government Regulations

Remarkable progress has been made in terms of improving pesticide use and safety. Many of the older persistent pesticides have been removed from the market place, and new pesticides have emerged that have much less effect on our environment. Although many of the rules and regulations regarding pesticide use were sorely needed, there is an imminent danger that too much unnecessary regulation will discourage investment in the search for new pesticides. The cost of generating the data to register a new compound has increased very rapidly, and the time required to generate the data has increased substantially. In some cases, half of the patent life of a new compound may have expired before it is available for use. This factor, combined with the possibility that the confidential data accumulated by a chemical company for pesticide registration can be released to other companies after the patent expires, is also discouraging to the patent holder. This would mean that the second producer of the compound would have a small investment in the generation of data for registration.

New herbicides will be much more costly to produce, and this high cost will be reflected in the immediate future.

Chemical weed control will remain as the most important component of weed control programs as long as it is economical. The potential constraints to herbicide use can be overcome with increased research.

HIGHER-PLANT PARASITES

More than 2,500 species of higher plants are known to live parasitically on other plants. Relatively few of the known parasitic higher plants cause economic loss to agricultural crops or forest trees. The botanical families and genera of the most common and serious parasites are shown in Table 1.

Witchweed

Witchweed is the most widespread and serious of all parasitic seed plants (8). In tropical and subtropical areas it attacks such crops as corn, sugarcane, rice, millets, and some small grains. Economic losses are particularly severe in India; in Africa it is reported to be the most important crop pest.

Four species of *Striga* are common in tropical areas of the world. They are *S. asiatica, S. hermonthica, S. gesnerioides*, and *S. euphrasioides*.

In 1956 *Striga* was discovered for the first time in the United States in North and South Carolina. It was identified as *Striga asiatica*. In 1957 federal and state quarantines were imposed, and in 1959 the USDA Witchweed Laboratory was established in North Carolina.

By 1959, 33 counties in eastern North and South Carolina were placed under quarantine. These measures have proven to be effective, and the size of the affected area has been reduced. From the standpoint of soils and climate, *S. asiatica* has the potential to infect all the corn-growing areas in the United States.

Witchweed seed germinates in response to stimulants exuded by the host plant. As soon as the witchweed rootlet comes in contact with the host root, it develops an haustorium that presses against the host root, and within 8–24 hours the tracheids of the witchweed root have reached the xylem and phloem cells of the

TABLE 1. The most common and serious higher-plant parasites

Family	Genera	Common names
Convolvulaceae	*Cuscuta*	Dodders
Loranthaceae	*Arceuthobium*	Dwarf mistletoes of conifers
	Phoradendron	American bine mistletoes of broadleaved trees
	Viscum	European true mistletoes
Orobanchaceae	*Orobanche*	Broomrapes
Scrophulariaceae	*Striga*	Witchweeds

host, where they absorb water and nutrients. Six to 8 weeks after the parasite attaches to the host plant, a stem develops and emerges from the soil. Within 90–120 days after germination, the parasite produces seed pods. A single plant can produce as many as 500,000 seeds.

Control programs for witchweed involve the prevention of seed production and the exhaustion of the seed reservoir in the soil. In the control program, catch crops such as small grains are planted to stimulate witchweed seed germination and then destroyed before witchweed produces seed. Trap crops are also used (9). These are legumes that will also stimulate germination but on which the witchweed is unable to complete its life cycle. Soil injections of ethylene have been particularly effective in stimulating germination. After the witchweed has emerged from the soil, 2,4-D and similar postemergence herbicides are effective in preventing seed production.

In 1979, *S. gesnerioides* was identified in Florida. Its present infestation is limited and is found to be associated with the host plant hairy indigo (*Indigofera hirsuta*). Weed scientists in the USDA believe that this species will not pose a threat to our crop production as compared to *S. asiatica*.

Dodder

Dodder is widely distributed in Europe and in North America. In the United States, it persists in fields where alfalfa and clover are grown intensively for seed. Other crops that suffer losses from dodder include lespedeza, onions, flax, sugar beets, potatoes, and several ornamentals. Three species of dodder are important in the United States. They are *Cuscuta indecora*, largeseed, *C. planiflora*, smallseed, and *C. campestris*, field dodder. Herbicides are used rather extensively for dodder control in the alfalfa seed production areas in the western states of the United States. Chlorpropham and DCPA are widely used, but dichlobenil, dinoseb, and pronamide are also effective. Recently Dawson (4) has found that glyphosate applied to alfalfa with attached dodder will give control without apparent injury to the alfalfa. Rates of glyphosate as low as 0.075 kg/ha have been effective. Recent U.S. delegations to the People's Republic of China (R. M. Goodman, University of Illinois, *personal communication*) have learned that dodder is a serious problem in soybeans in the soybean area of the northeast provinces. It apparently has increased in intensity over the past 30 years, and biological control research has been underway for the past 20 years. *Cuscuta chinensis* is the primary species, but *C. australis* is also reported to be present. This appears to be the first report of dodder being a serious problem on soybeans.

Mistletoes

The dwarf mistletoes occur in all parts of the world where conifer trees grow. In the United States, they are most serious in the western half of the country particularly along the Pacific Coast, but they also cause appreciable losses in the northeastern and southeastern states. The true or leafy mistletoes occur throughout the world, particularly in warmer climates. Although they attack primarily hardwood forest and shade trees, they can cause economic loss to common fruit and plantation trees such as apples, cherry, citrus, rubber, cacao, and coffee. Although these mistletoes are common in some areas, they are not nearly as severe as the dwarf mistletoes. All of the mistletoes spread by seed. In the case of dwarf mistletoes, the mature fruit develops internal pressure and when disturbed expels the sticky seed as far as 15 m. True or leafy mistletoes produce seed-containing berries that are eaten by birds. The birds excrete the sticky seeds in the tops of taller trees where they perch.

All of the mistletoes reduce tree growth, and heavy infestations weaken the trees and predispose them to wood-decaying and root pathogens. Infested trees are also more susceptible to insect and wind damage. The only control for the mistletoes is by physically removing the parasite. In urban areas this can be done by pruning infected branches or by cutting and removing entire infected trees. In forest management programs, controlled burning is used as a particular control.

Broomrape

The broomrapes are widely distributed in Europe and the United States and particularly in the tropical areas. Fortunately, in the United States they are a minor problem as compared to the dodders, mistletoes, and *Striga* species. They attack several hundred species of herbaceous crop plants including tobacco, potato, tomato, hemp, clover, and alfalfa. The parasite usually occurs in patches, but crop loss can be severe, depending upon the degree of infestation. The seed germinates only when roots of certain plants grow near them. Upon germination, the radicle grows toward the root of the host plant, becomes attached, and then penetrates and absorbs nutrients and water from it. Later the parasite develops a stem and it eventually appears above the soil to produce flowers and then seed.

Control of broomrapes depends on prevention of introduction of its seed to new areas, planting of nonsusceptible crops, and frequent weeding of the aerial portion to prevent seed production. Because broomrapes are a minor problem in the United States, very little if any herbicide control information is available. Recent reports from the Mediterranean area indicate that treatments of glyphosate on broad bean (*Vicia faba*) have given outstanding control of *Orobanche*.

Relatively good control measures are available for the dodder and *Striga* species found in the United States. With continued emphasis on research and education, these species should decline in intensity in the future. Since the mistletoes and *Orobanche* species are controlled by physical means, their status is unlikely to change very rapidly.

LITERATURE CITED

1. Batra, S. W. T. 1981. Biological control of weeds: Principles and prospects. Pages 45–59 in: Biological Control in Crop Production. G. C. Papavizas, ed. Allenheld, Osmun & Co., Totowa, NJ. 461 pp.
2. Buchanan, G. A., and Frans, R. E. 1981. The role of weeds in agro-ecosystems. Pages 46–49 in: Proc. Symp. Int. Cong. Plant Prot. 9th. Vol. 1. T. Kommedahl, ed. Burgess Publ. Co., Minneapolis, MN. 411 pp.
3. Charudattan, R., and Walker, H. L., eds. 1982. Biological Control of Weeds With Plant Pathogens. John Wiley & Sons, New York. 293 pp.
4. Dawson, J. H., and Shazhir, A. R. 1983. Herbicides applied to dodder (*Cuscuta* sp.) after attachment to alfalfa (*Medicago sativa*). Weed Sci. In press.
5. Fisher, R. F. 1979. Allelopathy. Pages 313–330 in: Plant Disease, An Advanced Treatise. Vol. 4. J. G. Horsfall and E. B. Cowling, eds. Academic Press, Inc. 466 pp.
6. Le Baron, H. M., and Gressel, J., eds. 1982. Herbicide Resistance in Plants. John Wiley & Sons, New York. 401 pp.
7. Miller, G. R. 1982. Integrated weed management systems technology for crop production and protection. Weed Sci. 30 (Suppl. 1). 54 pp.

8. Musselman, L. J. 1980. The biology of *Striga, Orobanche*, and other root-parasitic weeds. Annu. Rev. Phytopathol. 18:463-489.

9. Robinson, E. L., and Dowler, C. C. 1966. Investigations of catch and trap crops to eradicate witchweed (*Striga asiatica*). Weeds 14:275-276.

10. Unger, P. W., and McCalla, T. M. 1980. Conservation Tillage Systems. Academic Press, New York. 374 pp.

11. Walker, H. L. 1981. *Fusarium lateritium*: A pathogen of spurred anoda (*Anoda cristata*), prickly sida (*Sida spinosa*), and velvetleaf (*Abutilon theophrasti*). Weed Sci. 29:629-631.

CHAPTER 27

RECOGNITION AND SPECIFICITY BETWEEN PLANTS AND PATHOGENS

LUIS SEQUEIRA

The field of recognition between plants and pathogens, as an area of research, has finally come of age. Evidence for this is the upsurge of publications in this area during the past 10 years and the packed-house attendance at meetings, conferences, symposia, etc., where the subject has been discussed. It is useful to consider why there is so much interest in the subject.

Plant pathologists interpret recognition in the interaction between host and pathogen as an early specific event that triggers a rapid, overt response by the host, either facilitating or impeding further growth of the pathogen. If, indeed, recognition takes place at an early stage in the encounter of a potential pathogen with a host, this phenomenon may determine the ultimate fate of the interaction. Thus, the elucidation of mechanisms involved in recognition may provide answers to some of the key questions that plant pathologists have been exploring since the days of de Bary: 1) what is the nature of specificity? and 2) what is the nature of disease resistance? Although research on recognition is only in the preliminary stages, sufficient information has been obtained to provide a glimpse of what may lie ahead.

Interest in the study of recognition in plant-microorganism interactions was triggered about 10 years ago by Hamblin and Kent (17), who reported on the possible role of phytohaemagglutinins in the interaction of rhizobia with their hosts. A great flurry of activity on the possible role of lectins on adhesion of rhizobia to root hairs followed. Work with plant pathogens did not follow until later, but it was based on the same concept, namely, that recognition is the result of specific lock-and-key bonds that are established between complementary molecules on the surfaces of both host and pathogen.

The relationship between recognition and host specificity should be emphasized. Recognition generally is established after firm attachment of a potential pathogen to the host cell surface and is characterized by one of three host responses:

1. A compatible response that allows penetration and multiplication of the pathogen or the transfer of genetic information.
2. An incompatible response, characterized by rapid death of the host cells, as in hypersensitive interactions, or by the formation of chemical or physical barriers to further spread of the pathogen. All of these are inducible systems,

presumably the result of cascading biochemical events triggered by the interaction of preformed compounds on both cell surfaces.

3. No initial host response. This is characteristic of compatible interactions in which host cells in proximity to the pathogen are destroyed before they have an opportunity to respond to the pathogen. These three categories encompass most host-parasite interactions in which specificity is exhibited toward a particular tissue, plant part, or cultivar. They cannot illustrate all levels of specificity, however.

This chapter highlights some of the systems that best illustrate the principles that govern the initial interaction between host and parasite, and it points out some of the emerging challenges in the area of recognition. The discussion is limited to fungal and bacterial plant pathogens and to those that primarily attack leaves or roots.

THE GENETIC BASIS FOR RECOGNITION AND SPECIFICITY

Specificity in the interaction of plants and pathogens exists at many different levels. At one extreme are many omnivorous fungi and bacteria that attack many genera of plants. At the other extreme are obligate parasites that attack only a specific plant part of particular cultivars. Even pathogens with broad host ranges are unable to invade large numbers of plant species, however. Specificity is expressed as either resistance or susceptibility to infection.

From a biochemical standpoint, specificity can be understood only in cases where the resistance genes are superimposed upon a basic level of compatibility between host and parasite. In other words, a potential pathogen must have evolved mechanisms that allow it to utilize the nutrients available in host tissues and to overcome both chemical and morphological barriers to infection. Built upon this basic compatibility, however, is a superstructure in which specific genes for resistance in the host, usually dominant, interact with specific genes for virulence in the parasite, usually recessive. These gene-for-gene interactions have been studied most intensively with obligate parasites, but it is likely that they describe most of the host-parasite interactions where race or cultivar differences can be defined.

The simplest biochemical explanation for gene-for-gene interactions is that there is specific recognition between the products of the genes for resistance and avirulence. Since the product of the gene for avirulence of a race interacts only with the product of one specific gene for resistance in the host, it is likely that the interaction involves binding of a ligand to a protein receptor in a lock-and-key relationship. Ellingboe (13) has initiated the search for the products of a *P* gene in *Erysiphe graminis* by two-dimensional electrophoresis, on the assumption that mutations at this particular gene will show up as peptides that are displaced because of changes in net charge. Once the gene products can be identified, alterations in primary structure of the protein associated with mutations for increased virulence could be searched for. Should these attempts prove successful, one can predict, by analogy with a wide variety of systems for recognition in plants and animals, that the gene products will represent enzymes that are specifically involved in the synthesis of surface molecules of host and parasite.

Since recognition generally involves a surface polysaccharide or glycoprotein on one partner and a protein that binds specifically to sugar residues (a lectin) on the other, it seems likely that gene-for-gene systems involve a similar, complementary interaction. Albersheim and Anderson-Prouty (1) have suggested that the genes for avirulence code for glycosyl transferases, enzymes that add specific sugar residues at the ends of polysaccharide chains. Polysaccharides have enormous informational potential because the diverse types of linkages between sugar residues and the different steric arrangements that are possible provide an almost unlimited variety of configurations. This may explain why there appear to be no limits to the number of genes for virulence that may occur in one race of a particular rust. One may speculate that the genes for resistance, which generally occur as multiple alleles at relatively few loci, code for relatively few receptor proteins (lectins) at the target organelle. Mutations that affect the tertiary structure of these proteins would be expected to alter their ability to bind to specific sugar residues on the surface of the pathogen. One may speculate further that during evolution there has been selective pressure on the pathogen toward loss or alteration of specific sugar residues, leading to avoidance of recognition by the host. Similarly, there has been selective pressure on the host toward synthesis of proteins capable of specifically binding to the carbohydrates on the surface of the pathogen.

There is, of course, no reason why this model must be limited to a polysaccharide-lectin interaction or why the lectin should be present on the host cell surface. The genetics of the systems that have been studied, such as the *Melampsora*-flax interaction, favor this interpretation, but the location of the complementary structures may actually be reversed. Such is the case, for instance, in the highly specific interaction of certain strains of *Escherichia coli* with the intestinal mucosa (6). In addition, there are highly specific protein-protein or carbohydrate-carbohydrate interactions that may be important components in gene-for-gene interactions. Protein-protein interactions, for instance, appear to control the recognition system that prevents self-fertilization in sporophytically determined pollen-stigma interactions in plants. These interactions are controlled by multiple alleles of the *S* gene; this gene is thought to code for the synthesis of a stigmatic glycoprotein that effectively blocks attachment of self pollen containing a similar protein (14). Examples of polysaccharide-polysaccharide specific interactions can be found in certain bacteria that bind to plant cell walls (23).

The view that resistance and virulence are positive functions of the host and parasite, respectively, is not supported in host-parasite interactions that involve the action of host-specific toxins produced by fungi. In these systems, susceptibility usually is dependent on a single dominant gene, and, conversely, virulence is dependent on a single, dominant gene that controls the synthesis of the host-specific toxin (13). Thus, compatibility, rather than incompatibility, is the result of the apparent ligand-receptor interaction. In this instance, selection for resistance in the host has been for alterations in the putative receptor protein so that it can no longer bind the toxin. Admittedly, the evidence for the presence of such receptors is mostly indirect, but it seems likely that they are located on the target organelle.

The two genetic systems that we have described illustrate the basic principle that recognition in host-parasite interactions may be either for compatibility or incompatibility, depending on the adaptation of the parasite to the response of the host. In both instances, however, the mechanisms for recognition and specificity appear to depend on compounds (ligands) produced by the pathogen that are bound by proteins (receptors) on host organelles.

ADHESION AS A PREREQUISITE FOR RECOGNITION

The evidence that specific adhesion of the parasite to the surface of the host cell is essential for infection comes mostly from work with certain bacterial pathogens of mammals, which must adhere specifically to host cell membranes before they can cause disease (6). The same principle applies to plant symbionts, such as rhizobia, that attach to legume root hairs and then cause alterations in the host cell wall that allow infection thread formation. Such specific attachment appears to be rare in plant pathogens, however. Most plant pathogenic fungi attach nonspecifically to leaf, stem, or root surfaces before they invade the plant. Even fungi that produce toxins that act at a distance must first germinate and attach to the host surface via germ tubes or appressoria. The only apparent exceptions appear to be the nontumorigenic, plant pathogenic bacteria that initially remain free in the intercellular spaces or in the lumens of xylem vessels. There is evidence, however, that avirulent forms or saprophytic strains of certain bacteria become attached to the host cell wall (16,31).

Bacteria and fungi, both pathogenic and nonpathogenic, stick to many inert surfaces. It is likely, therefore, that the initial process of attachment to host cell walls is nonspecific and that it is mediated by hydrogen bonding, ionic interactions, and other surface forces that provide relatively loose binding. On the surfaces of host cells, however, attachment of the pathogen becomes irreversible within a short time and this indicates that additional, multiple bonds are formed between the host and parasite. Some of these additional bonds may be the result of ligand-receptor interactions of the type that we described previously. For example, Mendgen (26) has shown that germ tube walls of *Uromyces phaseoli* adhere strongly to host tissues but very poorly or not at all to nonhost tissues. The complementary molecules on host and pathogen surfaces must be arranged so that they are readily accessible and provide multiple bonding throughout the surface. Once tight bonding is established, the process is virtually irreversible.

Attachment may be a specific process, but is not necessarily related to host specificity. It is important that this distinction be made here. An example may suffice to illustrate the point. Furuichi and coworkers (15) have shown that the penetrating germ tubes of *Phytophthora infestans* traverse the cell wall and then bind to the plasmalemma of potato cells so firmly that upon artificial plasmolysis of the host cell the plasmalemma collapses except where it remains adhered to the invading fungus. Attachment to the plasmalemma appears to be specific because it can be inhibited by chitotriose, a hapten for potato lectin, which is the presumed receptor protein (28). Attachment of the fungus to the plasmalemmae of susceptible or resistant potato cells, however, is similar. Thus, the process of attachment is specific in the sense that it appears to be mediated by a ligand-receptor interaction, but the differential response of the host to attachment is associated with other unknown factors.

In certain cases, attachment is a prerequisite for induction of either compatible or incompatible interactions. In tumor induction by *Agrobacterium tumefaciens*, attachment to the host cell wall is necessary for transfer of the genetic information that will ultimately result in tumor formation (23). Transfer of information apparently occurs only on wound-exposed host sites, but the bacteria attach throughout the host cell surface. There must be specific features of the bacterium and unique sites on the host cell wall that are involved in attachment. Attachment occurs only with virulent strains and is maximal within about 2 hr of incubation (24).

With other plant-pathogenic bacteria only the incompatible or avirulent forms attach to host cells. When strains of *Pseudomonas solanacearum* that cause a hypersensitive reaction (HR) in tobacco leaves were labeled with ^{14}C and then incubated with washed tobacco suspension culture cells, a high proportion (up to 70%) of the bacteria were bound by the culture cells. In contrast, cells of the parental, virulent strain (K60) were bound less efficiently or not all (11). Attachment was reversible by high salt concentration within the first few minutes but was irreversible thereafter. This indicates that the initial reaction was probably ionic in nature but that additional and possibly specific bonds were formed soon after "docking" of the bacteria on the surface of the host cell.

There is no clear evidence at present that all incompatible bacterial strains must attach to the host cell wall before the HR is induced in the host. Nevertheless, when the intercellular spaces are maintained in a water-soaked condition, or if the bacterial cells are suspended in agar, the HR is not induced (32). This may be taken as indirect evidence that close proximity, if not attachment, of the bacteria to the host cell wall is necessary for induction of the host response.

Certain saprophytic bacteria have been shown to bind more efficiently to tobacco leaf cells than do HR-inducing bacteria (2). Although the data indicate clearly that saprophytes have very strong binding properties, induction of the HR by a pathogen may require efficient attachment by relatively few cells.

NATURE OF THE PATHOGEN COMPONENT INVOLVED IN RECOGNITION

The search for surface components that mediate attachment of plant pathogens to the host surface has centered almost exclusively on the bacteria. The possible reason for this is that most of the background information has been obtained with bacterial pathogens of animals or with *Rhizobium* spp. Such studies have emphasized the potential role that the surface components of bacteria, specifically the fimbriae, the extracellular polysaccharide (EPS), and the lipopolysaccharide (LPS), may have. It seems likely that one or more of these, and perhaps other components of the bacterial cell wall, play a role in attachment.

Fimbriae

The structures responsible for attachment of several Gram-negative bacteria to their hosts have been identified as the fimbriae or pili. These hairlike, proteinaceous appendages extend outward from the bacterial cell wall, sometimes for considerable distances. Some fimbriae bind specifically to carbohydrates and thus resemble lectin. Fimbriae enable many Gram-negative bacteria to agglutinate erythrocytes and play a role in adhesion to other types of cells of many mammalian species (6). The fimbriae of *E. coli* K88 cells, for instance, attach to the brush borders of intestinal mucosa cells of piglets. In piglets that are resistant to infection, however, the epithelial cells fail to bind the bacteria (6).

Fimbriae have been reported in species of *Rhizobium* and in several plant pathogenic bacteria (W. Stemmer, *personal communication*). As in mammalian systems, fimbriae may be involved in specific attachment to plant host cells. The traditional view, based on the model proposed for *Rhizobium*-legume systems, is that attachment results from the interaction of bacterial cell wall polysaccharides and a host lectin (5). This conventional view has obscured the possible importance of fimbriae in attachment. None of the data is inconsistent with the possibility that bacterial, rather than plant, lectins play a major role. For instance, 1) rhizobia attach on end to plant cell walls and some species of *Rhizobium* are polarly fimbriated, 2) heat treatment of *P. solanacearum* B1 cells markedly inhibits their ability to attach to cell walls (11), and 3) there is a strain-dependent, temperature-sensitive phase in the early stages of crown gall tumorigenesis (29). It remains a significant challenge for plant pathologists to conclusively determine whether fimbriae are essential structures in the recognition and attachment of bacterial pathogens.

Extracellular Polysaccharide

All plant pathogenic bacteria produce EPS either as a capsule or, more commonly, as slime. Some bacteria produce both forms of EPS. In general, EPS is a heteropolysaccharide consisting of repeating units of several hexoses and/or pentoses together with amino sugars and organic acids. EPS plays an important role in pathogenesis, particularly by wilt-inducing bacteria such as *P. solanacearum, Corynebacterium insidiosum, Erwinia amylovora*, and *E. stewartii* (3,7,19,34). EPS-deficient mutants of these bacteria are almost invariably avirulent.

Although the role of EPS in attachment of rhizobia to legume roots has not been resolved satisfactorily (because of contamination with LPS) recent work indicates that in *R. japonicum* recognition leading to nodulation in soybean is probably effected by EPS. It is not clear, however, just how or where in the process of nodulation EPS plays this role. Electron microscopic observations indicate that the binding site for soybean lectin, the presumed receptor, is the capsule of the bacterium (9,33). Although the relationship of capsule production to lectin binding ability is clear, Bauer (5) points out that there is no convincing evidence that a lectin-EPS interaction is involved in recognition. The role of EPS in recognition may be substantially different from that initially proposed. For instance, pretreatment of soybean roots with *R. japonicum* EPS enhances root hair infectibility (5). Precisely how this effect is mediated is not known, but it is unlikely that the same lectin receptor is involved since pretreatment with EPS should result in inhibition, not enhancement, of infectibility.

The rather confusing status of EPS as a recognition molecule in *Rhizobium* is further complicated by the evidence that capsule production is not required for nodulation (21,30). The emphasis on capsule formation, however, has caused researchers to disregard the fact that many bacteria produce both capsular and soluble forms of EPS. Law et al (21) make the important point that noncapsulated mutants of *R. japonicum* release EPS into the medium in amounts directly proportional to their ability to nodulate soybean.

A very different approach was used by Rudolph and collaborators to determine the role of EPS in infection of bean by *Pseudomonas phaseolicola*. Purified preparations of EPS from this bacterium caused persistent water-soaking of susceptible bean leaf tissues, favoring multiplication of the bacterium in the intercellular spaces (12). In resistant tissues, however, persistent water-soaking was inhibited due to enzymatic breakdown of EPS. It is difficult to see how the wide diversity of strain-variety relationships could be explained purely on the basis of the presence or absence of a specific enzyme capable of degrading EPS.

Lipopolysaccharide

The polysaccharide components of the outer membrane of Gram-negative bacteria determine specificity in many human infections (37). The highly variable nature of the O-specific antigen of *Salmonella typhimurium*, for instance, results in different antigenic properties that are used to characterize strains of the bacterium. Since antigen-antibody relations are not considered relevant for plant pathogens, it is not surprising that LPS has not been studied extensively in terms of specificity of these bacteria. One of the few studies with LPS was made by Whatley and coworkers (35), who determined that attachment of *A. tumefaciens* to a specific site on the cell wall was essential for tumor formation and that tumor formation was inhibited when the LPS of virulent strains was added before inoculation. The LPS preparations from *A. radiobacter* and from nonsite binding strains of *A. tumefaciens* were not inhibitory, however. Since the O-specific antigen (plus core components) prepared by acetic acid hydrolysis of LPS from virulent strains was equally as inhibitory as intact LPS, it was concluded that specific sequences in the O-polysaccharide chains probably were involved in attachment. More recent work (4) indicates that these sequences may involve *N*-acetyl galactosamine and β-D-galactose.

The possible role of LPS in attachment of *P. solanacearum* to intact tobacco leaf cells or cells from suspension cultures has been investigated in our laboratory. Attachment of avirulent (B1) cells of this bacterium to the host cell wall has been considered an essential step in the recognition phenomenon that leads to the HR. An initial study of a series of rough variants of the virulent strain, K60, led to the hypothesis that mutations that result in loss of the O-polysaccharide expose sugar residues in the core portion of LPS that are important in attachment and, ultimately, in induction of the HR (36). An avirulent, rough strain (B1) attached readily to tobacco cells or isolated cell walls, but a virulent strain (K60) showed no significant attachment over a 5-hr incubation period (11). Although the type of LPS may determine the initial ionic interaction with host cell walls, firm attachment is dependent on a temperature-sensitive factor that is probably enzymatic. The point needs to be emphasized again that attachment of bacteria is not a simple, one-step process; several pathogen and host factors need to be turned on in sequence to result in irreversible binding.

Cellulose

Whether LPS, EPS, or both play a role in initial adhesion of rhizobia and *Agrobacterium* to host cell walls, it is becoming increasingly evident that firm attachment may be mediated by additional pathogen factors. Among these is the production of cellulose fibrils that anchor bacteria to the host surface (25,27).

NATURE OF THE HOST COMPONENT INVOLVED IN RECOGNITION

The conventional view is that specific host responses are brought about by the interaction of host surface proteins that bind specifically to complex carbohydrate structures on the pathogen's surface. There is no reason, however, why specificity could not be associated with the interaction of pathogen proteins and host cell wall polysaccharides.

The hypothesis that host lectins are involved in recognition has gained support mostly from the work with rhizobia. There is considerable evidence, for instance, that soybean lectin attaches specifically to strains of *R. japonicum* but not to strains that nodulate other species (5). Similarly, a protein isolated from clover roots was shown to bind specifically to *R. trifolii* (10). Demonstration that these interactions occur in vitro, however, does not constitute proof that they occur in vivo. The host-lectin hypothesis was questioned by workers who noted that: 1) there is no incontrovertible evidence that these proteins are present on the surface of root hairs where infections generally take place, 2) soybean lines lacking soybean lectin are nodulated by *R. japonicum*, and 3) the binding of lectins from several legumes to rhizobia is not correlated with the ability of the bacteria to nodulate those plants.

The most convincing evidence that host lectins are involved stems from experiments in which the presumed haptens for the lectins have been shown to interfere with attachment of bacteria to the surface of legume roots (10). The lectin hypothesis is the only one that explains these results.

Plants contain a wide variety of proteins that agglutinate bacteria, but they do so nonspecifically, as opposed to the interactions with lectins. Of particular interest are the hydroxyproline-rich glycoproteins that are found in plant cell walls and thought to function primarily as structural components (20). An hydroxyproline-rich glycoprotein from potato tissue is an effective agglutinin and has been shown to bind selectively to avirulent mutants of *P. solanacearum* (22).

The idea that the receptor protein is located on the host's cell surface is at the heart of discussions on recognition phenomena. Plant pathologists have been reluctant to consider alternatives. One of these alternatives, for instance, is recognition of carbohydrate residues on the root slime by a receptor associated with the zoospore cell membrane (18).

THE CONSEQUENCES OF ATTACHMENT

Once the potential pathogen has established close contact with the host cell wall or membrane, the ultimate fate of the interaction is dependent on the response of the host. The host cell may not respond at all, it may grow abnormally or form structures or substances that inhibit growth of the pathogen, or there may be rapid death of the protoplast. It is not within the purview of this chapter to discuss the mechanisms that lead to these various responses, but it may be useful to speculate why they are dependent on attachment.

In recent years, considerable emphasis has been placed on the apparent association between the HR and the accumulation of phytoalexins. Of particular interest has been the demonstration by Albersheim and coworkers that certain fungi that degrade host cell walls release low molecular weight carbohydrates that are highly active as elicitors of phytoalexin accumulation in the host. Although there is no evidence that such elicitors are produced only in incompatible host-parasite combinations, substances that modify elicitor action ("suppressors") may eventually be shown to confer specificity (8). Neither elicitors nor suppressors are released until the host and parasite have come into close contact. Thus, attachment may be a necessary prerequisite for the HR because elicitors must be produced as the result of the breakdown of pathogen cell walls by host enzymes or of host cell walls by pathogen enzymes. In either case, the message for induction of phytoalexin biosynthesis is released as a result of the very close interaction of the pathogen with the host surface.

It is clear that in most host-parasite systems, whether they involve the production of toxins by the pathogen or the release of phytoalexin elicitors, host receptors for these molecules must exist. An important challenge for the future is to determine the nature of these receptors and of the "transducer" system that receives the message and relays the information to specific organelles where the host response is initiated.

NEED FOR INTERDISCIPLINARY WORK

It is in the very nature of science that major advances are made through the concerted action of scientists from different disciplines. My plea for the next decade is for more cooperation with biochemists and other scientists who could help us in the solution of problems pertaining to host-pathogen specificity. It remains an important challenge for plant pathologists, however, to demonstrate that the problems of recognition and specificity are of sufficient biological importance to deserve the attention of molecular biologists, geneticists, and biochemists. We cannot expect that this will happen without more effort on our part.

LITERATURE CITED

1. Albersheim, P., and Anderson-Prouty, A. J. 1975. Carbohydrates, proteins, cell surfaces, and the biochemistry of pathogenesis. Annu. Rev. Plant Physiol. 26:31-52.
2. Atkinson, M. M., Huang, J. S., and Van Dyke, C. G. 1981. Adsorption of pseudomonads to tobacco cell walls and its significance to bacterium-host interactions. Physiol. Plant Pathol. 18:1-5.
3. Ayers, A. R., Ayers, S. B., and Goodman, R. N. 1979. Extracellular polysaccharide of *Erwinia amylovora*: A correlation with virulence. Appl. Env. Microbiol. 38:659-666.
4. Banerjee, D., Basu, M., Choudhury, I., and Chatterjee, G. C. 1981. Cell surface carbohydrates of *Agrobacterium tumefaciens* involved in adherence during crown gall tumor initiation. Biochem. Biophys. Res. Commun. 100:1384-1388.
5. Bauer, W. D. 1981. Infection of legumes by rhizobia. Annu. Rev. Plant Physiol. 32:407-449.
6. Beachey, E. H. 1981. Bacterial adherence: Adhesion-receptor interactions mediating the attachment of bacteria to mucosal surfaces. J. Infect. Dis. 143:325-345.
7. Bradshaw-Rouse, J. J., Whatley, M. H., Coplin, D. L., Woods, A., Sequeira, L., and Kelman, A. 1981. Agglutination of *Erwinia stewartii* strains with a corn agglutinin: Correlation with extracellular polysaccharide production and pathogenicity. Appl. Env. Microbiol. 42:344-350.
8. Bushnell, W. R., and Rowell, J. B. 1981. Suppressors of defense reactions: A model for roles in specificity. Phytopathology 71:1012-1014.
9. Calvert, H. E., Lalonde, M., Bhuvaneswari, T. V., and Bauer, W. D. 1978. Role of lectins in plant-microorganism interactions. IV. Ultrastructural localizations of soybean lectin binding sites on *Rhizobium japonicum*. Can. J. Microbiol. 24:685-693.
10. Dazzo, F. B., Yanke, W. E., and Brill, W. J. 1978. Trifoliin—A *Rhizobium* recognition protein from white clover. Biochim. Biophys. Acta 539:276-286.
11. Duvick, J., and Sequeira, L. 1981. Binding of *Pseudomonas solanacearum* to tobacco and potato suspension culture cells. (Abstr.) Phytopathology 71:872.
12. El-Banoby, F. E., Rudolph, K., and

Mendgen, K. 1981. The fate of extracellular polysaccharide from *Pseudomonas phaseolicola* in leaves and leaf extracts from halo-blight susceptible and resistant bean plants (*Phaseolus vulgaris* L.) Physiol. Plant Pathol. 18:91-98.

13. Ellingboe, A. H. 1981. Changing concepts in host-pathogen genetics. Annu. Rev. Phytopathol. 19:125-143.
14. Ferrari, T. E., Bruns, D., and Wallace, D. H. 1981. Isolation of a plant glycoprotein involved with control of intercellular recognition. Plant Physiol. 67:270-277.
15. Furuichi, N., Tomiyama, K., and Doke, N. 1980. The role of potato lectin in the binding of germ tubes of *Phytophthora infestans* to potato cell membranes. Physiol. Plant Pathol. 16:249-256.
16. Goodman, R. N., Huang, P. Y., and White, J. A. 1976. Ultrastructural evidence for immobilization of an incompatible bacterium, *Pseudomonas pisi*, in tobacco leaf tissue. Phytopathology 66:754-764.
17. Hamblin, J., and Kent, S. P. 1973. Possible role of phytohaemagglutinin in *Phaseolus vulgaris* L. Nature 245:28-30.
18. Hinch, J. M., and Clarke, A. E. 1980. Adhesion of fungal zoospores to root surfaces is mediated by carbohydrate determinants of the root slime. Physiol. Plant Pathol. 16:303-307.
19. Husain, A., and Kelman, A. 1958. Relation of slime production to mechanism of wilting and pathogenicity of *Pseudomonas solanacearum*. Phytopathology 48:155-165.
20. Lamport, D. T. A. 1980. Structure and function of plant glycoproteins. Pages 501–542 in: Biochemistry of Plants. Vol. 3. J. Preiss, ed. Academic Press, New York. 644 pp.
21. Law, I. J., Yamamoto, Y., Mort, A. J., and Bauer, W. D. 1982. Nodulation of soybean by *Rhizobium japonicum* mutants with altered capsule synthesis. Planta 154:100-109.
22. Leach, J. E. 1981. Localization, characterization and quantification of a bacterial agglutinin from potatoes. Ph.D. thesis. University of Wisconsin, Madison. 111 pp.
23. Lippincott, J. A., and Lippincott, B. B. 1980. Microbial adherence in plants. Pages 377–397 in: Bacterial Adherence. E. H. Beachey, ed. Chapman and Hall, London. 466 pp.
24. Matthysse, A. G., Wyman, P. M., and Holmes, K. V. 1978. Plasmid-dependent attachment of *Agrobacterium tumefaciens* to plant tissue culture cells. Infect. Immun. 22:516-522.
25. Matthysse, A. G., Holmes, K. V., and Gurlitz, R. H. G. 1981. Elaboration of cellulose fibrils by *Agrobacterium tumefaciens* during attachment to carrot cells. J. Bacteriol. 145:538-595.
26. Mendgen, K. 1978. Attachment of bean rust cell wall material to host and non-host plant tissue. Arch. Microbiol. 119:113-117.
27. Napoli, C., Dazzo, F. B., and Hubbell, D. 1975. Production of cellulose microfibrils by *Rhizobium*. Appl. Microbiol. 30:123-132.
28. Nozue, M., Tomiyama, K., and Doke, N. 1980 Effect of *N,N′*-diacetyl-D-chitobiose, the potato-lectin hapten and other sugars on hypersensitive reaction of potato tuber cells infected by incompatible and compatible races of *Phytophthora infestans*. Physiol. Plant Pathol. 17:221-227.
29. Rogler, C. E. 1981. Strain-dependent temperature-sensitive phase in crown gall tumorigenesis. Plant Physiol. 68:5-10.
30. Sanders, R., Raleigh, E., and Signer, E. 1981. Lack of correlation between extracellular polysaccharide and nodulation ability in *Rhizobium*. Nature 292:148-149.
31. Sequeira, L., Gaard, G., and de Zoeten, G. A. 1977. Attachment of bacteria to host cell walls: Its relation to mechanisms of induced resistance. Physiol. Plant Pathol. 10:43-50.
32. Stall, R. E., and Cook, A. A. 1979. Evidence that bacterial contact with the plant cell is necessary for the hypersensitive reaction but not the susceptible reaction. Physiol. Plant Pathol. 14:77-84.
33. Tsien, H. C., and Schmidt, E. L. 1981. Localization and characterization of soybean lectin-binding polysaccharide of *Rhizobium japonicum*. J. Bacteriol. 145:1063-1074.
34. Van Alfen, N. K., and Turner, N. C. 1975. Changes in alfalfa stem conductance induced by *Corynebacterium insidiosum* toxin. Plant Physiol. 55:559-561.
35. Whatley, M. H., Bodwin, J. S., Lippincott, B. B., and Lippincott, J. A. 1976. Role for *Agrobacterium* cell envelope lipopolysaccharide in infection site attachment. Infect. Immun. 13:1080-1083.
36. Whatley, M. H., Hunter, N., Cantrell, M. A., Hendrick, C. A., Keegstra, K., and Sequeira, L. 1980. Lipopolysaccharide composition of the wilt pathogen, *Pseudomonas solanacearum*: Correlation with the hypersensitive response in tobacco. Plant Physiol. 65:557-559.
37. Wilkinson, S. G. 1977. Composition and structure of bacterial lipopolysaccharides. Pages 97–115 in: Surface Carbohydrates of the Prokaryotic Cell. I. Sutherland, ed. Academic Press, New York. 472 pp.

CHAPTER 28

CURRENT CONCEPTS OF DISEASE RESISTANCE IN PLANTS

J. M. DALY

The genetic manipulation of disease resistance has played an important role in the economics of plant production. It has been hoped that understanding the biochemical basis for resistance would improve the selection and/or deployment of genes for disease reaction in commercial crops. Despite intensive efforts, this expectation has not yet been realized. The failure may reside in the biological complexity inherent with plant disease, but technical problems also play a role. Unlike animal systems, resistance or susceptibility to disease is the property of individual cells, with resistance often expressed locally in just a few cells, thus making biochemical studies difficult. However, we also must acknowledge that our concepts of host-parasite interactions could be deficient, and therefore we have not always asked the appropriate questions experimentally.

CONCEPTS OF RESISTANCE

The term resistance is used so commonly in plant pathology that the variety of nuances it can convey often are overlooked. At a basic, operational level (for example, selection of progeny in breeding programs), the term is applied to plants that exhibit either less infection, or less damage, than plants subjected to equivalent inoculum loads in similar environments. There is an intellectual pitfall in the use of the operational term, in that, from common English usage, most observers almost instinctly infer a cellular mechanism; e.g., an active opposition to infection or disease. But the mere observation of less, rather than more, disease is not sufficient to establish a mechanism. For this reason, the above operational definition will be designated "resistance" in this section.

In cellular terms, true resistance can be defined as the innate ability of a plant to prevent the establishment, or limit the subsequent activities, of an invading pathogen (12). The ability may reside in passive (preformed) mechanical or chemical barriers, but it is widely held that resistance, in most instances, is a latent ability that becomes operative only after the act of invasion (induced, active, or metabolic resistance). Metabolic resistance can entail the development of new morphological barriers (e.g., papillae [1], cell wall thickening or strengthening [35]) that restrict or slow pathogen development. However, research on metabolic resistance has been dominated, almost since the establishment of plant pathology as a science, by the search for chemical events that lead to the death or inhibition of the invader.

The direct demonstration of pathogen death would be strong evidence for metabolic resistance caused by toxic chemicals produced by the host, but there is little unequivocal evidence available. The problem is best illustrated in rust diseases where there is abundant description of abnormal or necrotic development of haustoria in resistant tissue; yet when lesions on "resistant" plants are transferred to susceptible hosts, the pathogen resumes growth (9). Similarly, in plants where "resistance" is sensitive to temperature (30), development of the pathogen resumes readily when temperatures are changed to permissive conditions, even for diseases in which toxic metabolites were demonstrated to accumulate (45).

Although this view of resistance dominates current research, an alternate mechanism to explain differential disease reaction is possible. It may be the consequence of a passive or of an active, induced, metabolic susceptibility (11,13). Before the pathogen can develop, a set of favorable cellular conditions must be established. In a "resistant" host, the cellular environment is not conducive to pathogen development, but not necessarily because of the induction of an active mechanism that inhibits or kills the pathogen. Although a biochemical response in a "resistant" host may be observed, it may be a general, nonspecific response of cellular irritation, stress or injury, but not a *determinant* of disease reaction.

Which of these two major alternatives is the functional event for a specific disease is of little practical concern for the current strategies of disease control through deployment of "resistance" genes. The difference between them is not, however, mere semantics. The difference may be of considerable concern for future strategies (14). Every crop is exposed to a diverse array of potential fungal, bacterial, and viral pathogens. If future disease control requires the continual introduction of new genes for the production of diverse toxicants upon invasion of "resistant" hosts, is there a limit to the number of such genes that can be incorporated before crop yield or consumer qualities are seriously affected? On the other hand, if "resistance" is a "nonevent" biochemically, there should be no real penalty for continual incorporation of "resistance" genes (14).

Of more immediate concern are the expectations, now beginning to be generously funded by private sources, that molecular biology and genetic engineering will have novel application in crop production, including disease control through resistance. It should not be overlooked, however, that these research fields became possible only *after* the chemistry and properties of gene products (enzymes) were well understood. Attempts to clone genes for "resistance" against a specific disease will be disappointing if our basic assumptions about the nature of disease reaction are wrong or if we have not correctly identified the chemical mechanisms that account for metabolic resistance. What follows is an attempt to assess the current status of various alternatives upon which this future research will, of necessity, be predicated.

SOME BASIC PROBLEMS

The study of metabolic resistance is in transition from a qualitative stage (using descriptive terms and concepts) to a stage in which the events must be interpreted quantitatively in terms of isolated chemical and physical processes. It is important to recognize that the latter can not be achieved if our knowledge of the biological events in resistance is quantitatively imprecise (13). As of this date, the writer is not aware of any truly unambiguous quantitative measure of resistance. The investigator usually accepts a classification based on visual symptoms, which in themselves are somewhat variable. The extent of pathogen development in the tissue is not known. It is my belief that studies of resistance will not be placed on a completely satisfactory basis until quantitative measures of pathogen growth in, or disease development on, infected tissue is achieved routinely. There are a number of postulates concerning resistance that remain only postulates because such data are not available.

The lack of quantitative data on resistance limits the applicability of theoretical models formulated as though the symbols R and S represented well-defined biological properties. Theoretical use of Flor's gene-for-gene hypothesis (21), for example, often neglects to reflect that Flor's original analysis (23) required grouping of phenotypically distinct rust reaction types in order to obtain suitable overall genetic ratios and required the exclusion of certain intermediate reaction classes (13). At the other extreme, gene-for-gene control for reaction was invoked in order to explain the appearance of chlorosis or of necrosis around rust pustules that were all of a single susceptible (type 3) phenotype (29). Such disparate interpretations of gene-for-gene interactions might be reconciled if the extent of pathogen development were known.

Similarly, we need to refine to a much greater degree our observations on chemical events in resistance. The problem lies in the fact that often only a few cells are necessary to establish resistance or susceptibility. Further, the critical chemical events may be transient. There are relatively few available chemical or physical techniques that guarantee the analysis of events in only the affected cells, at the appropriate times, and on a statistically significant basis; this is a formidable handicap in the study of host-pathogen interaction.

EVIDENCE FOR METABOLIC RESISTANCE

Biological Observations

Frequently, disease development in a susceptible plant can be reduced by prior inoculation with unrelated pathogens or by avirulent (incompatible) strains of the same pathogen (challenge inoculations) (42). The fact that the response is time-dependent (usually 24 hr between inoculations) is presumptive proof that metabolic changes are required, but there is no clear chemical evidence that such changes are related to the natural, normal resistance to the pathogen of interest. The protection afforded by prior inoculations with avirulent strains is unlike normal resistance in two ways; it is nonspecific and frequently it becomes progressively systemic.

The hypersensitive response of resistant plants has been extensively examined for about 75 years because it is one of the earliest and most ubiquitous events found in resistance (46). It has become increasingly clear that further study of hypersensitivity along previous lines is unlikely to yield further useful information about the chemical basis for resistance. Recent studies (27) suggest that it probably represents a stress response of little functional significance, even in disease incited by obligate parasites (30).

Metabolic Changes

Attempts to implicate secondary plant products as factors in resistance to plants and pathogens have a long history, but the attempts have been, at best, inconclusive. The principal problem is the inability to explain the specificity of disease resistance. As a consequence, attention turned toward comparative studies of metabolism in resistant and susceptible plants. Initial studies supported the belief that resistance is characterized by an earlier advent of very high metabolic rates. The evidence for this belief is not overwhelming (13), and what evidence exists may be the consequence of the dissimilar genetic backgrounds in the resistant and susceptible hosts studied earlier. With lines of wheat near-isogenic for resistance or susceptibility (2), metabolic differences reported earlier for plants with dissimilar genetic backgrounds were not observed.

Increased activity of enzymes, particularly those associated with secondary metabolites, is suggestive of metabolic resistance triggered by protein synthesis. When tested critically, however, it is difficult to establish a causal role for enzymes in resistance (34) and they cannot be related directly to products of resistance genes (16). The same enzymes also arise under conditions of cellular stress or injury.

The apparent ability of certain inhibitors of ribonucleic acid or protein synthesis to cause reversion of incompatible to compatible reactions also is used as an argument for metabolic resistance (38,40). Interpretation of the data is difficult for several reasons (13). Why host, but not pathogen, protein synthesis is affected preferentially by such general inhibitors is not clear. Although earlier studies suggested that inhibition of synthesis of antifungal compounds resulted in susceptibility through inhibition of protein synthesis in resistant soybean hypocotyls, a later study found high, but equal, rates of synthesis of antifungal compounds in both resistant and susceptible tissues (48). Similarly, treatment of oat tissue resistant to crown rust with an inhibitor of protein synthesis permits colonization of a certain crown rust race. Simultaneously, however, the treatment caused considerable additional increase in the same enzymes that were activated in infected nontreated resistant, but not susceptible, tissue (38).

Postinfectional Inhibitors from Hosts

The most satisfactory evidence for metabolic resistance has been the demonstration of the postinfectional accumulation of compounds inhibitory to pathogen growth in vitro. Müller (31) coined the term "phytoalexin" for such compounds with the implication that they were *responsible* for resistance, but his experiments with potatoes infected with *Phytophthora infestans* did not prove their presence. It remained for Cruickshank and Perrin (10) to place phytoalexin theory on a reasonable basis with the chemical identification of several compounds that accumulated in the early stages of incompatible, but not compatible, interactions. Although the term phytoalexin currently is routinely and widely used in discussion of resistance, its original meaning as the *cause* of resistance often is forgotten. The demonstration of an inhibitory compound after infection, or its designation as a phytoalexin, does not establish a functional role for it in resistance.

The biological activity of known phytoalexins is of little help in establishing a role; they are nonspecific toxicants of low potency and can be induced nonspecifically by abiotic factors, as well as by pathogens. Although they accumulate at early stages of incompatible reactions, they also occur with compatible ones, but at somewhat later stages. These facts have led to suggestions that incompatibility is the result of differentially higher rates of synthesis of phytoalexins in resistant plants (4); that is, resistance is determined by a specific activation of phytoalexin metabolism in resistance rather than by the toxic properties of the phytoalexin. It also has been postulated that specificity in host-pathogen interactions could be achieved because of tolerance (41) of pathogens to phytoalexins, particularly through degradation by the pathogen.

These modifications of the original postulates of phytoalexin theory are reasonable possibilities, but they have not been conclusively demonstrated for any single disease. The central problem (11,12)—whether phytoalexins actually function in vivo to inhibit pathogen growth—is still unresolved. Attempts to correlate inhibition of growth with inhibitory concentrations of phytoalexins are limited and are flawed by a number of the quantitative uncertainties about pathogen development noted above. Neither can it be established that phytoalexins are in functional contact with pathogens and not sequestered from them. Even in those instances where apparent cessation of pathogen growth can be correlated with phytoalexin accumulation, the possibility remains that inhibition of the pathogen results from some other unrelated mechanism. Phytoalexin accumulation may be a secondary effect caused by cellular stress or injury. Necrosis almost inevitably accompanies production of phytoalexin, and necrosis may be essential for its appearance. It has been suggested that the hypersensitive (necrotic) response of the host arises as a consequence of pathogen death, not vice versa (26). Decompartmentalization of metabolic function in dying or stressed cells could lead to mixing of enzymes and substrates and the accumulation of phytoalexins, but the phytoalexins may have no real role in resistance.

Very recent experiments indicate that, with certain treatments, phytoalexins accumulate to apparently toxic amounts but have no effect on the degree of compatibility or incompatibility. For example, when soybeans are grown at 35 C or higher, susceptible cultivars form incompatible lesions that contain very high amounts of glyceollin. When temperatures are lowered, *Phytophthora megasperma* is able to begin ramifying tissue outside the lesion in an apparently compatible fashion, even though phytoalexin concentration does not change (44). Phytoalexin accumulation has been observed frequently in slices of resistant potato tubers infected with avirulent races of *P. infestans*. However, Bostock et al (5) have examined freshly harvested tubers at a number of stages of field maturation but could not detect phytoalexins even though tubers were characterized by their normal incompatible response to avirulent races. After a month of storage, tubers at all stages of maturation do exhibit a correlation between phytoalexin accumulation and incompatible reaction. This suggests that accumulation in infected tubers may be conditioned by handling.

Certainly some of the reservations given above will be answered satisfactorily with continued research on phytoalexins and resistance. To do so, research activities should center less on the demonstration and identification of phytoalexins and focus more sharply on proofs of their potential in vivo functions.

Recognition and Resistance

Because of the limited success in explaining the specificity of plant disease reaction through the production of phytoalexins, attention recently has been directed toward the possibility that there is an initial recognition between host and parasite that determines whether the interactions are to be compatible or incompatible. (The chemical mechanism actually causing resistance can be nonspecific.) It is postulated that the recognition mechanism for resistance determines specificity. It is logical to assume that recognition phenomena would be found most likely at the interfaces (cell walls or plasmalemma) between interacting organisms. This has lead to a search for pathogen components, particularly macromolecular entities of wall or of membrane, which can induce (elicit) the biochemical symptoms of resistance, i.e., phytoalexins, particularly. The results to date with fungal diseases are disappointing. Wall fragments of pathogens can be prepared and at low dosages can cause phytoalexin accumulation, but they show no specificity either relative to the host or to the pathogen from which they are obtained. Thus, the data are not definitive with respect either to their proposed function in recognition or to the postulated role of phytoalexins. There are recent data purporting to show that membrane preparations containing proteins from bacterial (6) and fungal (25) sources can induce phytoalexins in soybeans in a race-specific fashion, but the amounts required and the variability of the response are not encouraging.

The potential role of lectins as agents of recognition in bacterial infections is treated by L. Sequeira in Chapter 27 of this volume. Recently, there have been reports of apparently similar results with *Ceratocystis fimbriata*. Incompatible host species contain high molecular weight carbohydrates that cause agglutination of spores and inhibit the germination of certain incompatible fungal strains (28). Other than these findings, only host-specific toxins have roles in disease that might be construed to be involved in a formal recognition mechanism.

It is important to recall that, generally, the initial events of infections are similar in both resistant and susceptible hosts in that there is at least some limited development of the pathogen before resistance is expressed. With rusts, significant differences in pathogenesis on resistant or susceptible hosts are not observed for 48–72 hr. If a recognition device for incompatibility exists in such cases, its presence alone can not be responsible for resistance. In a thoughtful essay, Ward and Stoessl (45) have pointed to the logistic difficulties in applying the concept of recognition to incompatibility phenomena, which outnumber by far the examples of compatibility in disease.

Genetic Arguments

Flor's complementary gene-for-gene hypothesis can be used to create a quadratic scheme of host (Rr) and pathogen (Vv) genes in which the combinations R-v, r-V and r-v lead to susceptibility and only the R-V quadrant represents resistance. This has been interpreted to mean that resistance is the unique and specific feature of host-pathogen interactions (21) and that it results from the chemistry between a product of a dominant host gene for resistance and the product from an avirulence gene in the pathogen. In this view, susceptibility is the result of genes for basic compatibility in host species (21). Genes for resistance thus act as separate genetic controls superimposed on basic genetic compatibility.

As Ellingboe (21) has argued, genetic models are an important and powerful tool in the design, testing, and interpretation of experiments, but unfortunately the modes of inheritance by themselves do not solve questions about the biochemistry of host reaction. It seems premature, therefore, to extrapolate simple genetic ratios into molecular models, when the molecules themselves have not been identified. For example, primary gene products (i.e., polypeptides) of host and parasite could interact directly to form a dimer that determines resistance. Although this could explain specificity in disease reaction, the basic question still remains as to how the dimer functions biochemically.

Although the gene-for-gene relationship is considered to occur nearly universally, it has been demonstrated in only a limited number of known diseases, and not always with the rigor a major theory demands. The chief difficulty is in establishing the mode of inheritance of virulence in pathogens because of a failure to obtain sexual recombination. Even where the gene-for-gene hypothesis has been developed in considerable detail (rusts, mildews) there are puzzling facts. To assume that the genetic dominance for host resistance requires positive gene action (i.e., a unique gene product of the host) does not explain why certain genetic backgrounds cause some genes for rust resistance to exhibit dominance, whereas in other backgrounds inheritance of the same gene is recessive (20,27). It is also difficult to interpret the so-called mesothetic or intermediate reaction in rust disease (13), in which a single leaf supports infection sites ranging from highly compatible to highly incompatible. There is also at least one instance known (in powdery mildew of barley) of a dominant gene governing susceptibility (22).

It is recognized that inheritance of reaction for diseases in which host-specific toxins are important results in an inverse quadratic check. The unique quadrant in these diseases is for susceptibility, not resistance. This can be explained as the interaction of an active gene product (toxin) of the pathogen with a host receptor resulting from an active configuration of a host gene. Although it usually is assumed that active configurations exhibit genetic dominance, host inheritance is recessive in several such cases (15,47). This observation also is not in accord with the usual molecular interpretation of gene-for-gene systems.

Metabolic Susceptibility

There is a limited body of evidence to suggest that biochemical responses leading to compatibility may be the crucial event in determining specificity. That the evidence is limited should not be surprising. The notion of metabolic resistance has had an overwhelming influence on the direction of past research. With some exceptions, susceptibility appears to have been regarded as an inherently uninteresting condition of a host, not as a potentially unique property (11).

In the last decade, however, Ouchi and colleagues (33) have amplified earlier observations that prior inoculation with virulent pathogens can condition a host to support the development of incompatible strains of the same pathogen or even of species that ordinarily do not attack that host. Although the induction of susceptibility (accessibility [32]) thus appears to be nonspecific, the demonstration of this phenomenon with biotrophic organisms is of particular interest. It must involve the activation of certain biochemical events required to support growth and reproduction of pathogens ordinarily considered to be in fastidious metabolic balance with their hosts (13). This is in contrast to the situation with challenge inoculations involving the induction of resistance. A number of nonspecific general factors (e.g., pH) could prevent growth of a normally virulent pathogen in such instances, whereas the induction of susceptibility to a biotrophic organism would appear to require some selectivity in the pathways activated.

For biotrophic pathogens, it is clear that compatible interactions result in much more extensive changes in metabolism than do incompatible interactions. The changes encompass a full range of biochemical activities from nucleic acid metabolism to changes in translocation patterns of whole plants. Because of technological limitations noted earlier, such changes cannot be studied except in the later stages of pathogenesis, leading to the notion that they only represent the outcome of separate, earlier events determining resistance or susceptibility. There is no compelling evidence, or a priori reasoning, that this is true. As an example, diseases involving biotrophic pathogens are very often characterized by growth changes and hormonal imbalance. It is not illogical to suggest that hormonally induced metabolic changes limited to just a few cells in the very early stages of infection may be necessary for the initial establishment of a biotrophic pathogen, even in the absence of any morphological change (13). However, the induction of metabolic susceptibility obviously is not limited to hormonal control; regulation of biochemical pathways can be achieved in a variety of ways.

For diseases in which susceptibility is characterized by extensive cellular damage, the need for an alteration in host metabolism is much less obvious. Nonetheless, there are examples where prior inoculation with virulent races condition host cells for development of normally incompatible races (39,43). Even in necrogenic diseases, hormonal involvement cannot be ruled out. The degree of gene-specific resistance of tobacco tissue cultures to *Phytophthora parasitica* var. *nicotianae* appears to be regulated by the amount of kinetin supplied (24). It also is pertinent to note the growing list of diseases in which chemical compounds in hosts appear to enhance pathogen virulence. In most instances, these substances appear to function as nutritional factors, e.g., betaine and choline in wheat (37). A very interesting variant of this theme is the recent demonstration of host substances that transform the sporidial habit of *Ustilago violacea* in culture to a mycelial form, characteristic of its mode during parasitism (17). Only fungal cells possessing both mating types were affected. All of the host species tested contained such compounds, whereas most of the nonhost species did not.

It can be argued that pathogen production of compounds such as hormones, degradative enzymes, or toxins determines the degree of pathogen virulence and hence the degree of host susceptibility. The role of toxins as agents of virulence in disease has been increasingly less controversial, particularly for bacterial toxins, as their structure and cellular targets have been elucidated. There is little evidence that resistance stems from active mechanisms; for example, detoxification. Like phytoalexins, however, most of the known pathogen agents of virulence do not possess specificity. Important exceptions are the host-specific or selective toxins whose potential general importance has been recognized only in the last decade or so largely through the efforts of R. P. Scheffer and associates in documenting new examples. The number of examples is still small and centered in species of

Helminthosporium and *Alternaria*. This fact might indicate the evolution of a unique mode of pathogenesis in these genera or only that they are the most suitable species for the production of such metabolites in artificial culture (15). Extremely rapid strides in the sciences of chemical separation and identification may soon permit the use of diseased tissue in the search for toxins, tissues that are produced by pathogens but that cannot be produced by them in vitro.

Host-specific toxins have been considered as primary disease determinants because, in one instance at least (victorin affecting oats), they appear to be required for pathogen colonization (47). Victorin, therefore, appears to induce susceptibility in the usual sense of the word. In other examples (T toxin and corn), pathogens can colonize, to some degree, toxin-insensitive cultivars of the host. The toxin thus may act only as an additional factor for cellular damage, as do nonspecific toxins.

In either event, host-selective toxins provide an unusual opportunity for identifying the chemical bases of resistance and susceptibility. Because their effects on such processes as ion leakage can be seen in a relatively short time, the simplest explanation for their selectivity is that there is a constitutive gene product only in susceptible hosts with which the toxin interacts, whereas resistant cultivars lack the factor. This is the reverse of the usual interpretation of the gene-for-gene hypothesis. If, however, tissues are resistant because of their ability to degrade or detoxify toxin, the usual interpretation is permissible. Attempts to show degradation or detoxification have not been successful to date, probably because the attempts were made with toxin preparations of unknown chemistry and purity. In the last 5 years, structures have been assigned with some certainty (15) for host-selective toxins of *Alternaria mali, Helminthosporium maydis* race T, *Alternaria alternata* f. sp. *lycopersici* and, very recently, *Helminthosporium sacchari*. Earlier claims (36) for a structure of toxin from the last species were erroneous. With this new knowledge on structures, it should be possible to examine the potential for degradation or detoxification by more rigorous chemical and physical techniques.

The third alternative for explaining resistance to host-selective toxins, suggested by Wheeler (46), is that sensitive sites exist in both resistant and susceptible plants, but resistant plants have a mechanism for self repair of an initial metabolic lesion. This suggestion has been largely neglected but deserves careful attention in the future. Mechanisms involving detoxification or self repair would entail gene products for resistance that could be identified and isolated.

SUPPRESSION OF RESISTANCE RESPONSES

In the last few years, the somewhat divergent ideas of metabolic resistance and susceptibility have tended to converge with suggestions that susceptibility is the result of a specific suppression of a general, nonspecific, metabolic resistance. Specific induction of metabolic resistance would not be necessary; specificity would lie in the ability of the pathogen to overcome, or to prevent, events that lead to its inhibition. With *P. infestans*, mycelial slurries have been prepared that apparently prevent the development of hypersensitivity characteristic of resistance with late blight of potato, but the preparations do not show requisite race or varietal specificity (19). A claim for specificity in the suppression of the resistance of soybean cultivars to *P. megasperma* (43) has not been confirmed in recent studies in the same laboratory (18).

Bushnell and Rowell (8) recently have presented a genetic model that invokes race-specific suppressors of a general defense reaction and that can be aligned with the concepts of the gene-for-gene hypothesis. Two necessary postulates for the model are: the presence of constitutive genes for basic compatibility and the existence of a functional, general defense mechanism. Although genes for basic compatibility currently are in vogue for biotrophic diseases, none of them have been identified through conventional genetic techniques. As indicated earlier, the second assumption also is ambiguous. Although cytological and biochemical changes (e.g., phytoalexin formation) are observed in resistant plants, it has not been demonstrated convincingly that these changes are involved in metabolic resistance. The biochemical changes may be an inconsequential symptom of stress or cellular injury caused by foreign biotic or abiotic objects (12).

As presented, the model pictures a pathogen gene product with a specific configuration interacting with a specific gene product of the host. The resulting biological interaction is all-or-none; that is, either resistance or susceptibility occurs. It cannot easily account for the fact that, in many instances, disease reaction forms a continuum of phenotypic, and apparently genotypic, classes (in stem rust of wheat, classes 0, 0;, 1, 2 for resistance and 3, 4 for susceptibility). It is not clear from the model how these reaction classes are created through suppression of a general metabolic resistance. To permit pathogen development, suppressors of resistance must be present at the onset of pathogenesis. To explain reaction classes in which resistance can be high (immune infection type) or low (infection types 1 or 2), the model must be modified, either by postulating loss of pathogen suppressor production as a function of time or progressive loss of host sensitivity to suppressors. In either case, the biochemical mechanics for the model become complicated. The obvious alternative, that each reaction class is determined by individual suppressor structure or conformation, implies that each pathogenic race carries a very large complement of suppressors. The problems posed are similar to those raised by Ward and Stoessl (45) for recognition of resistance.

CONCLUDING REMARKS

In the very immediate future, efforts will be underway to use the techniques of molecular biology and genetic engineering to improve the protection of plants to disease. There has been considerable recent progress in characterizing some of the chemical events that occur in resistance, such as the accumulation of host-produced inhibitors of fungal and bacterial growth. These findings have not lead to completely satisfactory explanations for resistance, principally because they do not account readily for the specificity inherent in plant disease. Although logical schemes can be advanced to accommodate specificity, their usefulness is limited by the paucity of data about the biological events during infection. An urgent current need is better quantitative data on pathogen and disease development in both resistant and susceptible tissues.

There also is need to reexamine some of the basic postulates that determine the course of our research. The notion that certain hosts do not become diseased because of the existence of a unique, active resistance mechanism of that host has been a dominant theme for research in the past. The early science of plant pathology may have been influenced unduly by concepts from medicine emphasizing the active immune responses of vertebrates. In a review of evolutionary development of immunity, Burnet (7) noted that, although there has been far less research on invertebrates,

> any recognition is a recognition of self rather than the recognition of foreigness which is characteristic of vertebrate immune responses. In other words, the damaging effects occur when a positive recognition of a self component is no longer possible.

As with plant resistance, the damaging effects entail cellular necrosis (hypersensitivity?).

It may also be revealing that, despite many striking parallels in the biology of infection, resistance has not been a central theme in studies of symbiosis. A primary goal has been to understand compatibility, with encouraging, if not yet definitive, success (3). Similarly, with viral diseases of plants there has been less emphasis on active processes for resistance and more on the events of successful accommodation, even though with virus infection the full range of host response encountered in fungal and bacterial diseases occurs.

The presence of an active resistance undoubtedly is important in many host-parasite interactions, but the possibility that susceptibility is the inducible and unique event in host-pathogen interactions should not be ignored. Recognition can entail a variety of chemical mechanisms other than macromolecular recognition devices, such as low molecular weight effectors of metabolic regulation in the host. Signals of these sorts can provide for the specificity, as well as the biological diversity, of plant disease.

LITERATURE CITED

1. Aist, J. R., Waterman, M. A., and Israel, H. W. 1979. Papillae and penetration: Some problems, procedures and perspectives. Pages 85–98 in: Recognition and Specificity in Plant-Host Parasite Interactions. J. M. Daly and I. Uritani, eds. Jpn. Sci. Soc. Press, Tokyo.
2. Antonelli, E., and Daly, J. M. 1966. Decarboxylation of indole acetic acid by near-isogenic lines of wheat resistant or susceptible to *Puccinia graminis* f. sp. *tritici*. Phytopathology 56:610-618.
3. Bauer, W. D. 1981. Infection of legumes by rhizobia. Annu. Rev. Plant Physiol. 32:407-484.
4. Bell, A. A. 1969. Phytoalexin production and Verticillium wilt resistance in cotton. Phytopathology 59:1110-1127.
5. Bostock, R. N., Knuckels, E., Hensling, J. W. D., and Kuć, J. 1982. The effect of age and storage of potato tubers on sesquiterpene metabolites and glycoalkaloid accumulation. Phytopathology 73:435-438.
6. Breugger, B. B., and Keen, N. T. 1979. Specific elicitors of glyceollin accumulation in the *Pseudomonas glycinea*-soybean host-parasite system. Physiol. Plant Pathol. 15:43-51.
7. Burnet, F. M. 1976. The evolution of receptors and recognition in the immune system. Pages 35–58 in: Receptors and Recognitions, Series A. Vol. 1. P. Cuatrecasas and M. F. Greaves, eds. Chapman and Hall, London.
8. Bushnell, W. R., and Rowell, J. B. 1981. Suppressors of defense reactions: A model for roles in specificity. Phytopathology 71:1012-1014.
9. Chakravarti, B. P. 1966. Attempts to alter infection processes and aggressiveness of *Puccinia graminis* var. *tritici*. Phytopathology 56:223-229.
10. Cruickshank, I., and Perrin, D. 1960. Isolation of a phytoalexin from *Pisum sativum*. Nature 187:799-800.
11. Daly, J. M. 1972. The use of near-isogenic lines in biochemical studies of the resistance of wheat to stem rust. Phytopathology 62:392-400.
12. Daly, J. M. 1976. Some aspects of host-pathogen interactions. Pages 27–50 in: Physiological Plant Pathology. R. Heitefuss and P. H. Williams, eds. Springer-Verlag, Berlin. 890 pp.
13. Daly, J. M. 1976. Specific interactions involving hormonal and other changes. Pages 151–165 in: Specificity in Plant Diseases. R. K. S. Wood and A. Graniti,

eds. Plenum Publ., New York.

14. Daly, J. M. 1979. Basic mechanism of host/pathogen interactions. Pages 3–26 in: Recent Advances in Tobacco Science. Vol. 5. T. C. Tso, ed. Tobacco Chemists Research Conference.
15. Daly, J. M., and Knoche, H. W. 1982. The chemistry and biology of host-selective pathotoxins. Pages 83–138 in: Recent Advances in Plant Pathology. Vol. 1. D. Ingram and P. H. Williams, eds. Academic Press, London.
16. Daly, J. M., Ludden, P., and Seevers, P. 1971. Biochemical comparisons of stem rust resistance controlled by Sr6 and Sr11 alleles. Physiol. Plant Pathol. 1:397-407.
17. Day, A. W., Castle, A. J., and Cummins, J. E. 1981. Regulation of parasitic development of the smut fungus, *Ustilago violacea*, by extracts from host plants. Bot. Gaz. 142:135-146.
18. Desjardins, A. E., Ross, L. M., Spellman, M. W., Darvill, A. G., and Albersheim, P. 1982. Host-pathogen interactions. XX. Biological variation in the protection of soybeans from infection by *Phytophthora megasperma* f. sp. *glycinea*. Plant Physiol. 69:1046-1050.
19. Doke, N., Garas, N. A., and Kuć, J. 1980. Effect on host hypersensitivity of suppressors released during the germination of *Phytophthora infestans* cystospores. Phytopathology 70:35-39.
20. Dyck, P. L., and Samborski, D. J. 1968. Genetics of resistance of leaf rust in the common varieties Webster, Loris, Brerit, Cavina Malakof and Centenario. Can. J. Genet. Cytol. 10:7-17.
21. Ellingboe, A. H. 1979. Inheritance of specificity: The gene-for-gene hypothesis. Pages 3–18 in: Recognition and Specificity in Plant Host-Parasite Interactions. J. M. Daly and I. Uritani, eds. Jpn. Sci. Soc. Press, Tokyo.
22. Favret, F. A. 1971. The host-pathogen system and its genetic relationships. Pages 337–347 in: Barley Genetics II. R. A. Nilan, ed. Wash. State Univ. Press, Pullman.
23. Flor, H. H. 1947. Inheritance of reaction to rust in flax. J. Agric. Res. 74:241-262.
24. Haberlach, G. T., Budde, A. D., Sequeira, L., and Helgeson, J. P. 1978. Modification of disease resistance of tobacco callus tissues by cytokinins. Plant Physiol. 62:522-525.
25. Keen, N. T., and Legrand, M. 1980. Surface glycoproteins: Evidence that they may function as the race specific phytoalexin elicitors of *Phytophthora megasperma* f. sp. *glycinea*. Physiol. Plant Pathol. 17:175-192.
26. Kiraly, Z., Barna, B., and Ersek, T. 1972. Hypersensitivity as a consequence, not a cause, of plant resistance to infection. Nature 239:456-457.
27. Knott, D. R. 1981. The effects of genotype and temperature on the resistance to *Puccinia graminis tritici* controlled by the gene Sr6 in *Triticum aestivum*. Can. J. Genet. Cytol. 23:183-188.
28. Kojima, M., Kawakita, K., and Uritani, I. 1983. Studies on a factor in sweet potato root which agglutinates spores of *Ceratocystis fimbriata*, black rot fungus. Plant Physiol. (In press)
29. Loegering, W. Q., and Powers, H. R. 1962. Inheritance of pathogenicity in a cross of physiological races 111 or 36 of *Puccinia graminis* f. sp. *tritici*. Phytopathology 52:547-554.
30. Mayama, S., Daly, J. M., Rehfeld, D. W., and Daly, C. R. 1975. Hypersensitive response of near-isogenic lines of wheat carrying the temperature-sensitive Sr6 allele for resistance to stem rust. Physiol. Plant Pathol. 7:35-47.
31. Müller, K. O. 1956. Einige einfache Versuche zum Nachweis von Phytoalexinen. Phytopathol. Z. 27:237-254.
32. Ouchi, S., Oku, H., Hibino, C., and Akiyama, I. 1974. Induction of accessibility to a non-pathogen by a preliminary inoculation with a pathogen. Phytopathol. Z. 79:142-154.
33. Ouchi, S., Oku, H., and Shiraishi, T. 1982. Physiological basis of susceptibility induced by pathogens. In: Plant Infection—The Physiological and Biochemical Basis. Y. Asada, W. R. Bushnell, S. Ouchi, and C. P. Vance, eds. Springer-Verlag, New York.
34. Seevers, P. M., Daly, J. M., and Catedral, F. F. 1971. The role of peroxidase isozymes in resistance to wheat stem rust disease. Plant Physiol. 48:353-360.
35. Sherwood, R. T., and Vance, C. P. 1982. Initial events in the epidermal layer during penetration. In: Plant Infection—The Physiological and Biochemical Basis. Y. Asada, W. R. Bushnell, S. Ouchi, and C. P. Vance, eds. Springer-Verlag, New York.
36. Steiner, G. W., and Strobel, G. A. 1971. Helminthosporicide, a host-specific toxin from *Helminthosporium sacchari*. J. Biol. Chem. 246:4350-4357.
37. Strange, R. N., Majer, J. R., and Smith, H. 1974. The isolation and identification of choline and betaine as the two major components of anthers and wheat germ that stimulate *Fusarium graminearum* in vitro. Physiol. Plant Pathol. 4:277-290.
38. Tani, T., and Yamamoto, H. 1979. RNA and protein synthesis and enzyme changes in infection. Pages 273–287 in: Recognition and Specificity in Plant Host-Parasite

Interactions. J. M. Daly and I. Uritani, eds. Jpn. Sci. Soc. Press, Tokyo.
39. Tomiyama, K. 1966. Double infection by an incompatible race of *Phytophthora infestans* of a potato cell which has been previously infected by a compatible race. Ann. Phytopathol. Soc. Jpn. 32:181-185.
40. Vance, C. P., and Sherwood, R. T. 1976. Cycloheximide treatments implicate papillae formation in resistance of reed canary grass to fungi. Phytopathology 66:498-502.
41. VanEtten, H. D. 1979. Relationship between tolerance to isoflavonoid phytoalexins and pathogenicity. Pages 301–315 in: Recognition and Specificity in Plant Host-Parasite Interactions. J. M. Daly and I. Uritani, eds. Jpn. Sci. Soc. Press, Tokyo.
42. Varns, J., Kuć, J., and Williams, E. B. 1971. Terpenoid accumulation as a biochemical response of the potato tuber to *Phytophthora infestans*. Phytopathology 61:178-181.
43. Wade, M., and Albersheim, P. 1979. Race-specific molecules that protect soybeans from *Phytophthora megasperma* var. *sojae*. Proc. Natl. Acad. Sci. U.S.A. 76:4433-4437.
44. Ward, E. W. B., and Lazarovits, G. 1982. Temperature-induced changes in specificity in the interaction of soybeans with *Phytophthora megasperma* f. sp. *glycinea*. Phytopathology. 72:826-830.
45. Ward, E. W. B., and Stoessl, A. 1976. On the question of "elicitors" or "inducers" in incompatible interactions between plants and fungal pathogens. Phytopathology 66:940-941.
46. Wheeler, H. 1969. Genetics of pathogenesis. Pages 9–13 in: Potentials in Crop Production. N.Y. State Agric. Exp. Stn., Geneva.
47. Yoder, O. C., and Scheffer, R. P. 1969. Role of toxin in early interactions of *Helminthosporium victoriae* with susceptible and resistant oat tissue. Phytopathology 59:1954-1959.
48. Yoshikawa, M., Yamauchi, K., and Masago, H. 1979. Biosynthesis and degradation of glyceollin by soybean hypocotyls infected with *Phytophthora megasperma* var. *sojae*. Physiol. Plant Pathol. 14:157-170.

CHAPTER 29

MECHANISMS OF PATHOGENESIS

HARRY WHEELER and STEPHEN DIACHUN

Plant pathogenesis, which involves the initiation and development of disease in plants, may result from the effects of biotic or abiotic agents. In our view, only biotic agents are pathogens. Abiotic agents that cause disease are pathogenic, but this does not make them pathogens. This distinction is similar to that applied to toxins which are defined as injurious chemicals of biological origin. Injurious agents from other sources may be toxigenic, but this does not make them toxins.

Nearly all plant pathogens are heterotrophic, completely dependent for survival on energy supplied by autotrophs. In view of their numbers and diversity, together with their ability to reproduce and multiply, pathogens have the potential to wipe out all autotrophs and, in so doing, eliminate themselves and all other forms of life from this planet. Why then have so many virulent pathogens evolved and survived and so few organisms developed the mutualistic associations exemplified by mycorrhizae, lichens, and nitrogen-fixing rhizobia?

To digress for a moment, we cannot agree with those who would extend the concept of parasitism to include mutualistic associations. Complete dependence on a host for food, as in the case of rhizobia, does not make a symbiont a parasite. To extend the Horsfall-Dimond analogy: the penniless mother-in-law who cooks the meals, does the laundry, cleans the house, baby-sits, and mediates family quarrels is no parasite; she earns her bread and board. Parasite is a derogatory term that should be applied only to symbionts that are harmful to their hosts, not to those that are harmless or beneficial. In other words, all parasites are pathogens.

To return to the evolution of parasitism and pathogenicity, let us first examine the possibility that the question posed may be loaded. How do we know that parasitic associations greatly outnumber those that are mutualistic or commensalistic? Plant surfaces harbor a rich variety of microorganisms, and bacteria are present in low numbers in the interior of healthy plant parts, especially storage organs. Are these symbionts beneficial, harmless, or harmful and what prevents those in the interior from multiplying? Do we know that healthy plants are not hosts to beneficial or harmless viruses and mycoplasmas? In view of the prevalence of plasmids in bacteria and of viruslike particles in fungi, is it not possible that many higher plants harbor benign viroids? Perhaps the prevailing basic assumption, that most foreign biotic agents that colonize plants are pathogens, should be reexamined. As a case in point, wild fire bacteria

(*Pseudomonas tabaci*) can multiply and form colonies on the surface of roots of several unrelated species of plants without causing obvious visible injury. They can multiply, also, at least for several hours or days, inside leaves of resistant plants without inducing symptoms. Many pathogenic fungi and bacteria with host ranges thought to be limited may enter nonhosts and become established, at least temporarily, without causing disease. Perhaps the assumption that most pathogens cannot and do not attack most plants is the truth but not the whole truth.

Regardless of whether parasites outnumber other types of symbionts, many highly pathogenic parasites have survived and flourished. Biotrophic parasites, those that in nature complete their reproductive cycles only on a living host, usually do not rapidly destroy host tissue. To do so would open their habitat to swarms of organisms incapable of colonizing living tissues. The evolutionary advantage of the relative congeniality of biotrophs seems obvious. Equally obvious, at the other extreme, is the advantage gained by perthotrophs, which kill host cells in advance of colonization. This ability puts them in position to take the lion's share before the hyenas and buzzards move in. The paradox is posed by the hemibiotrophs: parasites that invade and colonize living tissues in the same way as biotrophs but continue to develop and reproduce after the tissue is killed. This killing seems senseless; as saprophytes most hemibiotrophs are poor competitors. Why do these organisms continue to "bite the hand that feeds" instead of evolving into more peaceful biotrophs? Does the advantage of a place at the head of the table more than offset the competition from scavengers? Is the killing required to enable the parasite to reproduce and multiply? Or, are such pathogens merely those in transition, the "slow learners" that have restrained their virulence enough to avoid self-extinction but have not yet acquired the skill to take the egg without killing the goose?

Leaving the question of why plants and pathogens have failed to evolve to a state of peaceful coexistence, we turn to the theme of this chapter which, in essence, is "know thine enemies." What weapons do they have, how and when do they use them, and how can we marshall defenses against them? In this attempt to answer these questions, we will examine pathogenesis primarily from the viewpoint of our enemies, the pathogens. We know the attribute they have in common is the ability to rob plants of the energy that pathogens must have to survive.

STAGES OF PATHOGENESIS

In general, plant pathogenesis progresses through a series of stages. First the pathogen makes contact with the plant surface, it then enters, establishes itself within the plant, induces disease symptoms, and finally reproduces a new generation. Only the final stage is essential for pathogen survival; the others may be combined or bypassed entirely. Some pathogens cause disease by producing toxic chemicals without making physical contact with the plant. Others, the sooty molds for example, grow over leaf surfaces without entering. A few, certain viruses, cause virtually no visible symptoms.

The stages of pathogenesis that have been outlined are most clearly defined in diseases caused by eukaryotic pathogens. Even with these, there is great variation in the time required to complete the progression and in that which separates the different stages. In the case of ectoparasitic nematodes, contact and entry may be virtually simultaneous, but most observations indicate that these pathogens do considerable exploratory probing before settling down to feed. In root hairs attacked by *Pythium* zoospores, the first four stages, cell death included, occur within 2 hours. In contrast, other fungal pathogens enter plant tissues and then remain dormant for months or, as some smut fungi, for years before completing the pathogenic progression. These variations suggest that eukaryotic pathogens differ widely in their ability to overcome obstacles at various stages of pathogenesis.

Stages in pathogenesis overlap or are fused in most diseases caused by pathogens that are not eukaryotic. Frequently, when a vector or wound is required for infection, contact and entry are reduced to a single stage. Unlike the eukaryotes, these pathogens often produce many generations of new infectious cells or particles before disease symptoms appear. In spite of this difference, which must be considered in devising methods of disease control, the overall course of pathogenesis caused by these agents often closely resembles that caused by eukaryotes.

We plan to examine each stage in pathogenesis with three questions in mind: 1) What obstacles do pathogens face? 2) How do pathogens overcome these obstacles? 3) How can we make these obstacles more effective barriers to the further progression of pathogenesis?

We hope to then project some directions for more efficient future approaches to disease control.

Contact

Most pathogens overcome the obstacle of making contact with susceptible plants simply by the production and dispersal of truly awesome quantities of inocula. Long ago, Gaümann calculated that the aeciospores produced in the spring on a single medium-sized barberry bush had the potential, by infecting and cycling on a cereal grass, of generating in a single growing season the astronomical number of 10^{46} uredospores with a total mass more than 1,000 times that of Earth. Some mobile pathogens, notably nematodes and fungal zoospores, are somewhat more subtle; they use tactic responses to plant exudates as guides to contact. Vectored pathogens, which take a short-cut ride past contact, will be considered later.

Prevention of plant-pathogen contact has been the objective of many past and present methods of disease control. Direct attempts to achieve this objective by quarantine, destruction of alternate hosts, fallow, flooding, fumigation, or crop rotation have often been ineffective or too costly to be practical. Prevention of plant-pathogen contact through the use of clean seed, disease-free seed stocks or seed treatment has provided effective control for many diseases but in some cases failed because pathogens, thought to require the continuous presence of a living host, survived epiphytically on nonhosts. Prevention of disease by trapping nematodes on plants on which they cannot reproduce has had limited success, partly, at least, because of the difficulty of maintaining pure, weed-free stands. In theory, the use of mixed cultivars or multilines would force the pathogen to waste most of its seed on infertile ground. Whether growers will accept almost certain slight to moderate losses to ensure against catastrophic epidemics remains to be seen.

Entry

After making contact, many pathogens enter plants through wounds or injuries caused by elements of the weather or by animals or by humans and their tools. Such entry seems simple, like a thief walking into a house through an open door. Elaboration is not needed.

In the absence of wounds, many pathogens enter through natural openings: stomata, lenticels, hydathodes. Others, chiefly fungi, enter by direct penetration. In the course of direct penetration, fungal spores themselves or their germ tubes or their appressoria become bound tightly to plant surfaces and then produce a fine penetration peg. Currently, opinion leans to the view that cuticular penetration is by a combination of localized pressure and enzymatic action. In penetration of cell walls, enzymatic action seems to be more important, localized at the hyphal tips in some cases, and less restricted in other cases, even at some distance in advance of penetrating hyphae.

Plant pathogenic bacteria are not as aggressive as plant pathogenic fungi. If leaf-spotting bacteria are on a leaf surface, they usually do not enter even if stomata are open. It was believed formerly that a bacterium could multiply over a stoma under humid conditions and that the ensuing colony could force its way into the leaf. That belief is not now widely espoused (except perhaps for a few leaf spot diseases). However, if the leaf is water-soaked or congested, naturally or artificially, the bacteria can get into the intercellular spaces by means of water films or channels that connect the water-soaked intercellular spaces with water on the outer surface.

Viruses are even less aggressive than bacteria; to cause disease, virus particles must be inside living host cells. Viruses enter through wounds caused mostly by the feeding of insects and other vectors.

Paradoxically, a high concentration of virus particles is necessary for mechanical inoculation to be successful, although it is believed that a single particle properly placed can initiate infection in a susceptible cell. Years ago it was suggested that only a few particles might be needed if a dilute virus suspension were brought in intimate contact with plasmodesmata. This suggestion seemed so logical to the junior author that he immersed a necrotic-spotting tobacco leaf in a suspension of TMV and stripped-off sheets of epidermis. He could envision the virus particles joyfully entering host cells by way of exposed plasmodesmata and he fully expected large numbers of necrotic spots to develop. Alas, none did. Later he rubbed TMV on leaf tissue with epidermis removed, as well as on adjoining leaf areas with epidermis intact. Necrotic lesions did not form on tissue without epidermis; numerous necrotic lesions developed on the adjoining areas. Perhaps plasmodesmata are not susceptible to infection by abrasion, or perhaps epidermal cells are necessary in such inoculations.

Entry by viroids, mycoplasmas, and spiroplasmas is essentially like entry by viruses. Entry by nematodes is a special situation. Indeed, the ectoparasitic nematodes do not enter at all, but feed on plants from the outside. Those nematodes that do enter plants (endoparasites) do so by pushing their way between cells or breaking in through cells, perhaps first softening the tissue by inserting their stylets and secreting saliva.

Do plants present obstacles or barriers to entry by pathogens? Yes; some real and others perhaps putative. Waxes, cuticles, epidermal cell walls, sunken stomata, and closed stomata are obstacles in the first line of defense. However, pathogens of various kinds obviously have developed ways to overcome such defenses, else all or most plant diseases would be caused by wound parasites.

Establishment

Many pathogens can penetrate into plants and even into cells of plants that are not susceptible without producing symptoms or causing disease. After a pathogen enters a susceptible plant, there is a period of time, the incubation period, during which there are no macroscopically visible signs of infection or pathogenesis. During this time the pathogen is becoming established. In some fungal diseases, especially powdery mildews and rusts, a specialized hypha (the haustorium) is formed. The haustorium penetrates the cell wall, invaginates the protoplast, and induces development of an encapsulating papilla. Presumably the haustorium withdraws, as sustenance for nourishment of the parasite, partially solubilized host cell walls, or nutriments from host cell vacuoles, or even parts of the host substance itself. In other diseases the parasite's hyphae remain intercellular.

In general, bacteria do not invade living host cells but remain in the intercellular spaces or in the vascular tracheal system. Current belief is that virulent bacterial leaf pathogens do not become attached, but remain free to multiply in so-called intercellular fluids. However, in such leaf spot diseases as wildfire and other leaf spots in which water-soaking at the margins of the spots does not occur or occurs only under very humid conditions, water in the intercellular spaces probably occurs mainly as thin films, perhaps molecular films, not as fluids. The multiplying bacteria can form discrete colonies called, in older literature, zoogleae. Under humid conditions, zoogleae resemble bacterial colonies on the surface of roots or on agar in petri dishes. In neither of these cases do fluids as such exist, although thin films of water with dissolved nutrients may be present.

What the forces are that cause bacteria to adhere to cell wall surfaces is not known. It may be that the bacteria develop pili, which interact with the fibrils postulated to be present on the surface of plant cell walls. In other cases, angular leaf spot of cotton, for example, the presence of intercellular fluid manifests itself as dark, "oily" water-soaking. In this and in other diseases characterized by localized water-soaking, it may be that bacteria do multiply in intercellular fluids and do not form zoogleae.

Whereas pathogenic bacteria become attached to cell walls of susceptible plants and then develop colonies, some nonpathogens induce a hypersensitive response by the host cells. Cells of incompatible plants are thought to engage in more or less violent warfare in an active process of resisting pathogens. Susceptibility occurs only after combat in which the pathogen breaches and overcomes host defenses.

Such active defense mechanisms to account for resistance to the establishment of pathogens may not be needed, at least not in all cases. Bacteria and fungi may enter leaves and begin to grow, as they can even on a clean glass slide, and then simply languish and eventually starve to death. Induction of susceptibility may be the key molecular event that allows a pathogen to become established. Resistance may be simply a nonhappening to be expected whenever the crucial specific susceptibility interactions do not occur.

Symptom Induction

The established pathogen must avoid being sealed off or restricted by the physical and chemical defenses of the plant. Biotrophic pathogens must also avoid suicidal, quick killing of host cells. These obstacles are the plant's final lines of defense. Once they have been overcome or bypassed, symptoms develop and pathogenesis is essentially completed.

In general, biotrophic pathogens succeed by not arousing the plant's last lines of defense, whereas nonbiotrophs overwhelm them with batteries of chemicals. Initial symptoms on susceptible plants are a reflection of these two modes of operation, mild in the case of biotrophs and severe in the case of nonbiotrophs.

Two mechanisms have been proposed to account for the success of biotrophic pathogens. One assumes that they don sheep's clothing so that plants will not recognize them as enemies until it is too late. The other proposes that biotrophs produce substances that suppress the plant's internal defenses. Evidence favoring the first mechanism is limited to a few cases in which pathogens have been found to have more antigens in common with susceptible than with resistant plants. Support for the second mechanism consists of evidence that hypersensitive responses and, presumably, the plant's internal defenses, can be suppressed by injection of killed cells or cell fractions of certain bacterial and fungal pathogens. The significance of these results is far from clear since the hypersensitive response can also be suppressed by a variety of other chemical and physical treatments, among which is the simple expedient of maintaining plant tissues in a highly hydrated state.

Toxins, growth regulators, and enzymes are the chemical weapons that many pathogens employ to overcome the plant's internal defenses. Attention has focused on toxins, in particular on those, termed "selective pathotoxins," that are selectively toxic to plants susceptible to the pathogen involved. Selective pathotoxins are not only the clearest examples of agents responsible for the specificity of plant-pathogen interaction; they have also provided model systems for studies of disease physiology and the nature of disease resistance. Some pathotoxins, tentoxin for example, apparently interfere with energy generation at a specific site in an electron transport system, whereas others, like victorin, cause decompartmentalization by disruption of membrane systems. In either case, the final result is the same; plant cells perish from the lack of a source of energy required for maintenance and repair of vital functions.

Galls, overgrowths, excessive elongation, epinasty, and witches' broom effects provide clear evidence that abnormal growth regulation plays a role in symptom development in many plant diseases. Unlike toxins, which are chiefly metabolic products of nonbiotrophs, excesses or imbalances in growth regulators are induced by both biotrophs and nonbiotrophs and, in most cases, appear to be produced by the plant. Galls induced by certain bacteria may be an exception to the general rule. In gall induction by *Pseudomonas savastanoi*, virulence appears to be dependent on genes for indole acetic acid synthesis carried by a bacterial plasmid.

Enzymes, in particular those that degrade cell walls, have long been recognized as the chief offensive weapons employed by nonbiotrophs that cause soft, watery rots. Pectic lyases and hydrolases, which macerate plant tissues and kill cells, are thought to spearhead the attack, but the origin of these enzymes in diseased plants is uncertain; they may be products of the pathogen, of the plant, or both. Since the lethal effects of pectolytic enzymes are suppressed by hypertonic solutions, it is thought that cell death results from uncontrolled swelling of protoplasts and rupture of compartmentalizing membranes.

Recently, attention has been drawn to the role of digestive enzymes in what have been aptly called replacement diseases. In these, host tissues are completely digested in precisely delimited areas and the space thus created filled by structures of the pathogen. Ergots, smuts, and rusts are typical replacement diseases caused by well-recognized biotrophs. Surprisingly, *Bipolaris* (*Helminthosporium*) *ravenelii*, a close relative of several pathotoxin-producing fungi, behaves as a typical biotroph in a replacement disease, false smut of smutgrass.

The most striking features of replacement diseases are the restriction of damage to specific host cells and the complete digestion and absorption of all host cell components, walls included. Although it has been suggested that the digestive enzymes are pathogen-produced, the whole process resembles microsporogenesis, during which cells surrounding developing microspores undergo autolysis and are absorbed and replaced by pollen grains. It may be that damage in replacement disease is sharply localized because the pathogen merely activates the host's own autolytic potential rather than secreting a battery of foreign enzymatic macromolecules.

Macroscopic symptoms usually reflect interferences with one or more vital plant functions and are, therefore, highly diverse. In contrast, a common pattern of physiological changes, termed the physiological syndrome, occurs in plants infected by a wide variety of pathogens. Embodied in the concept of the physiological syndrome is a series of pathological changes in plant metabolism triggered by some initial event. Currently popular as a candidate for the trigger is a change in permeability, which has been the first change detected in response to a wide variety of pathogenic agents. Although extensive investigations have failed to provide direct support for the physiological syndrome concept, the data that have accumulated suggest that during pathogenesis the plant's metabolic resources are diverted to the repair of damaged cells and tissues. Timely repair results in the suppression of symptoms or a resistant reaction. Faulty or slow repair puts an excessive drain on the plant's energy resources to the extent that defense mechanisms in uninvaded cells are rendered incapable of restricting the spread of the pathogen.

Control of plant diseases by blocking pathogenesis after pathogens have become established obviously depends on efficient utilization of the plant's internal defenses. Shortly after the turn of the present century, Noel Bernard demonstrated that infected plants may become resistant to reinfection by the same or a different pathogen. Since that time, resistance and, in some cases, apparent immunity to viral, bacterial, and fungal pathogens has been induced by inoculation with virulent or avirulent organisms, by injections of dead pathogens or fractions thereof, by infiltration of nucleic acids, and by prior treatment with pathotoxins. To date, however, these results have not led to practical measures of disease control. Instead, we have depended on the use of resistant plants to control established pathogens. Even here, lack of understanding of how plant internal defenses function has made the breeding and selection of resistant plants entirely empirical.

OUTLOOK

The challenge to predict how obstacles at each stage of pathogenesis can be utilized more effectively to control plant disease in the future carries with it the temptation to indulge in science-fiction fantasies based on space-age technology and recent advances in molecular genetics. In another century or two, present agricultural practices may seem as primitive as the hunting and gathering that preceded crop cultivation. Expanding human populations and withdrawal of fertile land from agriculture may require revolutionary new methods of food production. Visionaries see closed systems, floating in space like giant balanced aquaria, with everything but the sun's energy continuously recycled. Presumably photosynthetic organelles, genetically engineered to utilize all rather than only about 10% of the energy of sunlight, would provide the food. In such systems, pathogenesis would be blocked before it began by exclusion of all pathogens. Plant pathologists would be needed only to decontaminate systems that occasionally became invaded by an alien agent.

Looking more realistically at the next few decades, we see little prospect of more effective control measures based on prevention of plant-pathogen contact. On the other hand, certain recent developments make prospects quite promising for more effective utilization of obstacles at other stages of pathogenesis.

Many pathogenic fungi produce self-inhibitors of spore germination. A few, produced by certain rust fungi, have been identified as derivatives of cinnamic acid. Since these self-inhibitors are active at concentrations as low as 10^{-11} M, their potential as protective or, possibly, systemic fungicides is obvious. It is conceivable that, by genetic or chemical manipulation, plants could be induced to secrete sufficient quantities of specific cinnamic acid derivatives onto their surfaces to inhibit germination of spores of pathogens.

Certain other observations of spore germination and growth of germ tubes on plant surfaces suggest that obstacles prior to penetration can be utilized more effectively. For example, spores of *Peronospora tabacina* germinate in greater percentages on leaves of *Nicotiana debneyi*, which is immune, than on leaves of susceptible tobacco. Resolution of this paradox by some enterprising plant pathologist could lead to new methods of disease control. Germinating spores of certain fungi utilize tropic responses to ensure contact of germ tubes with stomates. If methods can be found to "confuse" the germ tubes by neutralizing these tropisms, they might reduce the fungi's efficiency in finding stomatal openings and thus reduce entry and infection.

Waxes and cutins are the first physical obstacles encountered by pathogens that penetrate directly. Infection by *Fusarium solani*, one of several fungi reported to secrete cutinase during penetration, was prevented by an antibody prepared against the enzyme and by a chemical inhibitor of cutinase. These results suggest that chemicals that inhibit fungal enzymes essential for penetration may provide a new class of fungicides.

Various types of structural modifications in plant cell walls occur during early stages of infection by many different kinds of pathogens. In some cases, treatments that inhibit these structural changes render plants susceptible to pathogens to which they are normally resistant. In other cases, induction of cell wall modifications by mechanical stress has rendered susceptible cells resistant to penetration. These results indicate that cell walls can be modified to provide effective barriers to pathogens. A better understanding of the nature and function of these modified cell walls may provide clues as to how they can be utilized in a practical way to control diseases.

Many pathogens enter and become established in resistant plants as readily as they do in susceptible ones. To effectively control these pathogens we must somehow bolster the plant's internal defenses. If we knew, at the molecular level, how these defenses function, present empirical methods of breeding for resistance could be replaced by methods designed to activate or stimulate specific metabolic pathways required for resistant reactions. The best hope for acquiring the knowledge needed to take advantage of new genetic technologies would seem to be continued work with model systems that have been developed over the past three decades to study the nature of plant-pathogen relationships.

Most work on plant-pathogen relationships has been based on one or the other of two theoretical models. One of these, often called induced resistance, grew out of the gene-for-gene concept. In essence, this model assumes that the product of a specific gene for avirulence in the pathogen must interact with a corresponding gene for resistance in the plant to produce a resistant reaction. Despite extensive searches, there is, at most, scant evidence for the existence of the highly specific gene products demanded by the gene-for-gene concept.

The second model, induced susceptibility, assumes that products of virulence genes in pathogens suppress or overwhelm defenses in susceptible plants. Results with pathogen-produced selective pathotoxins have provided strong support for the induced susceptibility model, but only in a limited number of diseases all involving nonbiotrophic pathogens.

In view of its theoretical and practical importance, a critical test of the gene-for-gene concept should have the highest possible priority. This concept holds that once resistance has been established through the interaction of any matched pair of genes for avirulence in the pathogen and resistance in the plant, symbolized A + R, resistance will be maintained regardless of any changes at other loci in the genomes of the pathogen or the plant. In other words, loss of resistance must involve a change at A or R or both. Since we know of no case of a plant resistant to a specific race of a pathogen becoming susceptible to that same race, changes in the pathogen must have been responsible for the losses of disease resistance that have often occurred. On the basis of the gene-for-gene concept, these changes must have involved either mutations to virulence or deletions at the A locus.

The frequency of spontaneous changes at single loci in fungi usually falls in the range of one in 10^6–10^8. Using a frequency of one in 10^7 as a reasonable average, and, for example, a wheat cultivar with a single R gene for rust resistance, the matching A gene in the pathogen should be lost in one of 10^7 spores produced, and, in view of the reproductive capacity of the pathogen, a new race virulent on this cultivar would be expected to arise quickly. As the number of A + R pairs in the pathogen and plant are increased, the number of spores that theoretically must be produced before a new virulent race can be expected increases geometrically. With two A + R pairs, 10^{14} spores are required; three, 10^{21}; four, 10^{28}; and five, 10^{35}. On the basis of these calculations, a cultivar with four or five R genes, each of which conditioned resistance to all races of the pathogen present in nature, would be expected to remain resistant for many years, perhaps for centuries, if the gene-for-gene concept is valid.

Cultivars of wheat, flax, and a number of other plants with single and multiple genes for resistance, which have been developed over the years, provide materials for a test of the gene-for-gene concept. The test would require that cultivars carrying one to five R genes be grown in strict isolation. Separate growth chambers, equipped to exclude cross contamination between chambers or entrance of spores from outside, should serve this purpose. Interplanted in each chamber, a cultivar with no R genes would be inoculated with a race of the pathogen to which all cultivars with R genes were resistant, to provide high quantities of inoculum. Generation after generation of the cultivars with R genes would be grown until some or all succumbed to the pathogen.

The outcome of the test just outlined should provide a guide to the future development of plants with lasting disease resistance. The gene-for-gene concept predicts that cultivars with one or two R genes should succumb quickly, whereas those with four or five should remain resistant for many plant generations. Although such a result would not constitute proof of the validity of this concept, it would justify undertaking the difficult and tedious task of finding new R genes, never faced by the pathogen in nature, and, in isolation, pyramiding them in a single cultivar that, when released, should have lasting resistance.

On the other hand, if cultivars with multiple R genes succumb to the pathogen as quickly, or nearly as quickly, as those with one or two R genes, the whole concept of induced resistance would be discredited. Such a result would indicate that single mutations to virulence in the pathogen are capable of overcoming the effects of multiple genes for resistance and that building gene pyramids would be futile. In this case, plant pathologists and plant breeders might do well to concentrate on developing tolerant cultivars that become mildly diseased rather than to persist in breeding highly resistant or immune plants.

ACKNOWLEDGMENTS

Space limitations precluded anything approaching complete documentation of this chapter so we have cited no references. It will be apparent to those familiar with the literature of our discipline that this chapter is mostly, if not entirely, a compilation of the results and ideas of many investigators. We acknowledge our debt to these unnamed colleagues. To them should go the credit for anything in the chapter that may be of value. We are especially indebted to Thomas Pirone and David Smith for valuable suggestions during the preparation of the manuscript.

CHAPTER 30

POTENTIAL OF MYCORRHIZAL SYMBIOSIS IN AGRICULTURAL AND FOREST PRODUCTIVITY

D. H. MARX and N. C. SCHENCK

It is necessary at the outset to point out that the potential management of mycorrhizal associations is not a panacea for correcting the numerous problems facing the world's people in the production of food, fuel, and fiber. Realistically, no one scientific specialty can solve all the problems. It is our belief, however, that many plant and soil scientists as well as most land managers are not sufficiently aware of the benefits of mycorrhizae. They do not know that on most plants mycorrhizae are as much a natural component of roots as chloroplasts are components of leaves and, therefore, this lack of understanding has limited serious exploitation of mycorrhizae to improve plant production. For this discussion we will examine only those aspects of mycorrhizae essential to a basic understanding. One can refer to the nearly 4,000 publications on mycorrhizae available in world literature for more information. Before we discuss the potential of manipulating mycorrhizae to improve crop and forest production, it is appropriate to understand the scope of the problem of supplying the world's needs for food, fuel, and fiber.

It has been estimated that the world's population will double from the present 4 billion to 8 billion within the next 50 years (38). Currently 500 million to 1 billion people in the world have less food than is necessary for basic survival (20) in spite of the use of high-yielding, disease-resistant crop cultivars and the widespread use of chemical fertilizer and pesticides. Worldwide fertilizer use has jumped from about 7 million metric tons in 1945 to more than 72 million metric tons in 1975 (44). This increase was accelerated as plant scientists developed new crop cultivars requiring high soil fertility for high yields. American farmers in 1981 used 4.9 million metric tons of phosphorus (P) fertilizer. According to the National Academy of Science, domestic supplies of P will be exhausted in a few years and, by 1990, three-fourths of the world's supply of P will come from the Middle East. Will this be another OPEC—Organization of Phosphate Exporting Countries? The demand of crops for nitrogen is projected to be 145 million metric tons by the year 2000 (15). Most genetically improved crop cultivars are inefficient in the uptake and use of fertilizer, especially nitrogen. Less than 50% of the applied nitrogen is taken up by nonlegume crops such as corn and wheat. Hardy and others (15) have estimated that more than $3 billion (1975 market) a year can be saved in the world's energy bill if efficiency in plant use of nitrogen fertilizer could be improved by 50%. In many parts of the developing world, soil management methods of any kind do not exist (13). Soil erosion, desertification amounting to 50,000 km^2 yearly, and soil salinization currently affecting nearly 1 billion ha are making many areas unusable for crop production (33).

Less than one third of the arable land on earth is classified as forest land. Forests are disappearing in the world at a rate of 1.2% annually with more than 11 million ha lost each year (9). This is a removal rate of more than 20 ha per *minute*. More than 25 million ha of deforested land exist now in Thailand, the Philippines, and Indonesia. Some countries, such as Haiti, have essentially eliminated all of their forests, whereas 17 other countries will reach total deforestation by 1990 at current removal rates (3). The destruction of tropical forests is proceeding at a tremendous rate (18) without even a backward glance at regeneration. The system of slash and burn migrant cultivation—shifting agriculture—has eliminated more than 3.6 billion ha of forests. Nearly 250 million people in developing nations, however, depend on shifting agriculture for their survival (1). Unfortunately, it has led to the destruction of many important watersheds which, in turn, has led to severe floods, soil erosion, and siltation of water resources. Based on current worldwide wood use rates, at least 24 million ha of tree plantations must be planted by the year 2000, which is 10 times the present rate of reforestation (9). Since 1925, the world's largest pine reforestation program has been conducted in 12 southern states of the United States by federal, state, industrial, and private land managers. However, only 11 million ha have been reforested in this 55-year period, and they were done with increasingly advanced technologies (43). Reforestation programs in the developing countries will have to be far more ambitious than this American effort if the deforested lands are to be reforested in the near future.

There is currently a strong and irreversible trend for use of the best forest land to produce food. The major challenge to forest science is to generate the biological technologies needed to grow productive forests on remaining lands now considered to be submarginal for economic production of either food or timber crops (45).

Only 10% of the projected global energy supply can come from renewable energy sources at the present state of biomass inventory. If used, this would cause a vigorous exploitation of all the biomass in Asia and Africa and much of it in Latin America. This prospect immediately raises questions of ecological stability, soil erosion, water requirements, and global climatic effects (38). Most people are not aware that nearly half of all the wood cut annually in the world goes for fuel, including 80% of all wood used in developing countries (9). Considering the past deforestation histories of these countries and the projected deforestation estimates for the future, there will be little value in improving local crop production or importation of food without a similar increase in availability of fuel wood. It is obvious that the people in these developing countries will not be able, economically, to substitute fossil fuels for fuel wood as has been done in developed nations. An exotic planned forest of pine or eucalyptus can produce up to 5–7 cords per hectare per year of wood in subtropical or tropical countries. In a 15-year rotation the BTU value of 1 ha of this green wood equals between 110 and 168 barrels of crude oil, or 56 and 83 metric tons of bituminous coal (4). Since these yields are based on stem wood and not on whole-tree harvest, these energy estimates are conservative.

Surface mining in the United States has disturbed 2.3 million ha of land. Nearly two thirds of this land (1.55 million ha) needs reclamation today. Land disturbance for all minerals, of which coal is the leading commodity, is projected to be nearly 115,000 ha per year by 1990 (2).

These gloom-and-doom statistics are real. The world's major problem is apparently a false philosophy of natural resource abundance due to a socially evolved myth of inexhaustibility of these resources. We must realize that the land and water resources of the world must produce enough food, fuel, and fiber in the next 25 years for as many people again as these resources produced in the whole history of humankind until now (26). The leadership in the governments of the world must make the right decisions concerning world stewardship of our plant and land resources during the next few decades to assure survival with an acceptable standard of life. New agricultural, forestry, and soil management practices must be developed if this is to occur. Our increasing population cannot survive by using the current technology for plant production on the decreasing inventory of productive lands. Basic and applied research on all aspects of crop and forest production on soils of different productivity potentials must be undertaken throughout the world on a large cooperative scale so that new biological technologies can be developed and implemented. A multidisciplinary approach to problem solving must include all aspects of plant growth, including the beneficial role of mycorrhizae and their possible manipulation to improve productivity of food, fuel, and fiber crops (36).

WHAT ARE MYCORRHIZAE?

The term mycorrhiza (fungus root) is used to describe a structure that results from a mutually beneficial association between the fine feeder roots of plants and species of highly specialized, root-inhabiting fungi. The fungi derive most if not all of their needed organic nutrition (carbohydrates, vitamins, amino acids) from their symbiotic niche in the roots. Evidence suggests that the mycorrhizal habit evolved as a survival mechanism for both partners of the association, allowing each to survive in the existing environment of low soil fertility, drought, disease, and temperature extremes.

Endomycorrhizae

This type of mycorrhiza is the most widespread and comprises three groups. Ericalean mycorrhizae occur on four or five families in the Ericales on such genera as *Calluna, Vaccinium, Rhododendron, Erica, Arbutus*, and *Pernettya*. Orchidaceous mycorrhizae are a distinct type that occur only in the family Orchidaceae. Since these two groups of endomicorrhizae may not play significant roles in solving any of the agricultural and forestry problems in the near future, they will not be discussed.

Vesicular-arbuscular mycorrhiza(e) (VAM) form the third group of endomycorrhizae. They occur on more plant species than do all other types of mycorrhizae combined. VAM occur in most cultivated crops such as cereals, corn, potatoes, beans, soybeans, tomatoes, and berry crops; in most fruit and tree-nut crops such as peach, cherry, pear, apple, citrus, grape, plum, walnut, and almond; and in most forest trees such as sweetgum, ash, maple, sycamore, elm, and mahogany. Most shrubs, climbers such as honeysuckle, grasses both cultivated and noncultivated, seashore plants such as sea oats, cacti in deserts, and a multitude of herbaceous plants all form VAM in natural environments. VAM are particularly widespread on plants growing in tropical and subtropical savannas and rain forests. They are also common on pteridophytes and bryophytes. Because of this broad host range, VAM fungi are ubiquitous in all natural soils throughout the world. Inoculum density and fungal species, however, differ in different soils. The fungi forming VAM belong to the class Zygomycetes and the family Endogonaceae, which includes the genera *Glomus, Gigaspora, Acaulospora, Sclerocystis*, and *Endogone*. These fungi have very little host specificity. The main characteristic of VAM is the presence of vesicles and/or arbuscules in the root cortex. The endodermis, stele, and root meristems are not invaded. Inter- and intracellular hyphae present in the cortex are connected to the external mycelium that spreads and ramifies in soil. Some VAM produce large sporocarps (5–10 mm in diameter) containing many spores, and others form large (100–600 μm in diameter), single, thick-walled spores on the root surface, in the rhizosphere, or in the root tissues. Early work with VAM dealt primarily with their anatomy and occurrence; they were virtually ignored by soil microbiologists, soil chemists, plant physiologists, agronomists, and all but a few plant pathologists. The main reasons for this were that VAM fungi cannot be grown on laboratory culture media and these fungi can be detected only under the microscope after root clearing and staining.

Ectomycorrhizae

This type occurs on trees belonging to the Pinaceae (pine, fir, larch, spruce, hemlock), Fagaceae (oak, chestnut, beech), Betulaceae (alder, birch), Salicaceae (poplar, willow), Juglandaceae (hickory, pecan), Myrtaceae (eucalyptus), Ericaceae (*Arbutus*), and a few others. Some tree genera such as *Alnus* and *Arbutus* will form both ectomycorrhizae and VAM, depending on soil conditions. Numerous fungi have been identified as forming ectomycorrhizae. In North America alone it has been estimated that more than 2,100 species of fungi form ectomycorrhizae with forest trees. Among the basidiomycetous fungi, species of Hymenomycetes in the genera *Boletus, Cortinarius, Suillus, Tricholoma, Laccaria*, and *Lactarius* and species of the Gastromycetes in the genera *Rhizopogon, Scleroderma* and *Pisolithus* form ectomycorrhizae. Certain orders in the Ascomycetes such as Eurotiales (*Cenococcum geophilum*), Tuberales (truffles), and Pezizales have species that form ectomycorrhizae on trees.

Ectomycorrhizae have intercellular hyphae surrounding cortical cells forming the Hartig net and several hyphal layers over the outside of the feeder root forming the fungus mantle. Ectomycorrhizal colonization normally changes the feeder root morphology and color. Like VAM, ectomycorrhizal colonization is limited to the primary cortex and does not spread beyond the endodermis or into meristem tissues. Unlike VAM, however, many ectomycorrhizal fungi can be grown routinely in pure culture. An important practical aspect of both VAM and ectomycorrhizal fungi is that neither group of fungi can exist saprophytically in nature without a plant-host association. Spores or resistant hyphae may survive long periods in soil without a plant host, but the fungi from these propagules will not grow as saprophytes.

Ectendomycorrhiza is a type that has some features of both ecto- and endomycorrhizae. They have limited distribution in forest soils and tree nurseries and are found on roots of normally ectomycorrhizal trees. Very little is known about their importance to trees, therefore, they will not be discussed.

BENEFITS OF MYCORRHIZAE TO PLANTS

Since the mid-1960s much research has been done throughout the world on VAM and ectomycorrhizae. Little research, however, has been done under field conditions. Most data have come from laboratory and greenhouse studies.

Vesicular-Arbuscular Mycorrhizae

The main value of VAM comes from their ability to furnish the plant with more nutrients, particularly P, Cu, and Zn. These elements are relatively immobile in soil and develop zones of depletion near feeder roots. The extramatrical growth of hyphae from VAM fungi can extend beyond the feeder roots and increase the volume of soil from which these elements can be absorbed. The additional nutrient absorption due to VAM fungi can result in severalfold growth increases in plants. The degree of plant benefit appears to be related to the plant's P requirement, ability to absorb nutrients from soil, amount of available P in soil, and the species of VAM fungus involved. Plants, such as grasses, with abundant fine feeder roots and root hairs are less dependent on VAM than are plants, such as sweetgum, citrus, and onion, that have fewer but larger feeder roots with fewer root hairs. However, even grasses benefit from VAM in soils with low amounts of available P. Baylis (6,7) has shown that the degree of root hair formation in many different plants was inversely correlated with an increasing dependence on VAM and directly correlated with the plant's ability to absorb P from soils containing little available P. Additions of available P to these soils eliminated the plant's dependence on VAM. Different species of VAM fungi may differ inherently in their effects on plant growth. All VAM fungi apparently use the same sources of soil P, but their effects on plants, in many instances, are not related to their degree of development or amounts of extramatrical hyphal growth. If soil P and N are raised to high amounts, VAM may not form at all. In regard to soil fertility, we can summarize that VAM-dependent plant species in soils of low and moderate P content will be stimulated by VAM. The fungus species of VAM can affect the magnitude of stimulation. Mosse and Hayman (31) have thoroughly discussed these aspects of VAM.

There are other significant benefits of VAM to plants. VAM are capable of reducing the effects of various fungal pathogens and suppressing the effects of parasitic nematodes on diverse agricultural crops (39). Not all reports are positive, however. Some reported that VAM had no effect on either the pathogen or the disease; others found that disease was increased. VAM have also been shown to enhance water uptake in soybeans, increase tolerance to toxic heavy metals, saline soils and drought, decrease transplant shock, and bind soil into semistable aggregates. Legume plants with VAM are often better nodulated than nonmycorrhizal plants, especially in soils with low available P.

VAM fungi may be eliminated or drastically reduced in soils that are fumigated or treated with certain pesticides to eliminate harmful pathogens and weeds. Large numbers of agronomic, forest, and horticultural crops use transplant production in fumigated nursery soils. These are citrus seedlings, hardwood tree seedlings, many vegetable (tomato, pepper), fruit (peach, apple, avocado, strawberry), and field (tobacco) crops. In the field, benefits from inoculation with VAM have been demonstrated for citrus, hardwood trees, and avocado, with growth increases exceeding 84-fold in certain instances. The addition of VAM to nonsterile field soil, especially soils containing low background populations of mycorrhizae, has been shown to be beneficial to corn, wheat, barley, cowpeas, soybeans, potatoes, white clover, and cassava, with growth increases from one- to sixfold (31). Reestablishment of various plant species on coal spoils has been improved also by enabling transplants to be colonized with VAM (19,21,35).

Ectomycorrhizae

Ectomycorrhizal fungi benefit tree hosts in a variety of ways. Some of these ways are increased longevity of feeder roots (growth-regulator function); increased rate of absorption of major and minor elements from the soil; selective absorption of certain ions from the soil; increased tolerance to soil toxins, low and high temperatures, and adverse soil pH; and resistance to certain feeder root pathogens (23).

Trees that normally have ectomycorrhizae require this symbiotic relationship to survive in natural ecosystems. This does not mean that trees such as pines and oaks cannot be grown for a few years from seed in aseptic seedling, hydroponic, or tissue culture. In these nonmycorrhizal conditions the seedlings, however, must be furnished with the factors (high nutrients, growth regulators, etc.) that are normally supplied by the symbiotic fungi and must be kept in a well-modulated environment or the seedlings will perish or not grow normally. The obligate requirement of pine for ectomycorrhizae in a natural environment has been clearly shown by numerous workers in tree regeneration trials (22,29). In many parts of the world ectomycorrhizal trees and their symbiotic fungi do not occur naturally. Forestation attempts with pine in these areas failed until ectomycorrhizal fungi were introduced. Most of the time, the essential fungi were introduced, often illegally, in a soil inoculum obtained from pine, eucalyptus, or oak plantations of other countries. The soil inoculum is used to inoculate seedbeds. Use of soil inoculum is still widespread in most developing nations where exotic plantations of pine, eucalyptus, and oak are established.

Soil inoculum has been the easiest and simplest method of inoculation but it has its limitations. Specific fungi in the mixture cannot be controlled nor can one be assured that it contains the most desirable fungi. Also, there is frequently a logistics problem in transporting large volumes of soil. The most severe limitation is that soil inoculum may contain a variety of microorganisms and noxious weeds that can be potentially harmful to the seedling crop and nearby agricultural crops. Due to these problems, a considerable amount of basic research has been done in various countries on selecting, propagating, and manipulating as pure cultures the more desirable ectomycorrhizal fungi in inoculation programs (22).

One recent and successful development came from research on the Gastromycete fungus, *Pisolithus tinctorius*. After techniques were developed (using spore and mycelial inocula) to "tailor" root systems of pine in nurseries with *Pisolithus* ectomycorrhizae, outplanting trials were set up. Dramatic improvements in survival and growth of various pine species with *Pisolithus* ectomycorrhizae in comparison to that of nursery-run (control) seedlings with naturally occurring ectomycorrhizae were reported from studies on acid coal-spoils in Appalachia, kaolin spoils in Georgia, severely eroded sites in the Copper Basin of Tennessee, borrow pits in South and North Carolina, and a prairie soil in South Dakota. Tree volume increases up to fourfold were not unusual on these hostile sites. Improvements in seedling performance were also reported on routine reforestation sites in Florida, Georgia, North Carolina, Oklahoma, Arkansas, and Mississippi. In parts of Africa, Asia, and South America, tropical pine seedlings with *Pisolithus* ectomycorrhizae also outperformed seedlings with ectomycorrhizae from soil inoculum. The greatest effects of *Pisolithus* ectomycorrhizae were on the more adverse sites such as acid coal-spoils and low-quality reforestation sites. The better the site, the better the control seedlings grew. In certain field studies on good-quality sites, control seedlings grew as well as seedlings infected with *Pisolithus* ectomycorrhizae. From an ecological view, this is understandable since *Pisolithus* was initially selected as a candidate fungus for inoculation programs due to its unique adaptation to adverse soil conditions.

PRACTICAL USE OF MYCORRHIZAE TODAY

Unfortunately, except for the following instances, the mass of information available on the benefits of mycorrhizae to plants has not been put to practical use.

Vesicular-Arbuscular Mycorrhizae

Since VAM fungi are obligate symbionts and cannot be grown in artificial laboratory media, it is essential to produce inoculum on plant hosts in pot cultures. There has been concern that pathogenic organisms also may be increased on the host at the same time the VAM fungi are being increased. Menge (27) and others devised a relatively safe procedure for preparation of "pure" VAM fungus inoculum. This method establishes the fungus on axenic plants, which are then used as inoculum for additional plants on which the VAM fungi are increased. The host used for increasing the VAM inoculum shares no major pathogens with the target plant to be inoculated in the field. Appropriate fungicides having no effects on VAM fungi are used to minimize the incidence of pathogens. Commercial concerns in California and elsewhere have employed these procedures to make inoculum of VAM fungi for use in commercial citrus nurseries. In England, a patented process has been devised to increase VAM fungi on plants in peat blocks, which then are ground up for use as inoculum (28). These procedures have made inoculum of VAM fungi available for use to a limited degree in commercial agriculture for the first time.

Although they have not been evaluated thoroughly, methods of application of these inocula usually consist of adding a dried spore-root-soil or peat-mixture to plants by placing the inoculum several centimeters below the seeds or seedlings by mechanical means so that seed and inoculum are added in one operation (28). Other means of incorporating inoculum of VAM fungi to soil include a mixture of soil, roots, and spores in soil pellets (14) and spores adhered to seed with adhesives (16). Incorporation of VAM fungi in soil using these latter procedures has not been attempted commercially.

Operational use of VAM inoculum is being made by a large paper company in

Mississippi to produce high-quality sweetgum seedlings with abundant VAM (R. C. France, International Paper Co., *personal communication*). A three-step method is used. First a specific VAM fungus is grown in standard pot cultures in the greenhouse on an appropriate agronomic host, such as millet. The second step involves spreading the contents of the pots (shredded roots and soil) onto recently fumigated soil and then mechanically incorporating the inoculum into soil. An agronomic crop is grown for 3–4 mo, and then disked into the soil. The third step is using this soil-root inoculum, after shredding, to infest fumigated nursery beds. This newly infested area is then sown to millet (or other crops) and grown for 2–3 mo. The area is disked and the plant material is allowed to decompose for an additional 2–3 mo. In the spring, after bed shaping, the beds are seeded to sweetgum and the resultant seedlings are grown for 9 mo. Seedlings grown with this method are superior in height, root-collar diameter, biomass, and amount of colonization by VAM fungi to seedlings grown in fumigated soil without inoculation and with moderate soil P. More than 250,000 sweetgum seedlings with abundant VAM were grown by this practical method in 1980. Although not an extensive commercial program, there have been successful attempts to utilize VAM in the revegetation of disturbed sites (21,27). Storage of topsoil before its replacement over mining spoils has been shown to decrease indigenous VAM fungal populations (11). Topsoil is now stored for shorter periods of time in order to maintain the highest possible VAM fungal populations that enhance revegetation efforts.

Ectomycorrhizae

The use of soil inoculum to form ectomycorrhizae on seedlings for establishing plantations of pine, eucalyptus, and oak, particularly in developing nations, is currently the main example of practical use of ectomycorrhizae in the world today. Although this method has several limitations, as mentioned earlier, it is effective.

In South Africa, Brazil, and Liberia, foresters collect basidiospores of *P. tinctorius* or *Scleroderma* sp. from basidiocarps in local forests and use them to inoculate container-grown pine or eucalyptus seedlings. In these countries, about 3 million seedlings are inoculated yearly with this method. Although research data are lacking, experience has shown these foresters that seedlings with ectomycorrhizae formed by these fungi survive and grow faster than seedlings having other ectomycorrhizae. In several pine nurseries in southern United States, nursery personnel collect basidiospores of *P. tinctorius*, mix them with mulch, and cover the seedbeds with this mulch after seeding. About 2 million seedlings are produced annually by using this method of inoculation.

A growing body of evidence shows the superiority of pine seedlings with *Pisolithus* ectomycorrhizae over seedlings with naturally occurring ectomycorrhizae on poor-quality reforestation, afforestation, and reclamation sites in the United States and elsewhere. This led to efforts to develop a commercial source of vegetative inoculum of *P. tinctorius*. After 5 years of intensive research by the Institute for Mycorrhizal Research and Development, USDA Forest Service in Athens, GA, and Abbott Laboratories, North Chicago, IL, a mycelial inoculum of *P. tinctorius* (MycoRhiz®) in a vermiculite-peat moss substrate was developed and is now commercially available in large quantities (24,25). A tractor-drawn machine has also been developed to inoculate fumigated nursery soils with MycoRhiz and seed the beds in one operation (8). For 1982, 1.5 million pine seedlings were produced with *Pisolithus* ectomycorrhizae from MycoRhiz in southern U.S. nurseries for operational outplantings.

During the research on MycoRhiz, which involved more than 100 experiments in 32 states on nearly 5 million seedlings of 18 tree species, numerous practical discoveries were made serendipitously. For example, in the past, one of the primary concerns of mycorrhizal researchers has been that the widespread use of fertilizers and pesticides would limit success of inoculation programs in tree nurseries. This concern was based on results from literally hundreds of publications showing how many of these chemicals inhibit ectomycorrhizal development on seedlings in either aseptic culture or greenhouse pot studies. In the MycoRhiz development program, however, use of these chemicals, as well as numerous other cultural practices, were found to have no effect on development of *Pisolithus* ectomycorrhizae or ectomycorrhizae formed by naturally occurring fungi. Neither total N ranging from 142 to 920 $\mu g/g$ and available P ranging from 13 to 54 $\mu g/g$ in soil nor 38 applications of various herbicides or 61 applications of various fungicides applied during one growing season significantly affected ectomycorrhizal development. Only two prerequisites appeared to be essential for successful development of *Pisolithus* ectomycorrhizae on bare-root pine seedlings—effective soil fumigation before inoculation and a highly viable inoculum.

FUTURE RESEARCH NEEDS

In all probability, more is known about the basic metabolic functions of mycorrhizae than is known about the metabolic functions of most medicines used to correct human maladies. Fortunately, these medicines are being used effectively on a broad practical scale even without knowledge of all the basic answers. We feel it is time to use the existing basic information on mycorrhizal associations and apply more research and development efforts to solving practical problems facing the world. We must concentrate more mycorrhizal research on "Does it work in the field?" rather than "How does it work in the laboratory?" The recently published text edited by Schenck (40) will contribute greatly to these expanded research activities.

Vesicular-Arbuscular Mycorrhizae

The potential uses of VAM in agriculture are now limited because an inexpensive, mass-produced inoculum of any VAM fungus is not available. Therefore, application of VAM to agriculture and forestry will have to be restricted to situations that require small quantities of inoculum. Certainly, the inoculation of plants in fumigated or treated seedbeds and nurseries and container-grown crops established in rooting media devoid of VAM fungi would fall into this category. The amounts of endomycorrhizal inoculum required in these situations could be supplied by the method devised by Menge, which is currently in use commercially.

The use of VAM fungi on plants intended for revegetation of waste lands and topsoil storage studies should be continued and expanded. Only minimum amounts of inoculum are needed if plants destined for use on disturbed sites are grown in container or bare-root nurseries.

Soilborne plant pathogens are extremely difficult and costly to control. VAM fungi that can reduce damage from a number of pathogens could be utilized especially on plants in nursery and container-grown crops. VAM fungal species active as biocontrol agents could be determined at the same time as evaluation of species for obtaining maximum crop response.

Use of VAM to increase salt tolerance of plants on high saline soils could aid in the reclamation of much of this land to crop production. In this same manner, VAM fungi capable of increasing drought resistance in plants could be utilized to increase plant productivity on arid soils. These possibilities exist if plants utilized in this process can be inoculated with VAM fungi in seedbeds or containers.

Obviously, priority should be given to using the above procedures on plants that are the most dependent on VAM for improved growth and yield and in soils containing low amounts of VAM fungi. The species of VAM fungi giving maximum plant response in a specific soil should be the ones used to maximize the response of crops on this soil. In addition, since differences in response can occur among cultivars of the same crop (5), a single species of VAM fungus should be compared on several plant cultivars. The variable reactions of cultivars suggest that geneticists, by selecting plant hybrids grown in soils of high fertility, may have inadvertently selected hybrids that have less dependence on VAM and more dependence on high soil fertility to satisfy their nutritional needs. Perhaps, plant-breeding programs are needed to reinstate genes to gain the benefit of mycorrhizae.

Although the addition of VAM fungi to extensive crop acreages is not currently possible, manipulation of cultural practices, cropping sequences, and fertilizer quantities should be tested to ascertain their impact on VAM fungi in the field. Crop sequences (32), pH (12), tillage practices (30), and fertility levels (17) may affect incidence of VAM fungi in field soil to some degree, but little is known about how to employ and manipulate these factors in an effective integrated program for VAM.

In addition to expanding the uses of VAM in commercial agriculture, there is a need to continue research to expand their potential uses. A search for alternative means of producing inoculum of VAM fungi should be increased so that quantities of inexpensive, viable inoculum can be used on large acreages. Fifty-one percent of the land in the tropics consists of soils with low amounts of available phosphorus, rendering this land unfit for agriculture without large applications of phosphate fertilizers (37). VAM as biotic fertilizers could be utilized to increase phosphorus absorption for plants grown in these areas of the world, allowing use of inexpensive rock phosphates rather than the concentrated phosphate fertilizers as sources of P.

Basic studies on VAM fungi should continue, especially on taxonomy and on factors affecting spore viability and germination, hyphal growth in soils, inoculum density, disease relationships, and the host-parasite relations. In all probability these basic studies would continue even without our encouragement since basic studies seem to attract the interest of more scientists, results are usually obtained faster, and the research is usually done in the comforts of the laboratory or greenhouse. However, there is a pressing need for more field studies where, unfortunately, the frustration threshold is high, current interest is low, and the work environment is frequently unpleasant and unpredictable.

Many crops in commercial agriculture can be grown more economically for the next few decades with their nutrients supplied almost exclusively by chemical fertilizers rather than by VAM (10). However, the incorporation of VAM into commercial agriculture could result in savings to the grower, acceptable yields with minimal amounts of fertilizers, and the establishment of a long-term sustainable agriculture that would reduce our dependence on dwindling reserves of fertilizer and other energy-expensive practices.

Ectomycorrhizae

Modern day operations of forest tree nurseries offer unique opportunities for basic and applied research on the ectomycorrhizal technology. In North America, nearly 2 billion tree seedlings are grown annually in nurseries for artificial regeneration programs. In most of these nurseries, fumigation of soil before sowing seed is a standard procedure. This creates a soil environment suitable for successful ectomycorrhizal fungus inoculations and the opportunity to regulate the physiological and ecological quality of seedlings with specific ectomycorrhizal fungi. The effects of other cultural practices, especially use of fertilizers and pesticides, root pruning, and lifting procedures, must be studied to determine their impact on specific ectomycorrhizae. It is wasted effort to produce ectomycorrhizae on seedlings in the nursery bed and then leave them in the soil because the lifting machine stripped the root system. The production of container-grown tree seedlings is rapidly becoming an important alternative to bare-root seedling production. The near-sterile rooting mixture used for container operations also offers a suitable environment for inoculation. Most container-grown seedlings are grown to the largest size possible in the shortest period of time with high rates of fertilizer, water, and supplemental light. These cultural variables must become part of the ectomycorrhizal research program.

In addition to the few fungal species being researched today, other species and strains of fungi must be identified and propagated for testing in various seedling production programs. With the exception of *Pisolithus tinctorius*, no other fungus has been intensively field tested for possible benefits to artificial regeneration programs. In the past, most researchers have been satisfied with studies on synthesis of ectomycorrhizae and have not extended their efforts to field studies. From a practical and economical point of view, results from studies on reclamation or reforestation sites are the only ones of importance. One major challenge facing the use of MycoRhiz is for it to gain acceptance by tree-regeneration specialists. This will come only after we make the transition from small research plots in the nursery and field involving only a few thousand seedlings to nursery production, lifting, and outplanting on an operational scale involving millions of seedlings. Unlike most research on annual agricultural crops, these field operational studies with forest trees must be monitored for 5–10 years to obtain useful results on survival and growth. These tests should also be replicated in time because environmental conditions encountered from year to year in the field are highly variable. In the field, various aspects of silviculture must be considered. Intensity of site preparation, site quality, tree genetics, fertilization, nutrient cycling, soils, disease and insect pests, competing vegetation and allelopathy, and tree spacing must be integral parts of the overall testing program. For instance, there would be little value in developing an ectomycorrhizal inoculation program that increased forest yields by 15% and then finding out later that improved growth on certain sites also increased incidence of fusiform rust disease that reduced yields by 20%. The program must be multidisciplinary in scope to have practical significance.

A research program dealing with the use of specific ectomycorrhizae in developing nations of the world deficient in these fungi must first train the local, qualified individuals who will eventually implement the programs. Such research programs should include basic studies on spread and persistence of specific ectomycorrhizal fungi as well as tree responses under the diverse conditions prevailing in these countries. Can a symbiotic fungus, such as *P. tinctorius*, persist and benefit tropical pines in a location experiencing five months of rain and seven months of drought? This must be studied because available isolates of this fungus have never been exposed to such moisture extremes.

Plant research has shown that pure culture inoculations are more successful if the soil is relatively free from other microorganisms. Soil fumigation can accomplish this objective. However, in developing countries fumigation of nursery soils is not widely practiced because of its expense and required technology. Other means of eradicating soil microorganisms, such as solarization, must be studied. Soil solarization, the process of heating soils via the "greenhouse effect" of transparent plastic tarping, has effectively eradicated many soilborne plant pathogens (34). If the technology can be developed, this procedure might furnish an economical means of eradicating these microorganisms before soil inoculation.

Nursery seedlings with a mixture of ectomycorrhizal fungus species should be tested in the field. Monocultures of introduced ectomycorrhizal fungi on seedlings is ecologically unsound especially in soils devoid of native symbionts. Field studies investigating the effects of several different fungi interacting in various combinations on seedling roots on diverse sites would furnish valuable information. Following the initial introduction of these fungi as pure cultures, they could be maintained in the various countries in "fungus gardens" on trees planted on the windward side of the operational nurseries. Not only would spores be blown onto the nurseries for natural inoculation of seedlings, but soil inoculum from under these trees could be used as an effective and economic means to introduce the selected fungi onto new seedling crops.

Inoculation of tree seed with spores of selected ectomycorrhizal fungi, in a manner similar to inoculation of legume seed with rhizobia, could be an effective method of introducing inoculum in nurseries and with seed sown directly on the site to be forested (direct seeding). Although methods of adding spores to naked seeds (41,42) and to the pelletizing matrix of encapsulated seeds (Marx, *unpublished*) have been developed, extensive field testing of these techniques is needed. Direct seeding, especially by aircraft, is unproved in the major deforested areas of the world (3). However, by adding spores to seed, then coordinating direct seeding with seasonal rains, this procedure may develop into an effective and economic way to rapidly reforest such areas.

CONCLUSIONS

Civilized peoples have been managing plants for thousands of years to satisfy their needs for food, fuel, and fiber. Agricultural and forestry research and development during modern times have shown that plants can be genetically manipulated, intensively cultivated, and protected from pests to increase yields dramatically. Without such progress many people in the world would have perished from exposure and hunger. However, with the world's population still increasing, with hunger persisting, and with a decrease in inventories of productive lands, we must seek, develop, and exploit additional biological weapons to combat these ever-increasing problems. Mycorrhizal technology is such a weapon, one that has great potential for combating the current and future worldwide problems of supplying enough food, fuel, and fiber for the expanding population.

LITERATURE CITED

1. Anonymous. 1975. Shifting agriculture in the world today. Editorial note. Unasylva 26:44.
2. Anonymous. 1980. Surface Mine Reclamation—A Position Statement. Soil Conserv. Soc. Am., Ankeny, IA. 6 pp.
3. Anonymous. 1981. Sowing Forests from the Air. Nat. Acad. Press, Washington, DC. 61 pp.
4. Anonymous. Wood Biomass for Energy—A Renewable and Expandable Resource. U.S. Dep. Agric. For. Serv., Washington, DC. 6 pp.
5. Azcon, R., and Ocampo, J. A. 1981. Factors affecting vesicular-arbuscular infection and mycorrhizal dependency of thirteen wheat cultivars. New Phytol. 87:677-685.
6. Baylis, G. T. S. 1970. Root hairs and phycomycetous mycorrhizas in phosphate-deficient soil. Plant Soil 33:713-716.
7. Baylis, G. T. S. 1972. Minimum levels of phosphorus for nonmycorrhizal plants. Plant Soil 36:233-234.
8. Cordell, C. E., Marx, D. H., Lott, J. R., and Kenney, D. S. 1981. The practical application of *Pisolithus tinctorius* ectomycorrhizal inoculum in forest tree nurseries. Pages 38–42 in: Forest Regeneration. Proc. Symp. Engineering Systems for Forest Regeneration, Raleigh, NC. Am. Soc. Agric. Eng., St. Joseph, MI. 376 pp.
9. Eckholm, E. 1979. Planting for the future: Forestry for human needs. Worldwatch Paper 26. Worldwatch Inst., Washington, DC. 64 pp.
10. Gerdemann, J. W. 1978. The practical application of research on arbuscular mycorrhiza. In: Proc. 2nd Woody Ornamental Disease Workshop. Missouri Acad. Sci. Occas. Pap. 5:1-3.
11. Gould, A. B., and Liberta, A. E. 1981. Effects of topsoil storage during surface mining on the viability of vesicular-arbuscular mycorrhiza. Mycologia 73:914-922.
12. Graw, D. 1979. The influence of soil pH on the efficiency of vesicular-arbuscular mycorrhiza. New Phytol. 82:687-695.
13. Greenland, D. J. 1981. Soil management and soil degradation. J. Soil. Sci. 32:301-322.
14. Hall, I. R. 1979. Soil pellets to introduce vesicular-arbuscular mycorrhizal fungi into soil. Soil Biol. Biochem. 11:85-86.
15. Hardy, R. W. F., Filner, P., and Hageman, R. H. 1975. Nitrogen input. Pages 43–61 in: Crop Productivity-Research Imperatives. Mich. Agric. Exp. Stn., East Lansing. 399 pp.
16. Hattingh, M. C., and Gerdemann, J. W. 1975. Inoculation of Brazilian sour orange seed with an endomycorrhizal fungus. Phytopathology 65:1013-1016.
17. Hayman, D. S. 1975. The occurrence of mycorrhiza in crops as affected by soil fertility. Pages 495–509 in: Endomycorrhizas. F. E. Sanders, B. Mosse, and P. B. Tinker, eds. Academic Press, London. 626 pp.
18. Howard, J. A., and Landry, J. P. 1975. Remote sensing for tropical forest surveys. Unasylva 27:32-37.
19. Lambert, D. H., and Cole, H., Jr. 1980. Effects of mycorrhizae on establishment and performance of forage species in mine spoil. Agron. J. 72:257-260.
20. Mahler, H. 1980. People. Sci. Am. 243:67.
21. Marx, D. H. 1977. The role of mycorrhizae in forest production. Pages 151–161 in: TAPPI Conf. Papers, Annu. Mtg., Atlanta. Tech. Assoc. Pulp Paper Ind., New York. 240 pp.
22. Marx, D. H. 1980. Ectomycorrhizal fungus inoculation: A tool for improving forestation practices. Pages 13–71 in: Tropical Mycorrhiza Research. P. Mikola, ed. Oxford Univ. Press, London. 270 pp.
23. Marx, D. H., and Krupa, S. V. 1978. Ectomycorrhizae. Pages 373–400 in: Interactions Between Non-Pathogenic Soil Microorganisms and Plants. Y. R. Dommergues and S. V. Krupa, eds. Elsevier Publishing Co., Amsterdam. 475 pp.
24. Marx, D. H., Cordell, C. E., Kenney, D. S., Mexal, J. G., Artman, J. D., Riffle, J. W., and Molina, R. J. 1983. Commercial vegetative inoculum of *Pisolithus tinctorius* and inoculation techniques for develoment of ectomycorrhizae on bare-root tree

seedlings. For. Sci. In press.

25. Marx, D. H., Ruehle, J. L., Kenney, D. S., Cordell, C. E., Riffle, J. W., Molina, R. J., Pawuk, W. H., Navratil, S., Tinus, R. W., Goodman, O. C. 1982. Commercial vegetative inoculum of *Pisolithus tinctorius* and inoculation techniques for development of ectomycorrhizae on container-grown tree seedlings. For. Sci. 28:339-366.
26. Mayer, J. 1975. Agricultural productivity and world nutrition. Pages 97–108 in: Crop Productivity-Research Imperatives. Mich. Agric. Exp. Stn., East Lansing. 399 pp.
27. Menge, J. A. 1983. Utilization of vesicular-arbuscular mycorrhizal fungi in agriculture. Can. J. Bot. In press.
28. Menge, J. A., Lembright, H., and Johnson, E. L. V. 1977. Utilization of mycorrhizal fungi in citrus nurseries. Proc. Int. Soc. Citricul. 1:129-132.
29. Mikola, P. 1973. Application of mycorrhizal symbiosis in forestry practice. Pages 383–411 in: Ectomycorrhizae: Their Ecology and Physiology. G. C. Marks and T. T. Kozlowski, eds. Academic Press, New York. 444 pp.
30. Mitchel, D. H., Schenck, N. C., Dickson, D. W., and Gallaher, R. N. 1980. The influence of minimum tillage on soil-borne fungi, endomycorrhizal fungi, and nematodes on oats and vetch. Pages 115–123 in: Proc. 3rd Annu. Southeast No-Tillage Systems Conf. Univ. Fla., Gainesville. 202 pp.
31. Mosse, B., and Hayman, D. S. 1980. Mycorrhiza in agricultural plants. Pages 213–230 in: Tropical Mycorrhiza Research. P. Mikola, ed. Elsevier Publishing Co., Amsterdam. 270 pp.
32. Ocampo, J. A., and Hayman, D. S. 1981. Influence of plant interactions on vesicular-arbuscular mycorrhizal infections. II. Crop rotations and residual effects on non-host plants. New Phytol. 87:333-394.
33. Oertli, J. J. 1980. Editorial. Reclamation Rev. 3:1.
34. Pullman, G. S., DeVay, J. E., and Garber, R. H. 1981. Soil solarization and thermal death: A logarithmic relationship between time and temperature for four soilborne plant pathogens. Phytopathology 71:959-964.
35. Reeves, F. B., Wagner, D., Moorman, T., and Kiel, J. 1979. The role of endomycorrhizae in revegetation practices in the semi-arid west: I. A comparison of incidence of mycorrhizae in severely disturbed vs. natural environment. Am. J. Bot. 66:6-13.
36. Ruehle, J. L., and Marx, D. H. 1979. Fiber, food, fuel, and fungal symbionts. Science 206:419-422.
37. Sanchez, P. A., and Buol, S. W. 1975. Soils of the tropics and the world crisis. Science 188:598-603.
38. Sassin, W. 1980. Energy. Sci. Am. 243:119.
39. Schenck, N. C. 1981. Can mycorrhizae control root disease? Plant Dis. 65:230-234.
40. Schenck, N. C., ed. 1982. Methods and Principles of Mycorrhizal Research. Am. Pathol. Soc., St. Paul, MN. 244 pp.
41. Theodorou, C. 1971. Introduction of mycorrhizal fungi into soil by spore inoculation of seed. Aust. For. 35:23-26.
42. Theodorou, C., and Bowen, G. D. 1973. Inoculation of seeds and soil with basidiospores of mycorrhizal fungi. Soil Biol. Biochem. 5:765-771.
43. Williston, H. L. 1980. A statistical history of tree planting in the South: 1925–1979. U.S. Dep. Agric. For. Serv. Misc. Publ. SA-MR-8. 37 pp.
44. Wortman, S. 1975. World productivity: Challenges to science. Pages 43–61 in: Crop Productivity-Research Imperatives. Mich. Agric. Exp. Stn., East Lansing. 399 pp.
45. Zobel, B. 1979. Growing more timber on less land. For. Farmer 38:15.

CHAPTER 31

USE OF MUTANTS TO STUDY SYMBIOTIC NITROGEN FIXATION

SHARON R. LONG

The generation and analysis of mutants has been a cornerstone of modern genetics. Availability of mutants has been of fundamental importance in functional studies of metabolic pathways and development. In turn, mutants with easily recognizable and definable phenotypes have provided systems where the nature of heritable genetic units—linkage, location, allelism, and regulatory relationships—could be studied. In examining the use of mutants to study the symbiosis of *Rhizobium* and plant hosts, we can keep these two approaches in mind: 1) the use of mutants to reveal, through phenotype, the inner workings of a bacterium-plant interaction and 2) the examination, via mutants, of the location, number, and relationship of the loci that carry symbiotic information.

It is clear from many lines of evidence that both bacterial and plant genes and gene products are required for symbiosis to occur. A number of plant genes have been described that affect nodule formation and function, and these are treated in several recent reviews (12,24). For the purposes of this chapter, I will discuss only bacterial mutants. My aim here is to present a particular view of studying symbiotic nitrogen fixation through genetics. The examples presented are few and only scratch the surface of the literature on symbiotic mutants of *Rhizobium*. A more complete treatment can be found in recent papers by Beringer et al (8), Schwinghamer (50), and Vincent (57).

NODULE DEVELOPMENT AS A PATHWAY

Rhizobium cells and their plant hosts are able to grow and reproduce independently of each other. When they come in contact, they establish a symbiosis through a series of interactions. Each partner exhibits changed morphology and biochemistry, and appears to induce such changes in the other partner. The properties of *Rhizobium* that we want to study genetically are part of this complex interaction. Therefore, mutant phenotypes must be described in the context of nodule development.

The way in which nodule development has been most thoroughly studied is with light- and electron-microscopic observation. The extensive background literature on nodule development has been reviewed by Libbenga and Bogers (31) and Dart (15), among others. The bacteria and plant recognize each other and form an attachment. A characteristic root hair curling is often seen in early stages of infection. Bacteria trapped in the crook of the hair proliferate and invade the root hair (13), and the plant deposits along the invasion route a layer of cell wall-like material that appears in the microscope as an "infection thread" (for recent reviews of these early steps, see references 2 and 17). The plant responds to this invasion by mitosis of cells in the root cortex; bacteria in the infection thread proliferate as the infection thread branches and penetrates through successive cells in the root. In certain cells, the infection threads release bacteria into the cytoplasm of the plant cell. Both bacteria and host cells undergo ultrastructural changes as they differentiate to produce an effective, nitrogen-fixing cell. The differentiated bacteria—enlarged, elongated, and/or branched in different symbioses—are referred to as "bacteroids." Plant cell nuclei in the nodule often become polyploid (31).

These structural studies form the basis for mutant analysis. Mutants that are originally classified as "Nod$^-$" (forming no nodules) or "Fix$^-$" (forming nodules that cannot fix nitrogen) can be further characterized according to the stage of development that is the last to be completed. This approach is analogous to classifying auxotrophic mutants with respect to the points in a metabolic pathway where a block occurs, by finding where in the pathway intermediate compound accumulate.

Vincent has presented an overview of these studies (57,58) and has suggested a phenotypic code for mutants, based on structural determination of stage of arrest (58, Table 1). This classification and code is a useful framework for comparing and describing *Rhizobium* mutants. It is probably an incomplete description at present; as more mutants are described, more steps in the pathway will be filled in (43).

The biochemistry of nodule development has also been studied by many groups. (These are reviewed in 4; for samples of recent approaches, see 16,20,29,38.) The sequence of biochemical events is much less complete than the picture provided by structural studies. Microscopic observations have described each step of the nodulation process; however, while there is some biochemical understanding of initial binding and of the final differentiation steps, the

TABLE I. Sample stages of symbiosis and phenotypic codes[a]

Stage	Code
Preinfection	
Root colonization	Roc
Attachment of root and bacteria	Roa
Root hair branching	Hab
Root hair curling	Hac
Infection and nodule formation	
Formation of infection thread	Inf
Nodule initiation (meristem formation, organogenesis)	Noi
Release of bacteria into target cell	Bar
Bacteroid development	Bad
Nodule function	
Nitrogen fixation	Nif
Complementary biochemical functions	Cof
Nodule persistence	Nop

[a]Adapted from Vincent (58).

intermediate steps of host cell invasion and stimulation of meristem activity are not understood at the molecular level. One difficulty is that nodule development does not occur synchronously. Also, each nodule is a mixture of cells at all stages of development. Therefore, biochemical studies on nodules can only correlate events with age; they cannot easily distinguish cause and effect.

For example, it is not known whether the high level of cytokinin in nodules is due to a stimulus from the bacterium to induce cell division in the host or whether it is a consequence of cell divisions (36). Biochemistry alone cannot unambiguously answer such questions. By using mutants that are known from structural studies to stop nodule development at particular points and by comparing the nodules biochemically, it will be possible to establish a better sequence of events and to build correlations between morphological and biochemical differentiation (37,40,56). As more is learned about the biochemistry of nodule development, it will be possible to use stage-specific biochemical events to score mutant phenotypes without always performing structural analysis on them. Recognition phenomena are discussed by Sequeira Chapter 27).

MUTANTS VERSUS VARIANTS

A number of naturally occurring variants have provided material for genetic analysis and have been exploited in research on symbiosis (3,8,50). Vance et al (55) and Vance and Johnson (54) have described and contrasted the Fix$^-$ nodules produced by ineffective bacteria on the one hand and by ineffective plants on the other. They pointed out the similarity between plant-determined ineffectiveness and some pathological reactions and suggested a role for plant defense compounds in the development of nodules (54). Studies such as these, comparing plant- and bacterium-determined ineffectiveness in the same system, should be useful in understanding the relative contributions of host and microbe at different steps of the symbiosis.

Naturally occurring variants are not necessarily comparable to mutants in the way they can be studied genetically. In natural populations, differences among individual genotypes are likely to occur at many loci. In addition, there are often many allelic forms of each gene available for reassortment in the population. Therefore, differences in phenotype may be due to the accumulation of changes at many loci, each of which may differ only slightly from other alleles at that locus.

Mutants, on the other hand, are new types that arise by a single "event" (which may be a complex rearrangement or a single base pair change). Abruptly different phenotypes that arise through mutation enable study of the one or few genes that have been changed. By generating and studying many mutants that affect related phenotypes, it is possible to delineate the role of each gene. This approach is especially helpful as we try to extend knowledge to the molecular level.

Variants can provide material for study of single genes as well, if genetic manipulation can be used to transfer genes from a variant, one or a few at a time, into a known background. This has not been exploited a great deal in *Rhizobium* due to the difficulty in genetics of this system. Instead, most genetic studies have focused on creating and analyzing mutants. However, a recent report by Brewin and colleagues (10) at John Innes Institute demonstrates the benefits of using genetics to study naturally existing strains. A particular cultivar of garden pea, *Pisum sativum* 'Afghanistan,' is not nodulated by most *Rhizobium leguminosarum* strains, although these same strains are capable of forming normal, effective nodules on other pea cultivars. However, there is one naturally occurring *R. leguminosarum* strain, TOM, that does nodulate Afghanistan pea. Brewin et al showed that transfer of a plasmid from TOM to a Nod$^-$ derivative of another *R. leguminosarum* strain, R1 300, resulted in R1 300 becoming able to nodulate Afghanistan peas. By this means, it was possible to narrow down the cause of changed host range from the whole genome of *R. leguminosarum* TOM to a plasmid. Additional studies may further delimit the genes responsible for the host range. It is not known whether one gene or a cluster is responsible for the altered host range. Further genetic and recombinant DNA studies will allow us to describe the genes for host range more exactly.

Detailed genetic study of natural variants may be especially valuable in investigating the combined genetics of host and bacterial partners. The type of allele that alters host range may not necessarily be generated in mutagenesis treatments of a given wild type. To make molecular studies on host range, we will need to take advantage of existing variants, identify the genes involved through crosses, and obtain the genes as recombinant clones for use in biochemical studies.

STUDIES WITH MUTANTS

There are two general ways to use mutants in the study of the symbiosis. One is to examine the symbiotic properties of strains with an altered free-living phenotype, e.g., auxotrophy, drug-resistance, phage-resistance, or colony morphology. The second is to generate and study mutants that have no apparent defect in growth outside of the plant but that fail to carry out a normal symbiotic interaction. I will give one or two examples of each approach, and discuss the advantages and disadvantages of some of the techniques used.

Free-Living Phenotypes and Symbiotic Properties

There have been many studies of the symbiotic effects of mutations that affect the phenotype of free-living *Rhizobium*. One widely obtained result is that within a class of mutants—for instance, those showing resistance to a particular antibiotic (1,39) or auxotrophy (18,39,41,47–49) or a change in cell surface properties (17,45)—some strains are symbiotically defective whereas others are normal. For example, a mutant of *R. leguminosarum* selected for its lack of exopolysaccharide (45) was found to be Nod$^-$ on its host plant. This pointed to a correlation between exopolysaccharide and host invasion and possibly specificity. However, when many different Exo$^-$ strains of several *Rhizobium* species were examined, it turned out that most of them were symbiotically normal (46). This complicates interpretation of the results, and in each case it is doubly important to confirm, through linkage and/or reversion studies, that the phenotype being scored (resistance, for example) is caused by the same mutation that causes the symbiotic defect. One possibility raised by these studies is that the various mutations or alterations represent lesions in different steps of a pathway, which occur either before or after a branch point that leads off to a specific symbiotic function. Figure 1 shows two sample pathways, one a metabolic pathway and the other a biosynthetic pathway for exopolysaccharide production, demonstrating this type of branch point.

Leu⁻ Auxotrophy in *Rhizobium meliloti*

As an example of combining observation with experimental studies on *Rhizobium* mutants, I will describe a series of studies by Jean Dénarié and his colleagues on symbiotically defective auxotrophic mutants of *R. meliloti*. In a discussion of the effect of mutations on the symbiosis (18), they pointed out that in vitro reversion could be studied more easily with auxotrophs than with antibiotic-resistant and phage-resistant mutants, thus establishing linkage of defect and mutation. Furthermore, they suggested that auxotrophic mutants with effects on particular stages of symbiotic nodule development may provide a useful tool for study of symbiotic development, since the metabolic defect may be repairable through in vivo experimental manipulation. Several studies of this kind have been reported in other systems (40,50).

Scherrer and Dénarié (47) found in a survey of various auxotrophs that among 11 Leu⁻ auxotrophs, all were symbiotically ineffective. Prototrophic revertants

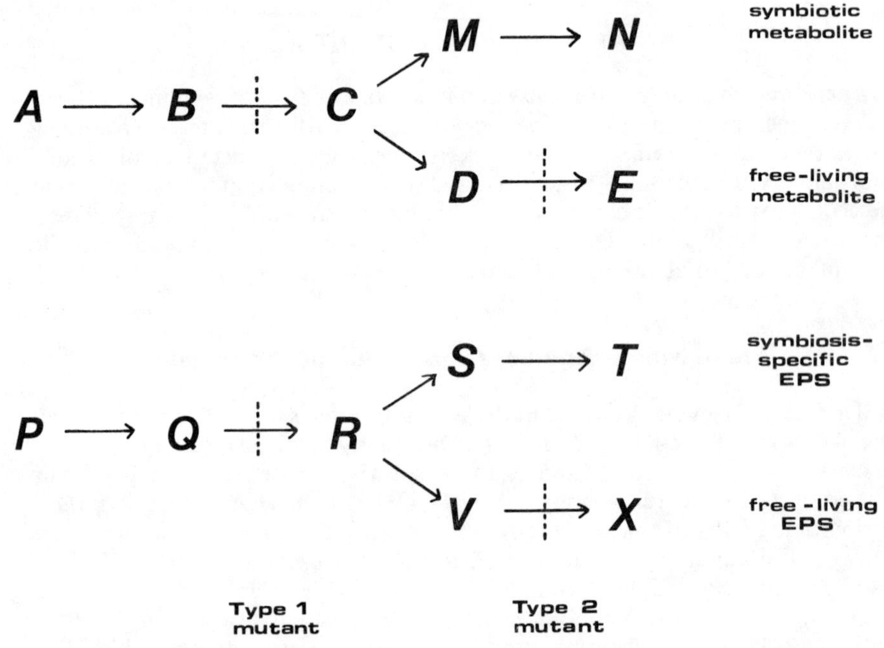

Fig. 1. Sample biosynthetic pathways with branch points leading to both a product found in free-living *Rhizobium* and a product unique to the symbiotic state. Mutants detected because they lack the "free-living" product (E^- or X^-) will fall into two classes. Type 1 mutants affect both the symbiosis and the free-living state because they control a step epigenetic to the branch point. Type 2 mutants affect the free-living phenotype, but these strains are symbiotically normal. The upper scheme would be exemplified by certain types of auxotrophy (E^-) that are sometimes Sym^+ and in other cases Sym^-. The lower pathway illustrates a possible mechanism for results such as those reported by Sanders et al (46) or Rolfe et al (42), who have found that some, but not all, extracellular polysaccharide (EPS) mutants (X^-) have a symbiotic defect.

of the Leu$^-$ mutants were completely normal symbiotically, as were revertants of other ineffective mutants characterized as Ade$^-$, Ura$^-$, and Ilv$^-$. Furthermore, when ineffective nodules formed by Leu$^-$ *R. meliloti* and alfalfa were supplied with leucine, they became symbiotically normal (51). These two lines of evidence show that the leucine auxotrophy, and not a cryptic second lesion in the various strains, was the cause of the symbiotic defect.

Truchet and Dénarié (51) used electron microscopy to determine the step at which nodule development had stopped, in order to more precisely define the phenotype of the mutants. Three different Leu$^-$ mutants had a similar phenotype: they induced the formation of small abnormal nodules ("nodosités") that were ineffective (no nitrogen fixation as measured by the acetylene reduction assay) and that showed little meristematic activity in the root cortex compared with nodules induced by wild-type *R. meliloti.* Infection threads had normal morphology at first and penetrated the root and cells of the nodules, but no bacteria were released from these infection threads into the cytoplasm of target cells. Bacteria within the threads showed precocious senescence.

Truchet et al (52) exploited the fact that leucine could restore abnormal nodules to the wild-type phenotype. Leucine-fed plants produce larger, normal nodules with a more extensive network of infection threads, and bacteria were released into target cell cytoplasm, where they fixed nitrogen. To determine whether part or all of this effect was due not to leucine itself but to improved nutrition from nitrogen fixation, they added low amounts of urea as a nitrogen source to Leu$^-$-induced nodules. Urea-fed plants produced nodules that were normal in overall size and morphology and had an actively growing meristem. However, there was still no release of bacteria from infection threads.

Since the abnormal nodule morphology appears to be a nonspecific consequence of starvation, it was possible to narrow the effect of the Leu$^-$ mutant to the release of bacteria from the infection thread. The fact that organogenesis continues normally, even without release of bacteria from infection threads, is strong evidence that organogenesis is triggered at a distance, presumably by a diffusible factor. Dénarié and colleagues hypothesize that such a factor exists and suggest it be called a nodule organogenesis-inducing principle (NOIP).

In a second set of observations in the same paper, Truchet et al found that in urea-fed Leu$^-$ nodules, where organogenesis is restored, the polyploidy found in normal mature nodules is absent. When leucine is provided, however, appearance of polyploidy accompanies the release of bacteria from infection threads. From these results, they conclude that organogenesis is not dependent upon preexisting polyploidization. The induction of polyploidy, rather, seems to depend upon the presence of bacteria in the target cell cytoplasm. They hypothesize action by a central tissue differentiation inducing principle (CTDIP) that does not act at a distance across cell boundaries.

NEED FOR CONDITIONAL MUTANTS

In this series of experiments, Dénarié and his group used the conditionality of an auxotrophic mutant to study control of nodule development. This approach partially overcomes one of the major drawbacks of nodulation as a developmental system, that is, the lack of synchronization. These experiments permitted them to rank in a hierarchy, events that had previously been confounded with each other.

Other types of conditional mutants are needed for experiments on nodule development. Temperature-sensitive mutants, for example, could be used to control activity of genes required at specific stages of development. These could be used to study genes with no apparent phenotype in free-living cells and would avoid the possible problem that auxotrophic mutants may affect the symbiosis through pleiotropic effects on bacterial or host cells.

Beringer et al (5) described the isolation of two temperature-sensitive symbiotic mutants of *R. leguminosarum* by screening nitrosoguanidine (NTG)-mutagenized cells on plants. At the permissive temperature, 13 C, both mutant strains induced the formation of nitrogen-fixing nodules on host plants of *Pisum sativum*. At 26 C, they were symbiotically defective: in one strain, there was no release of bacteria from infection threads, whereas in the other the bacteroids were often lysed. These two mutants appear to act at distinct stages of nodule formation.

Few groups of workers have tried to isolate such mutants; the large-scale screening necessary to find them is daunting. There may be ways of using technology to create these mutants in such a way that fewer screening trials need to be done. If a gene can be identified and cloned through other means, such as transposon mutagenesis (see below), then that gene can be artificially mutagenized in vitro. This could be accomplished either with chemical modifications or with enzymatic treatments that insert or misrepair nucleotides in the sequence. These altered, cloned genes could be tested for their ability to complement a strain defective for that function. This would provide a way of efficiently creating many types of mutations, since the target for mutagenesis is not the whole genome but a limited sequence known to contain genes of interest. This approach may be useful for creating functionally improved alleles as well as mutants.

Mutations Specific to the Symbiosis

To find symbiotic mutants, a population of wild-type cells is exposed to mutagens and individual colonies are screened on aseptically grown plants. A small proportion of mutants are likely to be defective in a gene required for the symbiosis. These mutants will either form no nodules (Nod$^-$) or will form ineffective nodules (Fix$^-$). Several techniques have been used to generate mutants for screening on plants. Beringer et al (5) screened survivors of NTG mutagenesis on pea plants and isolated a group of symbiotic mutants including the two temperature-sensitive mutants described above. Brill and co-workers have used a parallel approach to isolate mutants in *R. japonicum* and *R. meliloti* (30,33,38).

While most of the following discussion is aimed at the uses of transposon mutagenesis, rather than NTG or other point mutagens, it should be kept in mind that transposons have some disadvantages (see below) and that strains obtained by point mutagenesis are likely to provide us with valuable material for future experiments.

Transposon Mutagenesis

Principles

While transposons can be used to generate mutants with defects in various free living phenotypes, their usefulness is most evident when they are used to mark genes and/or effect mutation in genes uniquely expressed in symbiosis

Transposons, or transposable resistance elements, are discrete DNA sequences capable of transposing sequences, or relocating them by insertion, into other DNA sequences with which they have no homology. They are a subset of the more general group, insertion sequences. Transposable resistance elements are defined by their particular structure, which includes repeated sequences at the ends, and by the fact that they encode a drug resistance that marks their presence and position in the genome. These two properties—ability to transpose and a resistance marker gene—combine to make transposons a powerful tool for studying cryptic genes such as *Rhizobium* symbiotic genes.

When a transposon inserts itself into the coding sequence for a gene, it prevents expression and thereby creates mutation in that gene. At the same time, it marks its presence by its resistance function. Once a transposon-generated mutation is identified through screening as being in a function of interest, the inactivated gene can be located and cloned by using the drug resistance gene as a marker. In the case of symbiotic mutants, this means that potential mutants need be screened only once in association with plants; genetic studies can be done subsequently in the free-living state on plates.

This principle was first used in *Rhizobium* by Beringer and colleagues (6) at the John Innes Institute, who described in 1978 a means to obtain *R. leguminosarum* strains carrying transposon insertions. To introduce the transposon into the cells, they used a "suicide plasmid" (53), which is a plasmid able to conjugate into a recipient but unable to stably replicate. Descendants of *Rhizobium* cells that have received a transposon-bearing suicide plasmid will express the transposon's drug resistance only if the transposon has "hopped" into a *Rhizobium* DNA sequence. Such drug-resistant *Rhizobium* strains are therefore potential mutants, according to where the insertions have occurred.

Use in Marking Plasmids

The first, and still an important, use of transposons in *Rhizobium* symbiotic genetics was to mark a plasmid in *R. leguminosarum* and follow its transfer to new cells. Plasmids are defined both genetically, as a linkage unit independent of the chromosome, and physically by their differing properties of size, etc., that make them distinguishable from the chromosome. The correlation of the genetically and physically defined entities is not straightforward if no genetic markers are known for a particular plasmid. This is a problem with *Rhizobium* since many of the plasmids found in various strains at first had no known markers. A transposon insertion into a plasmid can be physically tracked because its DNA sequence is detectable by hybridization, and it provides a genetic marker for transfer or loss (curing) of that plasmid. This permits the correlation of a band seen in a gel or a gradient with other markers that are found to be transferred in linkage with the transposon.

By this means, Johnston et al (26) marked a plasmid in *R. leguminosarum* with transposon Tn5 and followed its transfer in conjugation with another *R. leguminosarum* strain. They showed that the transfer of the plasmid was accompanied by a change in nodulation properties in the recipient cell.

This experiment, and others that followed it, have made it possible to examine critically the behavior and role of intact plasmids in *Rhizobium* cells, using the transposons to track the plasmid as a unit as it is moved from cell to cell. Other markers are now known for various native *Rhizobium* plasmids; one is bacteriocin production (9,22,23), which can also be used to follow plasmids during genetic manipulation. A demonstration of the important role of plasmids in symbioses was recently published by Hooykaas et al (25), who moved a transposon-marked plasmid from *R. trifolii* into *Agrobacterium tumefaciens* and found that the *Agrobacterium* cells were able to form nodules on clover.

Use to Create Mutations

Moving beyond the function of entire plasmids, we come to the question of the nature of the individual genes that govern the symbiosis. Transposon mutagenesis can be used in two general ways in the study of individual gene location and action. First, as part of a random mutagenesis and screening experiment, it can be used to find the genes at large in the bacterial genome, in such a way that they can be mapped and cloned. Second, transposon mutagenesis can be used to probe the function of DNA sequences obtained as clones, by exposing one part of the sequence at a time to mutagens and then testing the mutated sequences in a known background.

Random transposon mutagenesis has been used by several research groups to generate *Rhizobium* symbiotic mutants (11,32,35,42,43). Rolfe et al (42) reported isolating mutants of *R. trifolii*, and Ausubel and co-workers (32,35) obtained symbiotic mutants of *R. meliloti*, using the "suicide plasmid" pJB4JI. The John Innes Institute researchers, in addition to using the original suicide plasmid method, found that Tn5 insertions already in a cell could be used as a source of new Tn5 insertions—"secondary transpositions." Starting with a *R. leguminosarum* strain that had a Tn5 insertion in the chromosome, they selected descendents that could transfer Tn5 at high frequency into a recipient cell. In these strains, Tn5 had transposed into a self-transmissible plasmid (11). A high proportion of these mutants were symbiotically defective.

So far, the use of transposon mutagenesis has been aimed at creating mutations and in mapping. Relatively few transposon-induced mutants have been studied at the structural level to determine their characteristic stage of arrest. Rolfe et al (42,43) reported that among their Tn5-mutated *R. trifolii* strains, there were several different stages that were blocked, including infection thread invasion. Hirsch et al (21) described the phenotypes of several transposon-induced mutants of *R. meliloti*, based on light- and electron-microscope observations. Two mutants are apparently defective in a gene that is required very early, because these mutants do not curl root hairs or form infection threads. Such examples show that transposon-induced bacterial mutants will probably be useful in the study of nodule development as well as in the task of locating and mapping the symbiotic loci.

Site-Specific Transposon Mutagenesis

A blend of recombinant DNA techniques and transposon manipulation have made it possible to do fine structure mapping of symbiotic genes in *Rhizobium* and should be applicable to studies of other genes such as those for virulence in plant pathogenic bacteria.

Recombinant DNA technology has greatly extended our ability to manipulate and study genes. At the heart of this field is the need for a *cloning vector*—a vehicle for carrying inserted genes or fragments and the limited availability of vectors for *Rhizobium* has slowed down the application of recombinant genetic techniques. Not many of the small, well-defined plasmids indigenous to *Escherichia coli* are able to replicate themselves in *Rhizobium* cells, so they cannot be used to stably maintain cloned genes in *Rhizobium*. There are several plasmids, some of them native to *Pseudomonas* species, that will transfer into and replicate in *Rhizobium*, along with many other Gram-negative bacteria. One of these, RP4 (very closely related to other plasmids, RK2 and R68.45), has been useful in *Rhizobium* genetics for its ability to mobilize the chromosome during conjugation. This has made it possible to construct chromosomal maps of several *Rhizobium* species (7,27,28,34). RP4 can be used as a cloning vector as well, because its ability to replicate allows it to carry insert DNA along with it and because it has unique restriction sites where DNA sequences to be cloned can be inserted enzymatically.

A disadvantage of using plasmid RP4 as a vector is its large size, almost 60 kilobase (kb) pairs. Such large plasmids break easily during preparation, and plasmids can be purified physically only if they remain intact; therefore it is difficult to prepare vector DNA quantitatively. Another drawback is that DNA molecules that are large are less efficiently transformed into recipient bacterial cells than are molecules that are small. An improvement in cloning technology for *Rhizobium* was made by Ditta et al (19), who developed a cut-down version of RK2 which they named RK290. This vector is 20 kb in size, one-third of the original plasmid size, and is therefore much easier to handle biochemically and physically. Although not self-transmissible, it is mobilized in the presence of a helper plasmid, and can thereby be moved into recipient cells by conjugation as well as by transformation. Because pRK290 retains the wide host-range of the parent plasmid, it can be used potentially in all Gram-negative bacteria.

By adding transposon mutagenesis to the versatile cloning system develped by Ditta and colleagues, Ruvkun and Ausubel (44) demonstrated a method for creating site-specific mutations in cloned DNA sequences and then "engineering" that mutant region back into a bacterium. This method can be used to demonstrate the in vivo function of a sequence by observing the cell phenotype when the sequence is inactivated.

The first step in the procedure is to expose an *E. coli* cell that harbors the cloned sequence to a transposon borne on an unstable carrier. The transposon will occasionally insert itself into the cloned sequence; these events can be detected and characterized by plasmid isolation and restriction mapping. Where the cloned region has a transposon inserted at a position of interest, clone plus insertion are placed into the vector pRK290 by means of restriction cutting and re-ligation.

At this point, the gene plus its engineered, mapped mutation is transferred into a *Rhizobium* cell. In some of those cells, a double recombination event places the "mutated" gene—transposon insert and all—into the *Rhizobium* genome, whereas the normal copy becomes combined into plasmid pRK290. The normal copy can then be eliminated from the cell by getting rid of the plasmid; this can be accomplished by moving in another plasmid that is incompatible with pRK290.

By this means, a series of mutations can be produced and tested along a defined DNA sequence, and the effect of each can be determined. Ruvkun and Ausubel (44) used this method to show required regions for symbiotic function in the vicinity of the *nif* genes of *R. meliloti*.

The significance of this technique is, first, that it allows a fine structure map to be made within a cloned gene because the position of transposon insertions can be determined so precisely. Second, it allows efficient, "shotgun" mutagenesis of defined regions that are thought to be of interest, such as DNA sequences in the vicinity of previously mapped genes. This may be useful for genetic study of the large plasmids that are indigenous to many *Rhizobium* strains (14,23).

Finding genes by random mutagenesis and screening is tedious, even with transposons; site-specific transposon mutagenesis amplifies the value of finding each symbiotic gene. From defined genes, it should be possible to move from each locus into adjacent regions, "walking" down the genome by cloning adjacent sequences, experimentally introducing mutations in each as it is cloned, and testing its phenotype. This will be an extremely useful approach to the genetics of the symbiosis.

NEED FOR NEW MUTANTS

Transposons, while handy in many respects, have several drawbacks. First, the addition of a drug resistance function into a cell may possibly have pleiotropic effects. Second, the insertion of a transposon has a polar effect on the adjacent region; that is, it affects the expression of genes downstream in the same transcription unit. Mutant phenotypes of transposon-containing cells may therefore be due to the inactivation of several genes at a time.

Finally, transposons can basically be used just to make "null" mutations of genes: lack of gene product or production of an incomplete product. Such mutants are useful in defining a gene's role. But for some experimental purposes, it is important to have mutants that have *altered* gene products. One example is in the use of in vitro reactions to establish gene function. To demonstrate that a gene produces a protein able to catalyze a particular step in a metabolic pathway, one type of proof is to show that a deletion of the gene—or transposon insertion into it—results in the lack of a particular protein and the lack of that activity. Another line of proof is to show that in a mutant with altered in vivo activity, such as temperature sensitivity, an altered product can be isolated whose ability to catalyze the reaction in vitro shows parallel properties. Such studies may be more useful for the functional dissection of metabolic pathways: the determination of rate-limiting steps or regulation by cellular environment.

Transposons and deletions can show where genes are and indicate what they do. But the use of mutants with altered gene products—found through genetic studies of natural variants, induced through chemical mutagenesis, or created through recombinant DNA manipulations—can help us to determine how the gene products work. The demonstration of function, always an elusive goal, is one of the great challenges ahead in the study of the symbiosis.

ACKNOWLEDGMENTS

I wish to thank Frederick Ausubel and Ann Hirsch for many helpful discussions of genetics, development, and possible mutants, and Gary Ruvkun and J.D.G. Jones for their suggestions on ways to make mutations in cloned genes. I am grateful to Thor Kommedahl for giving me the chance to present these ideas and for his encouragement and patience as I prepared this manuscript.

LITERATURE CITED

1. Amarger, N. 1975. Efficience symbiotique de mutants spontanes de *Rhizobium leguminosarum* resistants a la streptomycine, spectinomycine ou kanamycine. C. R. Hebd. Seances Acad. Sci. Ser. D 280:1911-1914.
2. Bauer, W. D. 1981. Infection of legumes by Rhizobia. Annu. Rev. Plant Physiol. 32:407-449.
3. Bergerson, F. J. 1957. The structure of

ineffective root nodules of legumes: An unusual new type of ineffectiveness, and an appraisal of present knowledge. Aust. J. Biol. Sci. 10:233-242.

4. Bergerson, F. J. 1977. Physiological chemistry of dinitrogen fixation by legumes. Pages 519–555 in: A Treatise on Dinitrogen Fixation. Vol. 3. Biology. R. W. F. Hardy and W. S. Silver, eds. John Wiley & Sons, New York. 675 pp.
5. Beringer, J. E., Johnston, A. W. B., and Wells, B. 1977. The isolation of conditional ineffective mutants of *Rhizobium leguminosarum*. J. Gen. Microbiol. 98:339-343.
6. Beringer, J. E., Beynon, J. L., Buchanan-Wollaston, A. V., and Johnston, A. W. B. 1978. Transfer of the drug-resistance transposon Tn5 to *Rhizobium*. Nature 276:633-634.
7. Beringer, J. E., Hoggan, S. A., and Johnston, A. W. B. 1978. Linkage mapping in *Rhizobium leguminosarum* by means of R plasmid-mediated recombination. J. Gen. Microbiol. 104:201-207.
8. Beringer, J. E., Brewin, N. J., and Johnston, A. W. B. 1980. The genetic analysis of *Rhizobium* in relation to symbiotic nitrogen fixation. Heredity 45:161-186.
9. Brewin, N. J., Beringer, J. E., Buchanan-Wollaston, A. V., and Hirsch, P. R. 1980. Transfer of symbiotic genes with bacteriocinogenic plasmids in *Rhizobium leguminosarum*. J. Gen. Microbiol. 116:261-270.
10. Brewin, N. J., Beringer, J. E., and Johnston, A. W. B. 1980. Plasmid-mediated transfer of host-range specificity between two strains of *Rhizobium leguminosarum*. J. Gen. Microbiol. 120:413-420.
11. Buchanan-Wollaston, A. V., Beringer, J. E., Brewin, N. J. Hirsch, P. R., and Johnston, A. W. B. 1980. Isolation of symbiotically defective mutants in *Rhizobium leguminosarum* by insertion of the transposon Tn5 into a transmissible plasmid. Mol. Gen. Genet. 178:185-190.
12. Caldwell, B. E., and Vest, H. G. 1977. Genetic aspects of dinitrogen fixation in legumes: The macrosymbiont. Pages 557–575 in: A Treatise on Dinitrogen Fixation. Vol. 3. Biology, R. W. F. Hardy and W. S. Silver, eds. John Wiley & Sons, New York. 675 pp.
13. Callaham, D. A. 1979. A structural basis for infection of root hairs of *Trifolium repens* by *Rhizobium trifolii*. M.S. thesis. Univ. Mass., Amherst. 41 pp.
14. Casse, F., Boucher, C., Julliot, J. S., Michel, M. and Dénarié, J. 1979. Identification and characterization of large plasmids in *Rhizobium meliloti* using agarose gel electrophoresis. J. Gen. Microbiol. 113:229-242.
15. Dart, P. J. 1977. Infection and development of leguminous nodules. Pages 367–372 in: A Treatise on Dinitrogen Fixation. Vol. 3. Biology. R. W. F. Hardy and W. S. Silver, eds. John Wiley & Sons, New York. 675 pp.
16. Dazzo, F. B. 1980. Lectins and their saccharide receptors as determinants of specificity in the *Rhizobium*-legume symbiosis. Pages 277–304 in: The Cell Surface: Mediator of Developmental Processes. S. Subtleney and N. K. Wessells, eds. Academic Press, New York. 374 pp.
17. Dazzo, F. B., and Hubbell, D. H. 1982. Control of root hair infection. Pages 275–309 in: Nitrogen Fixation. Vol. 2. *Rhizobium*. W. J. Broughton, ed. Oxford University Press, New York.
18. Dénarié, J., Truchet, G., and Bergerson, B. 1976. Effects of some mutations on symbiotic properties of *Rhizobium*. Pages 47–61 in: Symbiotic Nitrogen Fixation in Plants. P. S. Nutman, ed. Cambridge University Press, London 584 pp.
19. Ditta, G., Stanfield, S., Corbin, D., and Helinski, D. R. 1980. Broad host range DNA cloning system for Gram-negative bacteria: Construction of a gene bank of *Rhizobium meliloti*. Proc. Natl. Acad. Sci. U.S.A. 77:7347-7351.
20. Groat, R. G., and Vance, C. P. 1981. Root nodule enzymes of ammonia assimilation in alfalfa (*Medicago sativa* L.). Plant Physiol. 67:1198-1203.
21. Hirsch, A. M., Long, S. R., Bang, M., Haskins, N., and Ausubel, F. M. 1982. Structural studies of alfalfa roots infected with nodulation mutants of *Rhizobium meliloti*. J. Bacteriol. 151:411-419.
22. Hirsch, P. R. 1979. Plasmid-determined bacteriocin production by *Rhizobium leguminosarum*. J. Gen. Microbiol. 113:219-228.
23. Hirsch, P. R., Van Montagu, M., Johnston, A. W. B., Brewin, N. J., and Schell, J. 1980. Physical identification of bacteriocinogenic, nodulation and other plasmids in strains of *Rhizobium leguminosarum*. J. Gen. Microbiol. 120:403-412.
24. Holl, F. B., and Larue, T. A. 1976. Genetics of legume plant hosts. Pages 391–399 in: Proc. Int. Symp. Nitrogen Fixation, 1st. Vol. 2. W. E. Newton and C. J. Nyman, eds. Washington State University Press, Pullman. 717 pp.
25. Hooykaas, P. J. J., van Brussel, A. A. N., den Dulk-Ras, H., van Slogetern, G. M. S., and Schilperoort, R. A. 1981. Sym plasmid of *Rhizobium trifolii* expressed in different rhizobial species and *Agrobacterium*

tumefaciens. Nature 291:351-353.

26. Johnston, A. W. B., Beynon, J. L., Buchanan-Wollaston, A. V., Setchell, S. M., Hirsch, P. R., and Beringer, J. E. 1978. High frequency transfer of nodulating ability between strains and species of *Rhizobium.* Nature 276:1635-1636.
27. Kondorosi, A., Kiss, G. B., Forrai, T., Vincze, E., and Banfalvi, Z. 1977. Circular map of the *Rhizobium meliloti* chromosome. Nature 268:525-527.
28. Kondorosi, A., Vincze, E., Johnston, A. W. B., and Beringer, J. E. 1980. A comparison of three *Rhizobium* linkage maps. Mol. Gen. Genet. 178:403-408.
29. Legocki, R. P., and Verma, D. P. S. 1980. Identification of "nodule-specific" host proteins (nodulins) involved in the development of *Rhizobium*-legume symbiosis. Cell 20:153-164.
30. Leps, W. T., Brill, W. J., and Bingham, E. T. 1980. Effect of alfalfa ploidy on nitrogen fixation. Crop Sci. 20:427-430.
31. Libbenga, K. R., and Bogers, R. J. 1974. Root nodule morphogenesis. Pages 430–472 in: The Biology of Nitrogen Fixation. A. Quispel, ed. North-Holland Press, Amsterdam. 769 pp.
32. Long, S. R., Meade, H. M., Brown, S. E., and Ausubel, F. M. 1981. Transposon-induced symbiotic mutants of *Rhizobium meliloti.* Pages 129–143 in: Genetic Engineering in the Plant Sciences. N. J. Panopoulos, ed. Praeger Press, New York. 271 pp.
33. Maier, R. J., and Brill, W. J. 1976. Ineffective and non-nodulating mutant strains of *Rhizobium japonicum.* J. Bacteriol. 127:763-769.
34. Meade, H. M., and Signer, E. 1977. Genetic mapping of *Rhizobium meliloti* Proc. Natl. Acad. Sci. U.S.A. 74:2076-2078.
35. Meade, H. M., Long, S. R., Ruvkun, G. B., Brown, S. E., and Ausubel, F. M. 1982. Isolation of symbiotic and auxotrophic mutants of *Rhizobium meliloti* using transposon Tn5 mutagenesis. J. Bacteriol. 149:114-122.
36. Newcomb, W., Syono, K., and Torrey, J. G. 1977. Development of an ineffective pea root nodule: Morphogenesis, fine structure, and cytokinin biosynthesis. Can. J. Bot. 55:1891-1907.
37. Paau, A. S., Jing, Y., and Brill, W. J. 1981. Leghemoglobins in the alfalfa-*Rhizobium* symbiosis. (Abstr.) Plant Physiol. 67(Suppl.):92.
38. Paau, A. S., Leps, W. T., and Brill, W. J. 1981. Agglutinin from alfalfa necessary for binding and nodulation by *Rhizobium meliloti.* Science 213:1513-1514.
39. Pain, A. N. 1979. Symbiotic properties of antibiotic-resistant and auxotrophic mutants *Rhizobium leguminosarum.* J. Appl. Bacteriol. 47:53-64.
40. Pankhurst, C. E. 1974. Ineffective *Rhizobium trifolii* mutants examined by immune diffusion, gel-electrophoresis, and electron microscopy. J. Gen. Microbiol. 70:161-177.
41. Pankhurst, C. E., and Schwinghamer, E. A. 1974. Adenine requirement for nodulation of pea by an auxotrophic mutant of *Rhizobium leguminosarum.* Arch. Microbiol. 100:219-238.
42. Rolfe, B. G., Gresshoff, P. M., and Shine, J. 1980. Rapid screening for symbiotic mutants of *Rhizobium* and white clover. Plant Sci. Lett. 19:277-284.
43. Rolfe, B. G., Djordjevic, M., Scott, K. F., Hughes, J. E., Jones, J. B., Gresshoff, P. M., Cen, Y., Dudman, W. F., Zurkowski, W., and Shine, J. 1981. Analysis of the nodule forming ability of fast-growing *Rhizobium* strains. Pages 142–145 in: Current Perspectives in Nitrogen Fixation. A. H. Gibson and W. E. Newton, eds. North-Holland Press, Amsterdam. 534 pp.
44. Ruvkun, G. B., and Ausubel, F. M. 1981. A general method for site-directed mutagenesis in prokaryotes. Nature 289:85-88.
45. Sanders, R., Carlson, R., and Albersheim, P. 1978. A *Rhizobium* mutant incapable of nodulation and normal polysaccharide secretion. Nature 271:240-242.
46. Sanders, R., Raleigh, E., and Signer E. 1981. Lack of correlation between extracellular polysaccharide and nodulation ability in *Rhizobium.* Nature 292:148-149.
47. Scherrer, A., and Dénarié, J. 1971. Symbiotic properties of some auxotrophic mutants of *Rhizobium meliloti* and their prototrophic revertants. Plant Soil, Spec. Vol., pp. 39-45.
48. Schwinghamer, E. A. 1968. Loss of effectiveness and infectivity in mutants of *Rhizobium* resistant to metabolic inhibitors. Can. J. Microbiol. 14:355-367.
49. Schwinghamer, E. A. 1969. Mutation to auxotrophy and prototrophy as related to symbiotic effectiveness in *Rhizobium leguminosarum* and *Rhizobium trifolii.* Can. J. Microbiol. 15:611-622.
50. Schwinghamer, E. A. 1977. Genetic aspects of nodulation and dinitrogen fixation by legumes: The microsymbiont. Pages 577–622 in: A Treatise on Dinitrogen Fixation. Vol. 3. Biology. R. W. F. Hardy and W. S. Silver, eds. John Wiley & Sons, New York. 675 pp.
51. Truchet, G., and Dénarié, J. 1973. Ultrastructure et activite reductrice d'acetylene des nodosités de luzerne (*Medicago sativa*

L.) induites par des souches de *Rhizobium meliloti* auxotrophes pour la leucine. C. R. Hebd. Seances Acad. Sci. Ser. D 277:925-928.

52. Truchet, G., Michel, M., and Dénarié, J. 1980. Sequential analysis of the organogenesis of lucerne (*Medicago sativa*) root nodules using symbiotically-defective mutants of *Rhizobium meliloti*. Differentiation 16:163-172.
53. Van Vliet, F., Silva, B., Van Montagu, M., and Schell, J. 1978. Transfer of RP4:Mu plasmids to *Agrobacterium tumefaciens*. Plasmid 1:446-455.
54. Vance, C. P., and Johnson, L. E. B. 1981. Nodulation: A plant disease perspective. Plant Dis. 65:118-124.
55. Vance, C. P., Johnson, L. E. B., and Hardarson, G. 1980. Histological comparisons of plant and *Rhizobium* induced ineffective nodules in alfalfa. Physiol. Plant Pathol. 17:167-173.
56. Verma, D. P. S., Goodchild, B., Brisson, N., Thomas, D. Y., Haugland, R., Sullivan, D., and LaCroix, L. 1980. Expression of leghaemoglobin genes in effective and ineffective root nodules on soybean. (Abstr.) Plant Physiol. 65 (Suppl.):681.
57. Vincent, J. M. 1974. Root-nodule symbioses with *Rhizobium*. Pages 266–341 in: The Biology of Nitrogen Fixation. A. Quispel, ed. North-Holland Press, Amsterdam. 769 pp.
58. Vincent, J. M. 1980. Factors controlling the legume-*Rhizobium* symbiosis. Pages 103–129 in: Nitrogen Fixation. Vol. 2. W. E. Newton and W. H. Orme-Johnston, eds. University Park Press, Baltimore. 325 pp.

CHAPTER 32

DISEASE PREDICTION: CURRENT STATUS AND FUTURE DIRECTIONS

A. L. JONES

Disease prediction is the science of monitoring the physical conditions of the environment and declaring after "disease weather," but before symptoms are visible, that infection has occurred. Factors such as host susceptibility and inoculum availability may also be incorporated into a disease prediction program. Disease forecasting differs from prediction because forecasts are based on anticipated weather, not on past weather. Disease predictions are usually more accurate than forecasts and are used by growers to time or schedule control treatments. Besides the usefulness of disease prediction to growers, it can be used as a basis for planning research aimed at minimizing crop losses and pesticide usage.

The purpose of this chapter is not an in-depth review of the history of disease prediction, but rather a viewpoint of future directions in this field for the next decade or two. This will require some discussion of where disease prediction is today, of its importance both currently and potentially, and of problems that must be solved if it is to become more useful.

NEED FOR INTEGRATING DISEASE PREDICTION WITH CONTROL STRATEGIES

To control fungal and bacterial diseases on high value crops, fungicides and bactericides have been used to protect crops from infection and to reduce disease losses. Information on fungicides and bactericides for the control of diseases on fruit, vegetable, and some field crops is published in the form of spray calendars annually by many universities. This approach has been well accepted by growers and emulated by many commercial chemical companies who publish spray calendars outlining uses for their products. Control programs are based primarily on the characteristics of the control material, secondarily on the epidemiology of the disease or the host/parasite interaction. Routine calendar programs are designed to control disease in seasons of severe disease and result in excess use of chemicals in years of light or moderately severe disease. Except in wet years, chemical applications are costly to growers. Sometimes the routine use of highly effective fungicides in calendar programs has led to the development of resistant strains of the pathogen.

The timing of fungicide and bactericide applications based on current information about the climate in the pathogen-crop system has received attention in recent years as a method to increase efficiency of control and reduce dependence on calendar programs. Development of models, sufficient to identify when the weather is favorable for one or more stages of a pathogen's life cycle, is increasing nationally due in part to current emphasis on crop integrated pest management research. As the vast majority of these models key on time-wetness-temperature functions, pest managers must work with climatic events as they unfold. They must also work within a relatively short time constraint once key events are identified.

Predictive systems are highly dependent on having specific weather information available upon which to base a prediction. Delays in detecting an event and alerting pest managers takes time away from that available for implementing control strategies. Usually only a few hours to 2 or 3 days are available for carrying out the control procedures. Therefore, for the development of predictive disease systems the following factors are important: 1) identifying the criteria needed to predict stages in the life cycle of the pathogen susceptible to control, 2) having fungicides or some other strategy available for controlling the pathogen during the stage or stages of the life cycle being predicted, 3) having methods for disseminating disease predictions quickly so that control strategies can be implemented while still effective, and 4) understanding the economic benefits and risks of the predictive system.

IMPLEMENTATION SYSTEMS FOR DISEASE PREDICTION

Although the need for automatic acquisition of weather data for use in pest management and disease prediction has been recognized for about a decade (2,7,20), systems using central computer time-sharing facilities are currently experimental and practical systems for grower use have not been developed. System reliability and availability are severe limitations where the computer is not dedicated to disease prediction. Attempts to use a high-speed time-sharing computer to predict apple scab failed because the computer was not operational between 5 and 8 pm, when growers normally make control decisions (8). Such systems are currently not capable of dealing with local needs in an economic and efficient manner (2).

Because of the constraints associated with centralized communication networks, the development of on-site predictive systems is increasing. Two examples of integrating disease prediction and control strategies on-site are the control of scab in apple orchards east of the Rocky Mountains and the control of potato late blight in the northeastern United States. The experience and knowledge gained in developing these systems are similar in many aspects. I will discuss the apple scab system because this is one of the approaches to disease prediction under development at Michigan State University. Also, an in-depth review of the potato late blight system has been published by MacKenzie (14).

Research groups in many northeastern and northcentral states have given high priority to developing an improved system for implementing a predictive system for apple scab utilizing, in part, the early research of Mills at Cornell. The Mills system identifies putative infection periods based on duration of leaf wetness from rain and the average temperature during the wetness period (15). Environmental monitoring equipment specifically designed to record the time, wetness, and temperature data needed to predict scab was developed (13,16,17,21), as were methods and equipment for monitoring the maturation and the discharge of apple scab ascospores from pseudothecia in overwintering leaves on the orchard floor (6,18,19). Once infection is predicted, a suitable eradicant fungicide must be applied within 24–72 hr after the rain started.

In Michigan, a microcomputer was developed to alert apple growers of scab weather and to advise them of the most appropriate fungicide to use under current conditions (4,10). The microcomputer approach was selected because it can collect the data needed for making a prediction, then analyze and interpret the data for the grower. With previous instruments, the grower had to interpret the weather charts and then analyze the data using tables, charts, or figures. The unit is small and of rugged construction for placement directly in the orchard. It includes sensors for monitoring temperature, relative humidity, leaf wetness, and rainfall. A keyboard and display panel are used to retrieve data and predictions stored in the instrument. Units are commercially available (Reuter-Stokes, Inc., Cleveland, OH 44122).

Microprocessor units like the one developed for apple scab have considerable potential for greater reliability in disease prediction in the coming years. Such units are multipurpose and can be programmed with new models. They are designed with future improvements in mind. Recently, we programmed a unit with experimental models to predict apple scab and apple powdery mildew; forecast the development stages of codling moth, an important insect pest of apple; and predict growth stages of the apple tree itself. Based on our experience, it will be possible to develop a unit that can handle simultaneously several pest models associated with a single crop. Before all-embracing microprocessor units can be developed for disease prediction, certain technical problems must be solved. Current units are restricted in the amount of memory available to store data and computer programs. Increased memory will allow for better programming structures and will make it easier for pest managers to add new models. Moreover, as the capacity of units to store information increases, it will be possible to use higher level computer languages like BASIC or FORTRAN rather than assembler languages. When this occurs, biologists will be able to concentrate on developing and validating models with much less dependence on engineers. Progress should move quickly.

RISKS ASSOCIATED WITH DISEASE PREDICTION

Although it is possible with many diseases to reduce fungicide usage and the associated environmental problems by using predictive systems, this is often accompanied by greater risks for economic loss to individual growers. And if a program is accepted widely, losses from a control failure could extend across an entire region. Potential for loss is particularly high where production and harvesting costs involve high fixed investments in land and equipment. To protect their investments, growers are more willing to apply one or two extra sprays than they are to risk losing investments worth many more times the potential savings.

In practice, predictive systems only advise. It is the grower who must make the final decisions and implement the selected control tactics. For old established systems, the risks are fairly well understood, and experienced plant pathologists can alert growers to potential problems. In developing new models, simulation analysis may be useful for evaluating the risks. Through simulation, the model can be made more conservative, thus reducing the possibility of failure. Although more control treatments may be advised, they are advised only when changes of disease are likely.

NEED FOR CURATIVE FUNGICIDES IN PREDICTIVE SYSTEMS

There is a risk of a control failure associated with all fungicide programs. A protective program for apple scab may fail if an extended weather period suitable for severe infection comes without warning at a critical time in the spray schedule. This risk can be reduced by having eradicant fungicides available that will control infections from wetting periods starting a few days earlier. In predictive programs, sprays are not applied until after infection periods are identified. Once a prediction is given, control is critical and a day or two delay in spray application may result in no or reduced control. This is what makes after-infection programs inherently more risky.

New, highly effective fungicides that stop infection late in the incubation period but before symptoms develop, or that inactivate lesions if they should appear, reduce the risk of failure when sprays are delayed by bad weather or equipment problems. Similarly, these fungicides also reduce the risks associated with protective programs. Most of the sterol-inhibiting fungicides exhibit excellent curative properties against apple scab (11) and the acylalanine fungicides exhibit curative properties against potato late blight (5). Fungicides of this type are needed to avert the risks associated with predictive control programs; this is one of their important advantages.

Previous experience with highly specific curative fungicides indicate that strategies for their use should include techniques for avoiding problems with fungicide resistance; otherwise they may be lost as backstops to predictive programs. The development of metalaxyl-resistant strains of *Phytophthora infestans* in the Netherlands (3) and in Ireland (1) and of benomyl- and dodine-resistant strains of *Venturia inaequalis* in Michigan (9) have increased the risks associated with predictive systems for scheduling fungicides because alternative materials are currently not available or are not registered for commercial use.

Innovative control strategies are needed to reduce the risks of fungicide resistance while maintaining their potential for decreasing fungicide usage. The following illustrates how many apple growers in Michigan control apple scab economically and at the same time reduce the risks associated with prediction. Rather than spraying the entire orchard with each spray, a protectant fungicide is applied to alternate middle rows, as first developed in Pennsylvania (12). Sprays are applied with sprayers that have sufficient capacity to cover 80% or more of the tree from one side. The next half-spray is applied 1 week later, unless there is a predicted infection period. If the weather has favored infection, a protectant-eradicant fungicide is applied to the side of the tree that was not sprayed the previous time. If the grower is unsure about the effectiveness of the protectant spray because of rapid tree growth, the trees can be sprayed a second time. This system reduces fungicide usage and also reduces the risk of treating the whole orchard on an emergency basis. It may also delay the development of fungicide resistance because different fungicides are used for protection than for eradication. Risk is further reduced by using the "alternate-middle" procedure only in orchards where scab was well controlled the previous season.

SUMMARY

The development of more efficient control systems based on epidemiological concepts incorporated into predictive models has progressed in recent years. These systems have been stimulated by research on integrated crop management. Using these systems, control measures can be recommended based on the identification of "disease weather." The development of control systems has been stimulated by recent advances in electronics in which microcomputers with weather-monitoring capability have become available for making disease predictions on individual farms or in individual fields. Their acceptance will be led by a younger generation of farmers with training in computer technology. New fungicides with improved curative properties reduce the risks associated with predictive control strategies. This approach is expected to increase steadily in importance in the next 10–20 years as new models and improved implementation systems result in a range of comprehensive management models for a range of crops.

LITERATURE CITED

1. Cooke, L. R. 1981. Resistance to metalaxyl in *Phytophthora infestans* in Northern Ireland. Pages 641-649 in: Proc. 1981 British Crop Protection Conf. Pests and Diseases.
2. Croft, B. A., Miller, D. J., Welch, S. M., and Marino, M. 1979. Developments in computer-based IPM extension delivery and biological monitoring systems design. Pages 223-250 in: Pest Management Programs for Deciduous Tree Fruits and Nuts. D. J. Boethel and R. D. Eikenbury, eds. Plenum Press, New York. 256 pp.
3. Davidse, L. C., Looijen, D., Turkensteen, L. J., and van der Wal, D. 1981. Occurrence of metalaxyl-resistant strains of *Phytophthora infestans* in Dutch potato fields. Neth. J. Plant Pathol. 87:65-68.
4. Fisher, P. D., Neuder, D. L., and Jones, A. L. 1979. Microprocessors improve pest management practices. Pages 89-102 in: Microprocessor Applications: International Survey of Practice and Experience. J. Hosier, ed. Infotech Int. Ltd., Maidenhead, England. 359 pp.
5. Fry, W. E., Bruck, R. I., and Mundt, C. C. 1979. Retardation of potato late blight epidemics by fungicides with eradicant and protectant properties. Plant Dis. Rep. 63:970-974.
6. Gilpatrick, J. D., Smith, C. A., and Blowers, D. R. 1972. A method of collecting ascospores of *Venturia inaequalis* for spore germination studies. Plant Dis. Rep. 56:39-42.
7. Haynes, D. L., Brandenburg, R. K., and Fisher, P. D. 1973. Environmental monitoring network for pest management systems. Environ. Entomol. 2:889-899.
8. Jones, A. L. 1976. Systems for predicting the development of plant pathogens in the apple orchard ecosystem. Pages 120-122 in: Modeling for Pest Management. R. L. Tummala, D. L. Haynes, and B. A. Croft, eds. Mich. State Univ. 247 pp.
9. Jones, A. L., and Walker, R. J. 1976. Tolerance of *Venturia inaequalis* to dodine and benzimidazole fungicides in Michigan. Plant Dis. Rep. 60:40-44.
10. Jones, A. L., Lillevik, S. L., Fisher, P. D., and Stebbins, T. C. 1980. A microcomputer based instrument to predict primary apple scab infection periods. Plant Dis. 64:69-72.
11. Kelley, R. D., and Jones, A. L. 1981. Evaluation of two triazole fungicides for postinfection control of apple scab. Phytopathology 71:737-742.
12. Lewis, F. H., and Hickey, K. D. 1972. Fungicide usage on deciduous fruit trees. Annu. Rev. Phytopathol. 10:399-428.
13. MacHardy, W. E. 1979. A simple, quick technique for determining apple scab infection periods. Plant Dis. Rep. 63:197-204.
14. MacKenzie, D. R. 1981. Scheduling fungicide applications for potato late blight with BLITECAST. Plant Dis. 65:394-399.
15. Mills, W. D. 1944. Efficient use of sulfur dust and sprays during rain to control apple scab. N.Y. Agric. Exp. Stn. (Ithaca) Ext. Bull. 630. 4 pp.

16. Small, C. G. 1978. A moisture-activated electronic instrument for use in field studies of plant disease. Plant Dis. Rep. 72:1039-1043.
17. Smith, C. A., and Gilpatrick, J. D. 1980. Geneva leaf-wetness detector. Plant Dis. 64:286-288.
18. Sutton, T. B., and Jones, A. L. 1976. Evaluation of four spore traps for monitoring discharge of ascospores of *Venturia inaequalis*. Phytopathology 66:453-456.
19. Szkolnik, M. 1969. Maturation and discharge of ascospores of *Venturia inaequalis*. Plant Dis. Rep. 53:534-537.
20. Welch, S. M., Croft, B. A., Brunner, J. F., and Michels, M. F. 1978. PETE: An extension phenology modeling system for management of a multi-species pest complex. Environ. Entomol. 7:482-494.
21. Zuck, M. G., and MacHardy, W. E. 1981. Recent experience in timing sprays for control of apple scab: Equipment and test results. Plant Dis. 65:995-998.

CHAPTER 33

THE ROLE OF MATHEMATICAL MODELS IN PLANT HEALTH MANAGEMENT

RICHARD A. FLEMING and JOHN A. BRUHN

Control of crop diseases is a complicated venture that requires the manipulation of a complex of biological, chemical, and physical systems in an uncertain physical and economic environment. Field trials have generally been used to identify effective control practices. Unfortunately, this empirical and resource-intensive approach is only practical for testing a few possible alternative management practices over short time periods. Because of these limitations, crop disease management practices developed empirically have tended toward excessive control to minimize short term uncertainties. However, increasing production costs and increasing concern about the long term impact of control practices (e.g., on pathogen evolution and on environment) have complicated decision making and have produced a need for a more detailed understanding of plant disease management systems. Mathematical models are necessary, but by themselves not sufficient, to meet this need.

This chapter considers the role of mathematical models in improving the development of crop disease control policy. It emphasizes that mathematical models represent a single but important component of a variety of scientific methods available to help us understand, predict, and advantageously modify the behavior of plant pathogen systems.

MODELING

Modeling and simplification are essential aspects of the scientific process. The role of scientists is to observe and describe natural phenomena; to synthesize, summarize, and explain their observations; and to use their observations to predict the outcome of other, as yet uninvestigated, situations (16). Scientists deal directly with nature only during observation; all subsequent phases of the scientific process involve simplification in one form or another.

A model is a simplified representation of how we imagine reality. Ideally, a model provides the simplest representation of reality that retains all aspects of reality pertinent to the problem at hand. The optimal balance between the simplicity needed to allow easy understanding of the model and the complexity required to accurately describe how we perceive reality is dictated by the state of knowledge and the purpose of the model.

Since the objectives of the modeling effort determine the form of the model, model development is not a useful end in itself; rather, the model is a tool for fulfilling particular objectives. The best model for one purpose may be quite different from the best model for another purpose, even when they represent similar characteristics of reality.

Material and Conceptual Models

Generally, models used in plant epidemiology belong to one of two broad categories: material models, which represent nature in a physically simplified form, and conceptual models, which are abstractions of reality. Material models are used in field and laboratory experiments (26). It is assumed that the particular observations made in the experimental microcosm (the material model) are representative of occurrences in a real world of greater complexity and very different time and space scales.

Conceptual models are much more prevalent than material models. As diagrams they have helped in the synthesis and organization of observations concerned with life cycles and evolutionary relationships. Since language itself is effectively a collection of conceptual models, any communication between plant epidemiologists necessarily uses conceptual models. Furthermore, since any planned experiment requires that the experimenter first decide what to observe, when to observe it, and where to observe it, and since these decisions are necessarily based on mental conceptual models, it follows that experimental design is also in the domain of conceptual models.

MATHEMATICAL METHODS

Mathematical models are conceptual models that are represented symbolically and manipulated according to a set of rules derived directly from the logic of deductive reasoning. The power of mathematical modeling lies partly in the strict adherence of these rules to the laws of logic. Mathematics represents a kit of deductive tools for exploring unambiguously the logical consequences implicit in the assumptions of a model.

The mathematical notation has many virtues. First, it allows more concise representations of even the simplest relationships. For instance, suppose we are interested in the magnitude of some quantity (e.g., disease severity, inoculum density, or the number of atoms of a radioactive isotope) at time t. Then, if the variable x_t represents this quantity, the equation

$$dx_t/dt = mx_t \qquad (1)$$

expresses that its rate of change is equal to its magnitude multiplied by a rate factor.

Second, the mathematical notation exposes implicit assumptions and provides an economy of thought. For instance, equation 1 shows explicitly that the rate factor, m, is assumed to be independent of x_t and t. Moreover, since those familiar with equation 1 automatically associate the relationship

$$x_t = x_o \exp(mt) \qquad (2)$$

with it, equation 1 is readily recognized as describing exponential change.

Mathematical notation is also a common language for mathematically inclined scientists in diverse disciplines, and as such, it eases the transfer of ideas between disciplines and promotes the use of argument by analogy. For example, Yarwood and Sylvester (25) used equation 2 to describe the decrease of inoculum with time in poor environments (i.e., $m < 0$). Realizing that this equation had long been used to describe radioactive decay in atomic physics, they borrowed the concept of "half-life," the time taken for half of the atoms of an isotope to decay radioactively, and introduced it into plant pathology as the time taken for half of the inoculum to die.

However, mathematical notation also has a serious drawback. Scientists unfamiliar with it often find mathematical notation more confusing than helpful; this can lead to a communication gap between the mathematically skilled and the mathematically less skilled within a discipline (24).

The value of a mathematical model is determined by how well it fulfills its purpose and in how useful that purpose is. The degree to which a model fulfills its purpose is measured with respect to how well that purpose could have been achieved without the model (18). The usefulness of the purpose is much more difficult to assess. The model should be judged in the broadest of socioeconomic-scientific contexts: from the perspective of immediate practical application in particular circumstances to the perspective of long-term general theoretical insights with wide-ranging implications.

It follows that a mathematical model's ability to accurately describe and predict nature is not a universal criterion by which to judge its value. When the purpose is to test the logical foundations of a widely held hypothesis, the mathematical model should accurately reflect the hypothesis (not necessarily nature), and disproof may consist of demonstrating that the logical consequences of the model (and therefore of the hypothesis) are, in fact, at odds with observation. For example, to test whether the proportion of disease tissue, x_t, increases exponentially when x_t is large, one might adopt equation 2 with $m > 0$ and show that for $x_0 = 1$, $x_t > 1$. Since, by definition, $0 < x_t \leqslant 1$, the hypothesis is disproved.

Mathematical models should be robust: i.e., they should be insensitive to minor changes in their assumptions except where such sensitivity can be justified on biological grounds or by the purpose of the model. Fleming (5) discusses an example of a nonrobust model that describes aspects of host-parasite coevolution; changes in the sequence of host and parasite reproduction in the model can lead to vastly different conclusions, ranging from stable balanced coexistence to extreme fluctuations and eventual extinction.

Empirical Mathematical Models

Mathematical models can generally be classified according to their purpose as either empirical or mechanistic. Empirical models are motivated by a wish to provide a basis for prediction and sometimes control. They are exemplified by "curve-fitting" to reflect an observed relation among variables and are little concerned with the underlying mechanisms that caused the observed relation. Empirical models (e.g., regression models) are the product of experiment and observation; they provide a means of condensing vast quantities of data into a few statistical quantities (e.g., regression coefficients). Their usefulness in summarizing data has long provided an essential bridge between field observation and the development of mechanistic models. Empiricial models are based on inductive reasoning from particular observations to general relationships and are thus resumés of past experience. This inductive and largely descriptive foundation leaves them lacking explanatory ability and limits their predictive ability to occasions when the situation in the future will have changed little since the time the data were collected. Nevertheless, empirical models have often proved useful for short-term prediction; they are easy to develop and well suited to handling variability in the data through concepts such as confidence limits.

Mechanistic Mathematical Models

Mechanistic models are motivated by a wish to understand the processes causing observed relations among variables (17). They are often derived from theory and supposition about the underlying causal mechanisms by meticulously incorporating the known properties of the variables and their relationships. There is a change in emphasis from correlation to causation and from statics to dynamics (7). For instance, the difference in purpose between empirical and mechanistic models is embodied in the choice between determining the value of m in equation 2 as the coefficient of the regression of $\ln(x_t)$ against t (prediction) or measuring the rates of the processes (e.g., lesion expansion, sporulation) by which x_t (e.g., disease severity) changes (understanding). This attention to causation makes mechanistic models useful for explanation and potentially superior as predictors when one extrapolates beyond the data base on which the descriptive components were developed (20).

Empirical models and mechanistic models can sometimes be distinguished by comparing the level of biological organization at which the information used to construct the model was collected against the level of biological organization described by the model. Empirical models generally accept data and provide description at the same level of biological organization. Mechanistic models often describe biological organization at a level above that at which their data were obtained. Of those models that describe population processes, empirical models are usually derived from observations made at the population level, whereas mechanistic models often include empirical submodels derived from observations made at the level of the individual. The distinction between empirical and mechanistic models can become blurred since many mechanistic models include submodels that are empirical at the individual level and since successful empirical models are often developed from a structure that inherently reflects some of the mechanisms fundamental to the processes being described.

The many uses of mechanistic models in plant health management can be separated into three broad categories according to whether the model's main purpose is to synthesize existing knowledge, to explain observation and hypothesis, or to predict the future.

Synthesis

The ability of mechanistic mathematical models to logically synthesize existing knowledge is the basis for their largely unexploited but nonetheless potentially very important role as guides to the overall planning of research strategy (15). Jeffers (9) comments at length on the inefficiency of much scientific activity, inefficiency arising from poor experimental design, ineffective and often even invalid data analysis, and the failure to use available techniques to solve practical problems. Bigger data banks, built by uncritical information accumulation, are not necessarily better banks when their purpose is explanation, or prediction, or even description. The integration of mathematical models into the whole strategy of planning, organizing, and executing research in plant epidemiology has the potential to significantly reduce this waste of scientific effort. It could help to structure the processes of sampling and evaluation through the identification of crucial information gaps and the direction of research to efficiently fill those gaps.

In this sense, mechanistic mathematical models can provide a valuable complement to the more traditional experimental media of the field and the laboratory. Since field work is done under conditions approximating those on commercial farms, it ensures, as far as possible, that the results can be directly applied in farming practice. To manage effectively, the farmer must determine the effects of many possible combinations of treatment variations in assorted environmental conditions. But in the field it is rarely possible to do more than measure treatment inputs (e.g., times of fungicide application) and gross outputs (e.g., disease levels) in a limited range of environmental conditions (e.g., rainfall patterns). This can make it risky to extrapolate, or even to interpolate, the data (2). Mechanistic models provide a practical means of exploring beyond the treatment variations tested in the field (13).

For instance, suppose the effect of different schedules of fungicide application (the treatment inputs) on disease levels (the gross outputs) have been measured in one very "dry" and one very "wet" season. Since disease levels are likely to respond nonlinearly to "wetness," direct interpolation of the results to seasons of moderate "wetness" could be misleading. A mechanistic model incorporating the effects of wetness on crop growth, disease dynamics, and fungicide weathering could account for the nonlinearity and thus offer a potentially superior means of interpolation.

Laboratory work allows the controlled isolation of particular pathosystem components necessary for clarifying the mechanisms that govern overall system behavior. But again, much work is needed to evaluate each component, and one of the components most influential on system behavior may be among the last to be studied.

Sensitivity analysis is a means of identifying the pathosystem components seemingly most worthy of immediate investigation. It involves determing the sensitivity of the model's predictions to realistic variation in parameter values. For instance, suppose equation 2 accurately reflects the increase of a quantity x_t and suppose m and x_o have similar numerical ranges. Since the value of x_t changes more with changes in m than it does with similar changes in x_o, determining m is more important for estimating x_t than determining x_o. Hence, if x_t represents disease severity, laboratory and greenhouse workers might study the latent period (which affects m) before studying the overwintering capacity (which affects x_o).

Another characteristic of laboratory work is that interactions between components are usually omitted, leaving the properties of the pathosystem as a whole uncertain. Mechanistic models, as syntheses of scientific knowledge, provide a means of exploring these emergent pathosystem properties and testing scientific knowledge. For instance, if our best scientific knowledge indicates that equation 2 describes disease progress when disease severity, x_t, is small, measurements of sporulation rate and overwintering ability alone provide little information about x_t (the emergent property) until they are related to m and x_o in the context of equation 2.

Thus, mechanistic mathematical models could form a valuable link between field and laboratory in plant disease epidemiology. The field worker needs to reduce the complexity of the field to simpler components in order to understand the mechanisms responsible for what he observes. Laboratory work would benefit from a more directed goal than understanding the mechanics of a small assortment of these simpler components in isolation. A suitably formulated mechanistic model could meet both these needs by providing a common base for the development of a well-directed team approach.

Explanation

In plant health management, mechanistic models have been used principally for either explanation or prediction. Levins and Wilson (12) express concern that the explanatory role has been relatively neglected, leaving too narrow a theoretical basis for contemporary pest management.

Explanation has three facets. First, it identifies key processes that have determined the development of certain situations in the past. Second, it suggests how those processes might be affected by human actions and how those actions might ultimately influence the kind of situations that develop in the future. And third, explanation concerns hypothesis testing; the goal is to mathematically describe the hypothesis, to use the resulting model to explore the logical consequences of the hypothesis, and to compare the logical consequences with observation.

The assumptions of explanatory mathematical models are often deliberately oversimplified. This seems to cause a great deal of concern. The reason for this conscious over-simplification is to isolate key relationships in the hope that their explanation may be a useful first step in understanding more complex systems in which many factors interact. Although pathosystems as a whole have properties beyond those of their individual components, it is still of enormous value to understand the behavior of the basic elements. The philosophy is akin to that of laboratory and growth chamber work. The analysis of these idealized models provides standards of comparison and a framework in which to consider more complex systems (11). For instance, the classical theory of single locus Mendelian population genetics and the idealized Hardy-Weinberg equilibrium provide the almost universal basis for discussing the more complicated genetic mechanisms behind evolutionary change.

Pielou (14) argues that it is the discrepancies between models and reality, not their similarities, that lead to new discoveries. She cites Boulding (1) who claims,

> Knowledge increases not by the matching of images with the real world (which Hume pointed out is impossible), that is not by the direct perception of truth but by . . . the perception of error.

Oversimplification helps in identifying and interpreting such discrepancies. Pielou (14) concludes that such models "are as useful as realistic ones in contributing to the advance of knowledge."

In dealing with management questions, there is often a deliberate search for thresholds and nonlinearities in the pathosystems dynamics that can be exploited for disease control, e.g., Fleming (6). In this respect, explanatory models often provide a new and unique perspective and because of this, need not be closely tied to experimental work in the way predictive models must be. Furthermore, as Kranz (10) has noted, experimental testing may be impaired due to the limits of experimental accuracy or lack of appropriate techniques or apparatus. In applied ecology in general, the best explanatory models have become influential didactic tools, guiding research and determining control strategies (4).

Prediction

Although they may be the best tools available, mechanistic mathematical models are not as suited to prediction as they are to explanation (11). For the latter, the plant disease epidemiologist uses mathematics to focus ideas and to test preliminary hypotheses without resorting to new experiments. This purpose is distinct from the use of mathematics for prediction, and this difference is poorly understood, even by many mathematical modelers. Accurate prediction requires substantiated principles, but since plant disease epidemiology can offer only empirical generalizations to guide model construction, the predictive role of mechanistic mathematical models feels uncomfortable to purists.

In contrast to empirical models, the mechanistic models used for prediction are generally complex computer simulation models; to predict accurately they must describe the interactions between components as well as the components themselves. A structure is used that accurately describes the constituent biological processes (e.g., sporulation, infection, competition, dispersal) but since excessive complexity inhibits understanding, these models are ideally as simple as possible while maintaining their predictive reliability. To make the models of direct use to management, the influence of various management activities (e.g., fungicide application) are often included.

Many mechanistic predictive models in plant disease epidemiology have been used to forecast disease progress on a week-to-week basis. These models have the difficult job of contending in some detail not only with spatial and temporal heterogeneity, but also with weather, which tends to be a dominant factor in such forecasts. Since the reliability of even the most accurate predictive models depends upon the reliability of their input data, the prediction of disease incidence in the field is limited by the accuracy of initial inoculum estimation, parameter estimation, and weather predictions. Because of the prevalence of positive feedback relationships and nonlinearities, otherwise identical mechanistic models often quickly evolve large differences in the values of their state variables given very small initial differences. Hence small errors in determining the initial state of the pathosystem or small statistical fluctuations may drastically limit the ability of mechanistic models to predict particular events (3).

However, by including such uncertainties explicitly in terms of probability distributions, a mechanistic model might predict long-term averages reasonably well. The model could then be used as a laboratory world, to predict the distribution of pathosystem behaviors for each of a variety of possible management policies (8). Simulation then becomes an experimental method in which exactly the same experimental conditions (e.g., weather patterns) are available for successive experiments and in which a vast number of different treatment variations (e.g., fungicide application schedules) can be quickly tested. It provides an assessment of the relative performance (e.g., efficacy and reliability) of each possible alternative management policy (J. A. Bruhn and W. E.

Fry, *unpublished*). Thus, instead of using predictive mechanistic models to guide the tactics of week-to-week decision making, they would be used in a somewhat explanatory role as an aid to long-term planning and strategy development.

Methods

A variety of techniques and procedures are often required to develop and study a mechanistic model. Here we briefly comment on some selected aspects of these methods that are not always appreciated.

Techniques

Mechanistic mathematical models are generally investigated using either mathematical analysis or numerical simulation. Mathematical analysis produces general qualitative insights into the model dynamics in terms of algebraic relationships; numerical simulation produces particular quantitative summaries of the model dynamics in terms of numerical relationships. For instance, analysis of equation 1 produces equation 2; particular solutions are easily determined from this general solution by substituting the appropriate numerical values of m, x_o, and t. In contrast, simulation of equation 1 produces successive numerical values of the variables x_t and t for particular values of m and x_o. General statements based on simulation are logically weak because they require extrapolation or interpolation from outcomes of particular simulations. Only through mathematical analysis can logically well-founded general relationships be resolved.

Hence, mechanistic models are ideally explored as far as possible with mathematical analysis. This will often involve the analysis of simplified "special cases." But mathematical analysis quickly reaches its limits, and further investigation of a complex model may require numerical simulation. Computers are usually used to perform the repetitive calculations required in simulation. The numerical simulation is better directed within the framework established by the mathematical analysis of "special cases" than outside it, e.g., Fleming (5). Thus the distinction between models that do and do not require numerical simulation is dictated by the mathematical complexity of the model and not by its intended application.

Optimization

When a predictive model includes a large number of possible combinations of management activities, trial and error simulation can be impractical. For example, there are 4^T alternative schedules by which fungicide could be applied at one of four possible dosages in each of T time intervals. Optimization procedures are often used to overcome this problem (19).

The optimization procedure adopted depends on whether a model with only a few variables can be constructed to capture the essence of the behavior of the mechanistic predictive model. If this is possible, dynamic programming is generally applied since it can include observations of random events as they become available, include the fixed costs of management activities directly, and suggest policies over a range of initial values of the variables.

If the behavior of the mechanistic predictive model cannot be adequately described in a simple model of a few variables, then either simulation or deterministic optimization methods are needed. However, these methods require many calculations to suggest an optimal solution, and hence, can be expensive.

The solutions identified by the adopted optimization procedure are used to guide the search for good solutions in the complex predictive model. The goal is to point out policies that should be studied more thoroughly and to compare and combine alternate policies in order to determine the range and nature of available choices, not to identify some mystical "optimal policy." The efficacies of the "better" policies are then evaluated in the field.

Validation

The term validation is commonly used to describe the process of comparing the behavior and properties of mechanistic mathematical models with the behavior and properties of what they are intended to represent. Ideally, validation established the limits to the credibility of the model.

Although it has gained wide usage, the term validation is a misnomer. A model is essentially an explicit hypothesis of those aspects of pathosystem dynamics it is meant to depict, and, in keeping with the scientific method in which hypotheses cannot be proved but only disproved, a model cannot be validated but only invalidated (22). The only exceptions occur when mathematical models are used to determine the logical consequences of explicit hypotheses and it can be verified that the models accurately reflect the hypotheses.

To establish the limits to the credibility of the model, data from the extremes of pathosystem behavior (e.g., extreme weather of some special geographical area or of some past time) are compared (for techniques, see Teng [21]) with the model's behavior and properties for similar extremes. The more accurate the behavior and properties of the model at these extremes, the greater its credibility. The emphasis here is on attempting to invalidate the model with extreme data rather than on attempting to validate the model by tuning parameters to fit a given set of historical data.

Sometimes validation has consisted only of comparing a model's output with the observed behavior of the real system. However, as Solomon (20) emphasizes, several mutually contradictory sets of biological mechanisms can be incorporated in models that produce the same output. Hence, even when model output is consistent with observation, one cannot conclude that the assumed biological mechanisms on which the model is based are realistic. Accepting a model based on the wrong biological mechanisms could have disastrous consequences on a management strategy that attempted to exploit those imagined mechanisms for disease control. Thorough validation goes beyond the superficial level of comparing model output with observation.

A more common consequence of improper validation on management decision making may be the failure to use already adequate models (23). This occurs if attempts at invalidation are excessive and keep models in the "testing stage" longer than necessary. Ultimately, the degree of credibility required of a model is determined by its purpose.

CHALLENGE FOR THE FUTURE

Although mathematical models have substantial, but not unlimited, scope as tools for plant disease management, they are seldom utilized to their full capacity, and currently their full capacity is inadequate for many important problems. Correcting this situation is a major challenge for the future.

Better communication is needed to overcome the neglect of mathematical modeling in plant health management and to develop a general appreciation for the many situations in which it can be useful. Clearly this process must begin in the lecture hall, but just as clearly it must not end there. Both the inclusion of mathematical models in developing research strategy and the clearer, more concise, and more accurate presentation of mathematical models would improve communication between modelers and other scientists.

It is also important to extend the present potential of mathematical models in plant disease management. Better modeling methods and a coherent theory of plant disease epidemiology would help to achieve this. As discussed above, improved techniques are particularly needed (and may soon be devised) for validation and week-to-week prediction. In contrast, while the development of a strong theoretical foundation through the discovery of fundamental principles (e.g., analogous to Newton's laws) would provide a base from which nearly all else could be derived or predicted, such discoveries do not appear imminent.

Finally, we note that much information already exists that can help in extending the utility of mathematical models in plant health management. For example, in this chapter we have drawn on ideas, opinions, and techniques from a broad range of ecologically oriented disciplines. This emphasizes our belief that greater interdisciplinary communication in the future can only improve our use and knowledge of the tools of mathematical modeling.

ACKNOWLEDGMENTS

We thank Bill Fry and Ken Minoque, Department of Plant Pathology; Si Levin, Department of Ecology and Systematics; Dave Onstad and Chris Shoemaker, Department of Environmental Engineering; and Bob Seem, New York State Agricultural Experiment Station, all of Cornell University; and Carl Walters, Institute of Resource Ecology, University of British Columbia, for reviewing the manuscript. R.A.F. acknowledges partial support from the Environmental Protection Agency under grant CR-806227-020. The contents do not necessarily reflect the views and policies of the Environmental Protection Agency.

LITERATURE CITED

1. Boulding, K. E. 1980. Science: Our common heritage. Science 207:831-836.
2. Brockington, N. R. 1972. An agricultural research scientist's point of view. Pages 361–365 in: Mathematical Models in Ecology. J. N. R. Jeffers, ed. Blackwell, Oxford. 398 pp.
3. Clark, W. C., and Holling, C. S. 1979. Process models, equilibrium structures, and population dynamics: On the formulation and testing of realistic theory in ecology. Fortschr. Zool. 25:29-52.
4. Conway, G. R. 1977. Mathematical models in applied ecology. Nature 269:291-297.
5. Fleming, R. A. 1980. Selection pressures and plant pathogens: Robustness of the model. Phytopathology 70:175-178, 71:268.
6. Fleming, R. A. 1980. The potential for control of cereal rust by natural enemies. Theor. Pop. Biol. 18:374-395.
7. Gilbert, N., Gutierrez, A. P., Frazer, B. D., and Jones, R. E. 1976. Ecological Relationships. W. H. Freeman, Reading, U.K. 157 pp.
8. Holling, C. S., ed. 1978. Adaptive Environmental Assessment and Management. John Wiley & Sons, Chichester. 377 pp.
9. Jeffers, J. N. R. 1972. The challenge of modern mathematics to the ecologist. Pages 1–11 in: Mathematical Models in Ecology. J. N. R. Jeffers, ed. Blackwell, Oxford. 398 pp.
10. Kranz, J. 1974. Introduction. Pages 1–6 in: Epidemics of Plant Diseases: Mathematical Analysis and Modeling. J. Kranz, ed. Springer-Verlag, Berlin. 170 pp.
11. Levin, S. A. 1980. Mathematics, ecology, and ornithology. Auk 97:422-425.
12. Levins, R., and Wilson, M. 1980. Ecological theory and pest management. Ann. Rev. Entomol. 25:287-308.
13. Norton, G. A. 1977. Background to agricultural pest management modelling.

Pages 161–176 in: Conference on Pest Management. G. A. Norton and C. S. Holling, eds. Pergamon, Oxford. 352 pp.

14. Pielou, E. C. 1981. The usefulness of ecological models: A stock-taking. Q. Rev. Biol. 56:17-31.
15. Ruesink, W. G. 1976. Status of the systems approach to pest management. Annu. Rev. Entomol. 21:27-44.
16. Rutter, A. J. 1972. An ecologist's point of view. Pages 375–380 in: Mathematical Models in Ecology. J. N. R. Jeffers, ed. Blackwell, Oxford. 398 pp.
17. Schoener, T. W. 1976. Alternatives to Lotka-Volterra competition: Models of intermediate complexity. Theor. Pop. Biol. 10:309-333.
18. Shoemaker, C. A. 1980. The role of systems analysis in integrated pest management. Pages 25–49 in: New Technology of Pest Control. C. B. Huffaker, ed. John Wiley & Sons, New York. 500 pp.
19. Shoemaker, C. A. 1981. Applications of dynamic programming and other optimization methods in pest management. IEEE Trans. Automat. Control ac-26:1125-1132.
20. Solomon, D. L. 1979. On a paradigm for mathematical modeling. Pages 231–250 in: Contemporary Quantitative Ecology and Related Ecometrics. G. P. Patil and M. L. Rosenzweig, eds. Int. Co-operative, Fairland, MD. 695 pp.
21. Teng, P. S. 1981. Validation of computer models of plant disease epidemics: A review of philosophy and methodology. Z. Pflanzenkr. Pflanzenschutz 88:49-63.
22. Weigert, R. G. 1975. Simulation models of ecosystems. Annu. Rev. Ecol. Syst. 6:311-338.
23. Welch, S. M., Croft, B. A., and Michels, M. F. 1981. Validation of pest management models. Environ. Entomol. 10:425-432.
24. Wiens, J. A. 1980. Theory and observation in modern ornithology: A forum. Auk 97:409.
25. Yarwood, C. E., and Sylvester, E. S. 1959. The half-life concept of longevity of plant pathogens. Plant Dis. Rep. 43:125-128.
26. Zadoks, J. C. 1972. Methodology of epidemiological research. Annu. Rev. Phytopathol. 10:253-276.

CHAPTER 34

POPULATION GENETICS AND EVOLUTION OF HOST-PARASITE INTERACTIONS

CLAYTON PERSON and BARBARA CHRIST

As mentioned many years ago by Hudson and Richens (10), in their review of Soviet genetics during the influence of Lysenko, scientific hypotheses can be divided into two general categories according to whether they are verifiable or not verifiable, and those that are not verifiable can be further subdivided according to whether or not there is any likelihood of verification. Hypotheses for which there appeared to be no likelihood of verification were described as being "essentially unverifiable" (10).

It is our opinion that many of the biological hypotheses relating to evolution, and particularly those that deal with long-term phylogenetic processes, are essentially unverifiable. To take a single example, many textbooks present the hypothesis that life on this planet arose spontaneously out of nonliving matter, and many experiments have been done to demonstrate the plausibility of this hypothesis. But even if it were found possible through artificial synthesis in the laboratory to generate a living cell, the question "How did this occur thousands of millions of years ago in a totally lifeless environment?" would remain unanswered. It does not follow that hypotheses of this kind should not be formulated. It is only necessary to recognize them for what they are, viz., interesting explanations whose validity will forever remain in doubt.

MACROEVOLUTION

For plant pathologists and parasitologists, one of the key events in evolution was the adoption of the parasitic mode of existence by certain species. We should like to begin our discussion of host-parasite interaction by presenting an essentially unverifiable hypothesis that is intended to explain this event. In so doing we follow de Bary's (3) definition of symbiosis as the living together of organisms of two different species, and we regard parasitism as a form of symbiosis for which living together was an essential prerequisite.

Let us call the two species that live together "A" and "B," and let us imagine that the association together is initially without effect on either species. It would follow that in terms of natural selection it is of no selective advantage to either A or B to maintain the association. But if we now imagine that the association is selectively advantageous to one of the two species, say it is species A, selection will then discriminate against those of species A who do not associate with B and will favor those who do. Eventually, and providing the selective differential is maintained, an end point will be reached at which all members of species A are found to be in association with species B. From this point onward we can imagine that selection within species A will continue to operate in such a way as to maximize the reproductive benefit enjoyed through its association with B.

The hypothesis to this point requires only that one of the two species derive a reproductive advantage for a symbiosis to be initiated. When we now consider the other species, species B, there are clearly only three possibilities for the effect of the association on its reproductivity: 1) the reproductivity of members of B can also be enhanced, in which case all members of species B who associate with A will be selectively favored, a process that will lead to all members of B (as well as all members of A) being involved in the association; 2) the reproductivity of members of species B can remain quite unaffected by the association, and selection in species B, tending to affect the likelihood of the association being continued, will not occur; and 3) the reproductivity of those members of species B who are associated with A can be reduced, in which case natural selection in species B will favor those members whose reproductivities are least reduced, either through escaping the association or through reproducing in spite of it.

These hypothetical events are summarized in Fig. 1, where it is shown that the three possibilities just outlined fall into three familiar categories: mutualism, commensalism, and parasitism.

The hypothesis when extended further would predict that, for mutualism, selection would operate in both species to favor greater reciprocal benefit from the association as well as its continuation. As mentioned by Person (15), continuation of the association would be assured when the two species no longer reproduce independently of one another. With the two associated organisms reproducing as though they were a single unit, selection would act on the association as though it were a single biological species, and the mutualism would be further perfected. This hypothetical sequence of events would apply to the lichens and, providing the Margulis (12) hypothesis is accepted, to the existence of mitochondria-containing organisms as well.

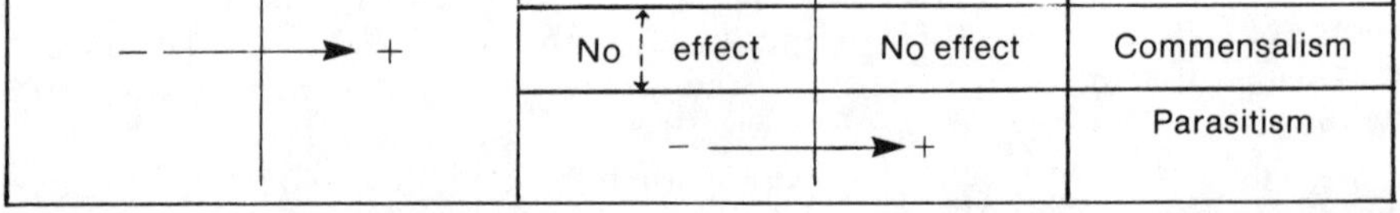

Relationship of				
Species "A" with "B"		Species "B" with "A"		
Not associated	Associated	Associated	Not associated	Classification
		+ ◄——	—— –	Mutualism
– ——	——► +	No ↕ effect	No effect	Commensalism
		– ——	——► +	Parasitism

Fig. 1. Outline of hypothetical events that would lead to three different kinds of symbiotic association. Plus and minus signs denote effect of association or nonassociation on reproductivity. Direction taken by natural selection is indicated by solid arrows; instability of commensalism is indicated by broken arrows.

For commensalism, where the reproductivity of one species is held to be totally unaffected by the continuing presence of the other, the expectation is that the association would be unstable. Any selective event in the benefiting species would have the possibility of affecting the reproductivity of the other species and, in consequence, of initiating selection in the direction of either mutualism or parasitism.

For parasitism, where continued association results in increased reproductivity of one and decreased reproductivity of the other organism, the directions taken by natural selection in the two associated species would be diametrically opposed. Selection would favor the parasite-free host on the one hand and greater parasitic efficiency on the other. Any selective event occurring in either of the two interacting species would initiate a selective response in the other (14).

As mentioned earlier, we regard hypotheses of the kind just outlined as being essentially unverifiable. Although they can lead to experimentation that would add plausibility to the hypothesis, they do not lead to experimentation that would result in rejection of the hypothesis.

The existence of complicated life cycles involving two or more hosts is another problem of long-standing interest to plant pathologists and parasitologists. It seems to us that any explanatory hypothesis that is based on natural selection would logically begin with a parasite, or potential parasite, whose life cycle includes two or more reproductive events. If we imagine, for example, an uncomplicated life cycle in which vegetative as well as sexual reproduction occurs, there would be an opportunity for competition between those members of the species that reduce their numbers of intraspecific competitors by moving one of their two reproductive phases to a new environment and those other members of the species that do not. In general, the mutant that is able to reproduce in the new environment would not be well adapted to it. But for an already parasitic species, the new environment may well be one that is not previously occupied by organisms of a different species and, in such a case, the mutant would be able to reproduce under conditions in which both inter- and intraspecific competition were absent, or at least minimal. Providing the mutant were able to return to the original environment to complete its life cycle, it would have opened the door to a new and previously untapped energy source for which there are few or no competitors. Selection would then discriminate between those members of the species that derive their energy entirely from a single environment and those that place their energy requirements partly in one environment and partly in another. We think it logical to conclude that, in some cases at least, the variant that spreads its energy requirements over two environments will enjoy a reproductive advantage over those other nonmutant members of the species that do not. Although we would regard this hypothesis as essentially unverifiable, we think it provides a plausible explanation of the fact that life cycles of parasites often involve two or more different host species.

MICROEVOLUTION

The term microevolution, as we intend to use it here, will refer to recent evolutionary events that have been directly observed and for which there exists a record of factual information that is known to be reliable. The selection of melanic forms of *Biston betularia* that has occurred during the past 130 years in the heavily industrialized areas of Britain is perhaps the best known example of microevolution. Documented examples of microevolution of more direct interest to plant pathologists would include the development (in treated fungal populations) of resistance to fungicides and, of course, the adaptation that leads (after a resistant cultivar has been grown) to the breakdown of resistance (15).

For the majority of cases, the breakdown of resistance can be attributed to the use of "major" resistance genes (R-genes) whose initial effect was to cause a striking reduction in the size of the pathogen population. The resistant cultivar when introduced also served as an efficient selective screen for the initially rare variants in the pathogen population that possessed a gene for virulence (V-gene) that "matched" the R-gene of the host. Continued selection in the resistance environment resulting in increasing numbers of "matching" pathogens accounted for the eventual breakdown of the resistance (11). According to Person (13) it also accounts for the origin of gene-for-gene relationships.

GENE-FOR-GENE RELATIONSHIPS

Flor's hypothesis of gene-for-gene relationships (6–8) was based on the finding that for crosses of flax cultivars that gave one-, two- or three-factor segregations in F_2 the appropriate crosses among flax-rust races also gave (respectively) one-, two-, or three-factor segregations. He found that resistance and avirulence segregated as dominant characteristics, i.e., that segregation at a single genetic locus in either the host or the parasite would generate two phenotypes. A single gene-for-gene relationship would therefore involve two phenotypes (resistant and susceptible) of the host and interaction of these with two phenotypes (virulent and avirulent) of the pathogen. The four phenotypes when brought together in all combinations produce the familiar "quadratic check," which is shown in Fig. 2a. Examination of the quadratic check reveals that resistance is expressed only when the dominant alleles for resistance and avirulence interact with one another. In other interactions, neither the dominant allele for resistance, alone, nor the dominant allele for avirulence, alone, will prevent compatibility. The dominant A- and R-alleles must therefore be regarded as "conditional" genes, since their phenotypic expression cannot be predicted unless it is known whether or not they are interacting with one another (16).

There is a remarkable similarity between gene-for-gene and antigen-antibody

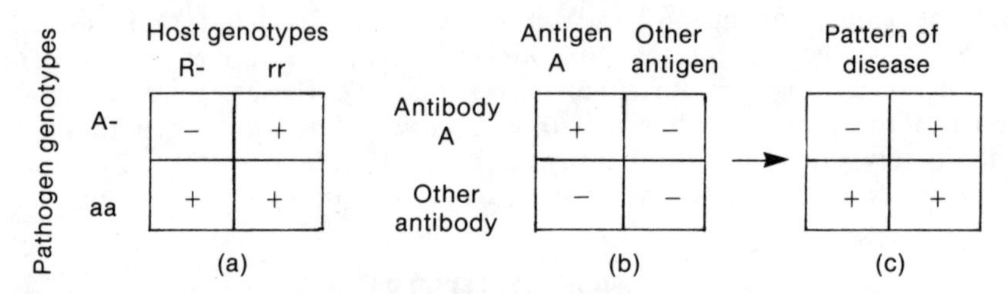

Fig. 2. Similarity of gene-for-gene and antigen-antibody interactions. **2a,** the disease pattern of the "quadratic check," where the specific reaction involves dominant alleles R (resistance) of the host and A (avirulence) of the pathogen. **2b,** interaction between antigen A and antibody A, with which it specifically interacts; here it is assumed that the "other" antigen and antibody are not mutually specific. **2c,** the "quadratic check" of disease development that would be generated by the interactions of Fig. 2b.

interaction. This is illustrated in Fig. 2b in which one of the cells of the antigen-antibody matrix also represents the specific interaction (designated by the plus sign) that takes place only when the antigen interacts with its "matching" antibody. Fig. 2c shows that when this matrix of interactions is expressed in terms of disease, the resulting pattern of disease development is identical with that of the gene-for-gene relationship. In both situations, the specific interaction of the quadratic check leads to inhibition or failure of disease development, and in both the development of disease is facilitated when specific interaction does not occur.

It will be observed that recognition of antigen-antibody specificity does not depend on genetic information, and Person (13) has pointed out that recognition of gene-for-gene (more accurately, "allele-for-allele") interaction can also be achieved even though genetic information is lacking. It will be evident also that a single quadratic check, together with knowledge of host and pathogen genotypes, provides all that is needed for a complete description of a gene-for-gene relationship. We do not understand the Vanderplank (25) argument that this is not possible.

There are other interesting comparisons between gene-for-gene and antigen-antibody interactions. The host's own antigens are genetically determined. Other antigens not produced by the host are recognized as "foreign," and these elicit the formation of specific antibodies. Parasites that succeed on particular hosts often do so because they form antigens identical with those of the host, a phenomenon termed "molecular mimicry" by Damian (2). Genetic polymorphism for antigen formation is also a common phenomenon. As pointed out by Snell (22), each existing host antigen serves as a protective umbrella for the fraction of the parasite population that happens to form the same antigen, and each new host antigen, entering the system via new mutation in the host population, enables its possessors to recognize as foreign *all* members of the parasite population. The new mutant allele, being selectively advantageous, is added to and becomes part of the genetic polymorphism of the host. The selective advantage of the new allele is progressively decreased as the pathogen adapts to its presence and forms "mimic" antigens. But during the process just described, the polymorphism has expanded to include the new allele and the new antigen whose formation it directs. The parallel with genetic resistance in plants lies in the fact that genetic polymorphism of R-genes has been reported for a number of pathogenic systems (21). Where there is a genetically heterogeneous host population, each of the several R-genes that are present confers resistance to all pathogens that do not have the matching allele for virulence. At the same time, each R-gene (or resistance genotype) can serve as a protective umbrella for each member of the pathogen population that does have the matching a-allele (or matching virulence genotype). As well, each R-gene that is entirely new to the host-pathogen system will be immediately effective against all existing pathogens and, through selection, will be added to the system. Snell's explanation of genetic polymorphism for host antigens is therefore applicable, and with minimal modification, to R-gene polymorphism in plants. Protein polymorphism in relation to the genetic interaction between host and parasite has been discussed by Clarke (1).

Another similarity becomes evident when gene-for-gene and antigen-antibody interactions are considered at the "systems" level. Where the host forms a number of different antigens, these must all be specifically matched if the infecting parasite is to successfully avoid recognition by the host. And where the plant host possesses a number of different R-genes, all of these must also interact with matching virulence alleles if the infecting pathogen is to succeed. In both cases, the success of the parasite or the pathogen leads to the conclusion that the host antigens or R-genes (as the case may be) have *all* been matched. Thus, where the constitution of the host is known, the success of the parasite or pathogen leads to definite conclusions concerning its antigens or v-alleles. On the other hand, the failure of disease development, although attributable to the failure of complete matching, is not greatly informative since it does not precisely indicate where the matching failed.

Finally it should be noted that as systems expand to include more units of interaction (antigen-antibody or gene-for-gene), the likelihood of "matching" through random contact is progressively reduced (20).

There is one fundamental difference between animals and plants. Animals are capable of forming acquired antibodies, whereas plants are not. This capacity on the part of the animals enables both individuals and populations of individuals to increase their resistance to disease without undergoing genetic change. Except for the formation of phytoalexins, and these are generally nonspecific, there appears to be no comparable phenomenon in the plant world. The changes in resistance and patterns of resistance that are achieved in the animal world through formation of antibodies are accomplished in the plant world through reproduction and formation of new genotypes. In either case, the pathogen is confronted with altered resistances for which the matching genotypes must be produced by natural selection. The indication from this is that plant and animal pathologists who are interested in epidemiology could benefit from each other's accumulated knowledge in this area of common interest.

MAJOR-GENE RESISTANCE

In agricultural crops, resistance to disease that is governed by major genes has often broken down. Recent interest has focused on polygenically controlled resistance, which is thought to be less easily overcome by the pathogen and therefore more stable. However, it is a matter of record that major genes for resistance can be obtained from natural (and presumably stable) populations of wild relatives of agricultural crops. The genes for resistance to oat crown rust obtained from natural populations of *Avena sterilis* represent just one of several such examples, from which it may be inferred that major genes do have a role to play in disease systems that are stable. It has to be admitted that very little is known about the mechanisms that contribute to the stability of natural systems, and it has to be recognized that the breakdown of major-gene resistance may be a direct result of knowing so little about the contribution that may be made by major genes to the processes that result in stability.

It has been shown mathematically that under certain assumed conditions the alleles involved in a gene-for-gene relationship will engage in a stable-limit cycle (5). An important assumption made by Fleming (5) and by others (see Fleming [5] for references) is that alleles for resistance and virulence (R- and a-alleles) decline in frequency when they are no longer needed. The same assumption had been made earlier by Person (14) in a speculative paper on R-gene polymorphism and by Vanderplank (23), who used the term "stabilizing selection" when referring to the selection that operates against unneeded genes. The evidence for declining frequencies of unneeded alleles appears to be inconclusive at the present time. Yet the theoretical studies make it quite evident that if stabilizing selection sensu Vanderplank does occur, it would serve as an important factor contributing to the stability of major-gene resistance. For example it could be an important contributor to the stability of multiline resistance (9), and it would offer the possibility of using and reusing R-genes in a continuous cycle. In order to determine whether any of these possibilities is real, it will be necessary to learn a good deal more about the genetic structure of natural populations in which the disease relationships appear to be relatively stable and, in particular, about the effect of selection on "unneeded genes."

It should be noted that the term "stabilizing selection" also denotes the kind of selection (operating at the two extremes of a normal frequency distribution) that maintains stability for polygenically determined characteristics. Because the term has two quite different usages, it should therefore be used unambiguously. It should also be noted that stabilizing selection sensu Vanderplank is not a process in which the frequency of an allele passively declines simply because it is no longer needed. It is more correct in our opinion to view the process as one in which the allele in declining frequency is being actively replaced by a needed allele that has greater fitness. Further, as pointed out by Person et al (17), the rate at which the replacement occurs is determined by the magnitude of the fitness differential that operates in favor of the incoming allele, and this in turn is a function of the proportion of total contacts of host and pathogen that involve "matching" of A- and R-alleles. The distinction made by Vanderplank (24) between "weak" and "strong" genes was based on an incomplete assessment of the factors involved.

POLYGENES AND HOST-PARASITE INTERACTION

When the expression of the phenotype being studied involves the orderly integration of several or many physiological processes (e.g., rates of growth and development; reproductivity; crop yield; meat, milk, or egg production; etc.), the analysis of recorded data usually reveals that the expression is variable, that the variability is continuous, and that the genetic component of the variability is attributable (in part at least) to the action of genes at several or many loci, that is, to polygenes.

Following successful infection, the pathogen must establish and maintain the supply of nutrient that supports it while it undergoes its growth, morphogenesis, and reproduction. A component of pathogenicity that is attributable to polygenes is therefore not unexpected.

There are relatively few reports of polygenically determined pathogenicity. For *Ustilago hordei*, it is known that pathogenicity involves both major genes and minor genes and that the minor genes can exert profound effects on the expression of the major genes (18).

There are many reports of polygenically determined resistance. Perhaps the best-studied example is the resistance of wheat to *Puccinia glumarum*, which Röbbelen and Sharp (19) have reviewed. But to the time of writing, there appears to be no report in which polygenes of *both* host and pathogen have been studied. The role of polygenes in host-parasite interaction is therefore unknown and therefore a subject of speculation. It is widely believed that polygenic resistance will be relatively more stable than major-gene resistance (4,20,23), but there is no consensus as to why this should be the case (18). This is obviously another area of research that deserves active and intensive study.

Johnson (11) pointed out 20 years ago that evolution in the rusts was "man-guided," and there are relatively few today who would disagree with the thesis that humans are guiding the microevolution of pathogens of many crops. But the basic question of how to manage the microevolution of pathogens to our best advantage remains unanswered. There are many problems yet to be solved!

LITERATURE CITED

1. Clarke, B. C. 1976. The ecological genetics of host-parasite relationships. Pages 87–103 in: Symp. Br. Soc. Parasitol. Vol. 14. A. E. R. Taylor and R. Muller, eds. Blackwell Scientific Publ., Oxford.
2. Damian, R. T. 1964. Molecular mimicry: Antigen sharing by parasite and host and its consequences. Am. Nat. 98:129-149.
3. De Bary, A. 1879. Die Eischeinung der Symbiose. Cassel, Strassburg.
4. Driver, C. M. 1962. Breeding for disease resistance. Scott. Plant Breed. Stn. Rep. 1-11.
5. Fleming, R. A. 1980. Selection pressures and plant pathogens: Robustness of the model. Phytopathology 70:175-184.
6. Flor, H. H. 1942. Inheritance of pathogenicity in *Melampsora lini.* Phytopathology 32:653-669.
7. Flor, H. H. 1947. Inheritance of reaction to rust in flax. J. Agric. Res. 74:241-262.
8. Flor, H. H. 1955. Host-parasite interaction in flax rust—Its genetics and other implications. Phytopathology 45:680-685.
9. Groth, J. V., and Person, C. O. 1977. Genetic interdependence of host and parasite in epidemics. Ann. N.Y. Acad. Sci. 287:97-106.
10. Hudson, P. S., and Richens, R. H. 1946. The new genetics in the Soviet Union. Imperial Bureau of Plant Breeding and Genetics, Cambridge. 88 pp.
11. Johnson, T. 1961. Man-guided evolution in plant rusts. Science 133:357-362.
12. Margulis, L. 1971. Symbiosis and evolution. Sci. Am. 225:48-57.
13. Person, C. 1959. Gene-for-gene relationships in host:parasite systems. Can. J. Bot. 37:1101-1130.
14. Person, C. 1966. Genetic polymorphism in parasitic systems. Nature 212:266-267.
15. Person, C. 1968. Genetic adjustment of fungi to their environment. Pages 395–415 in: The Fungi—An Advanced Treatise. Vol. 3. G. C. Ainsworth and A. S. Sussman, eds. Academic Press, New York. 738 pp.
16. Person, C., and Ebba, T. 1975. Genetics of fungal pathogens. Genetics 79:397-408.
17. Person, C., Groth, J. V., and Mylyk, O. M. 1976. Genetic change in host-parasite populations. Annu. Rev. Phytopathol. 14:177-188.
18. Person, C., Fleming, R., Cargeeg, L., and Christ, B. 1983. Present knowledge and theories concerning durable resistance. In: Proc. NATO/Adv. Study Inst. Symp. Durable Resistance in Crops, Bari, Italy. 1981. Plenum Press, New York. In press.
19. Röbbelen, G., and Sharp, E. L. 1978. Mode of Inheritance, Interaction and Application of Genes Conditioning Resistance to Yellow Rust. Verlag Paul Parey, Berlin. 88 pp.
20. Robinson, R. A. 1976. Plant Pathosystems. Springer-Verlag, New York. 184 pp.
21. Sidhu, G. S. 1975. Gene-for-gene relationships in plant parasitic systems. Sci. Prog., Oxf. 62:467-485.
22. Snell, G. D. 1968. The H-2 locus of the mouse: Observations and speculations concerning its comparative genetics and its polymorphism. Folia Biol. 14:335-358.
23. Vanderplank, J. E. 1968. Disease Resistance in Plants. Academic Press, New York. 206 pp.
24. Vanderplank, J. E. 1975. Principles of Plant Infection. Academic Press, New York. 217 pp.
25. Vanderplank, J. E. 1978. Genetic and Molecular Basis of Plant Pathogenesis. Springer-Verlag, New York. 167 pp.

CHAPTER 35

PLANT GROWTH MODELS AND PLANT DISEASE EPIDEMIOLOGY

DOUGLAS I. ROUSE

Throughout this chapter the term "plant growth model" will be used loosely to include models that dynamically relate the state of plant growth to environmental variables. The state of plant growth is quantified by dependent, or state, variables such as leaf, stem, fruit, tuber, and/or root biomass; leaf area; stem length; and root length. Environmental variables, also referred to as independent, driving, or forcing variables, may include almost any combination of physical, chemical, or biological components of the environment. In addition, dynamic plant growth models include rate variables that define how the state variables will change over time.

Plant growth models are simplified descriptions of plant growth representing statements of hypothesis, theory, or fact about the temporal behavior of plants or parts of plants at the individual or community level. The simplification inherent in plant growth models is facilitated by the use of mathematics and computer science since it is a quantitative understanding we desire, not a qualitative description. The necessity for simplification is twofold. First, we need to separate by exclusion from plant growth models relatively unimportant factors from factors of primary interest. Secondly, we must consciously recognize that our knowledge of the factors impinging on, or directly important to, the system is incomplete. Clearly, plant growth models (as is true of most models) are in a sense always wrong by virtue of their incompleteness or simplicity. Scientific method allows for this by requiring that the assumptions (including simplifications) be clearly and explicitly stated so that when the model does not appear to fit reality we know where to look for an explanation. Furthermore, a statement of assumptions allows us to take an experimental approach by indicating the specific experimental conditions that must be established to test or validate the model. The process of sorting through implicit and explicit assumptions can result in the identification of gaps in our knowledge, leading to the directed organization of experimental research activity rather than a piecemeal or haphazard approach to the acquisition of knowledge.

A clear understanding of the degree of knowledge we possess and its limitations allows the optimal practical application of knowledge. Plant growth models have a tremendous potential for practical application in agriculture and resource management by providing an aid to planning. This can occur in two ways. First, plant growth models can be used to examine various scenarios, to answer the question "What if?" For example, these scenarios or hypothetical situations may include particular types of growing seasons or particular fertilization regimes. Second, plant growth models can be used as predictive tools to assist in planning of marketing strategies or pest control strategies (forecasting). In the forecasting mode, models are fed data on actual past and present environmental and management variables. These data provide the basis for projections of future plant growth and yield.

Although the cynic may ridicule the formal mathematical plant growth model as incomplete and oversimplified in biological terms while being complicated and difficult to comprehend in mathematical terms, the preceding discussion suggests a substantial list of general reasons for developing and utilizing plant growth models. Models can: 1) provide a means for stating hypotheses in quantitative terms; 2) guide an experimental, or novel, approach or plan experiments while limiting the number of extraneous experiments; 3) clarify the current state of knowledge and identify research directions; 4) communicate knowledge clearly both in teaching and in research, and 5) aid in making decisions (planning or forecasting) important in crop management.

TYPES OF PLANT GROWTH MODELS

Thornley (28) utilized a common means of categorizing models by distinguishing between "empirical" and "mechanistic" plant growth models. Empirical models rely on the a posteriori quantitative summarization of data by means of statistical approaches such as regression analysis. This approach to modeling utilizes relationships established between variables at the time of data examination. Relationships are chosen for their mathematical simplicity and statistical appropriateness. When this type of model is utilized, there is no specific biological mechanism hypothesized a priori. Empirical models represent a statistical summary of what is observed.

Mechanistic models are defined as models specifically incorporating a priori knowledge or assumptions about the biological mechanisms that relate variables in the system. Statistical analysis is employed in conjunction with mechanistic models to test the null hypothesis that the model is inappropriate. Rejection of the null hypothesis does not imply that the a priori assumption of mechanisms was correct, only that the data were not significantly different from the proposed model. Although the term "validation" is used to describe the procedure of comparing model-generated information with independently derived data, in a real sense a mechanistic model is never fully validated.

Models with a mechanistic basis derived from knowledge of the underlying physiological processes of plant growth are referred to as physiological plant growth models. Physiological plant growth models may consist of a series of interconnected submodels that attempt to incorporate as fully as necessary the complex interactions of the system elements. Thus, a plant growth model may consist of a computer simulation linking together several computer subroutines that describe separate physiological processes, for example, assimilation, transpiration, respiration, and nutrient transport (37) in individual plant organs. Each of these submodels, in turn, may be made up of component models that eventually reduce at the lowest level to descriptive models. This type of plant growth model was reviewed by Loomis et al (20) and is referred to by them as explanatory models. An important feature of explanatory models is the interconnection via feedback loops of processes at multiple levels of organization. It is at this level of integration (i.e., making the connections between component models) that mistakes can be most easily made in the formulation of explanatory models.

Many different kinds of models describe various components of plant growth, as illustrated by the hierarchical nature of the submodels that make up explanatory plant growth simulators. There are, as well, many kinds of models, independently developed and reported in the literature, that can be grouped into a hierarchical scheme for defining plant growth models. Near the bottom of the hierarchy are models that describe specific physiological processes related to plant growth, e.g. photosynthesis or water uptake (8,36). Next are models that describe the growth of single plant organs such as leaves (29) or roots (4,14). Then there are whole plant models that attempt to quantify the growth of a single plant (28). Finally, there are true crop growth models that describe the growth of populations of plants (6,7,9,15,22). There are also microclimatological models that describe the variation in the physical environment around plant surfaces as affected by the plant or crop (11,34).

Over the last 15 years, plant scientists have shown increasing interest in various types of plant growth models, particularly physiological growth models of crops. This has been paralleled in plant pathology by an increased interest in modeling plant disease epidemics. What is the relationship between plant growth models and epidemic models?

A simplified conceptualization that illustrates the relationship between explanatory plant growth models and plant disease epidemic models is presented in Fig. 1. The focal point of this conceptualization is plant stress as it relates to plant growth and yield. It is based on the premise that most factors detrimental to maximum yield, whether biological (weeds, insects, or pathogens) or environmental (drought, heat, or nutrition), act by imposing measurable indications of physiological stress on the crop. Biological or environmental factors may cause plant stress in many ways: by reducing photosynthesis, nutrient uptake, and transpiration or interfering with water transport. The net result of plant stress regardless of cause can be observed by measuring its effect on plant growth and ultimately yield. The dashed line in Fig. 1 encompasses the components of the conceptualization that constitute most explanatory plant growth models. These plant growth models do not account for disease-related plant stress.

On the other hand, pathologists have developed a number of mathematical models describing pathogen population dynamics and disease progress which are dependent upon environmental factors. These are represented in Fig. 1 by the components of the conceptualization enclosed by the dotted line. These models may have the form of sophisticated computer simulations, but none of them have made use of explanatory plant growth models (25,35).

There has recently been an emphasis in plant pathology on yield loss modeling. Most yield loss models are empirical in their relationships between the quantity of signs or symptoms expressed by the host plant and final yield (16). This is represented in Fig. 1 by the thin solid line. Recently, there has been a growing interest by plant pathologists in mechanistic yield-loss models. These would be plant growth models that incorporate the impact of pests on specific physiological processes of plant growth and yield. Nevertheless, the epidemiologist has largely ignored the quantitative interactions between plant pathogen populations, disease, and host plant growth.

CHALLENGING PROBLEMS

The goal of each of the chapters in this text is to address challenging problems in plant health as related to the topic of that chapter. Two types of problems will be addressed in the remainder of this chapter: first, the problems associated with integrating the effects of plant disease into plant growth models and second, some of the future uses of plant growth models in plant pathology. This second question will be answered by identifying a series of plant pathological problems that require some degree of plant growth modeling to solve. The conceptualization in Fig. 1 indicates that a major problem confronting plant scientists is how to integrate pathogen-induced stress into plant growth models. An empirical approach could be taken by collecting data relating plant biomass, leaf area, or other plant growth descriptors to various amounts of disease and environmental factors over time. Statistical analysis of such data sets could lead to empirical plant growth models that incorporate disease as an independent variable. However, empirical models are limited in their statistical acceptability by the error associated with the estimation of each variable and by the correlation between variables. This results in the need for very large experiments when a large number of variables are being incorporated into a model. Thus, empirical plant growth models tend to include relatively small numbers of variables. This type of model also is limited in predictive value by the set of experimental conditions under which the data were collected since it is not known by what

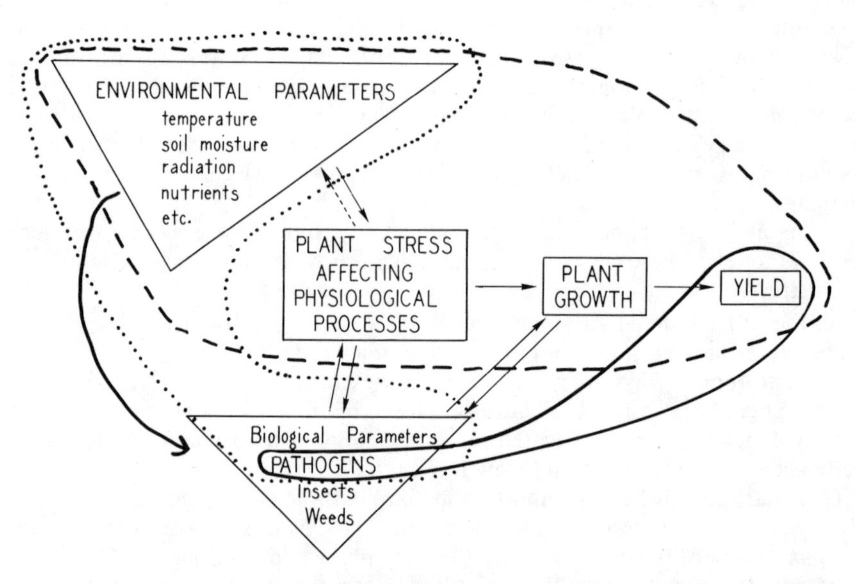

Fig. 1. Conceptualization of the relationships between current explanatory plant growth models (-----), disease dynamics simulators (····), and yield loss models (——).

mechanisms the variables are related. Mechanistic models, particularly explanatory models, are not as limited as the empirical models by constraints in number of variables. Although there may be a large number of variables in the whole model, only a small number of variables are simultaneously involved in any one statistical analysis. This is because subsets of the data base are analyzed for each submodel. Also, the final validation involves the model as a whole and an independent data set. There is, however, always the danger in large explanatory models that a number of errors occur that cancel each other and are, therefore, not evident in the models' output. Another possibility is that some components of the model have no effect on the outcome. These problems are dealt with by validating the component submodels within the main model and by sensitivity analysis to examine the contributing effects each of the variables has on the output. If the component mechanisms are correctly understood and incorporated, an explanatory model should have reliable predictive power within the constraints of its assumptions.

This approach to model building may provide a more efficient use of funds and research effort than does empirical modeling. Each of the submodel components in an explanatory model represents a unique contribution to an understanding of the specific physiological mechanisms governing plant growth. In other words, although a very large number of experimental results are incorporated into an explanatory model, the results of individual experiments are not invalidated by a poor fit of the model to an independent data set. In contrast, a large empirical model may yield less useful information if, for example, an important variable was not included, or, conversely, if a number of variables were confounded with a single relatively important variable. This speaks against haphazardly collecting data on a very large number of variables (with the concomitant expenditure of labor) and after the fact turning the multivariate statistical mill wheel in hopes of grinding out a clear result.

A physiological plant growth modeling approach that includes the impact of pathogens requires a quantitative understanding of the physiological effects of pathogenesis on plant growth. The effect of the pathogen on transpiration, respiration, and photosynthesis must be known if the plant growth model is driven by subroutines at that level. It is essential that this knowledge be quantitative. For example, although it is well known that powdery mildews and rusts increase host respiration and decrease photosynthesis after successful infection and colonization (24), little quantitative data relates the percentage of visual signs or symptoms to the degree of change in these physiological processes, particularly on a temporal basis.

The incorporation into physiological plant growth models of pathogen-induced effects on plant growth is further hindered by the physiological complexity of pathogenesis. A pathogen may produce adverse effects on the host by a combination of mechanisms such as acting as a nutrient sink, producing toxins, altering the plant hormone balance, and/or enzymatically degrading host tissue. There are a multitude of unresolved questions as to how the mechanisms of pathogenesis at this level relate to the general physiological processes of assimilation, transpiration, respiration, or translocation.

In contrast to pathogens, many insect pests, as a first approximation, can be assumed to affect plant growth simply by the removal of host tissue by ingestion. Thus, it is not surprising that more research has been done to incorporate into plant growth models the impact of insect pests than that of diseases. Gutierrez et al (12) recently reviewed these models in the context of applications of the systems approach to pest control. These models have not adapted the degree of explanatory nature that other physiological plant growth models have because their purpose is to define the role of defoliating insects on plant growth. Thus, while borrowing specific subroutines from physiological plant growth models, one can simplify the aspects of plant growth not directly affected by the insect pest. Some recent studies relate the effect of disease on transpiration, respiration, assimilation, or translocation to plant growth in a way that might be useful to a plant growth model. A couple of specific examples from recent literature will follow. Siddiqui (26) measured some effects of rust infection of sunflower on diffusive resistance and CO_2 assimilation. Plant growth parameters including total dry weight, green leaf area, and total leaf area were measured as affected by rust. Unfortunately, only two levels of rust infection were used in the study. Siddiqui's main objective was to examine the interaction of differing degrees of moisture stress and rust infection on the above-mentioned physiological processes and plant growth. Although a fully quantitative approach was still not used in this study, it contains useful information for a plant growth modeler because it links the effects of water stress and pathogen-induced stress. Considerable work with another group of pathogens, the vascular wilts, is related to water stress. Some of this work has directly related the effects of the pathogen on transpiration to plant growth factors (13).

The interaction of multiple pests with respect to their effect on plant growth and yield has also been studied via measurements of physiological processes in infected plants. Van der Wal and Cowan (32) quantitatively studied the influence of *Puccinia recondita* and *Septoria nodorum* on plant growth and transpiration. They measured the rate of dry weight accumulation, increase in leaf area, and change in transpiration. They found, not unexpectedly, that each of the pathogens, individually and combined, had unique effects on the factors measured. Specifically, they observed that the effect of both pathogens together on plant growth was greater than the sum of the effects of each pathogen separately.

A recent example of research related to plant growth modeling that successfully crossed traditional disciplinary lines involved the cooperation of an agronomist, agricultural engineer, entomologist, and plant pathologist in a study of the effect of Cercospora leaf spot *and* insect defoliation on peanut plant growth and yield (3). Assimilation of $^{14}CO_2$ was measured in the field as influenced by both the pathogen and insect. The correlation between pest effects on assimilation and plant growth depended on position in the canopy. The authors point out that this could have considerable effect on yield and pesticide control programs. The justification given for their study was that it would contribute toward the development of "... crop-pest models which can be used to evaluate and determine pest management decisions." In the future many more interdisciplinary studies will be required to bridge the gap between explanatory crop growth models and the impact of plant diseases and other pests.

A major problem that must be dealt with to integrate pathogen-induced stresses into plant growth models is the lack of communication between certain plant pathologists, specifically epidemiologists, and plant physiologists. On the one hand, the plant physiologist tends to have an interest in qualitative studies of mechanism and is, thus, not interested in going back to well-established procedures to work out a quantitative study. On the other hand, epidemiologists have not always adequately considered the importance of a knowledge of physiological mechanism and the role it could play in an understanding of disease progression at the population level.

EPIDEMIOLOGICAL PROBLEMS RELATED TO PLANT GROWTH MODELING

Host Growth Effects

The first obvious problem confronting an epidemiologist studying disease progression is how to account for plant growth when disease severity is expressed as percentage of total susceptible plant tissue affected by the pathogen. This is illustrated by the logistic equation utilized by Vanderplank (31) relating percent disease to time. The model does not separate the two ways that the percent disease could change with time. Disease could increase as a result of growth and sporulation of the pathogen on the host, leading to secondary cycles of infection, or as a result of growth of the host. Host growth is almost always ignored when disease progression data are collected and analyzed. The problem of accounting for host growth effects on percent disease could be resolved by expressing empirically the logistic equation as

$$\frac{dY}{dt} = rY\left(1 - \frac{Y}{K}\right)$$

where r = the apparent infection rate and K = carrying capacity. The parameter K could then be made a function of time to describe plant growth. Analytical solutions have been worked out for certain specific instances for variable K (30). A simple empirical plant growth model would suffice to estimate a function for K.

Vanderplank (31) proposed a very simple a priori model in rate equation form to account for the effect of plant growth on the apparent infection rate. The model asserted that the instantaneous rate of change of plant tissue (susceptible to disease) was proportional to the current amount of susceptible tissue. The model has neither an empirical nor a mechanistic basis, but may represent a useful first approximation for epidemics progressing during periods of rapid foliage growth. There is room for progress to be made to incorporate the effect of plant growth into simple theoretical epidemic models such as the logistic. These models have a purpose different from that of the large-scale epidemic computer simulator model since they are used primarily for statistical analysis of data and not as explanatory models of pathogen and disease dynamics as influenced by environmental conditions.

An interesting use of a plant growth model was made in the first dynamic simulation model of soilborne disease (2). This simulation contained a submodel describing the growth of Douglas-fir seedling roots. The simulator calculated the number of contacts between *Fusarium* propagules and the roots to estimate disease severity.

Host/Environment Interaction

Although the state of plant health as a result of its interaction with the environment may exert a significant influence on pathogen population dynamics, the epidemiologist rarely accounts for this interaction. The majority of disease simulators make the assumption that the plant or crop is healthy except for the single disease being simulated. The most sophisticated epidemiological models have gone no further than to include plant growth as a single variable that increases according to an empirical relationship with time. They otherwise essentially treat the plant as an inert substrate for the pathogen. There are a number of environmental factors that affect the pathogen only indirectly through their effect on the plant. One example is the effect of soil water potential on spore production of foliar pathogens. Soil water potential influences leaf water potential and transpiration, which have been shown to directly affect sporulation of leaf rust (33) and powdery mildew (1). Another example is the effect that photosynthesis may have on sporulation. Cohen and Rotem (5) used dichlorophenyl-dimethyl urea to inhibit photosynthesis and observed a direct relationship between photosynthesis and amount of sporulation of several fungal plant pathogens tested. Those types of indirect environmental effects, or environment-host interaction effects, on plant pathogens remain to be dealt with in future simulation models.

Plant microclimatological models will be particularly useful to epidemiologists in the future in refining explanatory epidemiological simulators. For example, it is known that radiation can have an effect on various plant pathogens. Currently, several crop canopy models can be used to quantify the effect of canopy structure and leaf area on incident radiation reaching leaves at different levels within the canopy (18,23). In a broad, complete perspective, radiant energy is only one component of models describing the energy balance of leaves which, along with micrometeorological variables, depends on variables derived from physiological plant growth processes such as transpiration. The energy balance of a leaf may result in temperatures at the individual leaf surface 8–10 C greater or 6–8 C lower than ambient air temperature. This range of temperatures may occur within a single crop canopy at one time. Quantitative disease progression simulators of the future will benefit by relating leaf temperature derived from energy balance submodels to component epidemiological processes such as germination, penetration, colonization, sporulation, and dissemination.

As indicated for leaf temperature, one of the features of microclimatological data is the extreme variability observed. Current explanatory plant growth models and disease progression models invoke the law of large numbers to bypass this apparent variability. However, that approach assumes that mean response of one variable is related to the mean stimulus provided by another variable. The possibility should never be ruled out without testing that a threshold level of one variable is required to elicit a response from a second variable. In that case, mean alone will not be an adequate predictor unless the degree of variability is constant. In the future, stochastic elements will certainly be included in plant growth models and in disease progression models to account for variability. This will require more complete data sets that describe the frequency distribution of variables.

Life History Strategies

An understanding of the evolutionary life history strategies of plant pathogens and the coevolution of host and parasite may also be enhanced with the aid of disease dynamic models that incorporate elements of plant growth. Ecological theories such as the theory of r and K selection suggest that there are trade-offs between efficient energy and nutrient transfer and maximum short-term reproduction (27). Plant pathogens tend to be colonizers that exploit relatively uncrowded niches, yet there appears to be considerable variation in the reproductive potentials of pathogens even within the same group (e.g., the powdery mildews). Is this related to the amount of disease (in terms of physiological or morphological damage to the plant) that these pathogens induce? How does the efficiency of substrate utilization by the pathogen relate to pathogen and/or host fitness? With an adequate data base providing a guide, physiological plant growth models that integrate the effects of pathogens on plant growth would allow the examination of the effects of various life history strategies on host pathogen. This approach might even be placed in the context of a population genetics model.

Disease Assessment

Accurate disease assessment has been identified as a major problem confronting plant pathologists. The usual methods, involving disease assessment keys or diagrams, are imprecise due to the subjectivity of the researcher and are inaccurate since they measure only visual effects of the pathogen's presence. The imprecision of visual assessment methods may be reduced in the future by the use of video cameras or other remote sensing technology in combination with analog-to-digital conversion and computer software. The problem of accuracy, however, remains since our understanding of the quantitative relationship between signs and/or symptoms of disease and the actual physiological effect on the plant is minimal.

The epidemiologist can too easily forget that there is not necessarily a one-to-one correspondence between the size of the pathogen population as measured by spore counts, signs, or visual symptoms and the amount of disease measured as the degree of physiological and morphological damage. Figure 2 represents several possible relationships that might exist between actual disease and a rating scale of visual disease severity. For example, Boote et al (3) provide data relating CO_2 assimilation to a visual disease scale for Cercospora leaf spot with a similar shape to curve A of Fig. 2. In contrast, Mignucci and Boyer (21) observed a photosynthetic-to-visual severity relationship similar to curve B in Fig. 2 for powdery mildew on soybeans. Examples could probably be found that are similar to surve C in Fig. 2. However, assimilation is only one of several physiological processes that may be adversely affected by disease. Plant growth models that incorporate disease stresses provide a means of estimating the degree of overall impairment of normal physiological processes that may then be related to visual scales of disease severity. Such models could provide this type of analysis for a range of environmental conditions.

Effect of Disease Severity on Yield

A directly related problem confronting plant pathologists is to improve understanding of the relationship between disease severity and yield. Gaunt (10) expressed this point succinctly:

> Percent disease severity based on lesion size and associated chlorosis represents a more satisfactory parameter, but it still ignores some of the effects of disease on the ability of the plant to grow and develop its yield potential.

Numerous yield-loss models have been developed that empirically relate visual signs or symptoms of disease to yield (16). On occasion, researchers have obtained significantly different yield-loss relationships for the same disease. This may be the result of differences in the relationship between visual disease severity ratings and amount of actual disease caused by the interaction of disease with environmental stress factors. Explanatory plant growth models that incorporate disease-induced stress may explain differences in empirical yield-loss results since they will include interconnecting submodels accounting for the interactions of various stress factors on a temporal basis. Loomis and Adams (19) provide a recent critical discussion of the potential of dynamic physiological models for crop loss assessment.

Because of the difficulty in clearly identifying symptoms and assessing damage to plants caused by air pollutants, considerable quantitative research has been done on the effects of air pollutants on physiological processes that are then relatable directly to plant growth and yield. Explanatory plant growth models have been developed specifically to quantify yield loss caused by air pollutant stress (17).

Integration of Results

The integrated application of epidemiological research results with similar results from other pest-related disciplines to obtain practical integrated pest management is another major problem confronting plant pathologists. An

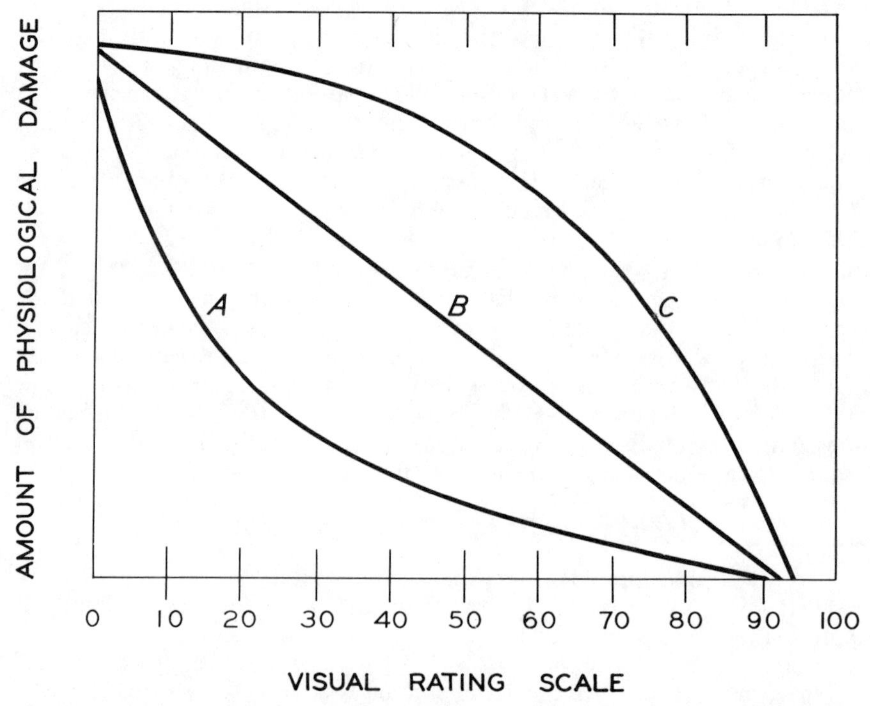

Fig. 2. Theoretical curves (A–C) relating amount of physiological damage caused by a pathogen to visual disease severity ratings.

integrated approach to optimizing crop production should cross traditional disciplinary lines to combine knowledge of the effects of environment, cultural practices, and pests on plant growth and resultant yield. A systems-oriented concept of plant growth and yield as depicted conceptually in Fig. 1 provides a framework for cross-disciplinary research leading to integrated pest management. Pest management models with a plant growth modeling basis have been developed (12). At least one model is being developed that reaches a level of integration including both insects and disease (A. P. Gutierrez, *personal communication*).

CONCLUSIONS

Disease progression is the result of the complex interaction between host, pathogen, and environment. The way to a more complete understanding of disease, disease progression, and its effects on yield is to integrate the efforts of the plant physiologist, disease physiologist, and epidemiologist. Dynamic explanatory disease models can provide a focus for research across these disciplinary lines while serving as a measure of the progress we have made toward our goal. The true disease progression model is the resultant of the interconnections at all levels between plant growth models and pathogen population dynamics models. The systems concept centering on plant stress, as it affects plant growth and yield, can be generalized to all pests. The application of this systems approach will enhance the implementation of integrated pest management.

LITERATURE CITED

1. Ayres, P. G. 1977. Effects of water potential of pea leaves on spore production by *Erysiphe pisi* (powdery mildew). Trans. Br. Mycol. Soc. 68:97-100.
2. Bloomberg, W. J. 1979. A model of damping-off and root rot of Douglas fir seedlings caused by *Fusarium oxysporum*. Phytopathology 69:74-81.
3. Boote, K. J., James, J. W., Smerage, G. H., Barfield, G. A., and Berger, R. D. 1980. Photosynthesis of peanut canopies as affected by leafspot and artificial defoliation. Agron. J. 72:247-252.
4. Brouwer, R., and de Wit, C. T. 1969. A simulation model of plant growth with special attention to root growth and its consequences. Pages 224-244 in: Root Growth. W. J. Whittington, ed. Butterworths, London. 450 pp.
5. Cohen, Y., and Rotem, J. 1970. The relationship of sporulation to photosynthesis in some obligatory and facultative parasites. Phytopathology 60:1600-1604.
6. Curry, R. B., Baker, C. H., and Streeter, J. G. 1975. SOYMOD. I. A dynamic simulator of soybean growth and development. Trans. ASAE 18:963-974.
7. Duncan, W. G. 1972. SIMCOT: A simulator of cotton growth and yield. Pages 115–118 in: Proc. Workshop on Tree Growth Dynamics and Modeling. Murphy et al, eds. Duke University, Durham, NC.
8. Duncan, W. G., Loomis, R. S., Williams, W. A., and Hanau, R. 1967. A model for simulating photosynthesis in plant communities. Hilgardia 4:181-205.
9. Fick, G. W., Loomis, R. S., and Williams, W. A. 1975. Sugar beet. Pages 327–355 in: Crop Physiology. L. T. Evans, ed. Cambridge Univ. Press, Cambridge, U.K.
10. Gaunt, R. E. 1981. Disease tolerance—An indicator of thresholds? Phytopathology 71:915-916.
11. Goudriaan, J. 1977. Crop Micrometeorology: A Simulation Study. Pudoc, Wageningen. 250 pp.
12. Gutierrez, A. P., Michele, D., Wang, Y., Curry, G. L., Smith, R., and Brown, L. G. 1980. The systems approach to research and decision making for cotton pest control. Pages 155–182 in: New Technology of Pest Control. C. B. Huffaker, ed. Wiley Interscience, New York. 155 pp.
13. Harrison, J. A. C. 1971. Transpiration in potato plants infected with *Verticillium* spp. Ann. Appl. Biol. 68:159-168.
14. Hayhoe, H. 1981. Analysis of a diffusion model for plant root growth and an

application to plant soil-water uptake. Soil Sci. 131:334-343.
15. Holt, D. A., Bula, R. J., Miles, G. E., Schreiber, M. M., and Peart, R. M. 1975. Environmental physiology, modeling and simulation of alfalfa growth. I. Conceptual development of SIMED. Res. Bull. 907. Purdue Agric. Exp. Stn., West Lafayette, IN. 26 pp.
16. James, W. C. 1974. Assessment of plant diseases and losses. Annu. Rev. Phytopathol. 12:27-48.
17. Kercher, J. R. 1980. Developing realistic crop loss models for air pollutant stress. Pages 90–97 in: Crop Loss Assessment: Proc. E. C. Stakman Commemorative Symposium. Misc. Publ. 7. Minn. Agric. Exp. Stn., St. Paul, MN.
18. Lemeur, R., and Blod, B. L. 1974. A critical review of light interception models for estimating the short wave radiation of plant communities. Agric. Meteorol. 14:255-286.
19. Loomis, R. S., and Adams, S. S. 1980. The potential of dynamic physiological models for crop loss assessment. Pages 112–117 in: Crop Loss Assessment: Proc. E. C. Stakman Commemorative Symposium. Misc. Publ. 7. Minn. Agric. Exp. Stn., St. Paul, MN.
20. Loomis, R. S., Rabbinge, R., and Ng, E. 1979. Explanatory models in crop physiology. Annu. Rev. Plant Physiol. 30:339-367.
21. Mignucci, J. S., and Boyer, J. S. 1979. Inhibition of photosynthesis and transpiration in soybean infected by *Microsphaera diffusa*. Phytopathology 69:227-230.
22. Morgan, J. M. 1976. A simulation model of the growth of the wheat plant. Ph.D. thesis. Macquarie Univ., North Ryde, N.S.W., Australia. 192 pp.
23. Ross, J. 1975. Radiation transfer in plant communities. Pages 13–56 in: Vegetation and the Atmosphere. Vol. I. Principles. J. L. Monteith, ed. Academic Press, New York. 278 pp.
24. Shaw, M. 1963. The physiology of host-parasite relations of the rusts. Annu. Rev. Phytopathol. 1:259-294.
25. Shrum, R. 1975. Simulation of wheat stripe rust (*Puccinia striiformis* West) using EPIDEMIC, a flexible plant disease simulator. Pages 1–41 in: Progress Rep. 347. Pa. Agric. Exp. Stn., University Park. 81 pp.
26. Siddiqui, M. Q. 1980. Some effects of rust infection and moisture stress on growth, diffusive resistance and distribution pattern of labelled assimilates in sunflower. Aust. J. Agric. Res. 31:719-726.
27. Stearns, S. C. 1976. Life history tactics: A review of the ideas. Q. Rev. Biol. 51:3-47.
28. Thornley, J. H. M. 1976. Mathematical Models in Plant Physiology. Academic Press, New York. 318 pp.
29. Thornley, J. H. M., Hurd, R. G., and Pooley, A. 1981. A model of growth of the fifth leaf of tomato. Ann. Bot. 48:327-340.
30. Turner, M. E., Blumenstein, B. A., and Sebaugh, J. L. 1969. A generalization of the logistic law of growth. Biometrics 25:577-580.
31. Vanderplank, J. E. 1963. Plant Diseases: Epidemics and Control. Academic Press, New York. 349 pp.
32. Van der Wal, A. F., and Cowan, M. C. 1974. An ecophysiological approach to crop losses, exemplified in the system wheat, leaf rust, and glume blotch. II. Development, growth, and transpiration of uninfected plants and plants infected with *Puccinia recondita* f. sp. *tritici* and/or *Septoria nodorum* in a climate chamber experiment. Neth. J. Plant Pathol. 80:192-214.
33. Van der Wal, A. F., Smeitink, H., and Maan, G. C. 1975. An ecophysiological approach to crop losses exemplified in the system wheat, leaf rust, and glume blotch. III. Effects of soil-water potential on development, growth, transpiration, symptoms, and spore production of leaf rust infected wheat. Neth. J. Plant Pathol. 81:1-13.
34. Waggoner, P. E. 1975. Micrometeorological models. Pages 205–228 in: Vegetation and the Atmosphere. Vol. I. Principles. J. L. Monteith, ed. Academic Press, New York. 278 pp.
35. Waggoner, P. E., Horsfall, J. G., and Luken, R. J. 1972. EPIMAY, a simulator of southern corn leaf blight. Bull. Conn. Agric. Exp. Stn., New Haven. 84 pp.
36. Wit, C. T. de, Brouwer, R., and Penning de Vries, F. W. T. 1970. The simulation of photosynthetic productivity. Pages 47–70 in: Proc. IBP/PP Technical Meeting, Trebon, Sept. 1969. Centre Agric. Publ. Doc., Wageningen.
37. Wit, C. T. de. 1976. Simulation of Assimulation, Respiration, and Transpiration of Crops. Simulation Monographs. Pudoc, Wageningen. 141 pp.

PART IV

Control of Biotic Pathogens

The day when chemical fungicides and insecticides would be relegated to the museum lay very far ahead, but if there was one line of attack which promised, in time, to outdistance all others, and to have an almost universal application against nearly all diseases, it was that of breeding disease-resistant varieties.

—E. C. Large, 1940

Many landmarks have been established in the realm of scientific agriculture during the past century. None of these rises more prominently on the plains of human endeavor than those that mark the forward progress in controlling plant diseases by chemicals. They stand as monuments to the research and ingenuity of many careful investigators.

—George L. McNew, 1958

The nurture of beneficial organisms within agricultural fields should become as central to crop production as the growth of crops themselves.

—Peter W. Price, 1981

CHAPTER 36

INTRODUCTION TO DISEASE CONTROL

WILLIAM E. FRY and B. G. TWEEDY

The challenge to suppress plant disease is more important during the latter part of the 20th century than ever before. The burgeoning world population and changing agricultural practices are largely responsible. Food supplies are now marginally adequate to meet needs (1,16), and, fortunately, world food production during the last three decades has matched increases in population (16). *However*, production of some important crops, such as grains in low-income, developing countries, is significantly less than demand; conversely, in developed countries, production is greater than consumption (1). *Consequently*, distribution of agricultural products is also an important concern (15). Predictions of continued increases in agricultural production are not optimistic because 1) yields per unit area are increasing less rapidly now than during the 1950s and 1960s and most high-quality arable land is currently being utilized, 2) in the absence of highly effective conservation practices, soil erosion will continue to significantly limit land productivity, and 3) the exciting promises of genetic engineering are many years from realization (2,15). *Consequently*, losses due to plant disease are expected to have a more critical influence on civilization during the next 20 years than they had in the previous several decades.

In addition to increasing the demand for food, increasing populations constrain disease control practices. In order to not overburden the environment, human activities including agriculture and plant disease control must have little detrimental impact on it. The permissible level of detrimental impact of any human activity declines as the population increases to a "crowded" status. In a crowded world the pollutant and hazardous effects of disease control procedures must be limited. In addition to the widely publicized direct and indirect effects of disease control chemicals, effects of some cultural practices such as burning, soil pH adjustment, and moldboard plowing must also be constrained (8).

As we intensify efforts to increase agricultural production, some new problems in plant disease control emerge as critical; some become more troublesome; and some persist as important problems. Fortunately, some previously primary problems become secondary. The increasing genetic homogeneity of crop plants and the increasing occurrence of pesticide-resistant pathogens are new problems. Rapid adoption of decreased tillage may introduce new problems. The need to decrease hazards to farm workers, to protect the environment, and to prevent hazards to consumers becomes more important as human populations increase.

The identification and use of durable resistance in plants is a problem that has been and remains a high priority. It is a challenge to plant pathologists to resolve these problems.

Genetic homogeneity of the host plants and resistance to chemicals in pathogen populations are two problems associated with the increased intensity of agriculture. In several respects, modern intensive agriculture increases the genetic homogeneity of crop plants (10). We depend now on fewer crop species for the majority of our food and feed needs than we did several decades ago. Additionally, the intraspecies genetic homogeneity is also increasing because modern intensive agriculture uses single cultivars or many cultivars with a common genetic background in monocultures over large areas. This genetic homogeneity makes a crop especially vulnerable to a pest problem. Susceptibility of corn hybrids with Texas male sterile cytoplasm to southern corn leaf blight in the United States in 1970 and susceptibility to tungro virus of some early IRRI rice cultivars are familiar examples.

The effect of monoculture on plant disease is well illustrated by the attempt to grow rubber trees in this manner in South America. Both rubber trees and a fungal pathogen (*Microcyclus ulei*) are indigenous to the Amazon basin. During the 19th century, latex was gathered from wild rubber trees in the jungle, and disease caused by *M. ulei* (South American leaf blight) was common but nondestructive. However, when rubber trees were cultured in plantations, leaf blight became very destructive and contributed to the failure of the plantations (13). Thus, a disease that was not particularly troublesome on plants growing in a heterogeneous population was very troublesome on plants growing in a homogeneous population. The increased genetic homogeneity of crop plants grown in monoculture increases the risk of devastating epidemics (10).

Pesticide-resistant pathogens are emerging as troublesome in modern intensive agriculture (4,6,7,11). As selective systemic pesticides increase in numbers and popularity, resistance to them will be a continuing threat. Plant pathologists are just now investigating pesticide resistance experimentally and theoretically (6,9,12). We need to develop technologies to suppress the occurrence or selection of pesticide-resistant individuals to prevent large-scale losses and to preserve the usefulness of some good pesticides. The use of plant resistance and of cultural and biological factors in combination with chemicals in an integrated control system (see Chapter 40) will probably contribute significantly to maintaining the long-term usefulness of some pesticides.

In addition to increased genetic homogeneity and increased pesticide resistance, other changes in crop production are likely to influence the risk of losses due to disease. The current trend of suppressing weeds with herbicide application rather than by mechanical tillage is likely to alter the spectrum of pathogens important on a given crop (5). For example, stalk rot of maize, caused by *Fusarium moniliforme*, has been lessened in fields tilled sparingly or not at all, but other diseases such as leaf blight and stalk rot induced by *Colletotrichum graminicola* may be favored by reduced tillage. Disease control practitioners must monitor crops to identify changes in the activities of important pathogens as production practices change.

As human populations increase, there is decreased tolerance for environmental or human hazards associated with specific disease control technologies. Society regulates pesticides to protect humans and the environment, and regulations of pesticide development and application are stricter now than in the previous decades (3,14,15).

The short-lived usefulness of some pathogen-resistant plants is a continuing problem that has led to losses due to disease and has demanded continuing effort on the part of plant breeders. Some investigators have suggested approaches for breeding cultivars with durable resistance. These predictions need to be evaluated critically. Only after several years of field evaluation will we be able to ascertain their accuracy.

Plant pathologists need to address these "new," "increasing," and "continuing" problems in disease control. Each of the five chapters in this section addresses these problems. Tweedy (Chapter 37) points to the new problems of pesticide resistance and the need for better pesticide application technology. He also addresses the continuing problem of ignorance among legislators, consumers, and some practitioners concerning the safe, responsible use of pesticides. Baker (Chapter 39) identifies disease control approaches appropriate for a crowded world—those biological and cultural controls with little or no detrimental environmental effect. Approaches include enhancing natural disease suppression, selective destruction of pathogens, and use of pathogen-free propagating material.

Andrews (Chapter 40) amplifies several of Tweedy's and Baker's points when he argues that integrated control is a necessary step in the development of future crop management and agroecosytem management systems. Integrated control focuses several approaches against a common problem (Chapter 43). Wolfe directs attention to the continuing problem of resistance durability and the emerging problem of genetic homogeneity. He describes host-parasite characteristics associated with resistance of limited durability, and he predicts the increased emphasis on some new techniques for breeding durably resistant plants. His suggestion for deploying resistant plants would aid in reducing genetic homogeneity of crops. Buddenhagen (Chapter 42) offers an explanation for the lack of disease-resistant cultivars—especially in developing countries. He suggests that breeding for resistance receives insufficient effort from agricultural scientists. He then argues convincingly that good progress is possible with intensive on-site breeding. Brewbaker (Chapter 41) sees an inexorable trend to plant breeding by industry and a few international institutes instead of by universities or government experiment stations. He discusses genetics of disease resistance from the perspective of the commercial plant breeder and thinks that monogene resistance will remain strategic to most commercial breeding programs against disease. On the other hand, Delp (Chapter 38) points out that breeding for resistance has been inadequate or ineffective for some diseases and that shifts in pathogen genotypes force changes in the use of host genotypes. He foresees increasing use of chemicals for disease control either singly or as a supplement to other control methods.

These chapters present all of us with challenges in future research and practice. They don't address all of the problems—there isn't enough space. You may not agree with all points of view, but the analyses provided by these experts are challenging and should stimulate productive effort to our task of suppressing plant disease.

LITERATURE CITED

1. Barr, T. N. 1981. The world food situation and global grain prospects. Science 214:1087-1095.
2. Brown, L. R. 1981. World population

growth, soil erosion, and food security. Science 214:995-1002.

3. Deck, E. 1975. Federal and state pesticide regulations and legislation. Annu. Rev. Entomol. 20:119-131.
4. Dekker, J. 1976. Acquired resistance to fungicides. Annu. Rev. Phytopathol. 14:405-428.
5. Doupnik, B., Jr., and Boosalis, M. G. 1980. Ecofallow—a reduced tillage system—and plant diseases. Plant Dis. 64:31-35.
6. Dovas, C., Skylakakis, G., and Georgopoulos, S. G. 1976. The adaptability of the benomyl-resistant population of *Cercospora beticola* in northern Greece. Phytopathology 66:1452-1456.
7. Gilpatrick, J. D. 1979. Contemporary Control of Plant Diseases with Chemicals: Present Status, Future Prospects, and Proposals for Action. Report prepared for EPA. Am. Phytopathol. Soc., St. Paul, MN. 170 pp.
8. Hardison, J. R. 1980. Role of fire for disease control in grass seed production. Plant Dis. 64:641-645.
9. Kable, P. F., and Jeffrey, H. 1980. Selection for tolerance in organisms exposed to sprays of biocide mixtures: A theoretical model. Phytopathology 70:8-12.
10. Marshall, D. R. 1977. The advantages and hazards of genetic homogeneity. Ann. N.Y. Acad. Sci. 287:1-20.
11. Ogawa, J. M., Gilpatrick, J. D., and Chiarappa, L. 1977. Review of plant pathogens resistant to fungicides and bactericides. FAO Plant Prot. Bull. 25(3):97-111.
12. Skylakakis, G. 1981. Effects of alternating and mixing pesticides on the buildup of fungal resistance. Phytopathology 71:1119-1121.
13. Thurston, H. D. 1973. Threatening plant diseases. Annu. Rev. Phytopathol. 11:27-52.
14. Wellman, R. H. 1977. Problems in development, registration, and use of fungicides. Annu. Rev. Phytopathol. 15:153-163.
15. Wittwer, S. H. 1979. Future technology advances in agriculture and their impact on the regulatory environment. BioScience 29:603-610.
16. Wortman, S. 1980. World food and nutrition: The scientific and technological base. Science 209:157-164.

CHAPTER 37

THE FUTURE OF CHEMICALS FOR CONTROLLING PLANT DISEASES

B. G. TWEEDY

The use of chemicals for controlling plant diseases dates back to Biblical times, when sulfur was applied to plants to control maladies. As the demands for food have increased and people have developed more intensive crop production, the demand for chemicals has also increased. The increased use of chemicals for controlling plant diseases has been most dramatic during the past 30 years. Some reasons for marked increase in use are the development of highly effective chemicals that are of relatively low cost in terms of financial benefits received, changing agricultural production practices that allow more effective chemical use, and an increasing food demand.

Of the three major types of pesticides (insecticides, herbicides, and fungicides), fungicides are the lowest in terms of worldwide sales (Table 1), but the recent growth rate of fungicides is greater than that of any other pesticide listed.

Previously, fungicides were applied mainly to high-income horticultural crops (apples, grapes, potatoes, etc.); however, their value has recently been demonstrated on agronomic crops such as peanuts, wheat, soybeans, and rice.

Data reported by Schwinn (15) show that more pesticides are used in North America than in any other continent (Fig. 1). When data are separated by type of pesticide, it is evident that fungicide use is greater in countries where highly intensive and diversified farming is practiced (Table 2). Approximately 75% of all pesticides and 85% of all fungicides are used in the developed countries. These data reflect primarily protective fungicides and not the addition of some of the new fungicides being introduced into the marketplace.

This brief review of pesticide use clearly shows that chemicals are currently very important throughout the world for controlling plant diseases and that the use of fungicides continues to grow in importance. The charge given to me in preparing this chapter is to look at the future role of fungicides in our plant disease control strategy in agronomic crops to ensure enough food to feed the world population.

CHEMICAL CONTROL IN THE PAST

To project the future, one should look at changes in past years that will have significant impacts on the future. Horsfall (12) in a recent article addressed the past as well as the present and future. He stated that "in 1776 when our nation was born, we only had two useful fungicides for controlling diseases on food crops, namely elemental sulfur and copper sulfate." We have come a long way since this time and most of the significant advances have been made during the past four decades.

In 1934, Tisdale and Williams of E. I. du Pont de Nemours & Company introduced the dialkyldithiocarbamates, which was the beginning of the organic fungicide era. According to Horsfall (11), the management of this company was

TABLE 1. Pesticides: world market

Pesticide Group[a]	Sales (billion U.S. dollars) 1978	1980	1984[b]	Percent Total Market in 1980	Percent Increase of 1984 over 1980
Herbicides	3.71	4.23	4.79	43.5	13.4
Insecticides	3.02	3.35	3.82	34.5	13.8
Fungicides	1.53	1.69	1.97	17.4	16.6
Soil fumigants	0.19	0.20	0.23	2.1	14.5
Others	0.22	0.26	0.30	2.6	14.9
Total	8.67	9.73	11.11	...	14.2

[a]Source: Schwinn (15).
[b]Projected.

TABLE 2. Pesticide use in six geographical areas (1979)[a]

Area	Herbicides (%)	Insecticides (%)	Fungicides (%)	Others (%)	Total (%)
North America	52	23	15	44	34
West Europe	21	13	37	38	22
Japan	8	13	20	7	12
East Europe	12	12	19	6	13
Rest of world	7	39	9	5	19
World total	100	100	100	100	100

[a]Source: Schwinn (15).

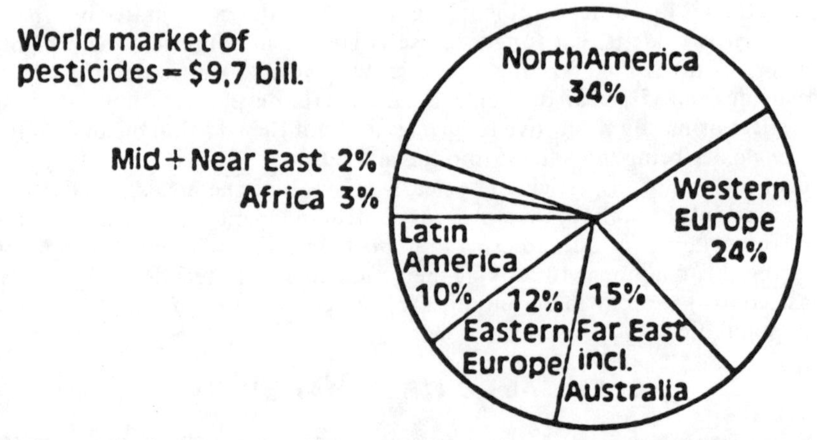

Fig. 1. Geographic distribution of pesticide market (1978), reprinted, by permission, from Schwinn (15).

reluctant to offer these compounds to farmers since it was during the depression; these new fungicides were expensive, whereas copper sulfate was available for only 6 cents a pound. Other fungicides such as chloranil, the ethylenebisdithiocarbamates (EBDCs) and captan were soon developed and were found to be very effective for disease control on many crops. The EBDC fungicides are possibly the most important group of organic fungicides ever developed. They have been and continue to be widely used throughout the world. In 1952, another significant contribution was made when captan was introduced. Captan and the EBDCs are relatively inexpensive, broad-spectrum, protective fungicides that must be applied to foliage before infection to be effective in controlling diseases. George L. McNew (*personal communication*, 1981) refers to this as the "Sherwin Williams approach," i.e., cover the world to protect it from the fungal pathogens. This pest control approach was later challenged by the environmentalists who thought that this was unnecessary because of potential detrimental effects to the environment. Federal regulations were soon passed, resulting in more restrictive uses, increased development costs, and delayed approval for use. However, a better data base for defining the safer use of all pesticides is one of the results of the new federal regulations.

During the 1960s, Horsfall and many other scientists proposed the "bullet" approach, chemotherapy, i.e., sending chemicals to the point of infection via the plant and making the chemicals more selective, thus controlling more specific diseases. In the 1960s and 1970s, several chemicals of this type were introduced such as carboxamides, pyrimidines, morpholines, phenylthioureas, and benzimidazoles. As pointed out by Horsfall (12), the use of plant disease chemotherapy has a built-in weakness that is not involved in the therapy of animal diseases. Whereas animals have phagocytes to clean up the left-over inoculum missed by the chemicals, plants unfortunately do not. Thus, chemotherapy added another tool for combating plant diseases, but provided another challenge to us, namely, resistance of pathogens to chemicals.

The highly selective approach of site-specific fungicides caused plant pathologists to join entomologists in combating chemical resistance. Horsfall (11) in 1956 did not list any fungi causing diseases of economic significance that had been cited as chemically resistant, and he stated that he "had difficulty identifying any resistant fungi." Within a decade the picture changed, and by the mid 1970s there were numerous examples of chemical resistance in plant pathology. In a survey made in 1976, 56 fungal species and 10 bacterial species were shown to have developed chemical resistance (FAO Plant Production and Protection Papers: a committee report, 1976). Until recently, reports of chemical resistance involved primarily the benzimidazole fungicides. During the past few years, reports of fungus resistance to several site-specific fungicides have occurred, thus indicating that the use of more site-specific fungicides has the potential for the rapid development of resistance to chemicals.

Except for a few examples in local areas, chemical resistance of plant pathogens has not caused serious crop losses affecting food production in the world. Fortunately, resistance problems were identified before they became too widespread, and other fungicides were used effectively. However, resistance severely limits use of some very effective chemicals, and in the example of metalaxyl, suitable alternative chemicals for controlling diseases caused by many of the pathogenic Oomycetes are not currently available.

The economics associated with farming operations influence plant disease and control practices. In the 1960s and early 1970s, pesticides were inexpensive and use directions were simple. However, prices for farm products have not increased in proportion to recent increases in costs of pesticide, fertilizer, labor, and other farm operations. The choices of pesticides are many, and the directions for use have become very complex. Farming has evolved from small individually owned and operated farms with few major decisions and relatively small investments to much larger, highly mechanized operations with very large investments and operating budgets. Farming is now a big business, and profits or losses can be large. Several farm economists giving seminars during past months state that farm profits have been hit harder during 1981–1982 than at any time since the early 1930s. Farming economics will certainly have a major impact on the direction of chemical use in the future.

Changing farm practices have caused cropping systems to change during the past 25 years. Farms have increased in size and there is less diversity in crops. We now have fewer crops grown in much larger fields; the monoculture extends both horizontally (across large areas) and vertically (through repeated seasons). This cropping system is necessary for more specialized and larger equipment, but it encourages the buildup of pests. Many of the more commonly used farming practices of the past, e.g., crop rotation and production of livestock of livestock on grain farms, have minimized pest populations, but these practices are not compatible with the current practice of monoculture of crops. Genetic vulnerability of major crops such as corn, wheat, soybeans, sorghum, cotton, etc., has become a big concern. Most commercial cultivars are genetically related and a change in resistance could be devastating without the availability of fungicides.

One last factor that I will discuss is a greater concern by the public for environmental quality; Horsfall (12) refers to this as the "Rachel Carson Syndrome." Carson's 1962 book (2) *Silent Spring* and subsequent publications related to environmental concerns had a tremendous impact on pesticide regulations. These regulations in turn increased registration requirements, time for registration approval, and development costs. The public has profited but also has paid for these sweeping changes. Although these changes are still occurring throughout the world, the pendulum appears to be swinging to a much more realistic position. Let us now look at some of the challenges we might expect.

SUMMARY OF COMMENTS MADE BY CHEMICAL CONTROL RESPONDENTS

Before preparation of this manuscript, I corresponded with more than 25 scientists who are directly or indirectly concerned with chemical control of plant diseases. Nearly all responded to my request for comments and I have summarized their comments in the following section.

Chemicals are and will continue to be an integral part of strategies for controlling plant diseases. The loss of chemicals for controlling nematodes has been disastrous to many growers, particularly in California. There is a real need to develop improved chemicals and novel approaches for controlling nematodes and other disease-causing organisms. It is generally accepted that our dependence upon chemicals for growing healthy crops will continue to be as great, if not greater, throughout the 20th century.

The two issues mentioned most frequently were 1) a need for educating the general public, EPA lawyers, etc., and other groups about the benefits as well as the costs of fungicide use and 2) the need to improve application techniques for pesticides. Exposure to the applicator, excessive use, and chemical misuse were closely related issues frequently listed. A few respondents indicated that industrial and extension specialists have generally done an inadequate job of demonstrating financial and other benefits from the use of chemicals and that this issue should be dealt with during the next several years.

The discovery of "new and better" fungicides was listed as a general need; higher priorities were placed on discovering new nematicides and bactericides than on new fungicides. The loss of the most effective nematicides put nematode control as one of the highest priorities for future research. Several respondents indicated a need to investigate the use of chemicals with other control measures (use in integrated pest management programs [IPM]) at an earlier stage in the development of a chemical than is currently being done. G. L. McNew (*personal communication*) described the ideal fungicide:

> the future will be served most effectively by seeking a general purpose fungicide that has strong ability to permeate cell membrane, remain viable and mature inside the plant tissues so it will progressively move toward the metabolite pools of the plant where protection is mandatory.

Several respondents indicated that government regulations are now reaching a point where registration costs and excessive time for approving registration are stifling the discovery and development of new biologically active chemicals. This is one of the challenges that needs to be addressed during the next few years.

Fungal resistance to fungicides is certainly a challenge to the plant pathologist during the coming years. Several approaches were suggested for investigation. Some respondents commented that strategies for combating resistance are being formulated with very little basic information; thus, there is a real need to obtain basic knowledge on the nature of chemical resistance and the epidemiology of the chemically resistant pathogen. Our goal is to strengthen strategies and hypotheses without permanently losing our most selective and valuable fungicides.

Another area needing attention is improved control of postharvest diseases. Even though progress has been made in this area, the losses are still staggering. The losses to postharvest diseases in some countries have been estimated to be as high as 25% of the total crop produced. The purchase and transportation of large quantities of wheat, corn, soybeans, peanuts, etc., from country to country have created a need for grain preservatives.

Last, but certainly not least, is the need for further investigation of host resistance induced by beneficial organisms or biochemicals extracted from the beneficial organisms. This approach has been investigated for several years, and a vast resource of information is available. What is needed at this time is to evaluate this information critically and follow it up with more complete studies (L. Sequeira, *personal communication*, 1982). This approach has merit and the advantage of using genetic materials that are commercially acceptable to growers.

For the sake of brevity, this is merely a consolidation of the many comments submitted to me by the respondents. Their names and affiliations are listed at the end of this chapter.

IMPORTANCE OF CHEMICAL CONTROL IN THE FUTURE

All of the respondents when asked about the future importance of chemicals for disease control indicated that chemicals will play an equally if not more important role during the next 20 years. Without chemicals there would be tremendous food production losses beyond what we currently encounter. The greatest increase in demand for chemical use will be in countries where intensive agriculture has developed during the last several decades. These countries include the United States, Brazil, and Argentina, as well as many of the less developed countries where highly commercial agriculture is just getting started. In the People's Republic of China and in many European countries where intensive and diversified agriculture has been practiced for centuries, disease pressures seem to have stabilized more or less. Major outbreaks are due primarily to changes in environmental conditions that favor specific pathogens over the host.

For decades we have emphasized plant resistance, sanitation, crop rotation, eradication, and quarantine as the main principles for controlling plant diseases, and certainly these are very effective tools. However, each has limitations and should be critically reviewed. For example, sanitation is an important concept but as pointed out by H. L. Bissonnette (*personal communication*, 1981):

> How can a cereal grain grower in the Great Plains area have a clean, trash-free field to control diseases and at the same time prevent his soil from eroding?

The grower must decide which is most important, and we are aware that erosion is a great problem. Programs of eradication and quarantine have been defined for many crops, but their effectiveness from the disease control standpoint are questionable, particularly eradication. The emphasis on a more integrated approach (IPM) is very important, and the use of chemicals should be considered as one of the necessary tools for a successful IPM program.

Specific diseases may change in importance, and these changes will become more of an international problem than a national or regional one. One excellent example of a minor disease becoming a major one is *Septoria* on wheat (H. L. Bissonnette, *personal communication*, 1981). In 1949, Septoria leaf spot was identified as a minor disease on spring wheat, whereas the rusts were very important diseases of wheat in the Great Plains. Several rust-resistant cultivars are now available commercially and the wheat grower has an effective means of coping with rust. Within 10 years, Septoria leaf spot became a major worldwide disease on wheat, and although now of less importance, it remains an important disease. Now we are observing rusts on our late-planted wheat and its importance may change. We are able to change the genetics of the host but we cannot control the environmental conditions nor the genetics of the pathogen. We have seen how quickly a minor disease can become nearly catastrophic to a crop, e.g., southern corn leaf blight in 1971. Diseases will continue to be serious problems in crop production, and fungicides will continue to be important for controlling them.

CHALLENGES OF THE FUTURE

Education

I believe one of the greatest challenges during the 1980s will be education—of the general public, university students, and pesticide users. All areas will be difficult, but by far the greatest challenge will be the education of the general public. Regardless of how difficult it may be, the education of these groups is essential for pesticide use to achieve maximum success in its role of improving crop production in the future. There are many misconceptions about chemicals ranging from "they are so toxic that you can't be safe in the same room with them" to "you have no worries at all because they are completely harmless." We must educate the general public on the costs, benefits, and relative importance of pesticides to the production of high quality food to feed a hungry world. We must inform the public of the magnitude of research being conducted by industrially and publically supported laboratories in order to define their safe use and the risks involved. The university graduate, particularly the agricultural college graduate, should be better educated on the safe use of pesticides and on pesticide toxicity.

Generally, high school and college graduates in agriculture lack pertinent information on all facets of food production and particularly plant protection. Where should we begin teaching the many factors involved in food production? How many youngsters grow up having little or no concept of the agricultural process and think that food simply comes from the local grocery store?

Users must be better educated in the safe use of pesticides. We are living in a time when more is known about proper chemical use than in any other time in history, yet misuse does occur and much of it because users are not properly informed.

Chemicals should be considered in every strategy for control of new pests, and the environmental concerns should be appraised. We need to educate everyone appropriately in order for this strategy to work with minimum impact to society.

Government Regulation

Regulation of pesticide use is needed and certainly serves a very useful purpose, but regulation must not be permitted to stifle the innovator, the scientist, the confidence of the consumer, or the protection of our food and fiber. Several of the scientists responding to my questionaire stated that overregulation of chemicals has become an acute problem. These were mainly scientists making recommendations to growers and who work with them on a day-to-day basis. Many of us in industry feel the pendulum is starting to swing to a more realistic position at the federal level, but the greatest concern now is registration at the state level. Can you imaging having different requirements for the federal government and for each of 50 states? Currently, registration at the international level appears to be developing into a more unified approach for many countries throughout the world, a trend that is badly needed.

When regulation is excessive and results in excessively high costs for compliance, then development of new products can be stifled. An American Phytopathological Society publication reported that representatives from 13 of 14 companies thought that increased costs of registration and other pesticide regulations severely limited availability of fungicides, viricides, and nematicides

(7). New pesticides come from industrial research, which is financed from sales of developed products. An incentive is needed by industry to continue discovery and development of these new products. The most negative impact on pesticide development by overregulation is development of chemicals for use on minor crops. The present average industry cost for the development of a new compound exceeds $10,000,000, and frequently total sales for a fungicide over a 5-year period would not exceed this value; thus development of a pesticide for only a few minor crops cannot be considered because of the very high development costs (7).

Chemical Resistance

One of the most urgent problems facing plant pathologists is chemical resistance by pathogens. Several recent reviews on this topic (3,4,6,13) have appeared. In 1978 resistance in one or more species had been found to 52 different fungicides and bactericides (FAO Plant Production and Protection Papers, committee report, 1978); most instances were cited for the benzimidazole fungicides. The list continues to grow. The most commonly recommended solution to stopping or delaying resistance is to apply a systemic with a broad-spectrum fungicide by mixing them in the spray tank or by using a commercially mixed product (prepack). Tank mixing is supposedly effective for preventing or at least delaying development of resistant strains of fungi normally controlled by benomyl (5,10). I do not believe this is the long-term solution. Marketing strategies, profitability to the pesticide manufacturer and the farmer, and good scientific practices are not always compatible. Strategies for coping with resistance are being developed before we really have obtained enough basic knowledge about chemical resistance.

Our knowledge relating to the development of resistant strains of fungi in the field is limited and we need to establish an appropriate research program before attempting to solve the problems of chemical resistance. We need basic information on the mechanism of resistance and the fitness of the pathogen, i.e., survival in the environment, competition with other strains and other microorganisms, etc. Are the tolerant strains already present but unable to compete with the wild type strains? How important is the environment in the development of high populations of the chemically resistant strains? Could we use beneficial bacteria or fungi in our fungicide application? Could we watch for the resistant strains and change chemicals when resistant strains appear? What can the grower and/or manufacturer afford? Epidemiological studies of the resistant strain are also badly needed. A screening mechanism for determining the potential for resistant strains at an early stage of chemical development is needed. Identifying the mechanism(s) of action will give a clue to the potential. Many conclusions are being drawn and strategies are being developed based on very few data.

Application

My major professor in graduate school, Dwight Powell, said to me "Bill, we are using more than twice the pesticides we need. Unfortunately, we aren't smart enough to say which half is needed and which half is not needed." I don't believe we have advanced very far from this point in the last 20 years, but there really has been little need. In most situations, the environment will accommodate chemicals placed in it when we use them according to direction. The costs have been relatively modest compared to the benefits received, yet so often cost-benefits are really not considered.

Basic knowledge is now being developed that will help solve problems in this area, particularly timing of applications where we are informed of what chemicals to use and when to apply them for control, e.g., scab on apples in Michigan and late blight on potatoes in Pennsylvania. We need to apply more basic knowledge to improve fungicide use. The delivery systems, i.e., the spray equipment, formulations, etc., still need improvement. We need to get a greater percentage of the pesticide to the target and less to nontarget areas. We need to have slow release chemicals so a more constant effective dose is present for protection. We now apply a high dose that rapidly decreases in concentration, making it necessary to reapply the chemical at a high rate. It is known that this is not an effective approach for controlling diseases in animals; therefore slow-release medications for controlling diseases of humans have been developed. I believe slow-release formulations should be more thoroughly investigated for controlling plant diseases. One excuse for not doing this is the high cost involved, but I really cannot accept this. I think much research needs to be done in the areas of 1) applicators (spray equipment), 2) formulations that reduce applicator exposure, increase the percentage of chemicals to hit the target, and are more effective in controlling plant diseases, and 3) better timing of applications.

INDUCED PLANT RESISTANCE

Induced resistance resulting from elicitation of the plant's own defense system has been the subject of several investigations (1,3,8,9,14). These studies show that plants have latent defense mechanisms that are effective against infection and that the defense mechanisms are more effectively elicited in some plants than in others. These defense systems can be "activated" by applying biochemicals that come from plant pathogens or by applying certain microorganisms. I will avoid a detailed review of the potential mechanisms; however, eliciting defense mechanisms certainly is a potential for chemical control of plant disease, and this approach will be covered briefly in subsequent paragraphs.

In recent years, there has been considerable interest in the use of fungal cell wall glucans that induce phytoalexin biosynthesis. These substances stimulate activity of many key enzymes in the aromatic pathway and lead to synthesis of numerous secondary metabolites, including phytoalexins. L. Sequeira (*personal communication*, 1981) states that this creates at least a potential problem in practical application of these compounds because of undesirable metabolites that accumulate in treated plants. Phytoalexins are also quite toxic to plant cells. For these reasons, the phytoalexin elicitors will probably have limited application. Nevertheless, this is an area that needs further evaluation for its practical application for disease control.

Inducers of systemic plant resistance that do not depend on phytoalexins could have a greater disease-control potential than is provided by the induction of the phytoalexins. Although this phenomenon can be induced for a wide variety of plant pathogens, little is known about what microbial components will induce systemic resistance in plants. The systems explored involve the use of live pathogens of different types, inactivated forms of these pathogens, or crude extracts. Bostock et al (1) have used both the fungal pathogens and extracts; Graham et al (9) have examined the use of bacteria lipopolysaccharides; Goodman (8) has used bacteria and bacterial lipopolysaccharides; and Ryan (14) has used cell wall components to induce reactions in several different hosts to protect plants from infection.

Goodman (8) suggests that for an efficient hypersensitive reaction to occur, the inducer or elicitor of resistance must be on the pathogen surface. He (*personal communication*, 1981) suggested later that these elicitors were biochemicals, specifically glycoproteins, which are comparatively small (5–15 thousand units) but complex enough to have a high degree of specificity.

This is a brief review of potential use of chemicals for induction of latent plant resistance systems. I believe that during the next few decades we will make some substantial advances in this chemical control area. Also, the application of genetic engineering could result in the use of chemicals to induce latent plant resistance as a singularly effective tool in dealing with control of plant diseases. A somewhat related approach is the integration of chemical with biological control such as shown in the work of Thomson et al (16) in California.

SUMMARY

Twenty-five years have passed since the golden anniversary for our Society and at that time the future looked bright for chemical disease control. During this quarter century, many changes have occurred and many concerns have been voiced, but the future of chemical disease control continues to look very promising. Most of the compounds used 25 years ago are still being used—the major exception being mercury compounds. We have seen the addition of many new chemicals in recent years, and these are much more active disease-control agents and most are systemic.

We do have new challenges ahead such as chemical resistance of the pathogen, induced resistance of the host by natural products, education of the public and the users, and increased government regulations. We are still lacking good bactericides to control plant bacterial diseases. We have lost the use of most nematicides, and we need to investigate new approaches to chemically controlling nematodes, such as antifeeding and/or hatching compounds and chemical confusants.

We should look forward to meet these challenges and to improve upon this very important tool for producing more and better food for a hungry world.

ACKNOWLEDGMENTS

I appreciate the help of these respondents: W. M. Dowler, USDA, Beltsville, MD; H. L. Bissonnette, University of Minnesota, St. Paul; L. V. Edgington, University of Guelph, Ontario; A. H. Epstein, Iowa State University, Ames; R. N. Goodman, University of Missouri, Columbia; K. D. Hickey, Pennsylvania State University, Biglerville; J. G. Horsfall, Connecticut Agricultural Experiment Station, New Haven; A. H. McCain, University of California, Berkeley; W. D. McClellan, ICI Americas Inc., Goldsboro, NC; A. A. MacNab, Pennsylvania State University, University Park; G. L. McNew, New Mexico State University, Las Cruces; J. M. Ogawa, University of California, Davis; A. O. Paulus, University of California, Riverside; C. Powell, Ohio State University, Columbus; J. N. Sasser, North Carolina State University, Raleigh; L. Sequeira, University of Wisconsin, Madison; M. C. Shurtleff, University of Illinois, Urbana; H. Sisler, University of Maryland, College Park; D. C. Torgeson, Boyce Thompson Institute, Ithaca, NY; S. D. Van Gundy, University of California, Riverside.

LITERATURE CITED

1. Bostock, R. M., Kuć, J. A., and Laine, R. A. 1981. Eicosapentaenoic and arachidonic acids from *Phytophthora infestans* elicit fungitoxic sesquinterpenes in the potato. Science 212:67-69.
2. Carson, R. 1962. Silent Spring. Houghton Mifflin Co., Boston. 368 pp.
3. Dekker, J. 1977. Resistance. Pages 176–197 in: Systemic Fungicides. R. W. Marsh, ed. Longman, London. 401 pp.
4. Dekker, J. 1980. Selective interference with the plant-parasite relation: Problems and prospects in plant protection. Meded. Fac. Landbouwwet. Rijksuniv. Gent 45(2):113-119.
5. Delp, C. J. 1980. Coping with resistance to plant disease control agents. Plant Dis. 64:652-657.
6. Georgogoulos, S. G. 1976. The genetics and biochemistry of resistance to chemicals in plant pathogens. Proc. Am. Phytopathol. Soc. 3:53-60.
7. Gilpatrick, J. D. 1979. Contemporary Control of Plant Diseases with Chemicals. Publication prepared for the EPA. Am. Phytopathol. Soc. St. Paul, MN. 170 pp.
8. Goodman, R. N. 1980. Defenses triggered by previous invaders: Bacteria. Pages 305–317 in: Plant Disease, An Advanced Treatise. Vol. 5. J. G. Horsfall and E. B. Cowling, eds. Academic Press, NY. 534 pp.
9. Graham, T. L., Sequeira, L., and Huang, T. R. 1977. Bacterial lipopolysaccharides as inducers of disease resistance in tobacco. Appl. Environ. Microbiol. 34:424-432.
10. Hickey, K. D. 1977. Fungicide tolerance and apple disease control. Pa. Fruit News 56:61-63.
11. Horsfall, J. G. 1956. Principles of Fungicidal Action. Chronica Bot. Co., Waltham, MA. 279 pp.
12. Horsfall, J. G. 1977. Fungicides—Past, present and future. Pages 113–122 in: Pesticide Chemistry in the 20th Century. Jack R. Plimmer, ed. ACS Symp. Ser. No. 37. Am. Chem. Soc., Washington, DC. 310 pp.
13. Ogawa, J. M., Gilpatrick, J. D., and Chiarappa, L. 1977. Review of plant pathogens resistant to fungicides and bactericides. FAO Plant Prot. Bull. 26:97-111.
14. Ryan, C. A. 1978. Proteinase inhibitors in plant leaves: A biochemical model for pest-induced natural plant protection. Trends Biochem. Sci. (Ref. ed.) 3(July):148-150.
15. Schwinn, F. I. 1982. Chemical control of fungal diseases: Importance and problems. Chap. 1 in: Fungicide Resistance in Crop Protection. J. Dekker and S. G. Georgopoulos, eds. Pudoc Publ., Wageningen, The Netherlands. 273 pp.
16. Thomson, S. V., Schroth, M. N., Moller, W. J., and Reil, W. O. 1982. A forecasting model for fire blight of pear. Plant Dis. 66:576-579.

CHAPTER 38

CHANGING EMPHASIS IN DISEASE MANAGEMENT

CHARLES J. DELP

As growers attempt to manipulate more of the interrelated factors in plant production, disease management is becoming increasingly complex. Individual methods of disease control will be blended with each other and with methods of production. Managers have always attempted to integrate some production factors, but many disease control methods are still used in relative isolation. These are likely to be put into a much broader context in the near future, for we are taking the next steps to holistic disease management, including the integration and optimization of all production factors. The development of these systems for growers will require an understanding of dynamic, artificial plant ecosystems and the epidemiology of major pathogens. Fortunately, computer technology will help to put these disease management strategies in concert with the entire production system. The challenge we face is for *cooperative* action to make the complex interactions most effective.

METHODS OF DISEASE MANAGEMENT

We've been in an either/or era, where recommendations were for a resistant cultivar *or* a spray regime *or* a cultural practice. The methods were frequently offered separately from different authorities and were received as almost competitive strategies. Now that situation is in rapid transition, toward integration of appropriate methods into the total production program. It is only for organizational purposes that the methods are described separately in the following paragraphs.

Plant Breeding (Genetics)

Host resistance to diseases will continue to improve production, reduce disease losses, and in some instances save crops from devastation. Despite the fact that plant breeding efforts have been intensive, there are many major diseases for which resistance breeding has been ineffective. Also, many pathogens eventually overcome resistance characteristics of hosts, leading to a need for continuously changing the host genotype to counter shifts in pathogen genotypes. Thus, there is a continuing need for new resistance genes and the use of strategies for their deployment such as mixed cultivar plantings.

There is hope that the emerging genetic engineering technology will be relevant in the provision of additional resources of host resistance for disease control. A challenge for plant pathology is the incorporation of resistance factors into breeding programs that frequently have broad objectives. Yield and other quality characteristics are given such high priority that disease resistance can get left out. The complementary effects of resistance and other methods of disease management must be thoroughly exploited in future production programs.

Chemicals

In coming decades, chemical control will be used with increasing intensity because of the growing population requirements for plant products and because of the proven effectiveness of chemicals. As a supplement to other methods, chemicals can provide predictable improvement in quality, yield, and profit. The discovery and development of more effective chemicals than currently available and the increased need for disease control on traditionally low value field crops and pastures will also result in increased use of chemical disease control agents. Intensified crop culture makes disease management with chemicals a profitable venture for the grower. Minimal treatment for broad spectrum disease reduction can provide consistent yield boosts even under conditions where total crop loss is not threatened. Such is the case with soybeans in southern United States and in an increasing number of other situations on cereals. Because of these expanded uses and also because pathogens have developed resistance to many of the highly effective fungicides, a large selection of chemicals is needed. This is a call for new chemicals with new modes of action. And they will come, but not enough, despite the tremendous industrial effort. It will continue to be a challenge to determine the risk/benefit facts and to use these facts to develop rational governmental regulations.

The problems associated with resistance to fungicides will complicate their use because strategies designed to prolong their effectiveness require consideration of many ecological factors and cooperative action. An indication of this cooperation was evident at a recent seminar on coping with fungicide resistance attended by representatives from 35 fungicide manufacturers. The problems of dealing with resistance are so complicated by cross-resistance among products and by the need for mixtures of unrelated products that scientists from these companies are looking for new ways to work together. To keep these most effective agents in the arsenal will also require the informed cooperative efforts of scientists, advisors, and growers.

Biological Control

Methods for biocontrol have been demonstrated as possible for at least six diseases (see Chapter 39), but there are few strong incentives to develop them to the extent that they have broad utility. Of course, many pathogens are restrained naturally by competitive or antagonistic organisms, and the basis for many cultural methods for disease control is biocontrol. The management of associated microbiota by the development of cultural practices requires a thorough understanding of ecological and epidemiological factors and their manipulation. In these studies, there is need for significant, practical discoveries. As these discoveries are made, it is possible that they will blend with other methods. A fascinating example is the control of *Eutypa* on apricot by applying a mixture of a fungicide plus a *Fusarium* resistant to the fungicide and also antagonistic to the pathogen. The temporary fungicidal effect gives the antagonist a chance to become established in the pruning wound. The rapidly emerging biotechnology industry may increase the likelihood of success, but there is little evidence indicating that biological control will be a major disease management tool in the near future except in very specific and limited instances.

Cultural Methods

Modifications of cultural practices are frequently the only practical means of disease management. Simple modifications of fertilization, crop rotation, cultivation, and soil moisture regulation can have dramatic results. For instance, raised beds can effectively control serious diseases. Unfortunately, some sanitation practices such as burning and deep cultivation are in conflict with air pollution concerns and soil erosion control, and no-till cultural methods may intensify disease pressure. The dynamics of the growing conditions of plants necessitate constant modifications in cultural practices. When used in conjunction with resistant cultivars and chemical treatments, they can be even more useful. These methods will continue to be developed in conjunction with growers' practices by knowledgeable field pathologists.

Biomonitoring Practices

Disease indexing, ELISA (enzyme-linked immunosorbent assay), and other monitoring techniques must be intensified to aid other control methods, especially disease-free certification programs. Their automation will be prompted by expanded use in contract clinics. Information from more reliable monitoring systems will also help identify sources of crop loss and give direction to management efforts. The untapped resource of technology from unrelated disciplines presents a challenge for exploration.

Computer Models and Programs

We are only seeing the first of the greatly expanded use of computers for the prediction of problems and integration of disease management and crop production components. Many desired cultural practices are in conflict with good disease management. These can be identified and evaluated in terms of cost/risk/benefit so that appropriate choices can be made. Epidemiological assumptions are being validated and modified to a useful state, but accurate determinations of pathogen population dynamics, disease intensity, and yield loss are major limiting components. The emphasis on disease management will center around rapid and effective development of these systems even for use in the most primitive agricultural areas. Of course, this is the hub where the efforts of so many interests merge, thus creating a dependence on cooperative action.

Changing Emphasis

I see a continuation of intensive efforts in plant breeding for disease resistance to keep up with past accomplishments and to provide growers with improved cultivars. Successes in genetic engineering could strengthen the emphasis on resistance but probably not in the near future. The use of chemicals will rise with the value of field crops and the discovery of more effective disease control agents. Biocontrol, the hope of many, will not have broad utility until a significant number of discoveries are made and developed. The increased development, integration, and flow of epidemiological information will strengthen the use of cultural practices and complimentary combinations of several disease management methods.

GROWER'S CHOICES

Production is not just doing those things that lead to highest output. Choices are necessary to suit individual needs. Each grower's situation is different, including resources and attitude. Therefore, there is no best production (disease management) system even for similar producers. For example, consider the following possible scenarios: Three cereal growers, who last season went all out for maximum production, were faced at harvest with severe losses due to lodging and low market prices. In response to these losses, each grower has an adjusted plan. This coming season, one neighbor decides to take the big risks and do it again—to maximize production. The high cultural inputs will make it necessary also to use the most intensive pest control and growth modification practices to attain maximum yield. The second grower, almost bankrupt by last year's disaster, is forced to employ minimal efforts and expenses. In this case, expensive crop treatments are not consistent with the objective of getting the most for the least. The third farmer elects to optimize all production factors for maximum profit. In all three cases, each production practice (including disease management) is considered in relation to the others but with different objectives. Because agroecosystems are in such a state of dynamic imbalance, it will be necessary to make constant observations and readjustments in the intensity of the different efforts.

Although the third program (maximum profit) makes good sense, it is not the best for every grower. There are many choices that will also have influences on a program, such as soil and energy conservation, or maximizing quality. I recently discussed the production management process with an agriconsultant firm in England and found that this firm is equipped to work with the cereal growers described above. Their sophisticated computer programs are regularly updated with new information from the technology of divergent disciplines as well as from their own field experiments and grower experiences. Their services are tailored to fit individual needs and local conditions.

So each of the three growers brings to the planning process an individual goal. With the help of a local agriconsultant (perhaps a "plant doctor"), each enters data into a computerized program designed to balance specific grower conditions and needs with available "tools." Analysis and integration of extremely complex and apparently unrelated factors produce guidelines for production. Each grower will plant a different cultivar at a different time on soil prepared in a different way and managed differently for yields consistent with that grower's situation. This yet imperfect process will improve as the complicated interactions are better understood.

This example of the three growers, of course, does not apply to all situations. Most of the growers in our world still cultivate a hectare or two with hand tools, and a large amount of the plant production in our world is under the control of governments and multinational businesses. But disease management practices in most of these situations will be increasingly integrated with each other and with the production methods of many other disciplines.

PEOPLE WILL MAKE THE CHANGES

Of course, there will be technical advances, natural disasters, wars, demands for food and energy, etc., that will influence the emphasis in disease management, but people will make the changes. The attitudes and actions of growers, grower-support groups, consumers, and business and government personnel are most likely to influence future disease management practices. So, let's take a look at some of these people.

Worldwide, at least 10 million families are engaged in agriculture and equipped with farm machinery. They are largely the modern growers at work in Europe, Israel, Australia, Japan, and North America. But what about the more than 300 million nonmechanized family farms? These relatively powerless peasant growers must be considered because they could be the source of the most radical changes. The interface between the hoe and the computer will become increasingly significant. Survival is the issue—not just increased profit or quality. The Green Revolution, which is buying time before the great famines, started the changes for peasant growers. Improved seed, fertilizer, and modified culture were a beginning. Some government agents and agrichemical representatives are committed to intensive village training programs. Other governments apparently must be faced with more critical starvation conditions before adequate resources are allocated to food production. Changing emphasis in disease management will be greatly influenced by the grower.

There are great contrasts between attitudes of the consumers in over-fed nations and the vast numbers of malnourished people in the world. Hungry people can't afford to demand excessive regulations that inhibit production.

On the other hand, sophisticated growers, frequently organized in cooperatives or businesses, are on the leading edge of new technologies. They are eager for the latest methods or tools to improve their productivity. To facilitate that process, there will be a rapid expansion in the use of agriconsultants or technical employees. A broad training in plant pathology supplemented with the practical field application of other disciplines is desirable for these farm advisors. It remains to be seen whether these plant doctors will be trained in departments of plant pathology or in newly created schools (see Chapters 5 and 48).

The innovative methods of the future will originate in increasing numbers from industrial and agribusiness scientists with strong motives to develop practical solutions to production problems. It is this group of people who will help growers put the integrated computerized programs into broad use. Of course, in countries with centrally planned economies, the government officials have control of these functions.

International communication, travel, and exchange of technology are in a rapid state of intensification. Language, territorial, and political barriers are crumbling under the overwhelming international interactions. This is especially true in agribusiness relationships. Agricultural produce frequently represents a primary source of foreign exchange and therefore national priority is given to staying abreast of the best international technology for crop production. Indonesia, for instance, has a major commitment to rice production, including strong grower subsidies and advisory services.

Many of us who think that the role of plant pathologists is to improve crop production are concerned about an emphasis and attitude that appears counter to that objective. We have come through a period when some basic scientific studies have received so much attention and support that the solution of practical problems was partially ignored and even looked down on. I sense a renewed acceptance of applied research in our profession but predict a growing segregation and conflict of interests as more pathologists are working as industrial scientists and plant doctors. And this separation is unfortunate, because there is so much to be gained by the cooperative interactions of basic and applied scientists.

CHALLENGES FOR THE FUTURE

- Plant pathology is a profession for practical plant health, and there is a continued need for consideration of the grower and consumer as well as of the basic science. Those developing fundamental knowledge are challenged to be in dynamic communication with those who can apply the information.
- The transition to the profession of plant health, in which interdisciplinary, basic research, and the plant doctor are complementary, is going to require bold determination. Clear guidelines for the process are needed. The challenge is to make it a cooperative effort.
- Plant breeders, even those using advanced genetic engineering techniques, must be persuaded that disease resistance characteristics deserve a high priority. This will require active participation by plant pathologists.
- The interface between the peasant's hoe and the computerized integrated production program will be increasingly significant. Will plant pathologists consider the common grower?
- Strong incentives to discover and develop new chemical and biological control agents will require more realistic government regulations. It is a challenge to determine the facts concerning safety and environmental impacts (risks) and to consider them in balance with the benefits.
- New discoveries and creative use of available technology are needed for the determination of epidemiological factors, crop loss, and the benefits of disease management. Once again, cooperative interchange will pay off.
- Political issues impact on the effectiveness of plant health activities, and plant pathologists should have an influence on the actions of their government. In addition to the local and national issues worthy of our attention, let's give serious consideration to world issues; perhaps starting with the challenge posed by Bill Paddock in Chapter 4.

CHAPTER 39

THE FUTURE OF BIOLOGICAL AND CULTURAL CONTROL OF PLANT DISEASE

KENNETH F. BAKER

Modern control of plant diseases is based on the principle that the ultimate sources of pathogens are previously infected plants and the soil, including its water and nonliving organic matter. Under the controlled conditions of commercial glasshouses, application of this principle by planting pathogen-free stock in pathogen-free soil, and practicing sanitation to prevent recontamination, has virtually eliminated many important diseases. Under the relatively uncontrolled environment of field plantings, however, the problem is more complex and disease control is therefore more complicated.

A second principle is that control of plant disease usually requires application of multiple integrated procedures, each operating in a different way or time to diminish disease, and collectively providing satisfactory economic control. Although this approach is much publicized in today's pest management programs, plant pathology has long depended on it, and it was there first clearly stated in 1882 by J. L. Jensen.

Pathogens tend to be selected for resistance to widely used chemical controls that may also be harmful to naturally occurring antagonistic microorganisms, and thus augment injury. The soil, water, and air may also be polluted by the chemicals applied. There is, therefore, increasing interest in alternative and supplemental control measures, such as altered cultural practices and biological control. These two general methods are properly grouped together in this book because they are often inextricably entwined. Biological control methods that operate through resident microorganisms must provide suitable environmental factors (e.g., minerals, organic matter, moisture, pH) for their realization. It is not accidental that most of the present effective large-scale commercial applications of biological control involve manipulation of cultural practices (6): e.g., take-all of wheat (caused by *Gaeumannomyces graminis* var. *tritici*); Phytophthora root rot of avocado and pineapple (caused by *Phytophthora cinnamomi*); Fusarium foot rot of wheat (caused by *Fusarium roseum* 'Culmorum'); Fusarium root rot of bean (caused by *F. solani* f. sp. *phaseoli*);

Armillaria root rot of fruit trees (caused by *Armillaria mellea*); potato common scab (caused by *Streptomyces scabies*). That there are examples of successful commercial use of introduced microorganisms in biocontrol (e.g., *Peniophora gigantea* against *Heterobasidion annosum* on pine, *Agrobacterium radiobacter* K84 against *A. tumefaciens* on fruit trees, *Fusarium lateritium* against *Eutypa armeniacae* on apricot trees, *Pseudomonas multivorans, P. fluorescens,* or *Enterobacter aerogenes* against *P. tolaasii* on mushroom) suggests that eventually the specific resident antagonists involved in many of the above examples may become known and effectively introduced. However, in that eventuality cultural practices will still need to be specifically favorable to the introduced antagonists if biological control is to be successful.

Since world food supplies and energy sources will continue to be in short supply for the foreseeable future, plant pathologists will be under increasing pressures to reduce field losses from disease. To the cost-benefit ratio of future disease controls, energy demand and pollution control must be added as other types of cost.

High productivity agriculture is high-energy agriculture, and humans may have to adjust their desire for high yields to the economics of energy cost (21). Obviously, if a 20% yield increase returns $100 more per acre but costs $110 for the various required forms of energy to do it, the grower will settle for lower yield. Thus, the high-yield-potential grains of the "Green Revolution," which require abundant fertilizer and irrigation for maximum performance, have proved less beneficial than expected. In addition, use of these cultivars in the developing countries has led to reduced genetic heterogeneity, with increased risk of catastrophic disease epidemics. Evans (13) pointed out that

> much of the increase in yield that has occurred . . . has been due to selection for better adaptation to those environments rather than to selection for greater yield potential.... If yields must be increased, so must [energy] inputs. It is futile—indeed harmful—to expect that yields can continue to be raised simply by breeding superior new varieties.

Concern about worldwide pollution of the environment can only intensify in the future, and chemicals for disease control inevitably will be subjected to further restrictions. If usual quick chemical control methods are applied against diseases, the resulting pollution may itself eventually reduce crop yields or prove so toxic to people that they will discontinue their use. If chemical application is restricted and alternative methods are used, the control may be too slow or the cost too high. Perhaps the only answer to this dilemma is the increased use of multiple indirect methods based on the specific ecology of the pathogen. This eventuality will pose less of an adjustment problem for pathologists, who have always emphasized such an approach, than for entomologists, who have emphasized application of chemicals.

BIOLOGICAL CONTROL

Microorganisms compete for nutrients, favorable sites, and oxygen, and are selected for tolerance of unfavorable conditions of carbon dioxide, pH, water, and other microorganisms. They secrete metabolic materials, some of which (antibiotics) inhibit other microorganisms; others stimulate microorganisms to form essential stages of their life cycles (e.g., *Pseudomonas* spp. on

Phytophthora cinnamomi). If a microorganism outstrips its nutrient supply, its population is reduced by starvation, antagonists, or both. The population density is thus maintained within definite limits, making for stability and biological balance.

An alien microorganism can establish itself in such a stable community only if it is better adapted to the ecological niche than are some of the residents, if it is introduced in such numbers as to "swamp" the residents, or if it, or some human disturbance, makes the environment more favorable to the alien.

In natural ecosystems plant disease epidemics are unknown or rare, pathogen suppression is usual, and health rather than disease is the norm. Biological control in this sense is the retention or restoration of such a disease-suppressive biological balance. This balance is achieved through antagonism, "the balance wheel of nature," that operates through antibiosis, competition, and parasitism.

The interlocking, flexibly buffered ecological niches may be modified by "nudging" with slight environmental changes, "swamping" with large numbers of some microorganism, or by a "shock" (e.g., chemical or heat treatment of soil, or addition to it of organic matter).

It is thus obvious that biological control is managed by manipulation of the *resident antagonists* through modification of cultural practices, or by the addition of *introduced antagonists*, and that biocontrol must, in any case, work within the limits of biological balance. Biologically "live" situations having an abundant, active, and heterogeneous microbial population are more likely to exert biocontrol of an introduced plant pathogen than are depauperate biologically "dead" niches.

Suppressive soils in nature have a balanced microbiota and probably cannot be fully explained in terms of a single antagonist. Populations antagonistic to a pathogen occur in soils biologically suppressive to it, but individual antagonists occur in many soils (6). A single antagonist may be effective against a single pathogen in a medium free from, or with diminished populations of, other microorganisms (e.g., in treated glasshouse soil), but its chance of success is less in the field, where mass transfer of the total antagonist population offers better prospects. It is much easier to obtain biocontrol of a pathogen in a one-on-one situation (e.g., one pathogen in a nearly sterile medium by one antagonist, as with *Peniophora gigantea* against *Heterobasidion annosum* in a fresh tree wound) than in multiorganismal interactions in field soil. It is, therefore, surprising and disappointing that so little use has been made of this method of disease control in glasshouse crops grown in treated soil, a very promising area for investigation and commercial application of biocontrol.

Biocontrol Strategies

A number of approaches to biocontrol that have proved fruitful and that offer promise for future exploitation are here considered.

Suppressive Situations

> Antagonists should be sought in areas where the disease . . . does not occur, has declined, or cannot develop, despite the presence of a susceptible host, rather than where the disease occurs.

This rather obvious dictum apparently has been one of the most valuable "take-home" messages in *Biological Control of Plant Pathogens* (6). Following this strategy, soils suppressive to wilt fusaria have been found in California and France; to *Rhizoctonia solani* in South Australia, Colombia, and Colorado; to root-knot nematode in California; to *Phytophthora cinnamomi* in Queensland; to *Sclerotium cepivorum* in British Columbia; and to cyst nematodes in England, among others. The strategy, although less exploited on aerial plant parts, has been used against *Botrytis cinerea* on chrysanthemum leaves, *Eutypa armeniacae* on apricot trees, *Poria carbonica* on Douglas fir poles, and *Heterobasidion annosum* on spruce logs.

Future investigations unquestionably will reveal many more suppressive situations. This is particularly needed for pathogens of aerial parts. For example, the striking persistent decrease of fire blight in New Zealand from disastrous to unimportance within a few years (6) still has not been investigated as a promising source of biocontrol microorganisms.

Once a biologically suppressive situation is found, several possibilities for exploitation follow: a) possible increase of effectiveness by modification of cultural practices; b) discovery of the microorganisms involved; c) possibility of transfer of the suppressive microbiota to other situations; and d) modification of cultural practices to favor the successful transfer of antagonists. Discovery of new suppressive situations and the above detailed studies of them will be an important area of research for some years.

Selective Treatments

Because living organisms differ in their tolerance of various environmental conditions, it is possible to use some factors selectively to inhibit or kill plant pathogens without injuring either the host plant or some saprophytic microorganisms.

Thermal. Plant pathogenic fungi, bacteria, nematodes, and apparently the mollicutes, are killed by moist heat at about 60 C for 30 min, but a substantial portion of the saprophytes are uninjured. It is therefore possible to eliminate pathogens from an infested suppressive soil so that it may be used in areas quarantined against the pathogen. Queensland soil suppressive to *Phytophthora cinnamomi* thus remained suppressive after treatment with aerated steam at 60 C for 30 min (8), and French soil to *Fusarium oxysporum* f. sp. *melonis* at 50 C for 30 min (22). Because antagonistic saprophytes exist in the tillage layer of many agricultural soils, it has been found that treatment of such soils with aerated steam at 60 C for 30 min in glasshouses will give a measure of biological control against reintroduced root pathogens. This protection depends on resident saprophytic antagonists in the soil, and it should be noted that it will not result from treatment of soils of near sterility, such as soil mixes containing perlite, vermiculite, or sand mined from deep deposits. Treatment of soil with aerated steam at 60 C, instead of the more usual 100 C steam, thus provides a useful measure of commercial biological control (3). This method of treatment is coming into wide commercial use, and will continue to be used in commercial glasshouse culture.

It is possible to heat soil in the field by solarization in areas with intense sunlight. The soil is moistened and covered with clear polyethylene sheets. The temperatures attained are lower (37–50 C) and are maintained for much longer periods (14 days or more) than with steaming methods. Pathogens are killed, or may be stressed and rendered more susceptible to antagonists (16). This treatment has tremendous potential for certain areas and will continue to be vigorously studied.

Chemical. Chemicals traditionally have been used to destroy microorganisms more or less indiscriminately, making little use of their potential for selective inhibition or killing. It is now clear that they can be used in conjunction with antagonists for effective disease control. Thus, *Eutypa armeniacae* on apricot is controlled in South Australia by application of a mixture of thiabendazole or benomyl and the nonsensitive antagonist, *Fusarium lateritium*, to fresh pruning wounds (10). The fungicide protects against the sensitive pathogen until the antagonist has established in the wound. Such an approach opens up new areas of *integrated control*, demanding new criteria of effectiveness of pesticides, and will undoubtedly be developed for many other pathogens in the future.

Bacterization

Bacterization (inoculation of propagules with selected bacteria prior to planting) was used in the Soviet Union during the Lysenko era, and was critically studied after 1959 in England, Australia, Finland, and the United States. This was approached from the standpoint of increased growth of the plant that was thought to come from hormone production by the bacteria, from nitrogen fixation (*Rhizobium* spp.), and from solubilization of soil minerals. These bacteria largely were obtained randomly from soil and selected by trial inoculations of seeds. Broadbent et al (8,9), in studies on biocontrol of seedling diseases, selected bacteria on the basis of their wide-spectrum antibiotic inhibition of plant pathogenic fungi and bacteria. They found that such inoculations markedly increased growth of seedlings even in the absence of known fungus pathogens, and postulated a direct relationship between antibiotic production and increased growth. They suggested that the effect was due to inhibition of nonparasitic root pathogens (subclinical pathogens, exopathogens), a conclusion supported and extended by Kloepper and Schroth (17), who showed that a nonantibiotic-producing mutant of an effective antibiotic-producing bacterium, as well as effective isolates grown under gnotobiotic conditions, did not give increased growth. Bacteriocins and siderophores produced by the inoculated bacteria were thought to inhibit harmful rhizobacteria. At least in part, bacterization is thus a form of biocontrol of rhizosphere microorganisms that inhibit plant growth without invading the plant.

Studies on bacterization are of great potential significance to agriculture and to plant pathology. They offer a biological means of increasing crop yields without increasing energy demands and without environmental pollution. They are expanding the scope of plant pathology beyond its traditional concern with parasites that penetrate the host and produce disease, for the field must now also include microorganisms in the rhizosphere, which decrease plant growth but rarely or never penetrate the roots. This research area is one of the most promising and exciting in plant pathology today. While it shares a continuum with mycorrhization by vesicular-arbuscular and ectomycorrhizae, it involves different microorganisms and operates through different mechanisms, and thus offers possibilities for studies on new aspects of microorganism interaction. It is to be expected that some of the bacteria used will also prove effective in biocontrol of root parasitic pathogens; this is already true for *Bacillus subtilis* A13 (8).

Genetic Control

Host Plant

The work of Neal et al (19) has opened new vistas of interaction between biocontrol and breeding for resistance to root pathogens. Root exudates largely determine the rhizosphere microflora (6,7), and, as would be expected, such exudation was shown to be under host genetic control. The specific alteration of host genotype so as to selectively manipulate rhizosphere exudates and microbiota is a largely unexplored but very promising potential means of biocontrol of root pathogens.

Antagonist

Another exciting research area today is the possibility of genetically designing antagonists for specific situations. We are now able to transfer genes for antagonism to a given pathogen, from a bacterium poorly adapted to the particular situation to a well-adapted but nonantagonistic bacterium. There are many laboratories working on this type of problem, and the future for such antagonist-designing is most promising. It is quite possible that we may be able to improve "natural" antagonists or increase their adaptability.

CULTURAL PRACTICES

Increased cost of labor and energy, with resultant mechanization, will bring about changes in cultural practices beyond those that would normally occur in agriculture. The use of no-till practices, mixed cropping or intercropping, drip or trickle irrigation, and mechanical picking or harvesting of the crop, examples of probable widely used future practices, will have an influence on biotic and abiotic plant diseases and their control (25). Seeding wheat in the stubble of the preceding crop without plowing will reduce both energy demands and soil erosion, but its effect on pathogens probably will be varied, increasing some and diminishing others depending on whether they use plant residues as a shelter or as a food base (11).

Some cultural practices reduce disease by reducing stress on the host plant. *Fusarium roseum* 'Culmorum' produces most injury when wheat is under water stress. Thus, delayed seeding and minimal use of nitrogen to reduce plant succulence and size, and growing water-efficient cultivars, all reduce water stress and root and foot rot (6).

Chemical or heat treatments may be used to "stress" the pathogen and increase effectiveness of biocontrol, as in soil treatment with carbon disulfide or heat to control *Armillaria mellea* (18). Mycelium of the pathogen so stressed produces less toxin inhibitory to the antagonist, *Trichoderma viride*, and therefore is attacked more vigorously by it. A similar stressful effect may be obtained by solarization of soil (16).

Determination of the proper cultural practice to use against a particular pathogen must be based on detailed knowledge of the ecology of host and pathogen, the culture of the crop, and the general agricultural problems of the area. To devise a successful cultural program thus requires a good deal of broad-based research, a generous dash of imagination, and much testing in field plots. This type of investigation may well become a dominant type of research in the future, as control comes to emphasize cultural practices, with or without chemical applications.

Many different types of cultural practices have been found to decrease disease incidence. Many of these have been reviewed recently (6,7,11,12,14,23,25), and there is no presently apparent means of increasing their effectiveness. Because of this and limited space in this book, discussion is here limited to those that appear most promising for either research or development. However, this does not imply that other cultural practices have limited use in the future.

Pathogen-Free Propagules

Use of pathogen-free propagules has become a very important means of disease control, and its importance will become even greater in the future. When infection of a plant propagule has occurred, the disease already is past the preventive phase. Thus, infected seeds are the crucial factor in outbreaks on lettuce mosaic of lettuce in California, even though the virus and its aphid vector may be present in weeds in adjacent fields (4). Plant propagule transmission of pathogens also provides an important automatic selective transmission of host-specific strains of a pathogen.

Selection Methods

Because of the great importance of infected plant propagules in perpetuating disease, development of methods for obtaining pathogen-free propagules will receive much attention in future research. The use of sterile explant cultures in plant introduction, started by R. P. Kahn in 1976 for asparagus, will greatly simplify intercontinental movement of propagative material. Such cultures are now accepted for introduction of propagules into Australia. The feasibility of the techniques considered here has been demonstrated, but much remains to be done in adapting them to a wider range of plants (15).

The smaller the plant part used for propagation, the better the chance of obtaining units free from plant pathogens, but the more complex and difficult the culture technique becomes. This line of work has proved extremely successful on numerous plants and will be extended to other vegetatively propagated crops in the future. Because it may provide propagules free from virus as well as microbial pathogens, it is the method of choice. However, each plant unit should be indexed for viruses before use in commercial propagation since some still may be infected.

The possibility of using such sterile cultures in place of the old mother blocks is being investigated by commercial propagators. The cultures may be held for surprisingly long periods in refrigerated chambers under low light intensity, without hazard of pathogen (particularly virus) spread. This fact, with the smaller space requirement, presents an exciting new prospect for propagators. The plants must, however, be checked for mutations and variability (24).

Apical Meristem Cultures. The apical 0.1–0.5 mm of the growing point of a stem of a healthy mother plant is aseptically removed and placed on a small strip of sterile filter paper arched over sterile nutrient solution or placed on nutrient agar in a test tube. When roots have formed, the tiny plant is transferred to a thumb pot of sterilized medium. This method is now widely used for producing plantlets, particularly of ornamentals and strawberries. Since each type of plant may require special handling methods or nutrient media, investigations will be needed to apply the technique to a wide range of plants.

Single-Cell Cultures. Separation of single cells from plant tissue and culturing them on special media is an even better, although still uncommon, method for obtaining pathogen-free plants. It was developed at Cornell University by F. C. Steward and associates about 1958, and has since been adapted for several kinds of plants.

Single-Protoplast Cultures. J. F. Shepard and associates reproduced tobacco and potato from single protoplasts at Montana State University and Kansas State University after 1975. In addition to obtaining plants free from plant pathogens, a surprising array of variant plant types appeared, some of them resistant to disease (24).

Treatment Methods

Heat Therapy. This treatment, based on the greater sensitivity of pathogens than the host to heat, has been used since its introduction by J. L. Jensen in 1887. It has been successfully used against fungi (particularly the Phycomycetes), bacteria, nematodes, viruses, and mollicutes. Treatments for all of these except the last two are for relatively short periods (30–60 min, depending on size of the treated material) and at temperatures of 43–57 C. Against mycoplasmas an extended treatment period (10–30 days) at 36–37.8 C has proved effective for fruit trees (20). Details of methods of handling material for treatment are given by Baker (2) and Nyland and Goheen (20). The effectiveness of the prolonged low-temperature treatment on bacterium-like mycoplasmas suggests that such treatments might be effective on more heat-resistant bacteria such as *Pseudomonas*.

Heat treatment of true seeds has been done with hot water, but the use of aerated steam has shown several advantages since its introduction in 1962 (2,4). It has more recently been shown (5) that holding the seed in polyethylene glycol following treatment with aerated steam gives opportunity for regeneration of metabolic systems of seeds injured by the treatment. This enables safe treatment at higher temperatures than before, and will probably be used extensively in the future. Prolonged heat treatment of virus-infected seed held in polyethylene glycol also has shown promise for inactivating the pathogen.

The recent suggestion that treatment at 5–8 C for 4–6 mo, followed by meristem culture, would free plants from a viroid should be investigated further.

Chemical Treatment. The emphasis in chemical treatment of propagules has been on eradication of the pathogen with minimal emphasis on phytotoxicity. There is already a trend toward using chemicals that inhibit the pathogen but not the accompanying antagonists, which then either eliminate the pathogen or inhibit it so effectively as to control the disease. Such combined chemical-biological control will come into wide use, requiring development of less potent and hazardous chemicals of marked specificity.

Certification and Standardization of Propagules

The production of pathogen-free plant propagules often, but not necessarily, leads to a certification program to identify such stock for the buyer. Experience indicates that such certification should be done by a state agency and that it should be voluntary rather than compulsory. Avocado and strawberry plants have been so handled for some years in California.

Numerous schemes have been devised for standardization of nursery stock, largely on the basis of plant size and form. However, a well-grown plant produced without check under consistently favorable conditions, and free from root pathogens, is a better buy than a larger specimen grown more slowly under intermittently unfavorable conditions and infected with root rot fungi but not yet showing disease symptoms. Standardization, to be useful, must therefore include certification for freedom from disease to provide a valid evaluation of growth potential of the stock (1).

LITERATURE CITED

1. Baker, K. F. 1959. Factors controlling plant standardization. Pac. Coast Nurseryman 13(9):31,32,70,71,74,75.
2. Baker, K. F. 1962. Thermotherapy of planting material. Phytopathology 52:1244-1255.
3. Baker, K. F. 1971. Soil treatment with steam or chemicals. Pages 72–93 in: Geraniums, 2nd ed. J. W. Mastalerz, ed. Penn. Flower Growers, State College, PA. 350 pp.
4. Baker, K. F. 1972. Seed pathology. Pages 317–416 in: Seed Biology, Vol. 2. T. Kozlowski, ed. Academic Press, New York. 477 pp.
5. Baker, K. F. 1979. Seed pathology—Concepts and methods of control. J. Seed Technol. 4:57-67.
6. Baker, K. F., and Cook, R. J. 1974. Biological Control of Plant Pathogens. W. H. Freeman, San Francisco. 433 pp.
7. Baker, K. F., and Snyder, W. C. 1965. Ecology of Soil-Borne Plant Pathogens. Univ. Calif. Press, Berkeley. 571 pp.
8. Broadbent, P., Baker, K. F., and Waterworth, Y. 1971. Bacteria and actinomycetes antagonistic to fungal root pathogens in Australian soils. Austr. J. Biol. Sci. 24:925-944.
9. Broadbent, P., Baker, K. F., Franks, N., and Holland, J. 1977. Effect of *Bacillus* spp. on increased growth of seedlings in steamed and in nontreated soil. Phytopathology 67:1027-1034.
10. Carter, M. V., and Price, T. V. 1974. Biological control of *Eutypa armeniacae*. II. Studies on the interaction between *E. armeniacae* and *Fusarium lateritium* and their relative sensitivities to benzimidazole chemicals. Austr. J. Agric. Res. 25:105-119.
11. Cook, R. J., Boosalis, M. G., and Doupnik, B. 1978. Influence of crop residues on plant diseases. Pages 147–163 in: Crop Residue Management Systems. W. R. Oschwald, ed. Spec. Publ. 31. Am. Soc. Agron., Madison. 248 pp.
12. Curl, E. A. 1963. Control of plant diseases by crop rotation. Bot. Rev. 29:413-479.
13. Evans, L. T. 1980. The natural history of crop yield. Am. Sci. 68:388-395.
14. Huber, D. M., Watson, R. D., and Steiner, G. W. 1965. Crop residues, nitrogen, and plant disease. Soil Sci. 100:302-308.
15. Ingram, D. S., and Helgeson, J. P. 1980. Tissue Culture Methods for Plant Pathologists. Blackwells, Oxford, England. 272 pp.
16. Katan, J. 1981. Solar heating (solarization) of soil for control of soilborne pests. Ann. Rev. Phytopathol. 19:211-236.
17. Kloepper, J. W., and Schroth, M. N. 1981. Plant growth-promoting rhizobacteria and plant growth under gnotobiotic conditions. Phytopathology 71:642-644.
18. Munnecke, D. E., Wilbur, W., and Darley, E. F. 1976. Effect of heating or drying on *Armillaria mellea* or *Trichoderma viride* and the relation to survival of *A. mellea* in soil. Phytopathology 66:1363-1368.
19. Neal, J. L., Jr., Larson, R. I., and Atkinson, T. G. 1973. Changes in rhizosphere populations of selected physiological groups of bacteria related to substitution of specific pairs of chromosomes in spring wheat. Plant Soil 39:209-212.
20. Nyland, G., and Goheen, A. C. 1969. Heat therapy of virus diseases of perennial plants. Ann. Rev. Phytopathol. 7:311-354.
21. Pimentel, D. 1980. Environmental risks associated with biological controls. Pages 11–24 in: Environmental Protection and Biological Forms of Control of Pest Organisms. B. Lundholm and M. Stackrud, eds. Ecol. Bull 31. Swedish Nat. Sci. Res. Counc., Stockholm. 171 pp.
22. Rouxel, F., Alabouvette, C., and Louvet, J. 1977. Recherches sur la résistance des sols aux maladies. II. Incidence de traitements thermique sur la résistance microbiologique d'un sol a la Fusariose vasculaire du Melon. Ann. Phytopathol. 9:183-192.
23. Schippers, B., and Gams, W. 1979. Soil-Borne Plant Pathogens. Academic Press, New York. 686 pp.
24. Shepard, J. F., Bidney, D., and Shahin, E. 1980. Potato protoplasts in crop improvement. Science 208:17-24.
25. Sumner, D. R., Doupnik, B., Jr., and Boosalis, M. G. 1981. Effects of reduced tillage and multiple cropping on plant diseases. Ann. Rev. Phytopathol. 19:167-187.

CHAPTER 40

FUTURE STRATEGIES FOR INTEGRATED CONTROL

JOHN H. ANDREWS

The term "integrated control" (IC), now used almost everywhere in the pest-oriented disciplines and in medicine (9), had its agricultural origins in entomology. "Integrated" strategies—initially restricted to the combination of biological control by parasites, predators, and pathogens, supplemented judiciously by chemicals—were devised to counteract problems of insecticide programs, such as the development of arthropod resistance, secondary insect outbreaks, resurgence, and presence of toxic residues (23). Since then, IC has attained remarkable scientific and political popularity; one consequence is that currently the meaning of IC is unclear. The concept has been broadened considerably and the term is often misconstrued. For example, depending on the author, IC, pest management, integrated pest management (IPM), integrated plant protection, supervised control, modified spray program, and bioenvironmental control can mean the same or quite different strategies. Therefore it is essential to establish clearly the essence of IC. To emphasize the importance of its component elements, I then orient discussion of future strategies around the definition of IC, rather than focusing on specific controls and how they may lend themselves to the concept.

THE CONCEPT OF INTEGRATION

While the descriptions of IC are as numerous as individuals willing to propound them, however defined, IC includes three essential facets: 1) diversified controls coordinated to achieve an additive or, preferably, synergistic, effect; 2) economic (including aesthetic) analysis to the extent that action thresholds are developed to distinguish biological damage from economic damage (i.e., the amount of injury justifying a given increment of control), and 3) ecological and environmental assessment to quantify, and subsequently minimize, a detrimental effect on nontarget organisms.

Two aspects relating to the IC concept need emphasis. First, despite traditionally diversified disease controls, IC *is* a new concept in plant protection generally and plant pathology in particular. Acquiring a detailed knowledge of pathogen life cycles and developing multiple controls are approaches used historically and successfully for combating plant diseases. They provide an excellent foundation for devising IC strategies. However, diversification is frequently confused with integration. The difference is analogous to a jigsaw puzzle, which in the former case exists as a stack of pieces and in the latter, an entity that conveys information as a whole. There also needs to be more attention paid to economic and ecological analysis to fulfill the spirit of IC. Secondly, IC should remain *distinct from* IPM (and the other terms noted above). IC is inherently and primarily a disciplinary undertaking, and one could theoretically devise programs for any specific pest. IPM, on the other hand, involves both disciplinary and methodological integration. IPM includes IC programs for various kinds of pests and additional elements such as consideration of societal values and impact on other resources. Thus there is a vast difference in hierarchy, scope, and complexity between the two strategies. This is inherent in the term "management," which implies a broader approach than "control," just as the term "financial management" refers to the flow of funds from an individual or corporation according to some (usually complex) plan, rather than merely the diversion of those funds to a savings account at 5% interest. As Coppel and Mertins (8) note, IPM is one element of overall resource management. It is the complexity of the IPM endeavor, and to a lesser extent of IC programs, that has necessitated use of modeling and general systems thinking. The interrelationships of these terms and the evolutionary sequence of agricultural pest control are depicted schematically in Fig. 1.

DIVERSIFIED, COORDINATED CONTROLS

New Components

Apart from regulatory actions, the three optional strategies available for control of any pest are cultural, biological (including genetic), and chemical. The total control effort for a specific pathogen, whether integrated or not, may be conceptualized as a pyramid wherein these three components comprise varying volumes. I start with the premise that it is desirable to build an "integrated" pyramid and one for which cultural control represents the foundation and major element, supplemented toward the apex by biological and chemical tiers, respectively. Currently, this conceptual pyramid is in fact often inverted, with chemicals or biological approaches (resistant cultivars) occupying the major portions. Constraints on space preclude a detailed discussion of each option; accordingly, I summarize tactics I consider to have the most potential.

Cultural Control

That the cropping system is the dominant feature of agroecosystems, and hence the major determinant of control strategies, is widely recognized (3,20). A key direction for future IC strategies will involve designing agroecosystems that minimize disease, rather than contending with consequences of crop production schemes, such as for ease of harvesting. This effort should be made on a grand, interdisciplinary scale and is more in the realm of agroecosystem design (Fig. 1) than IC. However, plant protection specialists need to provide the specific details pertinent to their respective areas of expertise before a coherent strategy can be developed for the entire system. For example, could novel orchard floor manipulations contribute to the decomposition of apple leaves or the prevention of aerial dissemination of ascospores of *Venturia inaequalis*? Can this approach form a major element in an IC program for apple scab and, if so, what will be its effect on weeds or predatory mites? The basic information from pathology must meld with that from complementary investigations by other specialists; one moves to the higher order of IPM, and ultimately, agroecosystem design.

The fundamental issue in agroecosystem design appears to be diversity vs.

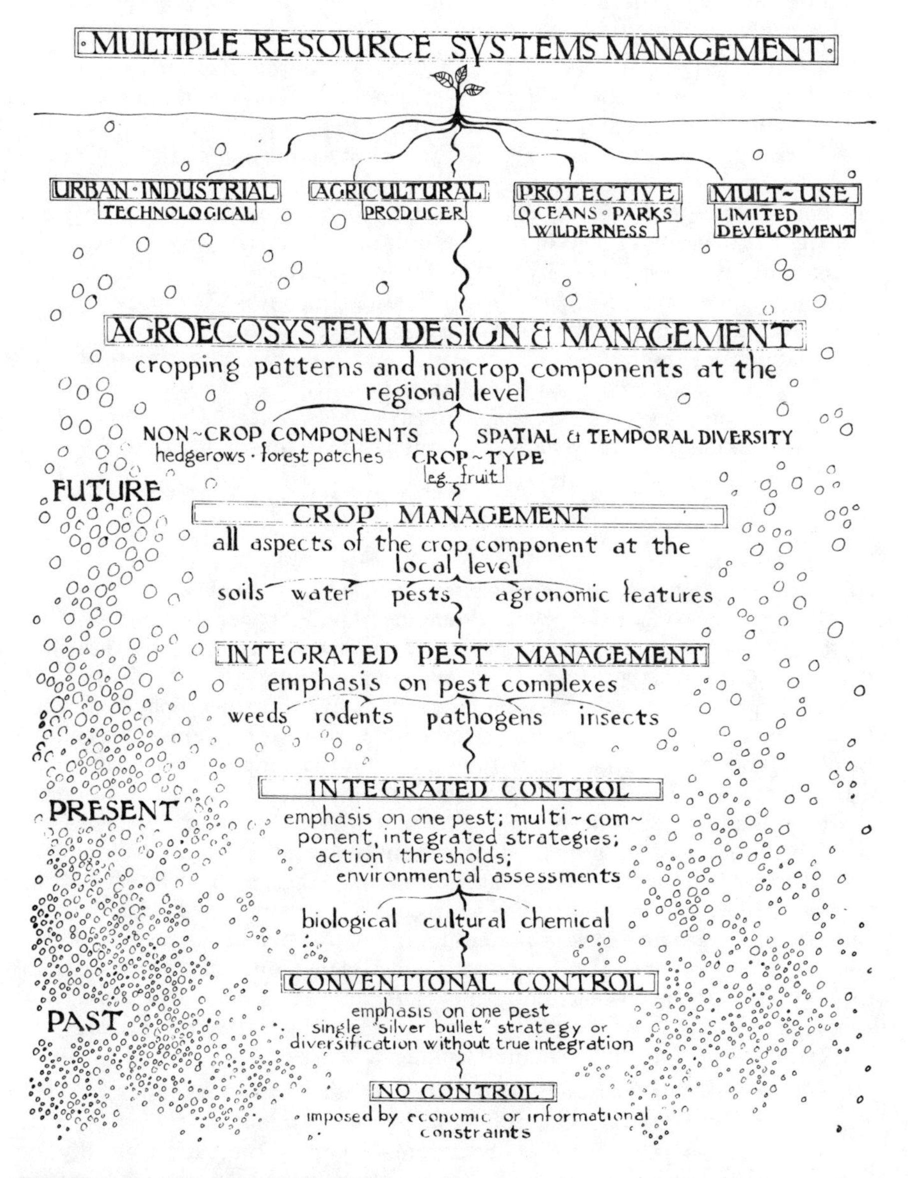

Fig. 1. Integrated control as it relates schematically to other levels of control and crop production. Hierarchical order and relative emphasis of strategies over time are depicted, not all possible interactions.

stability. This is because diversity, in terms of both habitat characteristics and richness of living organisms, is inevitably altered by agriculture. Although it now appears that the relationship between diversity and stability (i.e., the relative tendency of a system to remain near an equilibrium point) is parallel and not causative (19,24,27), certain key elements of diversity such as physical complexity can be stabilizing (19). What remains to be determined is the extent and kind of spatial or temporal diversity to seek for a particular pest-crop situation and the size of the geographic area over which adjustments are needed for a beneficial influence to be realized.

One way whereby pathologists can contribute to agroecosystem design is by evaluating the manipulation of factors influencing diversity and microclimate, factors such as row spacing and orientation; canopy modification by pruning, root stock selection, and trellising; and irrigation timing and method. Another possibility is to contribute through research on the relationship between disease and habitat complexity, as exemplified by cropping patterns ranging from monoculture to various forms of multiple cropping. Intercropping, for example, has reduced pest pressure on a variety of crop combinations in the humid and semiarid tropics (16,22,26). In general, this has not been used elsewhere for various reasons, among them abundance of land, accessible alternative controls, lack of labor, and emphasis on mechanization. To lay the foundation for future IC strategies, experiments need to be devised that address why certain natural and agrarian monocultures are conspicuously stable whereas others are not (22) and to what extent vegetative patterns, as one form of diversity, can provide the cultural basis of the control pyramid.

Biological Control

The most innovative and exciting component of biological control awaiting exploitation appears to me to be genetic engineering. Plant genetic manipulation to attain higher productivity, in part by transferring disease resistance, has attracted considerable attention (7). Gene cloning and gene transfer offer the prospect of imparting desirable characters while avoiding incompatibility that results from certain interspecific or intergeneric crosses. In an era when animal and plant genes have been shown to function within bacteria, and bacterial genes within plant and animal cells, the potential of gene-splicing methods for disease control is limitless. Among other agents that may be useful as "vehicles" to introduce exogenous "passenger" deoxyribonucleic acid (DNA) into plant host cells for transformation are plant pathogens (e.g., Ti plasmids or intact cells of *Agrobacterium tumefaciens* and DNA viruses such as cauliflower mosaic virus). Clearly the prospect now exists for identifying and transferring disease resistance genes by novel mechanisms and for altering these genes, once they are isolated in vitro, by various means, including use of exonucleases, chemical mutagens, ligases, or restriction enzymes; for reviews of recombinant DNA technology see Wu (28), Abelson and Butz (1), and Cocking et al (7).

Besides engineering the host plant directly, genetic alterations of the pathogen or potential microbial antagonists may be possible. In fact, it seems clear that recombinant DNA procedures will be developed considerably more rapidly for prokaryotes and microbial eukaryotes than for higher eukaryotes. Several plasmid vehicles for DNA cloning in *Bacillus subtilis* are known; genetic transformation of *Saccharomyces cerevisiae* has been demonstrated; and numerous fungal genes have been expressed in *Escherichia coli* (17). Indications are that many species of microorganisms may be useful as hosts for replication of recombinant DNA. These advances suggest the exciting idea of tailoring microbes that combine the traits of effective antagonism and competetiveness within the microbial community. One possibility is to splice genes regulating, for example, antibiotic production, into cells of species that are residents of the leaf or root surfaces. This approach would overcome several shortcomings of current microbiological control efforts by expanding the range of pathogens to which a given antagonist is effective, avoiding the rapid decline inherent with introduced antagonist populations, and decreasing the sensitivity of antagonists to pesticides.

Another option, engineering debilitation into the pathogen, offers two possibilities: 1) use of altered strains as vectors of foreign DNA, and 2) establishment of altered biotypes among other epiphytes to compete with or transfer hypovirulence into the wild-type strain. Recombinant DNA techniques may accomplish what the conventional "avirulent" strain approach often has not, namely, produce sufficient or stable avirulence and competetiveness of the biocontrol agent.

Overall, it seems clear that, analogous to advances in medical and industrial microbiology, plant pathologists will soon harvest the fruits of ongoing basic research efforts in genetic engineering for applied disease control strategies.

Chemical Control

Developing strategies in chemical control have attracted considerable attention. The subject is discussed by Tweedy in Chapter 37. Accordingly, chemicals will be considered here only briefly, and specifically from the standpoint of developments advantageous for IC. Broadly speaking, advances will be made in two related areas: 1) the types of chemicals or mechanisms of activity (e.g., viricides, antisporulants, antitoxins, promotors of host resistance) and 2) modes of application. The latter appear to offer the greatest promise for IC.

As an example of a new type of chemical, it may be possible to develop agents that are preferentially absorbed by the pathogen or the diseased tissue. For systemic therapy, these chemicals could potentially overcome two major current obstacles, phytotoxicity and massive dilution in the plant's transport system. Analogous advances are being made in medicine by the development of anticancer drugs that, bound to polymers, are less toxic than the free compounds and can be used in higher concentrations.

Innovations in modes of application will include increasing emphasis on development of action thresholds (see below) to facilitate timing; physical formulation (e.g., controlled release) or placement of the chemical to minimize adverse nontarget effects; and use of chemical combinations to obtain synergistic effects, combat mixed infections, and forestall development of resistance. For example, increased use of disease forecasting systems in IC will enable the emphasis to swing from a preinfection chemical strategy, often implemented on a calendar or phenological basis, to occasional postinfection sprays as needed. However, this approach will require development of chemicals possessing longer and better eradicative activity than most of those currently available have.

More Components

Within the three major control options, numerous specific tactics are potentially available. Although any given tactic is more or less feasible in a specific situation, I believe many possibilities have not been exploited in the past. Emphasis has been placed on a "silver bullet" approach. The alternative, deploying numerous tactics that each contribute a small increment, could meet our economic criteria and also offer long-term stability. A common statement in agriculture is "we have to use chemical A because no resistance is available in crop B to pest C." Paradoxically, resistant cultivars often were not developed *because* an effective chemical strategy was available. The rationale for this is that breeders should not waste their resources when a good alternative control exists. On the other hand, one should ask how long a specific control can exist effectively alone and to what extent can it be improved by integration with other, perhaps less obvious, components.

Integrating Components

The distinction between diversified control and coordinated IC was emphasized earlier. Development of truly unified control strategies hinges on identification and integration of appropriate components. "Appropriate" should involve both effectiveness and compatibility with other elements. A good example of the selection of complementary components is the beneficial combination of carbon disulfide or methyl bromide and *Trichoderma viride* for the control of *Armillaria mellea* in California citrus orchards (4,18). If benomyl is substituted, *Trichoderma* is impeded and *Armillaria* survives (3). Similarly, the combined biological and chemical control of *Eutypa armeniacae* is feasible because, fortuitously, the antagonist was tolerant to the fungicide (5). Elsewhere, work is in progress to incorporate fungicide tolerance into antagonists of the apple scab pathogen (Andrews et al, *unpublished*).

There have been few attempts to ascertain precisely the extent to which individual controls contribute to an IC program. Fry (12) quantified the relationship of fungicide dosage and polygenic resistance for the potato late blight disease. The greater polygenic resistance in the cultivar Sebago than in Russet Rural was equivalent in effect to about 0.5 kg (a.i.) of fungicide (Dithane M-45) applied weekly to Russet Rural. More studies of this sort are needed before it will be possible to rigorously select components, evaluate their interaction, and determine their relative contributions in IC programs.

ECONOMIC ANALYSIS

Economic threshold levels (ETL) have not attained nearly the prominence in pathology that they have in entomology. This is largely because of the inherent biological differences between pathogens and insects. Monitoring is the cornerstone of ETL, and pathogens are relatively more difficult than insects to monitor. More importantly, the damage caused by insects is often relatively easy to quantify, whereas pathogens typically cause physiological disorders (diseases) that are complex and separated in time from the incitant. Disease severity is therefore not necessarily directly relatable to the unit being monitored. It is thus more challenging to relate a given population of a pathogen to ultimate plant damage and to reduction of crop value. Finally, there is the question of systemic diseases wherein an ETL for the pathogen or vector is considerably different at the individual host and host population levels. For example, with aster yellows in Wisconsin, spring surveys have been valuable in predicting the date of arrival of leafhoppers immigrating from the south and therefore the time when insecticidal sprays need to be applied. Anticipated severity of the disease in any given year can be estimated from the number of arriving infectious vectors (6). For a field of carrots or lettuce, this system is analogous to an ETL. However, for an individual plant, one infectious leafhopper is sufficient theoretically to inoculate a systemic pathogen that, for all practical purposes, eliminates the economic value of the infected host.

Advances in modeling, disease forecasting, and crop loss evaluation methods (see chapters in parts II and III), coupled with development of chemicals that can be used effectively after infection, will accelerate use of ETL in pathology. A major effort is needed, with increased emphasis on sampling procedures and on determining the extent to which the ETL for a particular pest is influenced by other stresses such as various levels of different pests simultaneously attacking the host, crop age, and environmental factors.

The focus for IC in plant protection has been on determination of thresholds. The larger realm of cost-benefit analysis awaits more attention. It is emphasized that, although valuable, the ETL concept only allows an arbitrary control program to be implemented in an economically justified fashion. It implies *nothing* about *which* method of control is best. Cost-benefit analysis (CBA) permits different crop production schemes, including IC strategies, to be compared. The analysis also permits a level of optimization to be recognized. This is considered to occur when the difference is maximized between the crop value with specific controls in place and the cost of those controls (11,13). CBA is considerably more complex than determination of ETL and hinges on efficiency of control and the means to quantify it. There is scope for extensive economic research in this area of IC and a need for data on which to determine the relative merits of short- vs. long-term optimization.

ECOLOGICAL AND ENVIRONMENTAL ASSESSMENT

Documenting side effects of IC programs as a whole, and their component elements individually, and minimizing deleterious impacts are inherent in the IC concept. This philosophy carries with it the implication that IC strategies are needed that offer *both* synergistic control over conventional approaches and enhanced environmental quality.

Until the last decade there has been relatively little quantitative or systematic assessment of the nontarget effects associated with applications of fungicides, nematicides, or bactericides. Most reports concern exceptional observations, such as that a fungicide may control a specific disease while enhancing another (21). Other side effects, such as fungicide-induced delay of leaf senescence and concomitant increase in crop yield, e.g., Dickinson and Walpole (10), may be beneficial. Additional research is needed before the ecological effect of chemicals in disease control programs can be assessed adequately. Two issues are particularly important: 1) the possible interaction between chemicals resulting in synergistic toxicity and 2) persistence. For example, in the latter case, it is now thought that data that suggested a short persistence of certain insecticides were misleading because bound (unextractable) residues were not assayed. These residues were, nevertheless, available for biological uptake, e.g., Lichtenstein et al (15). Such studies raise fundamental questions about the concept of "persistence" and "nonpersistence"; one wonders also the extent to which information on the persistence of other agrochemicals may need to be reinterpreted. Although the environmental influence of the chemicals to control diseases, as a group, appears now to be relatively small compared to that of other classes of pesticides, side effects inevitably occur (2). Research in IC needs to give increased emphasis to the assessment of these side effects.

Even less is known about the ecological ramifications of biological and cultural controls because these effects are difficult to investigate and have been presumed to be comparatively harmless. Examples include various soil or plant refuse manipulations (e.g., tillage, flooding, burning) that influence the soil environment, incorporation of disease resistance genes that may be linked to genes influencing various factors such as susceptibility to other pests, and broad-scale application of antagonists that may be allergenic or alter nutrient cycling. In an era of increasing emphasis on nonchemical controls, it is important that more effort be made to quantify the costs along with the benefits of the biological and cultural components of IC programs.

It is debatable whether all external costs (i.e., indirect costs such as environmental impact) of any control strategy could ever be translated into realistic economic terms. However, insofar as possible, external factors must be identified and quantified so that the net social value of a specific policy can be assessed and resources can be allocated optimally. The issue is illustrated in Fig. 2, from Headley and Lewis (14), which represents a hypothetical agroecosystem receiving various amounts of pesticide. Without any pesticide, positive benefits equal to OQ_0 are realized. These benefits increase to Q_1 as pesticides are added to a level of P_0. If only direct costs are considered, the higher level of pesticide input, P_0, is preferable to the lower, P_1, because the distance $Q_1 - N_0$ is a maximum. However, if external costs are tallied, the net social value is maximized at P_1 because $Q_2 - N_2 > Q_1 + N_1$. The value of positive benefits from the last increment of pesticide does not exceed the total cost of that incremental input. This demonstrates the economic concept of marginality, which is central both to the rationale for and practice of IC and IPM. In essence the environmental and social issues of pest control are represented by the difference between the two cost curves. To choose and then implement a control strategy wisely, one must make marginal decisions as to whether, first, each extra dollar expended is met by a dollar returned and, second, which type of control technology to apply to realize the incremental return. More economic analyses, particularly with respect to indirect costs, need to be undertaken to provide a reliable basis for future IC strategies. These studies will hinge largely on a better understanding of environmental impact.

CONCLUSION

Watt (25) argued that pest control problems share the counterintuitive characteristic of complex systems in that application of an intuitively obvious control may exacerbate rather than improve the pest situation. In dealing with this attribute of pest systems, IC is at a key level in the hierarchical order (Fig. 1) ranging from single control elements through policy decisions affecting resource allocation at the agroecosystem level. This is because IC represents in effect a "staging area" where control tactics for a specific pest can be meshed in various combinations and assessed for their effectiveness, mutual compatibility, and environmental impact. It thus can account for the key interactions, while avoiding many of the complexities of the higher orders that would have to be dealt with in a simplified, abstract form. IC needs to be a major goal for pathologists in the coming decades. An outstanding base of fundamental and applied information exists for this endeavor. The future strategies that are developed will have major implications not only for disease control, but for the composition of IPM programs and the management of agroecosystems.

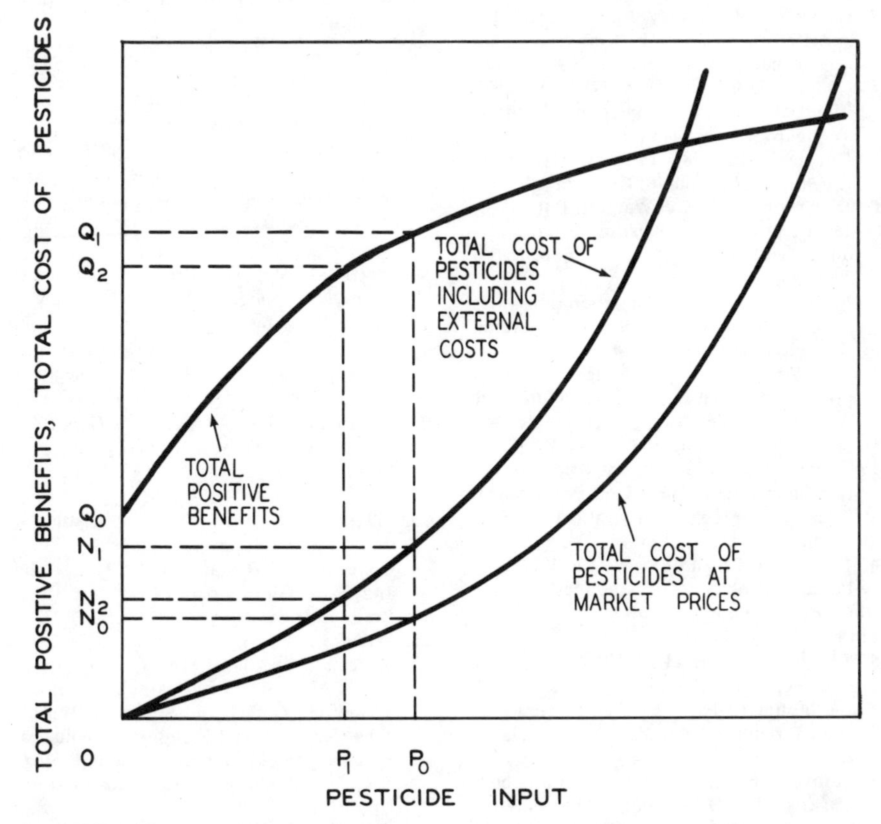

Fig. 2. Effect of external costs on determination of net social value of a control program. Net social value is maximized at the lower level of pesticide input, P_1, because $Q_2-N_2>Q_1-N_1$. Reproduced, by permission, from *The Pesticide Problem: An Economic Approach to Public Policy* by J. C. Headley and J. N. Lewis (14). Published by The Johns Hopkins University Press for Resources for the Future, Inc. 1967.

LITERATURE CITED

1. Abelson, J., and Butz, E. 1980. Recombinant DNA. Science 209:1319-1435.
2. Andrews, J. H. 1981. Effects of pesticides on non-target micro-organisms on leaves. Pages 283–304 in: Microbial Ecology of the Phylloplane. J. P. Blakeman, ed. Academic Press, New York.
3. Baker, K. F., and Cook, R. J. 1974. Biological Control of Plant Pathogens. Freeman, San Francisco. 433 pp.

4. Bliss, D. E. 1951. The destruction of *Armillaria mellea* in citrus soils. Phytopathology 41:665-683.
5. Carter, M. V., and Price, T. V. 1974. Biological control of *Eutypa armeniacae*. II. Studies of the interaction between *E. armeniacae* and *Fusarium lateritium*, and their relative sensitivities to benzimidazole chemicals. Aust. J. Agric. Res. 25:105-119.
6. Chapman, R. K. 1971. Prediction of aster leafhopper migrations and resultant aster yellows problems in Wisconsin. Proc. N. Cent. Br. Entomol. Soc. Am. 21:96-98.
7. Cocking, E. C., Davey, M. R., Pental, D., and Power, J. B. 1981. Aspects of plant genetic manipulation. Nature 293:265-270.
8. Coppel, H. C., and Mertins, J. W. 1977. Biological insect pest suppression. Springer-Verlag, New York. 314 pp.
9. Day, S. B., ed. 1981. Integrated Medicine. van Nostrand Reinhold, New York.
10. Dickinson, C. H., and Walpole, P. R. 1975. The effect of late application of fungicides on the yield of winter wheat. Exp. Husb. 29:23-28.
11. Ferris, H. 1981. Dynamic action thresholds for diseases induced by nematodes. Annu. Rev. Phytopathol. 19:427-436.
12. Fry, W. E. 1975. Integrated effects of polygenic resistance and a protective fungicide on development of potato late blight. Phytopathology 65:908-911.
13. Headley, J. C. 1972. Defining the economic threshold. Pages 100–108 in: Pest Control Strategies for the Future. Natl. Acad. Sci., Washington, DC. 376 pp.
14. Headley, J. C., and Lewis, J. N. 1967. The Pesticide Problem: An Economic Approach to Public Policy. Johns Hopkins Univ. Press, Baltimore, MD. 141 pp.
15. Lichtenstein, E. P., Katan, J., and Anderegg, B. N. 1977. Binding of "persistent" and "nonpersistent" ^{14}C-labeled insecticides in an agricultural soil. Agric. Food Chem. 25:43-47.
16. Monyo, J. H., Ker, A. D. R., and Campbell, M., eds. 1976. Intercropping in semi-arid areas. Int. Dev. Res. Ctr., Ottawa, Canada. 72 pp.
17. Morrow, J. F. 1979. Recombinant DNA techniques. Pages 3–24 in: Methods in Enzymology. Vol 68. Recombinant DNA. R. Wu, ed. Academic Press, New York. 555 pp.
18. Munnecke, D. E., Kolbezen, M. J., and Wilbur, W. D. 1973. Effect of methyl bromide or carbon disulfide on *Armillaria* and *Trichoderma* growing on agar medium and relation to survival of *Armillaria* in soil following fumigation. Phytopathology 63:1352-1357.
19. Murdoch, W. W. 1975. Diversity, complexity, stability and pest control. J. Appl. Ecol. 12:795-807.
20. Nusbaum, C. J., and Ferris, H. 1973. The role of cropping systems in nematode population management. Annu. Rev. Phytopathol. 11:423-440.
21. Nutman, F. J., and Roberts, F. M. 1969. The stimulating effect of some fungicides on *Glomerella cingulata* in relation to the control of coffee berry disease. Ann. Appl. Biol. 64:335-344.
22. Perrin, I. M. 1977. Pest management in multiple cropping systems. Agroecosystems 3:93-118.
23. Stern, V. M., Smith, R. F., van den Bosch, R., and Hagen, K. S. 1959. The integrated control concept. Hilgardia 29:81-101.
24. van Emden, H. F., and Williams, G. F. 1974. Insect stability and diversity in agroecosystems. Annu. Rev. Entomol. 19:455-475.
25. Watt, K. E. F. 1970. The systems point of view in pest management. Pages 71–79 in: Concepts of Pest Management. R. L. Rabb and F. E. Guthrie, eds. N.C. State Univ. Press. Raleigh. 242 pp.
26. Williams, P. H. 1979. Vegetable crop protection in the People's Republic of China. Annu. Rev. Phytopathol. 17:311-324.
27. Woodwell, G. M., and Smith, H. H. eds. 1969. Diversity and Stability in Ecological Systems. Brookhaven Symposia in Biology No. 22. Brookhaven National Laboratory, Upton, NY.
28. Wu, R., ed. 1979. Recombinant DNA. Methods in Enzymology, Vol. 68. Academic Press, New York. 555 pp.

CHAPTER 41

BREEDING FOR DISEASE RESISTANCE

JAMES L. BREWBAKER

Plant breeding is an ancient art that tends to plow ahead apparently oblivious to academic schemes for its edification. The academician can be annoyed to find so little impact on this ancient art by the intricacies of his quantitative or biochemical genetics, of his crop modeling and sink-source physiology, or of his epidemiological evidence and genetics of host-parasite relationships. Nonetheless, plant breeding works.

Breeding for disease resistance has an impact on crop production only when carried to completion in the form of acceptable commercial varieties. Thus it must be wed wholly to other breeding objectives—yield, quality, grower and consumer acceptability, insect tolerance, adaptability to site, climate, and management technology, etc. Control measures often lower the priority of disease resistance breeding far down this list of objectives. Disease resistance selections have nonetheless been commercialized for most crops, and the identification, introduction, and selection of resistant germ plasm is an adjunct to all breeding.

Some of the most notable academic achievements of this field, e.g., resistance to heterocaryotic basidiomycetes, have been marred by the insistent evolution of the pathogen. Breeders have been forced to retreat from simple gene-for-gene solutions.

The future will see more sophisticated gene technology and reliance on general or "horizontal" resistance. Despite common belief to the contrary, general resistance appears to be underlaid normally by single genes with adequate heritability for the breeder to make them the principal weapon in genetic advance against disease. Much plant breeding for resistance, however, can be expected to plow ahead in its peculiar way, relying on large segregating populations buffeted by disease under the artist's eye.

TRENDS IN CROP PRODUCTION AND IMPROVEMENT

Four crops account for two thirds of the world's food and feed grains: rice, corn, wheat, and barley. Together with soybeans and sorghum, they also embrace the majority of the world's plant breeders. The economic plants available are virtually numberless, if one includes everything from fuelwood species to algae of the sea. Although new crops will continue to appear, it is safe to predict a continuing trend toward fewer major crops and many minor ones—a global mirror of the modern corn belt farm, with its large acreages of corn and soybean and a small agroecosystem garden of highly diverse fruit trees, flowers, and vegetables. This trend leads indisputably in one direction—breeding resistance into a few major crops for their most suitable and highly tailored environment.

Intensification of agriculture changes pathogen spectra, effectively eliminating some species and enhancing others. Year-round production greatly increases the severity of insect-transmitted viruses and soilborne or weedborne pathogens, whereas irrigation and increased use of nitrogen fertilizer enhance diseases like rice blast and wheat powdery mildew.

Trends that promise to have a major effect on the breeding of disease resistance thus include:

- Consolidation of major crops into relatively few countries
- Consolidation of crop production into restricted regions and seasons of high yields and low incidence of pests
- Increased fertilizer inputs, better land preparation and weed management, and greater use of pesticides and biological control.

Plant breeding economics closely follows these trends and is expected to focus heavily on a few diseases that cannot be controlled cheaply on a few crops in their most productive environments. Global import-export patterns for major cereal crops begin to reflect these trends. Few diseases then assume the commercial importance to justify a breeding program.

Who will be doing this breeding? The trend is inexorably toward commercialization of all plant breeding, plus contributions from a few major international institutes with a few mandate crops. A refreshing aspect of disease resistance breeding has been its historical development in universities, with their urgency to publish and share results, and to determine the mechanisms underlying resistance. As plant breeding evacuates the universities, we can expect this to change to a more-focused, less-explained, but probably more cost-effective progress in disease resistance breeding.

MODE OF REPRODUCTION

The methodology of disease resistance breeding is directly tailored to the mode of propagation of the species, whether by self- or cross-pollination and by seed or vegetative reproduction. The economics of seed production has driven seed-reproduced crop marketing toward hybrids, over which the seedsman has complete control. This should ultimately occur for all cross-pollinated, seed-reproduced species, and for many others as well. At the same time, plant protection legislation and patenting have encouraged the marketing of clones and strengthened the economics of pure-line production. Some impacts of these trends in crops of different propagation methods are discussed.

Vegetative Propagation

Perennial crops that can be propagated clonally appear to have survived the rigors of natural selection in part due to high levels of general pest and disease resistance. This has been associated increasingly with high concentrations of toxic alkaloids, terpenoids, plant cyanogens, pigments, and other chemical and morphological barriers. Such traits often involve many gene loci in biosynthetic sequences, i.e., they are complex in inheritance. Perennial crops also include an unusually high proportion of polyploids that undoubtedly exploit multiple pest resistance loci from the parent species and the multiple allelism available at resistance loci.

The risk of genetic disaster due to disease is popularly believed greatest for clones propagated on a wide scale. The epiphytotics of Irish potato blight (1850s) and coffee rust (1900s) are examples. Both, however, involved narrow plant germ plasm bases (a single clone in coffee). Vegetatively propagated plants also present the greatest problem for germ plasm transfer from one country or region to another in order to stem epiphytotics, a current major problem with many root crops.

Important clonally propagated species are all perennial, largely cross-pollinating, and highly heterogeneous genetically. They include sugarcane,* coffee, cacao, tea, orchids and many perennial ornamentals, apples, pears, almonds, peaches, cherries, grapes, strawberries and other berries, pineapples, bananas, macadamia and other nuts, Irish* and sweet* potato, taro, cassava,* rubber,* and gliricidia. Large scale disease-resistance breeding and major efforts to establish disease-free clones for international transfer of germ plasm have been confined largely to the species asterisked. Very limited resistance breeding occurs in most other species and is often confined to single institutions and scientists on limited budgets. In general, the disease problems of the tropical perennial species tend to exceed those of temperate species, whereas institutional support lags far behind in the tropics.

Cross-Pollinated Seed-Grown Crops

Seed dealers increasingly market hybrids of cross-pollinated crops, exploiting cytoplasmic and genetic male sterility, sex chromosomes, and self-incompatibility—nature's devices for promoting heterozygosity. Where these species exist in nature, it is as highly heterogeneous populations often displaying wide variations in genetic disease resistance, and they are normally grown in developing countries as open-pollinated cultivars. In many cases there are no natural populations left of these crops, and the breeder must turn to related species for exotic disease resistance.

The search for resistance in cross-pollinating populations is tedious, especially if homozygosity is essential for expression of resistance and if gene frequencies are low. Major investments are involved in the establishment of stable lines for general resistance or of homozygous resistant parents for hybrid production. As a result, the breeding of genetic resistance has been confined largely to hybrids of a few crops (asterisked below), primarily under high-input agriculture with high supplemental control measures, and a few hybrids have come to dominate large acreages.

Important cross-pollinated seed-grown crops include corn,* sunflower,* cabbage* and cole crops, alfalfa,* red clover, white clover, rye, spinach, papaya, passionfruit, sugarbeet,* safflower, cucumber,* celery, onion, pumpkin, squash, petunia, snapdragon, oil palm,* coconut, kenaf, pine tree,* eucalyptus, carrot, and acacia.

Marketing of broad-based composite or synthetic lines in these species provides a genetic base minimizing evolutionary advance of pathogens. However, such breeding is more tedious and expensive than that of hybrids, which if not superior in yield are almost always preferred by growers for their uniformity in appearance, maturity, quality, and other important traits.

The wisdom of hybrid development has been questioned, especially for developing countries, as it can markedly restrict the genetic base of crops and increase disease risks. Against this stands the logic of the marketplace, since seed production of outcrossed species has not been economic without hybrids, over which the seed dealer has complete control. Hybrids must thus be considered inevitable in most of these crops, and resistance breeding can focus on resistances with high combining ability, where only a single parent of the hybrid need carry the requisite genes.

Self-Pollinated Seed-Grown Crops

Gene-for-gene resistance had its birth in crops such as wheat, flax, and rice. It suggests an historic pattern for self-pollinated annual species, in which a high degree of homozygosity and regular displacement of one plant population by that of its offspring drives the coevolution of specific resistance. This is especially evident as single modern crop cultivars have come into wide commercial production.

Self-pollination is a genetic luxury that most plants cannot afford except as a rare event, but which has markedly increased under pressure of human activities. Many of our self-pollinated annual crop plants are derived from cross-pollinated wild relatives (e.g., tomato, tobacco, soybean, bean, and peanut). The self-fertility that arises with autopolyploidy in most self-incompatible plants has rarely survived the rigors of natural selection.

Important self-pollinated seed-grown crops include wheat,* rice,* barley,* soybean,* bean,* pea, cowpea, mung bean and other grain legumes, sorghum,* lettuce, tomato, pepper, flax, peanut,* tobacco, leucaena, and cotton (partially outcrossed). Extensive resistance breeding occurs in the species asterisked, and it tends to dominate breeder's attention in many self-pollinated crops. An exception is the highly polyploid perennial leucaena, where general pest resistance appears to be very high.

Breeding advance for resistance in self-pollinated crops has relied heavily on massive screening of germ plasm (pure lines) and exploitation of monogenes that are easily transferred to commercially acceptable cultivars. Techniques like disease resistance backcrossing, as in the development of isogenic multilines for specific resistance, have been too expensive for wide application. Incorporating polygenic resistance and achieving homozygosity in self-pollinated species is a stimulating academic objective for classroom exercises but largely impractical for the commercial breeder.

Few self-pollinated species are sold as hybrids (tomato, pepper, sorghum) exploiting heterozygosity for resistance loci. Hybrid dimeric and tetrameric enzymes may also be involved in the specific combining ability for disease resistance, a phenomenon again restricted in its application to such hybrids.

EPIPHYTOTICS

Failures in disease resistance breeding are attributable as often to inadequate epiphytotics as to inadequate genetic resources. Diseases such as Fusarium wilt of cotton can defy the breeder for decades, due to difficulties in epiphytotic establishment. Diseases resulting from complexes of different rotting or viral pathogens add to this problem. The partnership of breeder and pathologist in the field is thus commonly in evidence where resistance breeding has proceeded with ease.

Natural epiphytotics may suffice in early stages of breeding when resistant lines are few. Spreader rows of highly susceptible lines are valuable to enhance epiphytotics and may be inoculated. Field inoculation of all lines is often essential to avoid escapes. Methods for machine inoculation have been developed for many fungal diseases. Spore or mycelial suspensions can be machine-sprayed or applied to sorghum seeds dropped in the whorl; engineering to automate such infections increasingly is critical to the economics of large-scale selection programs and often involves several pathogens applied simultaneously.

Environmental management can enhance the severity of diseases like leaf blights and rusts, as by the extension of dew periods through irrigation. Stage of field inoculation is critical for the expression of most resistance genes; young seedlings may be fully resistant (as in bean bacterial blight) or may only show resistance after a few weeks of development (as in corn downy mildew).

Insectborne and soilborne pathogens pose special problems to epiphytotics. Most viral and bacterial pathogens kill even the most resistant host if applied in very high concentrations. Epiphytotic severity is particularly difficult to control for insect-transmitted viruses. The control of soil moisture and the incorporation into the soil of diseased tissues may prove effective in enhancing soilborne fungal diseases. Only too often these problems prove insurmountable to any but the most dedicated breeder or pathologist.

An inevitable requirement in breeding is that of large segregating populations, as disease resistance will be only one of many goals. Plants that escape infection and are misclassified as resistant can prove to be very expensive errors. While multiple inoculation can help avoid escapes, the breeder normally plans for a certain frequency of them. Clonal or selfed progeny normally must be evaluated under an epiphytotic to reduce the chance of such errors.

SOURCES OF RESISTANCE

Increasingly exotic sources of resistance are exploited in breeding, although it remains most economic when the resistance is identified in adapted germ plasm. Major crops of the United States were often introduced from abroad with limited initial germ plasm—e.g., soybean, sorghum, sugar beet, sugarcane, potato, corn, tomato, wheat, barley—and explorations in the center of species diversity have been essential for resistance breeding. The screening of such germ plasm is almost overly routine in many institutions, notably for self-pollinated crops (e.g., 70% of 30,000 tested rice cultivars resist tungro virus). Genetic variation in the pathogen and cross-pollination or polyploidy of the host greatly complicate the screening process, and its effectiveness necessarily depends on excellent epiphytotics.

Transfer from Wild Relatives

Natural populations of species related to crop plants often resist a wide spectrum of diseases in nature; examples would include all of the crops listed above. In some instances these resistances are not present in the crop species, e.g., resistance to curly top virus of sugar beet, rust of soybean, or tobacco mosaic virus of tomato. Genetic transfer from wild relatives presents at least three major hurdles—the species hybridization itself, the chromosome homology necessary for gene transfer, and the ability to derive the disease resistance locus free from undesirable linkages. Transfer of polygenic complexes is virtually unmanageable.

Exotic hybridization can be aided by hybrid embryo culture, by attention to the genetics of self- and cross-incompatibility, and by generous survey of parental germ plasm in crosses. Close linkages are a bane in most plant breeding and are always to be expected from wide species crosses. Extended backcrossing to adapted species is usually required, with F_2 populations interpolated to allow identification of crossover homozygotes. Thus the pragmatic breeder stays "close to home" for a desired gene until that possibility has been exhausted.

Resistance Induction and Genetic Engineering

The induction of resistance by mutagens generally has been overrated and appears to have led to no commercially important varieties. Most induced mutants have problems associated with cytological irregularities. In combination with cell or tissue culture, however, chemical mutagens may prove more effective notably in selecting toxin-resistant cell lines.

Genetic engineering (Chapter 13) involving the viral transfer and cloning of resistant genes from one unrelated species to another clearly deserves generous evaluation. Optimism must be tempered by the recognition that only single deoxyribonucleic acid sequences can be transferred, and the proteins they code for must cope with a generally hostile new cellular and tissue environment. A promising possibility is to transfer genes between near-relatives that cross, if at all, with difficulty—e.g., corn, sorghum, sugarcane, sudangrass, johnsongrass.

GENETICS OF RESISTANCE

The genetics of disease resistance is dealt with fully elsewhere in this volume and is addressed here from the perspective of the commercial plant breeder. The identification and use of resistance by the breeder often precedes by many years the knowledge of its genetic basis and can be unaffected by such elucidation. It is often of greater value to refine the methodology for superb epiphytotics that clarify resistance segregations (i.e., thereby increasing heritability values), notably where insect vectors or modification by environmental agents affect disease expression.

Heritability and Dominance

Heritability and resistance expression in hybrids ("dominance") are major genetic questions from the breeder's perspective. Heritability is the proportion of phenotypic variance that can be ascribed to genetic control, and it directly predicts rates of genetic advance. Generation mean and diallel cross analysis provide accurate statistical estimates of heritability, but it can often be estimated with surprising accuracy lacking statistical analyses by an experienced breeder. Heritability can be very low for a monogenic resistance under a poor epiphytotic, like bacterial wilt for many crops. It can be very high for oligogenic or polygenic resistance with extremely low environmental influences, like horizontal rust or blight resistance in several cereal crops.

The expression of resistance in a hybrid determines its ease of genetic transfer in backcrossing and its utility to the hybrid breeder. The concept of dominance is applied loosely in this context to mean expression in the hybrid, and "incomplete dominance" implies a partial resistance (which may or may not be economically marketable). At the level of gene action, it is probable that gene-for-gene resistance usually involves toxins or antibiotic proteins synthesized by a resistance gene, whether present singly or multiply. In contrast, most resistance appears to involve genes acting through enzymes that synthesize products like pigments or alkaloids produced in direct quantitative relationship to gene dosage ("incompletely dominant"). From the breeder's perspective, a good resistance is dominant, i.e., expressed in the hybrids of susceptible and resistant parents.

Specific and General Resistance

Specific or "vertical" resistance applies to host-pathogen gene-for-gene relationships, detailed in Chapter 34. Rust, blast, and smut diseases provide the classic examples, with clear distinction of resistant plants (usually disease-immune or with hypernecrotic lesions). Resistance is displayed by heterozygotes of the host, and virulence is recessive in the pathogen (often heterokaryotic). Plant genomes may abound with such loci; e.g., the three genomes of hexaploid wheat carry at least 35 known *Lr* loci for leaf rust and 15 *Sr* loci for stem rust.

Breeding for specific resistance is about as routine as the coevolution of the pathogen. Our Hawaii conversion of corn in Hawaii to allele Rp-d for hypernecrotic common-rust resistance was followed within 4 years by new virulent heterokaryons. Allelic replacements Rp-g and Rp-Td survived less than 2 years, leaving us no resistance among the six resistance loci and 30 alleles available in the world germ plasm. This situation is exacerbated in Hawaii by year-round sexual cycling of the pathogen. In most gene-for-gene relationships, the host resistance gene is effective in double or single dose (i.e., "dominant"), and few have proven commercially effective.

General or "horizontal" resistance applies to host-pathogen relationships that are not gene-for-gene. The original definition included genetic assumptions regarding the host that have not proved widely applicable—namely, polygenic control and additive gene action. General resistance often involves single genes, which, for instance, delay establishment or sporulation of most gene-for-gene rusts, blights, or blasts. Gene-for-gene relationships are rare or unknown for diseases such as those attributed to viruses, bacteria, *Fusarium*, and other organisms that cause rot, for which host-resistance monogenes, however, are commonplace. The concept of dominance is generally inapplicable to alleles of such loci, where heterozygote reaction may range from resistance to susceptibility depending on virulence of pathogen or titer of inoculum.

Polygenes are too easily inferred where disease reactions vary in a continuous spectrum, whether due to host or pathogen genetic variability or to incomplete epiphytotics. Resistance in corn to maize mosaic virus, downy mildew, and corn streak virus all have appeared polygenic in quantitative genetic analyses of open-pollinated materials; major resistance genes appeared only under careful epiphytotics applied to homozygous inbreds and their progenies.

Polygenicity is to be expected in polyploid crops like wheat, sugarcane, alfalfa, mustards, yams, cotton, tobacco, coffee, potato, sweet potato, plums, berries, etc. Careful analysis can reveal underlying monogenes, as it has for alfalfa bacterial wilt and sugarcane stalk rot. Allopolyploids like wheat can incorporate identical resistance loci from two or more ancestral diploids. When gene-for-gene in nature, these loci can act independently, but if not they may appear to be additive polygenes, as suspected for "slow-rusting" or "slow-mildewing" genotypes in hexaploid wheat. A high proportion of crop plants are polyploids, so these genetic complications are very familiar to breeders.

BREEDING STRATEGIES

Disease resistance breeding presents no unique challenges to the breeder, once epiphytotics work and a sense emerges of heritability and dominance of the resistance. The principal tools of the breeder are pedigree selection and backcrossing, but on occasion all available genetic strategies are used. Rarely can the strategy be confined to a single pathogen, but often it involves several at once. Usually the breeder seeks commercially acceptable levels of resistance that are far short of immunity, on a scale related to cultivars of the crop grown in the field.

Pedigree and Mass Selection

Identification of segregating plants or progenies under an epiphytotic and growth of their progeny is fundamental to most resistance breeding. Selfed lines of uniform resistance may be used directly as parents. Inbreeding to homozygosity proceeds slowly in polyploids and must be extended in the event of escapes.

Mass selection among heterogeneous populations can range from ineffective roguing to major improvements in resistant gene frequencies, but it is automatically a part of all breeding research. The method is commonly modified by growing single-plant progenies under an epiphytotic, allowing favoritism of progenies whose female parent combines well for resistance.

A particularly powerful method for population improvement in crops like corn is S_2 selection, in which two selfed generations are grown from individual plants under an epiphytotic. The S_2 progenies showing homogeneity for resistance and high yield are recombined to establish the next cycle of selection. The extraordinarily high-yielding, downy mildew resistant tropical corn, Suwan 1, resulted from four cycles of 1% selection of this type.

Backcrossing

Introduction of a gene into adapted germ plasm by backcrossing is probably the most predictable of a breeder's methods, and exotic disease resistance has often been introduced in this fashion. As noted previously, the dominant corn rust gene Rp-d was bred by six backcrosses into 60 inbreds of maize in an elegantly simple program in Hawaii. Three nurseries per year were grown with 30 segregating plants per cross, and susceptible plants were removed before flowering to allow backcrossing of a few resistant plants. Homozygotes for the dominant gene were established in two additional selfed generations. Regrettably, the pathogen evolved virulent strains shortly after this was all completed! Backcrossing recessive and incompletely dominant genes is slightly more tedious, and the method is greatly complicated by polygenic inheritance of polyploidy.

Multiline Cultivars and Hybrids

The backcross incorporation of diverse monogenes in the gene-for-gene system can create multiline cultivars of reasonably homogeneous morphology but high genetic diversity for disease resistance (Chapter 43). The strategy has significant implications for pathogen evolution but is difficult to sell to administrators of commercial seed programs.

It is commonplace to have only one parent of a hybrid provided with the requisite disease resistance, avoiding cost of converting both parents. Hybrid breeding in crops like tomato, pepper, cucumber, and cole crops capitalizes as much on this covering-up of deficiencies of one parent as it appears to on hybrid vigor per se, i.e., it is a highly economic way to breed resistance adequate to consumer needs.

The incorporation of two diverse genetic sources of resistance into parents of a hybrid is an attractive theoretical goal. Tobacco mosaic virus resistant tomatoes reputedly often have a resistance source from one wild relative on one side of the cross and from another relative on the other. The evolution of more virulent tobacco mosaic virus strains appears to be stalled by this strategy.

The eventual life of a commercial cultivar rarely exceeds a decade, as improvements arise in quality, yield, and grower and consumer acceptability. The breeder is thus always under pressure to pyramid the achievements, including new sources of disease resistance. Polygenic resistance can be retained with adequate backcrossing, but clearly it is an impediment to rapid breeding. Despite a general concensus that monogene resistances evoke high risks of pathogen evolution, they can be expected to remain strategic to most commercial breeding programs against disease. And although no one would accuse breeders of generating problems, there is a certain comfort in knowing that the evolving diseases and insects of crops ensure a healthy future for the art and science of resistance breeding.

REFERENCES

Brewbaker, J. L. 1964. Agricultural Genetics. Prentice-Hall, Englewood Cliffs, NJ.

Hooker, A. L., and Saxena, K. M. S. 1971. Genetics of disease resistance in plants. Annu. Rev. Genetics 5:407-424.

National Academy of Sciences. 1972. Genetic Vulnerability of Major Crops. Nat. Acad. Sci., Washington, DC.

Nelson, R. R., ed. 1973. Breeding Plants for Disease Resistance. Pennsylvania State Univ. Press, University Park. 413 pp.

Russell, G. E. 1971. Plant Breeding for Pest and Disease Resistance. Butterworth, London.

Simmonds, N. W. 1979. Principles of Crop Improvement. Longman, London.

Vanderplank, J. E. 1968. Disease Resistance in Plants. Academic Press, New York.

CHAPTER 42

AGROECOSYSTEMS, DISEASE RESISTANCE, AND CROP IMPROVEMENT

IVAN W. BUDDENHAGEN

The development of new cultivars of crops is an enormous worldwide activity, split among thousands of individuals in state, federal, and commercial organizations in dozens of countries. An accurate comprehensive overview of this activity and its implications for the future of agriculture and humankind is hard to envision. But, the importance of crop diseases today and, say, 25 years from today, is largely determined by those who decide from which plants a new crop cultivar will be generated—and by those who decide which cultivar they will grow. All subsequent agricultural research by pathologists, physiologists, agronomists, and others will be influenced, and even determined, by these earlier decisions.

Imagine a crop cultivar that has been so successfully bred that it is both high-yielding and so resistant year after year to all potential pathogens that it has no significant disease losses. Does such a situation exist? Probably yes in localized agroecosystems, for instance, for sugarcane in Hawaii. But if such success were common and widespread, it would result in the ending of many jobs for plant pathologists. Whether or not they realize it, plant pathologists in general have a vested interest in a continued lack of success in developing sufficient or durable disease resistance. The pathologist who has a real involvement in development of new disease-resistant cultivars through direct participation toward that goal is an exception. At least this is so for most of those in plant pathology departments in both the developed and less developed parts of the world. These are strong statements to make and difficult for most people to accept. Moreover, plant breeders also have a vested interest in developing cultivars that will later need to be replaced; insufficient or impermanent resistance is a help in this need. It makes new cultivars with "new" resistance easier to promote. This is especially true for commercial breeders. In unguarded comments I have been told as much. On the other hand, it is in commercial seed companies in high-technology countries that plant pathology comes closest to direct participation in crop improvement. In Third World countries this involvement is minimal, and it is here where it is most needed. Such realizations have been slow to come to me (and hard for me to admit) as I have probed why diseases (and insect pests) remain so important in the face of all the apparent efforts to make them insignificant through breeding.

Contrarily, there are rewards in developing superior and more durably disease resistant crop cultivars. A new cultivar (licensed under breeders' rights) that really is disease resistant and remains so (and is also high yielding) gives a tangible economic reward to the developer. This can balance the opposite situation, of needing a cultivar that requires replacing due at least in part to inadequate or impermanent resistance. But, we should face the reality of the basic schism between most plant pathologists who want "important" diseases (i.e., high cultivar susceptibility) to investigate and plant breeders who (especially now with breeders' rights) want greater resistance in their new embryonic cultivars.

HIGHLY BRED CROPS

If there still are important plant diseases in crops where new cultivars are continually being developed, one can ask why.

My analysis reveals the following possibilities:

First, the breeder has not really come to grips with this part of his crop improvement needs. A series of potential steps fail to be taken. The reasons for this are varied, with some mentioned above.

Second, the breeder does come to grips with the resistance needs, but cannot follow through because of inadequate plant pathological input. The reasons for this are varied, with some mentioned above.

Third, with both breeder and pathologist, recognition, cooperation and input progress remain low. The reasons for this are varied: 1) the system is intractable, with sufficient resistance for normal genetic manipulation just not available; 2) the breeding system and the many objectives do not enable high disease resistance to be incorporated easily with the other required characters; 3) the conceptual basis for the employed methodology is inadequate; or 4) the conceptual basis is adequate but the methods used are inadequate.

The ideal situation would be for the breeder and pathologist to come to grips with the problem as a team, use adequate concepts and methods, and develop new cultivars with high and stable resistance. Initially this will be economically rewarding, although long-term support may dwindle due to success, with lowering of perception of future need.

We should be able to agree that the development of such cultivars, with high and durable resistance, is the desired goal in this increasingly food-deficient world and that much productivity is lost in many crops because this goal is not attained. This loss of productivity should justify greater investment and effort to reach the goal. Then careful analysis of the situation in crop improvement should reveal which of the three items above require change to improve chances for success.

PERENNIAL CROPS

The foregoing applies to most field and vegetable annual crops, where plant breeding and varietal development are so active. A different situation exists for many clonal perennial crops, such as orchard and small fruits, where breeding cycles are long or difficult and fruit quality is paramount. With bananas, pineapple, citrus fruits, and some tropical root crops, nearly all production is from varieties developed in antiquity, long before modern science evolved. Even for many of our temperate fruits such as grapes, apples, peaches, etc., a similar situation exists, combined with newer cultivars selected by chance or by amateurs or horticulturalists concerned mainly with quality. For all of these crops and for most of our forests, the level of disease susceptibility is a result only of natural selection, modified for the former only slightly by primitive people's selection for suitability in the local environment, where disease resistance may or may not have been important. In modern times, since fungicides have been developed, selection has been based largely on quality, enabling us to ignore disease susceptibility, possibly even permitting an increase in susceptibility.

For high-quality, high-value perennial crops we may generalize that disease susceptibility has largely been ignored in relation to breeding and it is accepted that disease will be reduced to tolerable levels by other means—there are some exceptions. A large investment in plant pathology research is made in this area of "other means."

It follows then that for perennial, orchard-type crops there should be a large gap between susceptibility of existing cultivars and susceptibility of cultivars that might be developed by taking a new look at disease resistance when plants are bred. Possibly it is for such crops that the greatest progress could be made, since little has been done and genetic reservoirs are very large. The major inhibition would seem to be lack of realization of opportunities combined with the slow traditional breeding cycle and the inadequate methods available in assessing small differences in disease resistance that could be effective in a plant population. New methods in recombination and transformation techniques and in in vitro culture and selection could radically change the present stagnation if the opportunities were recognized.

RECENT HISTORICAL PERSPECTIVE

At the Golden Jubilee meetings of the American Phytopathological Society (APS) in 1958, a symposium was held on the genetics of disease resistance (16). The bias of the times (toward physiological host-parasite interaction) was inherent in the title: "Genetic approach to elucidation of mechanisms governing pathogenicity and disease resistance." There was really no evaluation nor review of what I consider the really important question: how much progress is being made in developing cultivars that get less disease—and stay resistant—and how can this progress be accelerated?

Only 4–5 years had passed since major epidemics of stem rust had devastated durum and bread wheats in the United States. Cultivars previously considered to be resistant on the basis of major hypersensitive genes were spoken of as having "broken down." The previous resistance, although still there, was no longer effective against races fostered by the resistance genes so carefully incorporated into the new cultivars. For potatoes, 25 years of work on *Phytophthora* specialization and the major R genes responsible had gradually been revealed to be a dead end, with fungal races readily overcoming all R genes in all combinations. "Field resistance," coined for the residual non-R resistance factors, was described only in 1957, but even so, the advice was not to neglect the dominant R-genes; the new ones provided "untapped sources for broadening the basis of resistance" (7). Time has not proved this to be so and most breeders now want to eliminate the R genes that earlier had been so important as to merit incorporation into cultivars. But in any event, is there substantial acreage of potatoes today in high-technology countries of cultivars that have higher degrees of blight resistance due to conscious selection than did the potatoes of 50 years ago? I doubt it.

At the time of the Golden Jubilee meetings no one had heard of "vertical" and "horizontal" resistance. These terms were introduced 5 years later in 1963, by Vanderplank (18), based largely on analytical appraisal of data previously published by various workers, along with a quantitative and mathematical approach to the study of the development of disease epidemics. This was a major break with tradition that has had a striking influence on studies of plant disease epidemics. Soon there was also considerable interest, mostly by plant pathologists, in "horizontal" resistance and in a more critical and logical analysis of resistance in relation to epidemiological factors. But the marriage of epidemiology and disease resistance breeding was never fully consummated. The then new gene-for-gene concept of Flor (6) and the elegance of classical genetic manipulation for major resistance (and avirulence/virulence) genes were too strong for the abstruse and often conflicting characteristics and definitions surrounding horizontal resistance. Attempts to clarify and promote horizontal resistance by Vanderplank (19,20), Robinson (14), and others led to rebuttals (5,12), compromises (13), and new terms (9). Polarization of views by the advocates did not aid in winning adherents, and the traditional way of looking at and carrying out disease resistance breeding has changed much less in the last 20 years than one might have expected. A plethora of terms qualifying types of resistance has emerged; slow-rusting has been resurrected and examined (21); and an evolutionary viewpoint of diseases has been offered in relation to breeding (2). But no generally accepted cohesive, logical, and analytical pattern has emerged in relation to either understanding resistance and susceptibility (8) or to breeding and selecting for stable resistance. Cultivars released as resistant often became susceptible or were soon revealed as susceptible to something else that was not considered important during the selection process. Retrospective studies on inheritance of resistance dominate thesis research, and major genes are found, for obvious reasons. The marriage of the vast field of quantitative genetics and resistance breeding is seldom made. Most plant breeding in university departments involves plant pathologists and diseases and insect pests and entomologists much less than it does biometricians and biometric manipulation of data. Change comes hard. With computers and mathematics, why do we need to examine the biological basis for variable data in the field?

Most plant pathology departments give short shrift to really productive involvement with varietal development. They are indeed organizations wherein the pathology of plants is studied. They are aptly named. The objective of plant health, the title of this volume, is much less in evidence than we would like to believe. A lot of good research is done—that is not the point nor the criticism. Some of it is geared to control—control of a disease after it occurs and is well studied. The reality of existing disease and existing crop cultivars is accepted as the base from which studies are to begin. The idea of influencing that base through new research that will result in lower incidence of disease to start with hardly occurs. The recently published abstracts of the 1981 Pacific Division APS meetings reveal that only two of 51 were concerned with disease resistance. Such meetings are a forum for recent research, often by young people who will be the plant pathologists of the future. Disease resistance is hardly there. This is the trend of our times and our field. It has not always been so.

L. R. Jones very early on established for the Wisconsin Department of Plant Pathology a direct involvement in developing cultivars that would suffer much less disease. He started with cabbage plants that he observed as resistant in farmers' heavily infested fields, an approach that is less often used today. This direction was continued there by J. C. Walker and many others to the present day. But this direction for plant pathology at Wisconsin did not take firm root in many other places. In recent years, with the great expansion of knowledge of the complexity of plant diseases, there has not been a concomitant feedback expansion into plant breeding. When L. R. Jones started breeding for resistance it seemed simple. We now know that manipulation of pathogen evolution through breeding in the host involves many complex interactions. The complexity is not easily solved by plant breeders who have many objectives and who are often only superficially trained in understanding disease complexity.

One may argue that we are plant pathologists and that varietal development is for plant breeders. But developing new cultivars is essentially an attempt to produce maximum suitability of a plant genotype for an ecosystem. And an ecosystem contains many microorganisms, viruses, and insect pests ready to interact with a new cultivar in complex ways. To understand such a system enough to be able to develop enough stable disease (and insect) resistance in new cultivars is an enormous task. It needs many disciplines and substantial teamwork in breeding and selection. It is not enough to leave the task to breeders or only to supply them with inoculum or disease scores. I recently attended the 1981 American Society of Agronomy meetings (where breeders, among others, give papers) just to see how the crop improvement field appears from "inside." I was struck by the scarcity of papers in which disease resistance was a major effort and the generally superficial treatment of the disease (and pathogen) in those few papers. It appears to me the important and complex task of developing stable disease-resistant cultivars is often being given short shrift from both disciplines.

TRAINING AND THE THIRD WORLD

If one takes a world view of plant pathology and of varietal improvement, the separation of the two fields becomes more critical than if one looks only at the United States and Europe where commercial breeding operations largely make restitution and dominate varietal development today. Diseases and insect pests are major contributors to production (and postproduction) losses in the Third World, especially in the tropics. In developing countries, resistance is the ideal and often the only means to reduce disease losses. Agriculture is shifting from old landraces to recently bred cultivars with higher yield potential. The record of new problems generated thereby does not speak well for good teamwork and understanding among disciplines or for balancing ecosystem and pathosystem knowledge with varietal development. An examination of plant pathology departments in Third World countries (and in their federal and state ministries) reveals usually young and small copies of the U.S. (or British or European) departments where some of the staff members were trained. There is little direct involvement with breeders and crop improvement. One should not expect anything else, but let us look at our foreign student numbers in U.S. pathology departments—often 25% or more of postgraduate students—and ask ourselves if they are getting the training they need to help raise their national food production. We may not need to be more practical and realistic because our predecessors were. But our drift away from plant pathology involvement in varietal improvement does a disservice to the needs of research in the Third World through its influence on those we educate. A real opportunity exists to redirect our approach for these students and thereby help redirect some of our own research to the long-range benefit of the science of plant pathology and to agriculture. We may not always have the largesse for our present large proportion of esoteric research; it would be well to prepare for that time, even for the United States.

THE CHALLENGE

Diseases have evolutionary temporal and geographical antecedents and ecological reasons for being. Diseases are complex interactions of many environmental forces acting on specific past evolutionary events. Why do we not spend more effort to understand and unravel these complexities and then use this knowledge to reduce future diseases by manipulating host evolution appropriately, in breeding? Either we think it is not important to do so, we think we cannot do it, or we really do not want less disease in our crops.

One can argue, contrarily, that this view is wrong, that much useful disease resistance is being incorporated into new cultivars, and that there is much progress. Of course there are many useful major resistance genes and of course progress in lessening disease through resistance breeding is occurring. But the very existence of this large meeting of plant pathologists and the many chapters in this volume must mean that many diseases remain important and thus have not been solved through breeding. Some, such as southern corn leaf blight in 1971, even became major diseases after an unimportant past (due to breeding) or, like rice blast in Korea in 1978–1979, became major after years of unimportance (due to breeding). Progress is fitful and often seems due more to luck than to design. Failure likewise seems unpredictable.

Great potential vulnerabilities crucial to national economies and even to international politics and technological stability are essentially ignored until it is too late; witness coffee rust in tropical America since 1970 and the vulnerability of rubber in Asia and Africa to the not-yet-introduced destructive South American leaf blight. What really is being done about this threat to rubber?

Wolfe (Chapter 43) recommends varietal mixtures to reduce disease, and peasant agriculture uses species mixtures to reduce both disease and insect damage. These are practical and innovative approaches, but when needed for recently bred cultivars they represent failure of disease control through breeding. In Western Germany it is now accepted that yield trials for wheat before varietal release may be conducted under fungicidal treatments. Thus, yield superiority of susceptible cultivars may be demonstrated (when protected) and they then can be grown widely. This is a clear case of failure in breeding and a bad portent for the future. What direction are we going? Potato growing was the basis of the evolution of the Incan empire. In that homeland of potatoes, seed certification schemes have not been used. In advanced-technology countries potato culture could not continue without seed certification schemes. Why the difference? Surely it is related to the differences in tolerance or resistance levels among the cultivars grown. The scientifically bred cultivars have less resistance, requiring more sanitation and fungicides. Try to suggest (as I have done) that we breed potato cultivars so tolerant to disease that disease would become unimportant, and one is met with resistance and dismay. "What would happen to our seed potato industry? We need susceptibility for that industry as well as for making certification easy!"

Epidemics of tungro virus disease of rice have occurred in India and other tropical Asian countries in recent years. A recently conducted 3-year study of field resistance to tungro (A. Anjaneyulu, *personal communication*) revealed that even with new varietal releases vulnerability continues. The 89 high-yielding rice cultivars released by state and federal agencies in India and by the International Rice Research Institute (IRRI) were evaluated for field resistance, using growth reduction and yield depression as criteria. Only 7 cultivars were ranked resistant, 27 were intermediate, and 55 were susceptible. Thus, tungro, a widespread virus disease of explosive potential is being maintained as a threat to the major food crop in a chronically food-threatened country. Surely there are lessons to be learned from such a story.

All of the foregoing is to emphasize the great challenge that lies before us to reduce diseases through plant breeding. The challenge is enormous, but the hints on its complexity are revealing for the unwary.

The road ahead is filled with greatly unexpected psychological, traditional, organizational, and discipline-rigidity boulders—above and beyond the practical biological ones of how to get the job done.

NARROWING THE GAP

How to enhance the development of cultivars that have better resistance—better here meaning higher levels of and more stable and durable resistance—is a complex issue (3,4). But some key points can be raised.

Any lack of sufficient resistance in a cultivar reflects either selection of plants known to have insufficient resistance from the start or a difference in performance (or perception of performance) in breeder plots from performance in farmers' fields in relation to disease.

Thus, a major problem is the cryptic error represented in the performance in breeders' (and yield trial) plots. Reducing this cryptic error is a major *resolvable* contribution that can be made by a breeder/pathologist team. When this error is reduced to nearly zero, new cultivars will have sufficient resistance.

How can such cryptic error between breeders' plots and farm performance be lessened? It is easy to compare genotypes agronomically in breeders' plots or yield trials. It is more difficult to know the relationship between genotype performance in plots and disease levels that may occur in farmers' fields. This is because the breeders/pathologists are seldom observing or researching in farmers' fields, and the different pathogens' variability, ecological requirements, and epidemiological potentials are little studied. Circumscribing a target area for a new cultivar and studying the crop throughout the area can help greatly in breeding so as to prevent the ascendancy of pathogens now-minor as well as to suppress the continuance of the now-major ones. Picking and using selected trial sites in the target area that have different ecological influences on different pathogens will enable selection of more stable resistance. On-site breeding and selection can contribute greatly toward developing cultivars with ecosystem balance to all local stresses and with stable disease resistance.

One should not, for instance, expect rice cultivars developed in and ideal for Latin America to have sufficient resistance to the many indigenous rice pathogens of Asia to which they have not been exposed. Asian rice cultivars should not be expected to be ideal in West Africa, where they are faced with local African pathogens and pests not present in Asia, etc. While this may seem self-evident to some, it is surprising how much effort is expended to test and select finished cultivars in ecosystems very different from those in which they were bred. Initial yield comparisons mean little since only after being widely grown are their deficiencies revealed by the buildup of pathogens and pests heretofore minor on the local cultivars.

Many commercial breeding operations are located distant from major production areas, in locations such as California, where a dry climate makes easy the growing of high quality and pathogen-free seed. In this same climate, large segregating populations are grown to permit selection and development of a few finished cultivars. Although these are then tested comparatively in the target-growing area, for instance, the midwestern United States, this approach may not reveal initially the vulnerability and insufficient resistance in promoted cultivars.

The major rice diseases of tropical Asia in the period after the Green Revolution (tungro virus and bacterial blight) were not even known to be present or were not considered important when the first high-yielding dwarf cultivars were promoted there. These diseases were indigenous but were elevated into major significance by the large cryptic error between breeder-plot performance and the millions of acres in Asia containing these pathogens at a low incidence in old cultivars and landraces. Breeder plots were then often sprayed with insecticides (a practice not yet extinct), thereby preventing us from detecting virus as well as insect vulnerability. There sometimes has been a tendency to *increase* the potential cryptic error in breeder plots (such as with insecticides), and the reasons given are that otherwise one would not see the yield potential of the different lines. It should be self-evident that no one harvests "yield potential," but rather they harvest "yield." However, the goal of yield potential has often overshadowed the goal of harvestable yield and stability of that yield. It is more exciting to tell about 10-ton plot yields than to tell how to raise maximum yields, i.e., rice yields above 2 tons.

Rice bred in the West African rice stations is never tested for resistance to the major rice bacterial pathogen, which has only recently been discovered in West Africa. This pathogen exists in Africa only in the savanna-sahelian zone of wild rice evolution, north of the rice breeding locations. Thus, vulnerability and losses should be expected when southern-bred cultivars are moved north or when the disease spreads to the south. The converse is true for maize (the foliar blights and rusts seldom occur in the savanna-sahel); therefore cultivars bred there cannot compete further south due to disease vulnerability. Moreover, southern-bred cultivars do better even in the north, over the years, since their inherently higher resistance is sometimes needed even there.

The ecosystem determines the amount of resistance needed and should determine where it is most useful to breed and select for the most stable resistance. We think of American corn hybrids as a great success story in breeding, which includes many resistances, along with high yield. Except for the 1970 leaf blight debacle and vulnerability to some newly introduced viruses, this appears to be so. But this success is highly circumscribed by the ecosystem. Those same hybrids are devastated (and yield is nil) when grown in tropical Africa, and by the same pathogens that occur in the United States. Their resistance is not nearly sufficient where the environment increases the epidemiological potential of these pathogens. If we look wider, at exotic African pathogens such as maize streak virus, other viruses, and downy mildew, the vulnerability of the American corn hybrids is revealed as very great.

The largest gap between performance in trials and in agriculture is imprecisely known as the "breakdown" of resistance. Actually it means the absence of a test of the resistance one sees in plots against sufficient variability of the pathogen population. No resistance ever "breaks down"; it was not there in the first place but was not perceived to be absent. It is for such variable pathogens where breeding concepts and selection methods should be examined and alternatives to "breakdown" type resistance genes should be sought—hence the stress on trying approaches toward horizontal and durable resistance (3,9). But much needs to be done. It was only in 1977 that slow rusting for stem rust of wheat was shown to be location specific (21). For blast of rice the first attempt to determine whether the existing horizontal resistance was location specific was done in 1981. Using concepts of pathosystem evolution and horizontal resistance breeding and selection, coffee berry disease in Ethiopia, recently devastating, has been reduced to nonsignificance (17). The development of methods to breed and select wheat for horizontal resistance to many disease problems in difficult areas of Zambia (11) and southern Brazil (1) appears to be successful. Such activities are generally outside the mainstream of communication on plant breeding for disease resistance because of their remoteness and radical approaches. Attempting such solutions to impermanence of resistance in traditional programs in North America and Europe has not been supported. One wheat program in Montana, however, has utilized some of the approach and found transgressive segregation for resistance in advanced generations, that is beyond the parental and F_2 levels (10). But in general, crop improvement workers have been loath to try innovation, in spite of obvious deficiencies of existing approaches.

But let us go back to the first level of the gap between plot performance and agriculture, that of perceived susceptibility in plots; it is being dismissed as unimportant. This occurs more often than one would expect. It is an area little researched. *Incidence* may be so low that the problem is dismissed on the presumption of low transmission or low inoculum concentrations, both remarkably plastic characteristics and influenced greatly by weather and crop population size. Or severity in small plots may be low because of low amounts of inoculum that is adapted initially to the new genotype at that moment, or because of a short period of weather suitable for epidemiological expansion. Both potential reasons are dangerous things to dismiss as unimportant in agricultural production. This type of gap is part of the larger question of the relationship of trial yields to agricultural yields, now being questioned by Simmonds (15).

The solution is a much greater concern for delimiting a target area for the new cultivar and a much greater examination of that ecosystem, the crop diseases in it, and their elasticity with normal environmental swings from year to year. The pathogen's variability and plasticity need to be appraised in a practical manner (not trapping of prevalent races; prevalent races are only a reflection of present crop genotype). Agricultural disease assessment needs to be clearly evaluated by the breeder/pathologist team (not just by extension personnel) in relation to incidence and severity. Both incidence and severity need to be understood as moving variables and largely a reflection of present crop genotype; a new cultivar can magnify either or both.

CONCLUSION

If we view crop disease as preventable by breeding and find this is not readily occurring, we must be able to recognize both the need and possibility of change. If we compare disease performance of cultivars on farms with performance (or perceived performance) of that same cultivar in breeder and trial plots, we can determine what changes are needed to produce cultivars with less disease. The gap in disease performance between those two situations will be either large or small. However large, it will exist for specific reasons, which can be identified, researched and resolved. But first we will need to understand well the conditions at the source (the plots where decisions are made) and the conditions in the target area—the complete ecosystem of the farmers' fields. The challenge is there and it needs taking up in much greater depth in the next 25 years. A proper marriage of pathology and breeding new cultivars through teamwork with an understanding of the ecological, epidemiological, and evolutionary aspects of crop disease is badly needed for stable harvests from bountiful progenies in the future.

LITERATURE CITED

1. Beek, M. A. 1983. Breeding for disease resistance in wheat: The Brazilian experience. In: Durable Resistance in Crops. J. M. Waller, N. van der Graaff, and F. Lamberti, eds. Plenum Press, New York. In press.
2. Buddenhagen, I. W. 1977. Resistance and vulnerability of tropical crops in relation to their evolution and breeding. Ann. N.Y. Acad. Sci. 287:309-326.
3. Buddenhagen, I. W. 1981. Conceptual and practical consideration when breeding for tolerance or resistance. Pages 221–234 in: Plant Disease Control: Resistance and Susceptibility. R. C. Staples, ed. John Wiley and Sons, New York. 339 pp.
4. Buddenhagen, I. W. 1983. Crop improvement in relation to virus diseases and their epidemiology. In: Plant Virus Epidemiology. R. T. Plumb and J. M. Thresh, eds. Blackwell Scientific Publications, Oxford. In press.
5. Crill, P., Jones, J. P., and Burgis, D. S. 1973. Failure of "horizontal resistance" to control *Fusarium* wilt of tomato. Plant Dis. Rep. 57:119-121.
6. Flor, H. H. 1959. Genetic controls of host-parasite interactions in rust diseases. Pages 137–144 in: Plant Pathology, Problems and Progress 1908–1958. C. S. Holton, ed. Univ. Wis. Press, Madison. 588 pp.
7. Gallegly, M. E., and Niederhauser, J. S. 1979. Genetic controls of host-parasite interactions in the *Phytophthora* late blight disease. Pages 168-182 in: Plant Pathology, Problems and Progress 1908–1958. C. S. Holton, ed. Univ. Wis. Press, Madison. 588 pp.
8. Heath, M. S. 1981. A generalized concept of host-parasite specificity. Phytopathology 71:1121-1123.
9. Johnson, R. 1981. Durable disease resistance. Pages 55–63 in: Strategies for the Control of Cereal Disease. J. F. Jenkyn and R. T. Plumb, eds. Blackwell Scientific Publications, Oxford. 219 pp.
10. Krupinsky, J. M., and Sharp, E. L. 1979. Reselection for improved resistance of wheat to stripe rust. Phytopathology 69:400-404.
11. Milliano, W. A. J. de. 1983. Breeding for disease resistance in wheat: The Zambian experience. In: Durable Resistance in Crops. J. M. Waller, N. van der Graaff, and F. Lamberti, eds. Plenum Press, New York. In press.
12. Nelson, R. R. 1978. Genetics of horizontal resistance to plant disease. Annu. Rev. Phytopathol. 16:359-378.
13. Parlevliet, J. E., and Zadoks, J. C. 1977. The integrated concept of disease resistance; A new view including horizontal and vertical resistance in plants. Euphytica 26:5-21.
14. Robinson, R. A. 1976. Plant Pathosystems. Springer-Verlag, Berlin. 184 pp.
15. Simmonds, N. W. 1979. The impact of plant breeding on sugar-cane yields in Barbados. Trop. Agric. (Trinidad) 56:289-300.
16. Snyder, W. C., Chairman. 1959. Symposium on genetic approach to elucidation of mechanisms governing pathogenicity and

disease resistance. Pages 135–218 in: Plant Pathology, Problems and Progress 1908–1958. C. S. Holton, ed. Univ. Wis. Press, Madison. 588 pp.

17. Van der Graaff, N. A. 1981. Selection of arabica coffee types resistant to coffee berry disease in Ethiopia. H. Veenman and Zonen, Wageningen. 110 pp.
18. Vanderplank, J. E. 1963. Plant Diseases, Epidemics and Control. Academic Press, New York. 349 pp.
19. Vanderplank, J. E. 1968. Disease Resistance in Plants. Academic Press, New York. 206 pp.
20. Vanderplank, J. E. 1975. Principles of Plant Infection. Academic Press, New York. 216 pp.
21. Wilcoxson, R. D. 1981. Genetics of slow rusting in cereals. Phytopathology 71:989-993.

CHAPTER 43

GENETIC STRATEGIES AND THEIR VALUE IN DISEASE CONTROL

M. S. WOLFE

A resistant cultivar provides an empty niche for a pathogen variant able to overcome the resistance. The probability of occurrence of such a pathogen variant, and of its accidental migration into the niche, depends on various features of the life style of the organism. Consequently, host resistance lasts longer against some pathogen species than against others.

Within a single pathogen species, pathogenic variants occur at different frequencies. Consequently, it is also observed that the host resistances that do or do not happen to match these variants will also differ in their durability (17). The larger and more uniform the niche provided by the resistant cultivar, the greater will be the probability of accidental migration of a matching pathogen variant into it. Reducing the individual niche size, and arranging a changing pattern of different niches, will reduce the probability of accidental migrations and thus extend the useful life of each host resistance.

Genetic strategies of disease control involve the interaction of these two elements, inherited durability of resistance and the system of cultivar use to affect the niche size. For convenience, they are discussed separately, but in the practical application of genetic strategies they should be used together as much as possible.

The accidental migration of a pathogen variant into an empty niche may occur between leaves, fields, or continents. It can occur passively, for example, by wind or rain-splash, or actively, by human interference. In this sense there is no fundamental difference between a new rust race overcoming a new cereal cultivar and the sudden increase of Dutch elm disease in western Europe. Such events are largely unpredictable in their timing and scale: there is no absolute certainty of continuation of the durability of a particular resistance, but we can try to improve the probability that it will so continue.

Host resistance varies continuously in effect from qualitative to quantitative; so does pathogenicity. Specificity is often discernible between particular hosts and matching pathogen variants, again varying continuously from large and obvious effects to those that may be of no practical consequence, such as the varietal adaptation in *Phytophthora infestans* observed by Caten (7). Often specificity is not discernible, but its absence cannot be proved. For example, it has been widely held that *Septoria nodorum* shows no specific interaction with cultivars of wheat. Recent observations suggest that this is incorrect, at least for some environments (29).

It is, of course, more likely that if a host resistance is highly specific, the accidental migration of the appropriate pathogen variant may occur more rapidly, because the matching pathogenicity may be under simple genetic control and cause little disadvantage to the pathogen that carries it. Nevertheless, if a large area of resistant host is available, it may also become occupied accidentally by a pathogen variant that may not show specific pathogenicity for that host. For example, there is no evidence that the damaging new form of *Ceratocystis ulmi* in western Europe is specifically adapted to the particular range of elm genotypes that occur in the area.

The theoretical concept of vertical and horizontal resistance provided by Vanderplank (37) and elaborated by Robinson (28) suggests a division between forms of resistance that are either within or beyond the pathogen's capacity for change. This seems most likely, but in practice it is only possible to prove that resistance is vertical, i.e., within the pathogen's capacity for change. The occurrence of horizontal resistance cannot be proved since it cannot be demonstrated. It implies permanence and denies the occurrence of future evolutionary accidents. It seems more logical therefore to retain this concept for theoretical use rather than for practical application.

DURABILITY OF RESISTANCE IN DIFFERENT HOST-PATHOGEN SPECIES

Outbreeders are likely to be more durably resistant to variable pathogens than are inbreeders, partly because of the diversity of resistance that may occur within the population and partly because of the ability of the population to respond through changes in frequency of the components. Beyond this general difference, comparisons between host species are difficult to make because an unknown number of the many differences that occur between species are involved in disease resistance.

Comparisons between different species of pathogens are simpler because the majority of characters involved affect the ability of the pathogen to overcome resistant host cultivars. For example:

1) biotrophy: an obligate biotroph is intimately compatible with its host. If it does not produce variants that overcome host resistance, it may become extinct. A facultative necrotroph may need to adapt to both parasitic and saprophytic environments and thus be less well able to overcome host resistance.
2) dissemination: pathogen variants that are spread by airborne spores are more likely to migrate into matching resistant hosts than are those that are restricted to the soil.
3) generation time: the shorter the generation time, the more quickly will the pathogen spread in a new niche.
4) sexual stage: an effective sexual stage increases the rate of production of new pathogen variants. Lack of a sexual stage may be compensated for, to some extent, by somatic hybridization (5).
5) continuity: pathogenicity matching a particular host resistance will increase most rapidly where the pathogen is endemic and suitable susceptible host

plants are continuously available. The ultimate continuity in annual crops may be in the Philippines, where it is proposed that a few rows of rice plants be sowed on alternate days the year round; rows that are ripe would be harvested on the intervening days (L. T. Evans, *personal communication*). Year round survival for any pathogenic organism would thus be ensured.

6) epidemiological competence: some pathogens are better adapted than others to the environment in which they occur; they are thus more likely to survive any change in host resistance.

By considering these features of the life style, it is possible to make an approximate prediction of the relative rate at which different pathogens may overcome host resistance and therefore to decide on the most appropriate control strategies. Because different kinds of pathogens may be important, however, the strategy that is finally adopted may be determined by the pathogen for which durable control is least easily obtained.

RECOGNITION OF INHERENT DURABLE RESISTANCE

Within any one host species there may be differences in the durability of resistant cultivars against a particular pathogen. However, Johnson (17) pointed out that durability could only be recognized retrospectively following large scale exposure of the host varieties in environments favorable for the pathogen.

Presumably some resistances may be more durable than others because matching pathogenicity is impossible, or rare, or because it can be assembled only from the effects of a number of genes. If host resistance is inherited polygenically it is usually regarded as potentially durable on the assumption that matching pathogenicity will also be under polygenic control (31,33). Although this may be generally so, it cannot be excluded that the matching pathogen response may be more simply inherited and thus emerge more quickly than anticipated. In practice, it is also evident that some host resistances that are assumed to be under polygenic control are actually controlled by few genes that show a large environmental interaction and that may be easily matched by the pathogen. Moreover, if a single host with polygenically inherited resistance is over-exposed, sufficient opportunity may be provided for the emergence of the matching polygenic pathogenicity.

Differences in durable resistance between cultivars of outbreeding crops may be recognizable, but the reasons for the differences may be more complex than in inbreds. It is likely that durability in an outbreeder is a function of the range of resistance genes, their effects, and their frequencies.

INHERENT DURABLE RESISTANCE IN PRACTICE

The approach taken by Johnson (16,19) for control of wheat yellow rust is to further exploit in breeding programs cultivars that continue to exhibit durable resistance. Progeny of crosses involving durably resistant parents are exposed to pathogen isolates known to produce the highest amount of infection on the parent so that similar or better progeny can be selected. Although technical care and precision are required, evidence so far suggests that inheritance of such resistance is not necessarily complex, which simplifies handling. There is no certainty that the progeny will have the same durability as the parent, but the probability that it will do so is high. A similar approach is used by Parlevliet (25) and Clifford (9) for barley brown rust, although the choice of parents is not so rigorous because less use is made of historical data in the selection of parents. Only cultivars are used that lack hypersensitive resistance to the pathogen, since this character is often associated with qualitative specificity of host and pathogen.

The disadvantages of these methods are that they require a considerable investment of pathological testing in the breeding programs (which may increase with success as appropriate pathogen test isolates become more difficult to find), and they introduce some conservatism into programs in which improvements in yield and quality are being sought. For a durable cultivar to be successful it also needs to be durably resistant against other major pathogens. For example, the considerable success of the winter wheat cultivar Maris Huntsman, in Europe, depended on a combination of durable resistance to *Puccinia striiformis, Erysiphe graminis* f. sp. *tritici, Pseudocercosporella herpotrichoides* and *Septoria nodorum.* Unfortunately, the variety had poor grain quality and attempts to improve that character disrupted the remarkable assembly of disease resistances.

Where known durably resistant cultivars are not available, it is probably wisest to base the strategic program on the best locally available resistant sources and to import foreign sources only as a last resort.

Robinson (28), however, suggested that breeding for permanent (horizontal) resistance should be started by intercrossing local susceptible lines. The rationale is that in such lines the pathogen will have matched all of the resistance genes that it is capable of matching; any that remain will be beyond the pathogens' capacity for change. So, by using the same matching pathogen race, recurrent selection should allow the accumulation of such residual genes for resistance into a single host genotype to the level at which they will provide adequate, and permanent, disease control. However, it is most unlikely that the pathogen isolates used will represent the sum of present variation and future evolution, so that the durability of any new cultivar cannot be predicted (19). The technique is being used by Scott (32) to try to accumulate resistance in wheat to *Gaeumannomyces graminis* since resistance to this pathogen seems to be otherwise unobtainable in *Triticum.* If it can be obtained, there is a high probability that the resistance will be durable, if for no other reason than that the pathogen, a soilborne necrotroph, is unlikely to be able to respond quickly.

Simply inherited qualitative resistance is easy to handle in breeding programs and provides complete disease control. It has fallen into disfavor with pathologists, although less so with breeders, because it is often associated with poor durability, although this may only be applicable for some obligate biotrophs. This reputation has been earned, to some extent, by uncritical introduction and use of such resistance. For example, in Germany in the 1930s when Honecker introduced genes of major effect into barley for control of powdery mildew, he was aware of the existence of uncommon pathogen genotypes able to grow on resistant cultivars, but, because of their relative rarity at the time, he considered them unimportant. Further, because of their initial effectiveness, cultivars with simply inherited qualitative resistance rapidly became popular and extensively used, so maximizing the probability of coincidence of resistant cultivar and matching pathogen genotype. The large scale, rapid failure of the "R" genes from *Solanum demissum* to control *Phytophthora infestans* in cultivated potato was a further, unfortunate example. It does not necessarily mean, however, that other resistance genes of large effect but different mode of action, properly used, would fail so quickly in this or other host species.

An unfortunate consequence of such failures was the attempt to recombine simply inherited resistances that had been previously exposed alone to the pathogen. For example, the wheat cultivar Mendos, introduced into eastern Australia in the 1960s, combined the previously exposed genes for black rust resistance, *Sr*11, *Sr*17, and *Sr*Tt. It rapidly became susceptible (22) because of the preadaptation of the pathogen population. In a similar example, the barley cultivar Impala, which combined two previously used genes for mildew resistance (Mlg; Mla6), became highly susceptible during its first year of extensive commercial use in the United Kingdom (42). Following the introduction of single resistance genes, such fragile combinations are inevitably and continuously produced by chance and released from breeding programs because breeders need to exploit other characteristics of the cultivars involved.

However, in recent years in Australia, pyramidal accumulation of resistance genes in wheat has been carefully controlled by analysis of the reponse of *Puccinia graminis* f. sp. *tritici* to new cultivars. If necessary, pathogen isolates are selected after artificial induction of mutants for screening host progeny with combinations of resistance genes not matched by naturally occurring pathogen genotypes (23).

Pyramidal accumulation of major genes for black rust control in wheat has also been practiced in Canada. Because primary infection in Canada is due almost entirely to immigrant spores, which come from other cultivars, it is argued that pathogen genotypes with appropriate combinations of matching genes are likely to be rare. Pathogen genotypes that do survive on Canadian wheat are unlikely to be highly adapted to wheat cultivars grown elsewhere and are unlikely to contribute to overwintering populations in the south. Experience shows, however, that some combinations of resistance genes are more durable than others, and there does not yet seem to be a satisfactory way of predicting these differences (18).

COMBINING QUALITATIVE AND DURABLE RESISTANCE

Qualitative resistance may be combined with potentially durable resistance either by design, to take advantage of the complete disease control and to provide a safe background for the qualitative resistance, or by necessity, because of the particular parents used in a breeding program. The appropriate selection is difficult but feasible (19). If a cultivar with simply inherited qualitative resistance is crossed with a durably resistant cultivar, progeny with the durable resistance phenotype can be selected if a race is used that overcomes the qualitative resistance. If such a race is not available, segregating families can be selected in which the most susceptible segregants resemble the durably resistant parent. The difficulty of selection increases with increasing numbers of specific resistance genes. A variation of this method was used in the selection of the potentially durable wheat cultivar Bounty (19).

As an alternative, Clifford (9) described ways in which temperature-sensitive resistance genes, switched on and off by appropriate temperatures, could be used to reveal the phenotype of the host background in segregants that possessed the resistance genes.

To summarize, the potential for using inherited durable resistance can be gauged initially from the biology and ecology of the host-pathogen system involved. For many pathogens that are unable to respond easily and quickly to large scale exposure of a resistant host, all that may be required on the part of the breeders is ". . . a strategy of watchful neglect" (33). Even under these circumstances there is no certainty that the resistance will remain effective; a more pathogenic form may arise by local selection, immigration, or as a result of some change in agricultural practice.

For potentially responsive pathogens, or to allow greater exposure of individual host cultivars, a more rigorous approach is required, preferably utilizing resistant cultivars whose durability has already been proved. Successful exploitation requires a high investment and the simultaneous incorporation of durable resistance to several diseases.

The advantages of simple inheritance and complete disease control provided by some qualitative resistances should be exploited in combination with known durable resistance, or against pathogens unlikely to respond quickly, or in systems designed to promote durability.

SYSTEM DURABILITY

The basis of system durability is to reduce exposure of each resistant host to the pathogen, and, by increasing the range of different resistances used, to further slow down the overall rate of pathogen response. It is essential to know that the diversity between hosts is relevant to disease resistance. The range of maize hybrids used in the late 1960s in the United States no doubt encompassed a considerable diversity of characters. The epidemic of southern corn leaf blight in 1970 revealed, however, that too little of that diversity was relevant to infection by *Helminthosporium maydis*: the almost universal use of Tms cytoplasm provided a crop medium that was effectively uniform to the pathogen.

SYSTEMS FOR IMPROVING DURABILITY

Between Crops: Deployment in Time

The commonest system is the "reserve strategy" (11). It has never been used in a controlled manner, and probably never will be. The obvious reasons for cultivar replacement are either that an existing cultivar has become or is becoming susceptible to a disease or pest or that it is being superseded by a new cultivar improved in other characters.

However, it has been suggested that used resistance genes should be recycled following a period of nonexposure to allow a decline in frequency of matching pathogenicity. Such a strategy may be feasible only if the resistance genes involved could be removed from cultivation before there was any obvious pathogen response; the motivation for such an undertaking would probably be small and impossible to enforce. It would also be made more difficult by the rapid dissemination of the resistance genes among various breeding programs, perhaps in neighboring regions. For this same reason, the delay before a used resistance gene is finally eliminated from an area is likely to be great. Under these conditions, matching pathogenicity might become fixed in the pathogen population, so that the resistance would not become effective again. On the other hand, it might decline rapidly to an undetectable frequency, the rate being dependent on the particular array of nonmatching hosts that are used. Unfortunately, with currently available methods, such an apparent disappearance may reflect poor sampling rather than low frequency. Further, the real rate of decline tends to slow down as the character becomes more infrequent.

A more useful form of temporal deployment occurs where a crop is grown more than once in a year, with separate or partially overlapping seasons. Under these conditions, it is likely that different groups of cultivars will be bred or selected for the different seasons because of different environmental requirements. Under these circumstances, the resistances that are used at different times should also be separated. For example, in recent years in western Europe, there has been a rapid upsurge in the cultivation of winter barley, so that the areas of winter and spring-sown crops are roughly equal. As a consequence, Schwarzbach (31) has taken the initiative of persuading breeders to use particular sources of mildew resistance only for the winter crop, which he and others have tested and provided, and to prevent, or to eliminate by selection, the exchange of resistance genes between the two crops. The strategy is unlikely to be completely successful because breeders are interested in exchanging other characters between winter and spring crops, which will carry over associated resistance genes.

Separation of winter and spring crops in this way imposes disruptive selection on the pathogen migrating from one crop to the other, to the benefit of both. If this strategy is not used, then the crop continuity that is provided for the pathogen may accelerate its response to resistant cultivars.

Between Crops: Deployment in Space

Spatial deployment of host resistance on a regional basis is valuable for areas in which a pathogen does not overseason so that infection at the beginning of the epidemic is entirely by immigration from distant sources. If cultivars in the receiving area are resistant to all but occasional variants in the incoming population, and if the spore clouds arrive late in the growing season, then it is unlikely that a serious epidemic will ensue. Large-scale regional deployment of resistance genes has been discussed for the "Puccinia Path" in North America and for the distribution of cereal rust diseases on the Indian subcontinent (24). As a principle, it seems unlikely that such strategies will work if restrictions are placed on breeders and growers at all stages along such pathways. The onus must be on those at the receiving end of the system to use cultivars or, if necessary, species that are resistant to alloinfection from distant sources.

Within a region, spatial deployment may be more easily introduced where growers are particularly concerned with major diseases. However, if diseases are of major significance it often implies that monoculture of a particular species is already well established, so that there is little scope for effective disposition of fields of different species.

Spatial deployment for a single species becomes feasible if similarly acceptable cultivars are available that have different resistance genes. For example, following the introduction of diversification schemes for yellow rust and powdery mildew in wheat, and for powdery mildew in barley (27), there was a considerable voluntary response from wheat growers because of recent memories of severe problems with wheat yellow rust (26).

Diversification schemes have a potential value in the insurance that they provide against failure of particular cultivars. They may also provide a direct benefit in slowing down large-scale spread of windborne pathogens from field to field. This assumes that alloinfection provides an important contribution at least in the early stages of the epidemic and that the differences between the cultivars used cause directional or disruptive selection on the immigrant spores. Ideally, the cultivars involved should be completely different in their qualitative resistances, but such distinctions are impossible to maintain because of deliberate intercrossing of new and old cultivars for other reasons.

A basic assumption in diversification is that the pathogen population selected on one cultivar will be adapted to it and will be less well able to grow on other cultivars. It is also assumed that selection against the favored pathogenicity characters will be similar on all other nonmatching cultivars. In practice, this may not be so. For a number of pathogenicity characters in barley mildew, selection on nonmatching hosts was found to be similar, but for others it was greater or less than expected (45). Consequently, diversification would be expected to be more effective between some neighboring cultivars than between others.

The evidence for deviations in selection against nonmatching pathogen populations on particular hosts in the field may be difficult to obtain. Wolfe and Knott (46) pointed out that the evidence can be gained only from defined, random mating populations. Unfortunately, a number of conclusions in the literature are incorrectly drawn, such as, for example, the suggestion that the gene combination in *Puccinia graminis* necessary for overcoming *Sr*6 and *Sr*9d in the host occurs at a lower frequency than expected (38,39). It may do so, but this cannot be determined from the evidence and calculations provided.

Persistent deployment of particular sets of cultivars may lead to a loss of effectiveness in diversification because of selection for complex pathogen races able to infect all cultivars, even though such selection operates only during alloinfection (47). Some evidence of such a change came from observations in Austria over three years of an increasingly high frequency of phenotypes of *Erysiphe graminis* f. sp. *hordei* able to attack two different groups of resistant cultivars (21). This may occur rapidly if very few cultivars are used, but if no single cultivar occupies more than, say, 10% of the crop area, the change is likely to be acceptably slow.

Within Crop: Deployment in Space and Time

Outbreeding crops that are utilized as heterogeneous populations should have an obvious advantage. However, breeding methodology and the search for increasingly higher yields have combined to limit the variability that occurs within and between cultivars, and the variation that is present is largely unknown and uncontrolled. Factors other than disease may be more important in the selection that acts on the population, and the range of variation that does occur may not be relevant to disease control. For example, it is likely that the recent rapid epidemic development of *Erysiphe betae* on sugar beet in the United States occurred because of the absence of relevant variation in resistance in the cultivars developed in the southwestern United States (30).

For inbreeders, the advantage of within-crop diversity for disease control was probably first pointed out by Targioni Tozzetti in 1767, quoted by Groenewegen and Zadoks (13), who observed a reduction in black rust on wheat in a wheat-vetch mixture when compared with the infection in wheat grown alone. Darwin (10) also pointed to the advantages of growing mixtures of different wheat cultivars although he did not single out disease as a contributory factor. Since then, numerous papers have recommended heterogeneous cropping in terms of disease and pest control, more efficient exploitation of the environment, and buffering of the crop against environmental stresses. Interest has grown in recent years as attention has been focused on some of the limitations and disadvantages of extensive monoculture.

Apart from reductions in disease levels, the advantages of heterogeneous cropping are not always evident in the published papers, although disadvantages apparently occur only rarely. This may be because cultivars that are successful in monoculture have been selected for wide adaptability, so that if they are mixed, there is little scope for demonstrating mixture advantage unless comparisons are made over a wide range of environments. Unfortunately, the experiments that are reported often tend to make use of a relatively small number of cultivars tested at one or a few sites over only one or a few years, e.g., Trenbath (36). More extensive studies, particularly with lines that are appropriately different, should reveal a more obvious benefit.

Disease development in mixtures compared with that in pure stands has been examined recently by several authors (1,4,6,8,14,15,20,43). Essentially, it involves a reduction in fitness of pathogen genotypes best adapted to individual host components. This occurs because, in the mixture, unlike the pure stand, autoinfection and alloinfection at the individual plant level may have a different outcome, analogous to the difference between autoinfection and alloinfection at the field level.

Epidemic development thus proceeds at a slower rate than in pure stands so that, for a wind-disseminated pathogen, a mixture of three cultivars with different resistance genes may halve the amount of infection expected from observation of the pure stands (44). Where there is no evident pathogen specialization toward different host cultivars, epidemic development in a mixture is likely to be slowed down by differences in the levels of host resistance, essentially because increase in infection of the resistant components tends to be less than the decrease in infection of the susceptible components (14,15).

In a cultivar mixture, the reduced fitness of pathogen genotypes best adapted to individual components means that more widely adapted genotypes, which are usually slower-growing on the hosts to which they are compatible, may now increase, relatively, more rapidly, and so become more frequent than simpler genotypes. This does not necessarily mean that the absolute population size, and thus the amount of disease, will increase. Increase in disease may be dependent on selection for increased rate of reproduction among the widely adapted pathogen genotypes.

To reduce selection for increased fitness in widely adapted pathogen genotypes, different mixtures should be deployed in space and time. This can be achieved easily with simple mixtures: new resistant and improved cultivars can be immediately incorporated into the system, providing protection for the variety and improvement in mixture performance. Ordered rotation of components between mixtures is also feasible if the supply of new, different cultivars is limited. Particularly for barley mildew, the advantages of the mixture system have been repeatedly demonstrated by experiment in terms of control of powdery mildew and other foliar diseases, increased yield, and greater yield stability (35,40,44). However, the large-scale, long-term test of the durability of the system has not yet been performed, nor has the flexibility of response of the system been tested against large-scale changes in the pathogen population.

Because of its simplicity, which requires mixing to be done only at the final stage of seed preparation, there is no difficulty in adding the system to existing legislative procedures for cultivar registration, recommendation, and production. It has the further advantage that management of disease control is removed or lessened for the grower; it is possible to control the whole system within the area of seed production. As a result, commercial development is now well in hand in the United Kingdom and Denmark, with increasing interest elsewhere.

The mixture system may be widely advantageous for other crops and pathogens, for example, in control of rice blast (K. M. Chin, *personal communication*) and powdery mildew of swede (L. Sitch, *personal communication*) and in many forestry systems, where disease resistant selections may be available but cannot be easily bred, and which therefore require protection. The argument can be further extended to species mixtures, upon which evolving agriculture has depended for 5,000 years, and which would doubtless repay more careful investigation and exploitation. Mixtures of different species have additional advantages in that they provide predictably better durability of disease control and a quality of product not easily achieved otherwise. For example, the various possible cereal-legume mixtures can provide a combined output of carbohydrate and protein that is impossible to obtain from either crop alone.

SOME CONCLUSIONS

Genetic strategies can provide both control of disease and durability of that control. The value of the control can be measured, but that of the durability cannot. In general terms, the advantages are obvious. Effective strategies increase the probability that crop production and quality can be maintained at a high level continuously and over a wide area. Avoidance of local and large-scale fluctuations provides continuity of income for the grower and the prospect of continual upward progress in production for the country. The frequent changes in cropping pattern and agricultural practice that occur can be accommodated with greater safety.

However, against the benefits, one has to balance the cost of each strategy itself, which is considerable, at all levels of introduction and use. Resistance genes in a host cultivar may be energy-demanding in their activity (34), and they may replace genes that could provide higher yield and quality. Introducing resistance genes into a new cultivar is an expensive process, particularly so if it detracts from more productive aspects of crop breeding. At the production level, it may be difficult to evaluate different strategies against the, probably, short-term gains from maximum utilization of a particularly high-yielding single cultivar.

Because of the range of considerations involved for any crop, its pathogens, and the possible solutions that may be available, a different answer will be appropriate to each different situation, whether it be inherent durable resistance or pyramidal resistance in some inbreeders or spatial diversification in forest trees; as Barrett (2) pointed out, it is unlikely that a single approach will suit all crops in all environments. Nevertheless, a major defect in current disease control strategies, whether by resistant cultivars or by pesticides, is the widespread opportunity provided for the controlled organism to overcome the control measure. The advantages provided by the enormous acceleration in monoculture and crop homogeneity over the last hundred years have lain in the uniformity of high-yield characters and in simplicity of handling and processing. The uniformity has frequently been embarrassing, however, because of epidemic development of pathogens and pests and its sensitivity to unpredictable environmental stresses. Heterogeneous crops, on the other hand, may have the strength of diversity for dealing with pathogens, pests, and environmental stresses, but may also be criticized for their lack of uniformity in characters such as crop maturity and ease of handling.

An important attempt to resolve this dilemma was the development of multilines (3), in which the lines that formed the cultivar were uniform in all respects except for genes for resistance to a particular disease. This compromise system in its classic form did not become widely popular, partly because the system was balanced too far toward uniformity and lacked the necessary heterogeneity to deal with a wide range of variable pathogens and environmental stresses. A second disadvantage was the limitation imposed by breeding technology: breeders regarded the introduction of limited heterogeneity into a cultivar that was becoming outclassed as being too conservative.

However, we may now be able to see more clearly those characters for which uniformity is advantageous and those for which appropriate heterogeneity would be more suitable. We are also being provided with more rapid and efficient plant breeding methods, such as doubled haploids and single seed descent, and with much improved equipment. For the future, therefore, it seems both possible and desirable to synthesize new cultivars that have the required levels and advantages of both homogeneity and heterogeneity.

This chapter has been deliberately concerned only with genetic strategies for disease control and the integration of inherent and system durability. Such strategies can obviously be improved by the integration of other methods such as crop sanitation, fungicide use, and biological control, dealt with elsewhere. For some host-pathogen systems, integration of genetic with other strategies may be extremely valuable if not crucial. For example, host resistance genes and fungicides can act together to provide high levels of disease control (12) and a system that is more difficult for the pathogen to overcome than can either host or fungicide alone. This simple concept can be extended (41) and should be explored more fully to preserve the effectiveness of both host resistances and fungicides. Exciting developments in all aspects of fungicide application and disease forecasting should provide the means for effective and cheap integrated strategies.

To provide the necessary background for these and other developments presents several key challenges. We need more intensive studies of the ecology and epidemiology of hosts and their pathogens to evaluate the relative contribution of different features that allow the pathogen to respond to the use of host resistance. We must also learn more of the reasons why some resistant cultivars are more durable than others. A third challenge is to integrate more closely studies of the population ecology, genetics, and epidemiology of major systems to reveal and evaluate critically those strategies that will provide the greatest benefits. Finally, the challenge of genetic manipulation in its new sense must be taken up as soon as it offers the opportunities for introducing novelty into inherited resistance and for increasing the rapidity and ease with which diversity can be gained and dynamically maintained.

LITERATURE CITED

1. Barrett, J. A. 1980. Pathogen evolution in multiline variety mixtures. Z. Pflanzenkr. Pflanzenschutz 87:383-396.
2. Barrett, J. A. 1981. The evolutionary consequence of monoculture. Pages 209–248 in: Genetic Consequences of Man-Made Change. J. A. Bishop and L. M. Cook, eds. Academic Press, London.
3. Browning, J. A., and Frey, K. J. 1969. Multiline cultivars as a means of disease control. Annu. Rev. Phytopathol. 7:355-382.
4. Burdon, J. J. 1978. Mechanisms of disease control in heterogeneous plant populations—An ecologist's view. Pages 193–200 in: Plant Disease Epidemiology. P. R. Scott and A. Bainbridge, eds. Blackwell Sci. Publ., Oxford. 329 pp.
5. Burdon, J. J., Marshall, D. R., and Luig, N. H. 1981. Isozyme analysis indicates that a virulent cereal rust pathogen is a somatic hybrid. Nature 293:565-566.
6. Burdon, J. J., and Shattock, R. C. 1980. Disease in plant communities. Appl. Biol. 5:145-219.
7. Caten, C. E. 1974. Intra-racial variation in *Phytophthora infestans* and adaptation to field resistance for potato blight. Ann. Appl. Biol. 77:259-270.
8. Chin, K. M. 1979. Aspects of the epidemiology and genetics of the foliar pathogen, *Erysiphe graminis* f. sp. *hordei* in relation to infection of homogeneous and heterogeneous populations of the barley host, *Hordeum vulgare*. Ph.D. thesis, University of Cambridge. 137 pp.
9. Clifford, B. C. 1983. Combining different resistances to barley brown rust. Proc. Int. Barley Genet. Symp., 4th. In press.
10. Darwin, C. R. 1872. On the Origin of Species by Means of Natural Selection, 6th ed. Murray, London. 458 pp.
11. Duvick, D. N. 1977. Major United States crops in 1976. Ann. N.Y. Acad. Sci. 287:86-96.
12. Fry, W. E. 1975. Integrated effects of polygenic resistance and a protective fungicide on development of potato late blight. Phytopathology 65:908-911.
13. Groenewegen, L. J. M., and Zadoks, J. C. 1979. Exploiting within-field diversity as a defense against cereal diseases: A plea for 'poly-genotype' varieties. Indian J. Genet. Plant Breed. 39:81-94.
14. Jeger, M. J., Griffiths, E., and Jones, D. G. 1981. Disease progress of non-specialised fungal pathogens in intraspecific mixed stands of cereal cultivars. I. Models. Ann. Appl. Biol. 98:187-198.
15. Jeger, M. J., Jones, D. G., and Griffiths, E. 1981. Disease progress of non-specialised fungal pathogens in intraspecific mixed stands of cereal cultivars. II. Field experiments. Ann. Appl. Biol. 98:199-210.
16. Johnson, R. 1978. Practical breeding for durable resistance to rust diseases in self-pollinating cereals. Euphytica 27:529-540.
17. Johnson, R. 1979. The concept of durable resistance. Phytopathology 69:198-199.
18. Johnson, R. 1981. Durable resistance: Definition of genetic control, and attainment in plant breeding. Phytopathology 71:567-568.
19. Johnson, R. 1983. Genetic background of durable resistance. Proc. NATO Adv. Study Inst. on Durable Resistance in Crops. Italy, 1981. In press.
20. Leonard, K. J. 1969. Genetic equilibria in host-pathogen systems. Phytopathology 59:1858-1863.
21. Limpert, E., and Schwarzbach, E. 1983. Virulence analysis of powdery mildew of barley in different European regions in 1979 and 1980. Proc. Int. Barley Genet. Symp., 4th. In press.
22. Luig, N. H., and Watson, I. A. 1970. The effect of complex genetic resistance in wheat on the variability of *Puccinia graminis* f. sp. *tritici*. Proc. Linnean Soc. New South Wales (Aust.) 95:22-45.
23. McIntosh, R. A. 1976. Genetics of wheat and wheat rusts since Farrer. J. Aust. Inst. Agric. Sci. 42(Dec.):203-216.
24. Nagarajan, S., and Singh, H. 1975. The Indian stem rust rules—An epidemiological concept on the spread of wheat stem rust. Plant Dis. Rep. 59:133-136.
25. Parlevliet, J. E. 1981. Deploying rust resistance. Pages 149–152 in: Proc. Symp., IX Internat. Congr. Plant Prot., Vol. 1. T. Kommedahl, ed. Burgess Publ. Co., Minneapolis. 411 pp.
26. Priestley, R. H., and Bayles, R. A. 1980. Factors influencing farmers' choice of cereal varieties and the use by farmers of varietal diversification schemes and fungicides. J. Nat. Inst. Agric. Bot. 15:215-230.

27. Priestley, R. H., and Wolfe, M. S. 1977. Crop protection by cultivar diversification. Proc. Br. Insecticide Fungicide Conf., 9th 1:135-140.
28. Robinson, R. A. 1976. Plant Pathosystems. Springer-Verlag, Berlin. 184 pp.
29. Rufty, R. C., Hebert, T. T., and Murphy, C. F. 1981. Variation in virulence in isolates of *Septoria nodorum*. Phytopathology 71:593-596.
30. Ruppel, E. G., Hills, F. J., and Mumford, D. L. 1975. Epidemiological observations on the sugar beet powdery mildew epiphytotic in western USA in 1974. Plant Dis. Rep. 59:283-286.
31. Schwarzbach, E. 1983. Progress and problems with breeding for disease resistance. Proc. Int. Barley Genet. Symp., 4th. In press.
32. Scott, P. R. 1981. Variation in host susceptibility. Pages 219–236 in: The Biology and Control of Take-All. M. J. C. Asher and P. J. Shipton, eds. Academic Press, London. 538 pp.
33. Simmonds, N. W. 1979. Principles of Crop Improvement. Longmans, London. 408 pp.
34. Smedegaard-Petersen, V. 1980. Increased demand for respiratory energy of barley leaves reacting hypersensitively against *Erysiphe graminis, Pyrenophora teres* and *Pyrenophora graminea*. Phytopathol. Z. 99:54-62.
35. Stølen, O., Hermansen, J. E., and Løhde, J. 1980. Varietal mixtures of barley and their ability to reduce powdery mildew and yellow rust diseases. Yearbook. Royal Danish Veterinary Agricultural College, pp. 109-116.
36. Trenbath, B. R. 1974. Biomass productivity of mixtures. Adv. Agron. 26:177-210.
37. Vanderplank, J. E. 1963. Plant Diseases: Epidemics and Control. Academic Press, London. 349 pp.
38. Vanderplank, J. E. 1975. Principles of Plant Infection. Academic Press, London. 217 pp.
39. Vanderplank, J. E. 1983. Should the concept of physiological races die? Proc. NATO Adv. Study Inst. on Durable Resistance in Crops. In press.
40. Wolfe, M. S. 1981. The use of spring barley cultivar mixtures as a technique for the control of powdery mildew. Proc. Br. Insecticide Fungicide Conf., 11th. 1:233-239.
41. Wolfe, M. S. 1981. Integrated use of fungicides and host resistance for stable disease control. Philos. Trans. R. Soc. London, Ser. B. 295:175-184.
42. Wolfe, M. S., and Barrett, J. A. 1976. The influence and management of host resistance on control of powdery mildew on barley. Proc. Int. Barley Genetics Symp., 3rd, Garching. pp. 433-439.
43. Wolfe, M. S., and Barrett, J. A. 1979. Disease in crops: Controlling the evolution of plant pathogens. J. R. Soc. Arts 127:321-333.
44. Wolfe, M. S., and Barrett, J. A. 1980. Can we lead the pathogen astray? Plant Dis. 64:148-155.
45. Wolfe, M. S., Barrett, J. A., and Slater, S. E. 1983. Pathogen fitness in cereal mildews. Proc. NATO Adv. Study Inst. on Durable Resistance in Crops. In press.
46. Wolfe, M. S., and Knott, D. R. 1982. Populations of plant pathogens: Some constraints on analysis of variation in pathogenicity. Plant Pathol. 31:79-90.
47. Wolfe, M. S., and Schwarzbach, E. 1978. Patterns of race changes in powdery mildews. Annu. Rev. Phytopathol. 16:159-180.

PART V

Socioeconomic Factors Affecting Plant Health Maintenance

It is the task of the day and the responsibility of the younger farmers to stride out briskly, to tie together science and practice, and to take full advantage of the results of the former to perfect the latter, all to our own profit as well as that of our fellow man.

—Julius Kühn, 1858

I have become convinced that the resistance of the farmer to accepting changes in crop production methods with any given crop, even when research data are available and clearcut, is greatest in the area of the center of origin and domestication of that crop. The resistance to change in cultivation practices is tied directly to the cultural patterns and traditions of the people. Life itself in these areas has depended upon the "success" or failure of the one basic crop.

—Norman E. Borlaug, 1965

More than ever we need leadership. In plant pathology as in all fields, it is the job of the teaching institutions to attract good students and to develop them into competent plant pathologists to serve society anywhere in the world.

—William C. Snyder, 1971

CHAPTER 44

ECONOMICS OF PLANT DISEASE CONTROL

G. B. WHITE

New problems and concerns in pest management have made economic input important and perhaps even essential in the development of strategies for plant disease control. The concern in the last three decades about environmental quality has forced attention on policies and programs to balance the need for an adequate food supply at reasonable prices with the need to reduce the use of pesticides, especially those with more lasting environmental effects. Rising energy prices (relative to other costs of production and commodity prices) since the 1973 Arab oil embargo have made pest control more expensive by increasing pesticide application costs and, in some cases, increasing the cost of pesticides. Regulation of the registration of new chemicals has made it increasingly expensive and time-consuming for new compounds to reach the market. All of these factors have focused attention on the efficient allocation of pest management resources.

Consideration of the economics of plant disease control could be directed toward several different groups or levels within the economic and political system. Most frequently, attention has been directed to investigation of resource allocation decisions at the individual producer level. Other areas of research of interest to economists include the macroeconomic (national) or regional effects of pesticides or various pest management strategies, the efficient allocation of research dollars, or the effects of alternative regulatory schemes. All of the latter topics are important, but this chapter is limited to a consideration of resource allocation decisions at the individual producer or the pest management delivery system level. The emphasis is on making efficient pest management decisions at the farm level.

PEST MANAGEMENT INPUT DECISIONS

Several alternative pest management inputs are available to producers for plant disease management. These include pesticides, resistant cultivars, certified seed, sanitation measures, cultural control, and information. All of these inputs have a cost, although the cost is differentially shared by producers and society; some of the inputs also have a cost in terms of external environmental effects.

A common goal of producers is to maximize profits. There is a unique amount of each of the inputs mentioned above that will maximize profits. The conceptual tool used by economists to identify combinations of inputs that maximize profit is marginal analysis, or comparing incremental returns and costs from small increases in inputs. At least two slightly different forms of marginal analysis have been used to study the optimal use of pest management inputs. For reasons I will discuss later, there are problems in applying these marginal analysis techniques in the management of plant diseases. However, marginal analysis is useful in providing a framework for understanding the economic factors involved.

Economists have traditionally used a production function to solve for the optimal level of inputs. A production function describes the relationship between the use of all inputs in crop production and the resulting product. Input prices, product prices, and the production function are assumed to be known with certainty. The profit-maximizing level of all inputs is then determined. External effects on other producers and on consumers are not accounted for in the functional relationship. (For a more detailed treatment of maximization using the production function approach, see, for instance, references 5 and 6.)

The production function approach to resource allocation—while useful in a general sense and for inputs such as fertilizer—has proved less useful for consideration of pest management problems. First, the functional relationship describes a planning process where inputs are applied in anticipation of certain prices and yields. Pesticide applications, at least for many insecticides and fungicides, are in reaction to conditions that develop as the growing season progresses or are applied according to a calendar schedule. While many inputs are committed at the time of planting, pesticides generally are applied in reaction to potential pest damage. Secondly, there is an intermediary that stands between the use of the pesticide and the resulting yield and quality of product. The effect of the pesticide is exerted through its effect on the pest population, which exerts a direct effect on yield or quality. A third problem might be that the production function is not a useful frame of reference for communicating with scientists in the crop protection disciplines. All of these difficulties were addressed by Headley (15) in his economic formulation of the economic threshold (ET).

Headley's formulation considers three variables—damage (a constant price × yield change), pest population, and time. A value of production function and a cost function are expressed in terms of the effects of the pest population. The ET occurs at the pest population where the rate of change of the value of production equals the rate of change of the cost function. The ET approach is consistent with the traditional approach; useful results that emerge from both approaches include the following: 1) maximum yield or revenue (i.e., zero damage) is not optimal unless pest management inputs have zero prices; 2) the optimal pest population becomes smaller (i.e., the level of pest management inputs becomes larger) as the value of product increases (or as product price increases), and 3) increases in the unit cost of control result in higher optimal pest populations. Other researchers (12,19,24) have modified Headley's model, or developed other models, to account for some of the many complexities of pest management decisions using marginal analysis.

Examples of the use of marginal analysis for determining fungicide applications may illustrate why, for example, high-value crops receive more fungicides than do small grains. It has been estimated that 80% of the fungicides used on growing crops is applied to vegetables, fruits, and other high-value crops (1). In contrast, relatively little is applied to small grains, although the use of fungicide-treated seed is a common practice. In 1976, it was estimated that only

1% of the wheat acreage and 2% of the oats, rye, and barley acreage were treated with fungicides, excluding seed treatment and treatment of stored grain (8). It is often presumed that high-value crops are treated more frequently because of the value of the crop alone, but this is an oversimplification. The economical use of fungicides for a high-value crop that requires frequent applications of fungicides could be viewed as choosing the optimal number of applications for an average growing season. In New York, apples for fresh market typically receive about seven primary fungicide applications for a protectant schedule against apple scab (W. C. Stiles, *personal communication*). If the choice were between no sprays and seven, the obvious economic choice would be seven sprays since scab can lead to nearly 100% reduction in fruit meeting fresh market requirements. An appropriate question for marginal analysis, however, is whether four, five, six, seven, or more applications are optimal in an average growing season.

The marginal returns of each successive additional application could be compared with the incremental cost of an application (i.e., the variable cost of one spray, which includes labor, spray materials, and fuel costs). For small grains, the development of resistant cultivars and the use of fungicide-treated seed have been the main approaches to disease management, with applications of fungicides directly on the growing crop being used infrequently. This may be due to the marginal cost of an application of fungicide being high relative to the expected value of damage prevented. For example, wheat yields in New York during 1978–1980 averaged 26.4 quintals per hectare and prices averaged $13.85 per quintal (17). Chemical costs for wheat production in the Northeast, excluding seed treatment, were virtually zero (7). Triadimefon, a fungicide that is currently being tested, costs about $29.60 per hectare at rates currently being tested, and custom aerial application rates are $14.80 per hectare; thus if the material is applied in a single application, the cost is about $44.40 per hectare. The breakeven yield response would be 3.21 quintals if the expected price is $13.85 per quintal. If a producer expected anything less than an additional 3.21 quintals of wheat from an application of the fungicide, the treatment would not be justified economically. It is my hypothesis that the cost of materials plus application is high relative to the expected gain, and this has inhibited the market potential and hence the registration of fungicides for foliar applications to small grains. The fact that most wheat is grown in the Midwest, where disease pressure and base yields are lower than in the Northeast, lends additional support to this hypothesis.

The various approaches to determine the optimal amount of pest management inputs, all of which involve marginal analysis, have been difficult to apply. The concept of the ET as formulated by Headley has been useful in facilitating communication with crop protection scientists but has been elusive to apply in research and extension programs. In particular, plant pathologists have not found the ET to be a useful concept for managing plant diseases (2). Although the same can be said for the other marginal analysis approaches, I will discuss the lack of acceptance by plant pathologists from the frame of reference of the ET.

Two difficulties have emerged in applying the ET. The first lies in the simplifying assumptions made in the analyses. Conditions that violate the assumptions of marginal analysis underlying the ET include combined applications of insecticides and fungicides, the effect of pest populations this year on the yield and quality of the crop in subsequent years (especially for perennials), the buildup of resistance by pests to pesticides, and current federal regulations prohibiting the use of pesticides at any dosage other than the recommended rates (20). Each of these problems can be handled by adjustments in the marginal analysis, but each also complicates empirical application of the ET concept.

The second difficulty in the applicability of Headley's definition is that the ET was formulated specifically for insect pest populations. While the same relationships are relevant for disease control (a damage function, pest population growth, yield and quality, and a cost function), there are important differences. First, many diseases have been difficult to successfully monitor for inoculum, in contrast to monitoring for insects. Once initial lesions appear, damage is irreversible. Another difficulty has been the extreme variability in yields where pest control is not practiced. Even when the presence of inoculum can be monitored, the development of diseases can appear erratic, with extreme variability due to environmental conditions. Partially for this reason, Fry (10) suggested that for many diseases, monitoring of environmental conditions is the cornerstone of disease pest management.

This large potential for variability in the damage resulting from a given pest population at a given time introduces explicitly the importance of risk and uncertainty for pest management resource decisions. Other authors (3,4,18) have addressed this issue. Norgaard (18) has noted a widely held view that a substantial portion of total pesticide applications occur for insurance purposes. The effect of this is to cause producers to sacrifice income on the average to attain greater stability of income. It has been shown theoretically (5,9) that when the outcome of a particular action is uncertain and producers are risk averse, the optimal use of risk-reducing inputs such as pesticides can be higher (i.e., the optimal pest population density is lower) than is justified when the outcome is known with certainty.

Still another problem in applying the ET to disease management is the need for producers to anticipate the timing so as to employ the pest management resource at the optimal time. Diseases that have the potential for destroying a high percentage of the crop, or that move very rapidly, leave little detection time and little time for fungicides to act. Consequently, producers often rely on the precautionary use of fungicides.

Given these difficulties in applying marginal analysis to plant disease management, the best approach to an economical use of pesticides for crops such as fruits, vegetables, and peanuts, which receive frequent applications of fungicides during the growing season, may be to monitor for the environmental conditions that would lead to development of the disease. This could lead to a reduction in pesticides applied in some years when the weather is unfavorable for disease. As knowledge of epidemiology improves, perhaps disease forecasting could be linked to ET criteria; however, risk aversion may preclude this approach for many diseases.

These difficulties in implementing the ET in no way invalidate its conceptual value. Rather the difficulties point to the gaps in empirical knowledge regarding the relationship between the pest population, the value of production, costs, and the impact of other pests, parasites, and weather. These difficulties also point out rather large gaps between the conceptual formulation of optimal pest management input use and actual decisions made by producers or recommendations made by consultants.

ROLE OF INTEGRATED PEST MANAGEMENT IN THE MANAGEMENT OF PLANT DISEASES

If producers had more information about the relationships between pest populations, the value of production, costs, and the impact of pests, parasites, and weather, they could make more efficient pest management decisions. Integrated Pest Management (IPM) delivery systems typically have as their main objective the providing of information to producers.

IPM is

> the use of multiple tactics in a compatible manner to maintain pest populations at levels below those causing economic injury while providing protection against hazards to humans, domestic animals, plants and the environment. . . . Tactics include chemical, biological, cultural, physical, genetic, and regulatory procedures (2).

The economic rationale for IPM has been developed elsewhere (2,18,25). Benefits of IPM generally are based upon some type of monitoring or scouting program in which information is substituted for prophylactic applications of pesticides when that substitution is economically advantageous to producers. Specialization of labor and economies of size in monitoring may imply more efficiency for a collective IPM program than for a situation in which each producer monitors components individually. The wide variety of tactics and the regional collective control potentially provided by an IPM program enable more efficient management of target pest resistance through time. Finally, there are usually environmental benefits, as most IPM programs result in less pesticide being applied. Proponents believe that use of IPM results in a net economic gain for society.

It has already been suggested that because of the importance of environmental conditions and the difficulty in monitoring inoculum, indirect approaches to disease forecasting, based on weather effects or on host resistance, have been used to guide disease management. These indirect approaches typically assume the pathogen to be present (11). While disease forecasting may be less precise than monitoring and predicting insect and weed populations, several interactions and interdependencies suggest that IPM has an important role in plant disease control.

First, there are interdependencies in monitoring and scouting costs. If a scout is already present in the field to sample insect and weed populations, there is a relatively small additional cost to monitor disease incidence or the environmental conditions that are criteria used in disease forecasting. Second, there are interdependencies in application costs since insecticides and fungicides are frequently applied together. The incremental cost of including an extra chemical in the spray tank may be relatively small if the spray rig is in operation to apply another chemical. Third, there are many interactions between plant diseases and insect and weed pests that can best be managed by an integrated approach. These interdependencies and interactions suggest that, although monitoring and scouting activities may not be economically justified for plant disease control alone, they may be justified for an integrated approach to pest management.

ROLE OF COMPUTERS IN PEST MANAGEMENT DECISIONS

The use of computers should enhance the storage, retrieval, and processing of information generated by IPM research and extension programs. Pest management practices have not changed appreciably in recent years. A mail survey of 513 tree fruit growers in the northeastern United States confirmed the fact the producers use a blend of extension recommendations, personal experience, and chemical company retailers or representatives as primary sources of information for pest related decisions (Table 1). Consultants or advisors paid for by subscription was listed as a source for information by only 9% of the respondents.

With the advent of IPM delivery systems, the possibility arises that computer programs from the land grant university or other computer systems can be made available to producers to provide information for pest management decisions. Decision models could incorporate pest population dynamics, crop growth, and economic criteria to indicate optimal pest management decisions when data relating to weather, input and output prices, and pest monitoring counts are entered for a specific producer. Indeed, with the recent explosion in small computer technology, it is currently within the means of many producers to purchase a small computer. A recent survey of deans and directors of the Cooperative Extension Service revealed that there were more than 800 farmers in the United States in the summer of 1980 who were known to own small computers (23). Various models with sufficient accessible memory (20–64K) are available for $3,000–10,000. In this section, the potential for IPM delivery systems with computer decision models will be explored.

Various simulation models of plant disease have been formulated. Modeling and simulation efforts may provide the impetus to attain the multidisciplinary cooperation necessary for IPM. Simulation has already proved useful to test theories and to do experiments that are too expensive or complex to do in the field. For the most part, these models do not include economic criteria, nor are they linked with other plant disease, insect, or weed pest models to provide an

TABLE 1. Proportion of producers using alternative sources of information for pest related decisions (New England and northeastern United States, tree fruit production, 1980)

Sources of Information[a]	Percent
Extension personnel	84
Personal experience	79
Label or container	59
Chemical retailer	52
Representative of chemical manufacturer	47
Farm magazines	41
Neighbors, friends, relatives	32
Sales leaflets, etc.	28
Publications	22
Radio	14
Processor fieldmen	13
Consultant or advisor paid for by subscription	9
Newspaper	5
Television	2

[a] Unpublished data from a study by Peter Thompson, Department of Agricultural Economics, Cornell University. The survey excluded Wayne County, NY, where there is an extension-operated tree fruit pest management program with 23 participating farms.

integrated approach to crop protection. Current research by the Consortium for Integrated Pest Management groups for alfalfa, cotton, soybeans, and apples will be an important test for the feasibility of crop modeling that includes components from all of the plant protection disciplines as well as having economic input. Thus far, simulation has played an increasingly important role in research.

There remains the challenge of translating simulation and other research models into a form that can be used to aid in management decisions by producers or IPM consultants. The models formulated by researchers are inherently basic and complex, and in addition are often poorly documented. They tend to be used for a particular research project and then are filed with no further improvements, validation, or adaptations by other researchers in other states or regions.

It will be a challenge for plant pathology researchers and extension workers to develop software that would enable full potential use of the current hardware. Hardware configurations that are currently feasible include 1) terminals at field locations linked to mainframe computers at the land grant institutions, 2) small on-farm computers, or 3) a linked network system that includes on-farm as well as mainframe computer capability. The latter configuration makes possible the use of the full power of simulation models in theory—and even in practice if the model is not too expensive to run.

The development of pest management software has, however, been limited. It was reported that there are currently nine mini- and microcomputer programs related to pest management currently in use at land grant universities, and another eight programs are planned or are being developed (23). There were no known commercial software vendors currently supplying pest management programs. A similar list, being prepared for programs on mainframe computers, is not yet available. However, a partial list would include seven programs related to pest management in the Michigan State University Telplan system (14) and a number of programs available on the SCAMP system in New York, including three predictive models (21). Ten years of experience of extension agricultural economists with computer decision aids and other information systems have highlighted the difficulty in making high quality programs available to producers. The greatest difficulty in developing programs for extension has been that few individuals have really been interested in developing computer programs into extension subject matter programs. (For a discussion of some of the problems and successes, see references 16, 22, and 26.) An effort was made by agricultural economists at several institutions (e.g., Virginia Polytechnic Institute, Michigan State University, and the University of Nebraska) for sharing a computer network so that programs written at one institution would be available on mainframe computers to others. Communication systems with remote terminals are being used to attain the objective of easier access. However, many believe that the small on-farm computer offers the greatest potential breakthrough in utilizing computer technology (13). The chief limitation in their adoption is the development of software and making data bases from mainframe computers available to the smaller storage of the small computers (13).

Many operational problems face those who would make plant disease decision models utilizing small on-farm computers alone or in networks. The first problem is that software is not usually adaptable from one company's equipment to another, so the choice of hardware by the plant pathologist limits the choice for producers. This is becoming less important with the development of operating systems that enable programs to be interchanged among computers. Second, once a program is written, there is the question of need for interaction by the user with others (the author of the program, extension agents, or IPM consultants) for interpretation, discussions of limitations of these decision models, etc. A third problem is program maintenance, a difficult problem even with mainframe computers. With 20, 30, or 100 programs in the field in the hands of producers, updating data and making corrections when programming errors are discovered becomes an even greater problem. Private vendors of software could circumvent many of these operational difficulties, but this raises the problem of linking the field models to research since the research base resides at universities and research institutes.

MAKING PLANT DISEASE MANAGEMENT DECISIONS IN THE FUTURE

In this chapter I have discussed the efficient use of pest management inputs through marginal analysis. I have also discussed the role of IPM and computers in providing easier, more timely access to information that would improve efficiency in using pest management inputs. These considerations lead me to the following view of the potential for plant disease management in the next decade.

1. Economies of size in delivery of pest management information; specialization of labor in monitoring; and biological, monitoring, and control cost interactions imply that IPM delivery systems will play a much larger role in plant protection. The extent of that role depends upon the pest pressures on a particular crop and on the extent to which scientists in the plant protection disciplines within a given state are able to develop interdisciplinary cooperation. The scouts and consultants in such a system will maintain close coordination and working relationships with research and extension specialists at land grant universities.

2. Pest management options in the field under such a scheme will always involve a final decision by the individual producer. Increasingly, however, scouts will collect data on pest population densities and consultants will make recommendations to producers based on weather conditions, expected pest population dynamics, and economic criteria. Such recommendations may have to include probabilistic estimates, given the uncertainties inherent in pest problems.

3. Simulation and modeling will be used much more extensively by researchers. The models developed may be too complex and basic for use directly in the field, but they can form the basis for simplified field decision models. These models will also help to define data gaps, test theories, and make experiments that are too expensive or too complex to do in the field.

4. Computer technology will play a much larger role in pest management decisions. To facilitate access to the research base in plant protection, most hardware systems used will probably be terminals and/or small computers linked to mainframe computers at land grant institutions. Some states may be successful in developing pest management software for small computers, but these computers may reside with consultants or extension agents, not with producers. Computers will also play a role in the collection and processing of data for disease management decisions, with the potential for linking monitoring devices (particularly weather monitoring) to computers for storage and retrieval.

5. Producers for some commodities where disease pressures are low, and who are remote from other producers, will continue to make pest management decisions as they have in the past. Main sources of information will be extension agents and chemical field representatives. In these instances, costs of providing IPM delivery are high relative to expected benefits.

Decision making by the producer has been explored in this chapter, and the variables that affect the economic decision for plant disease management have been highlighted. A rationale was presented for IPM as a superior means for making plant disease management decisions when compared with current disease control practices and decision making. The potential and problems in applying computer technology to plant disease management have also been reviewed. In my view, delivery systems that incorporate economic criteria, approach pest management from an IPM perspective, and rely upon decision models in or resulting from computer analysis, represent the optimal potential for plant disease management.

Attaining this optimum is not a foregone conclusion. It depends upon several factors, including 1) successful integration of economists in research and extension pest management projects, 2) the implementation of research projects that include the yield and quality effects of various pest management tactics, 3) effective multidisciplinary research with full participation by plant pathologists, entomologists, weed scientists, nematologists, economists, and systems scientists, and 4) a commitment to overcome some of the problems in the development of and access to software for computerized decision aids. From this viewpoint, the challenge for plant pathologists, scientists of the other plant protection disciplines, and social scientists is to rethink our traditional ways of doing research and extension so that this potential can be attained.

ACKNOWLEDGMENTS

I gratefully acknowledge P. Arneson, J. Conrad, G. Fohner, W. Fry, W. Knoblauch, and R. Milligan, colleagues at Cornell, who provided constructive criticism and helpful insights in reviewing earlier drafts of this chapter.

LITERATURE CITED

1. American Phytopathological Society. 1979. Contemporary Control of Plant Diseases with Chemicals. Report prepared for the U.S. Env. Protect. Agency. St. Paul, MN. 170 pp.
2. Apple, J. L., Benepal, P. S., Berger, R., Bird, G. W., Maxwell, F., Ruesink, W. G., Santelmann, P., and White, G. B. 1979. Integrated Pest Management: A Program of Research for the State Agricultural Experiment Stations and the Colleges of 1980. Study by the Intersociety Consortium for Plant Protection for the Experiment Station Committee on Organization and Policy. N.C. State University, Raleigh. 190 pp.
3. Carlson, G. A. 1970. A decision theoretic approach to crop disease prediction and control. Am. J. Agric. Econ. 52:216-223.
4. Carlson, G. A., and Main, C. E. 1976. Economics of disease-loss management. Annu. Rev. Phytopathol. 14:381-403.
5. Dillon, J. L. 1977. The Analysis of Response in Crop and Livestock Production, 2nd ed. Pergamon Press, New York. 213 pp.
6. Doll, J. P., and Orazem, F. 1978. Production Economics: Theory with Applications. Grid Publishing Inc., Columbus, OH. 406 pp.
7. Economics and Statistics Service, USDA. 1981. Costs of producing selected crops in the United States—1978, 1979, 1980, and projections for 1981. Prepared for the Committee on Agriculture, Nutrition and Forestry, U.S. Senate, U.S. Govt. Printing Office: Washington, DC. 79 pp.

8. Eichers, T. R., Andrilenas, P. A., and Anderson, T. W. 1978. Farmers' use of pesticides in 1976. Agric. Econ. Rep. 418. USDA, Economics, Statistics, and Cooperatives Serv. U.S. Govt. Printing Office: Washington, DC. 58 pp.
9. Feder, G. 1979. Pesticides, information, and pest management under uncertainty. Am. J. Agric. Econ. 61:97-103.
10. Fry, W. E. 1980. Management of late blight. Paper presented at a Conference on Potato Pest Management, Rutgers University, New Brunswick, NJ. 18 pp.
11. Fry, W. E. 1982. Principles of Integrated Plant Disease Management. Academic Press, New York. 378 pp.
12. Hall, D. C., and Norgaard, R. B. On the timing and application of pesticides. Am. J. Agric. Econ. 55:198-201.
13. Harsh, S. B. 1978. The developing technology of computerized information systems. Am. J. Agric. Econ. 60:908-912.
14. Harsh, S. B. 1980. M.S.U. index of Telplan programs. Mimeographed. Michigan State Univ., East Lansing. 9 pp.
15. Headley, J. C. 1972. Defining the economic threshold. Pages 100–108 in: Pest Control Strategies for the Future. Nat. Acad. Sci., Washington, DC. 376 pp.
16. Infanger, C. L., Robbins, L. W., and Debertin, D. L. 1978. Interfacing research and extension in information delivery systems. Am. J. Agric. Econ. 60:915-920.
17. New York Crop Reporting Service. 1981. New York Agricultural Statistics, 1980. Albany. 75 pp.
18. Norgaard, R. B. 1976. Integrating economics and pest management. Pages 17–27 in: J. L. Apple and R. F. Smith, eds. Integrated Pest Management. Plenum Press, New York. 200 pp.
19. Regev, U., Gutierrez, A. P., and Feder, G. 1976. Pests as a common property resource: A case study of alfalfa weevil control. Am. J. Agric. Econ. 58:186-197.
20. Reichelderfer, K., and Bender, F. 1978. A simulative approach to controlling Mexican Bean Beetle on soybeans in Maryland. Misc. Publ. No. 935. Md. Agric. Exp. Stn., College Park, MD. 52 pp.
21. Sarette, M., Tette, J. P., and Barnard, J. 1981. SCAMP—A computer-based information delivery system for Cooperative Extension. Food and Life Sci. Bull. 90. N.Y. State Agric. Exp. Stn., Geneva. 8 pp.
22. Shaffer, R. E. 1978. The developing technology of computerized information systems: Discussion. Am. J. Agric. Econ. 60:913-914.
23. Strain, J. R. 1980. Survey of availability of micro and mini computer software. Staff paper 165. Food and Resource Econ. Dep., Univ. Fla., Gainesville. 39 pp.
24. Talpaz, H., and Borosh, I. 1974. Strategy for pesticide use: Frequency and applications. Am. J. Agric. Econ. 56:769-775.
25. Thompson, P., and White, G. B. 1979. Linking economic and action thresholds in pest management: Implications for advisory services. Acta Hortic. 97:97-104.
26. Walker, H. W. 1978. Interfacing research and extension in information delivery systems: Discussion. Am. J. Agric. Econ. 60:921-928.

CHAPTER 45

THE ROLE OF PLANT QUARANTINE IN PLANT HEALTH

MARK A. SMITH and GEORGE H. BERG

The term "quarantine" originally meant 40 and was applied to the period of detention (40 days) for passengers and crew of ships arriving from countries subject to epidemic diseases such as bubonic plague, cholera, and yellow fever. The original numerical significance soon disappeared and the word came to connote only the detention feature and associated practices. In all of human, animal, and plant quarantines, the fundamental concept that a public authority is empowered to establish protective barriers against the dissemination of injurious pests and pathogens has persisted over the years. However, greater knowledge of pests and pathogens in all of these fields as well as increased public awareness of their potential threat has resulted in expanded quarantine activities (9). *The term "plant quarantine" has come to include an increasing range of procedures and practices needed in this protective effort and will be used in this context here.*

The goal of plant quarantine activities is to protect agriculture and the environment from avoidable damage by hazardous organisms introduced by human activities and thus to contribute to improved plant health. The goal is achieved by government-established legal restrictions on the movement of plants, plant products, soil, cultures of living organisms, packing material, and commodities as well as their containers and means of conveyance to prevent or inhibit the establishment of plant pests and pathogens in areas where they are not known to occur (6,14).

Certain basic criteria should be used to establish a plant quarantine: 1) the pest or pathogen must present a demonstrated and/or expected threat to one or more plant species, 2) the organism must have little or no chance to spread rapidly without the assistance of humans, 3) no less disruptive substitute action must be available, 4) the stated objectives of the quarantine must be reasonably possible to attain, and 5) the economic gains from implementation must substantially outweigh the costs of administration and disruption of normal trade activities (14).

The establishment of a plant quarantine can have substantial impact on human activities and the environment. Thus it should come as no great surprise that there is significant debate as to the value of plant quarantine among quarantine officials and plant scientists. Some officials and scientists think that plant pests and pathogens will eventually occupy the geographic range of the host species because quarantine action can not permanently stop but can only delay the spread of these organisms. Albeit possible, this hypothesis remains unproved. Many plant pests and pathogens are still not distributed throughout the geographic range of the host species even though people have been moving about the world for thousands of years (6). For example, it would only be through humans functioning as the primary mover that the blue mold of tobacco (*Peronospora tabacina*) could become established in East Africa, that the plum pox virus could gain access to North America, or that the downy mildew of corn (*Peronosclerospora philippinensis*) could gain access to the Western Hemisphere. Nevertheless, even if the hypothesis proved to be correct, agriculture can derive substantial benefits from plant quarantine if the spread of plant pests and pathogens is delayed by these efforts while additional control strategies are being developed and deployed.

The role of plant quarantine in plant health should be viewed as one component of a comprehensive pest and disease management strategy and not relied upon as an exclusive control measure over an extended period of time. Effective plant pest and disease management programs should integrate a number of strategies, of which plant quarantine is one. Also, plant quarantine systems to control pests and pathogens should be examined over time and altered to address changing situations or discarded altogether when they are no longer effective.

Given that a great many plant pests and pathogens still are not distributed throughout the geographic range of their hosts, an effective deterrent against pest and pathogen introduction and spread is needed now and will be needed even more as the world population grows and the demand for food and other plant products increases accordingly.

Thus, effective plant quarantine systems are needed to function as a part of an integrated pest and disease management effort to protect crops and improve plant health. The remainder of this discussion will focus on 1) national plant quarantine efforts, with the United States as an example, and 2) international plant quarantine efforts currently under way and planned for the future.

NATIONAL PERSPECTIVE

Responsibility of Federal and State Agencies

The U.S. government's primary responsibility in plant quarantine activities is to prevent the introduction and spread of exotic plant pests and pathogens, whereas individual states have similar responsibilities on an intrastate basis. In practice, federal and state officials work closely together on the establishment and execution of plant quarantine activities (12). One key forum used to enhance communication and cooperation is the National Plant Board and four Regional Plant Boards, whose annual meetings bring together leading federal and state quarantine officials to discuss current and future strategies and programs.

Federal plant quarantine responsibilities are planned and executed by the Plant Protection and Quarantine (PPQ) programs of the Animal and Plant Health Inspection Service, U.S. Department of Agriculture (USDA). The basic statutes under which the USDA acts to prevent the introduction or spread of plant pests and pathogens as well as to run control and eradication programs are:

the Plant Quarantine Act of 1912 as amended in the Organic Act of 1944, the Federal Plant Pest Act of 1957, and the Noxious Weed Act of 1974 (14).

Many states have broad plant quarantine authorities under their individual statutes. When the federal government does not take action against a particular pest or pathogen, many individual states have the authority to impose actions they deem appropriate under their own laws and within their own borders.

International Activities—Part of a National Program

Traditionally the United States has concentrated on three strategies to protect its crops from the introduction and spread of new plant pests and pathogens: exclusion, detection, and action programs. With greatly increased worldwide movement of people and commodities, emphasis also has been placed on the international aspects of plant quarantine activities. The purpose of this emphasis is to reduce at the country of origin the risk of pest and pathogen introduction into the United States and to gain additional information concerning the occurrence and distribution of pests and pathogens in foreign countries. The international activities strengthen the U.S. plant quarantine effort as well as the commitment to the International Plant Protection Convention. *Thus the United States now employs four basic strategies or lines of defense in its overall plant quarantine effort: international activities, exclusion, detection, and domestic action programs* (13).

The USDA's PPQ is reorganized to emphasize international activities as the nation's first line of defense against the introduction of new plant pests and pathogens. The realignment includes the establishment of an assistant deputy administrator position for international and emergency programs. This administrator is responsible for existing and proposed foreign programs and three international regions: Region I—Latin America; Region II—Europe including all of the USSR and Africa; Region III—Asia, Australia, and Oceania. Long-term plans call for the establishment of two additional international regions and a redistribution of the areas for which each region is responsible, thus giving more comprehensive worldwide coverage (13).

Each regional office serves as a liaison with countries in the region and with regional plant protection organizations. The purpose of the liaison is to exchange information of plant quarantine significance, thus reducing the pest and pathogen risk associated with the movement of people and goods among cooperating countries and the United States. In some countries the liaison also may include PPQ assistance in pest and pathogen survey and control programs.

Exclusion

Agricultural Quarantine Inspection

PPQ's agricultural quarantine inspection program, which is designed to exclude plant and animal pests and pathogens from entry into the United States, is the second line of defense. At international air, sea, and border ports of entry, PPQ inspectors working with the U.S. Customs Service and other federal agencies examine incoming baggage, cargo, mail, and carriers (aircraft, ships and vehicles) for prohibited fruits, vegetables, plants, meat, and animal products. PPQ officers also work in some foreign countries to secure the preclearance inspection of agricultural commodities for plant pests and pathogens before shipment to the United States (13).

To illustrate the volume of the inspection activities, one can cite the 200,000 commercial aircraft carrying 16 million passengers who entered U.S. international airports in 1980. Several hundred thousand lots of prohibited agriculture items were confiscated from travelers as a result of baggage inspections. Projections indicate that the number of passengers will continue to increase at an annual rate of 5–10%. Consequently, promising inspection techniques such as the use of detector dogs, chemical and mechanical sniffers, and X-ray devices are being evaluated to improve detection capabilities and to reduce passenger delays.

PPQ has increased its emphasis on detection of plant pests and pathogens in maritime cargoes and at Mexican border sites. A computer-based system that facilitates the timely exchange of information concerning the occurrence of plant pests and pathogens in maritime shipments has been field tested at several ports of entry and may be implemented nationwide. Inspectors at Mexican border clearance facilities provide agricultural clearance for baggage, hand-carried groceries, automobiles, and trucks and railroad cars of Mexican produce destined for U.S. markets.

PPQ preclearance operations aim to reduce plant pest and pathogen risk to U.S. agriculture at the country of origin. Preclearance of agricultural commodities under cooperative agreements with foreign exporters or U.S. importers requires either fumigation, cold treatment, inspections, sanitation in production, or a combination of these before shipment to the United States. The cost of this service is borne by the exporter or importer. PPQ also cooperates with the U.S. Department of Defense and the Customs Service in the predeparture inspection of military personnel and cargoes being shipped from overseas military installations to the United States.

PPQ plant inspection stations are located at major U.S. international airports and maritime ports. Personnel at these stations screen plant material that enters the United States from foreign countries for presence of plant pests and pathogens. The largest station, located in Miami, processed 135 million plant units during 1980. Plant inspection station facilities and equipment were upgraded in 1980 in anticipation of even greater volumes of plant materials in the future.

Plant Importation and Exportation Activities

The USDA, in cooperation with other federal and state agencies, provides the phytosanitary services and safeguards necessary to minimize the risk of pest and pathogen movement associated with the import and export of germ plasm to and from the United States. The Plant Germplasm Quarantine Center at Beltsville, MD, and the Plant Quarantine Facility at Glenn Dale, MD, are maintained and staffed by the USDA to provide quarantine services, including virus indexing, required for the import and/or export of a wide variety of fruit, vegetable, and ornamental planting material. In addition, quarantine services are provided for wheat, corn, rice, sorghum, sugarcane, citrus, and grape plants through cooperative programs with the Agriculture Research Service, USDA, as well as through some agriculture experiment stations at land grant universities (13).

The USDA under the authority of the Plant Quarantine Act and the Federal Plant Pest Act, issues permits to individuals, companies, and organizations authorizing the importation of plants, seeds, plant products, and soils as well as cultures of organisms that meet the requirements of quarantine regulations. Scientists may be issued permits with specified safeguard requirements for importation of articles with high pest and pathogen risk.

The United States continues to be the world leader in the sale of agriculture commodities. Export sales of U.S. agricultural goods rose from about $15 billion in 1973 to about $40 billion in 1980 (5). The rapidly expanding export pace has created a corresponding increase in requests for phytosanitary inspection and certification. Nearly 120,000 federal phytosanitary certificates were issued during 1980. Even greater demands for phytosanitary certification are foreseen as agricultural exports continue to increase in the future. To meet these demands as well as the requirements of the importing countries, the USDA, in cooperation with various states and agribusinesses, is developing uniform phytosanitary inspection guidelines for many of our major export crops such as corn, soybeans, and wheat.

Detection—Pest and Disease Survey and Emergency Programs

Detection is the nation's third line of defense against the establishment and spread of new plant pests and pathogens. Under this strategy, cooperative federal/state surveys are implemented to detect foreign plant pests and pathogens early enough to institute eradication or containment programs. Cooperative survey programs of this type also monitor important endemic plant pests and pathogens and supply information useful in integrated pest management.

Beginning in 1917, the USDA, working with the states, developed a cooperative plant disease survey of major agricultural crops. The results were published in the *Plant Disease Reporter*. In the early 1920s, the USDA also began a cooperative federal/state insect survey of agricultural crops. In 1951, the insect survey was reorganized and expanded into a nationwide effort. Results were published regularly in the *Cooperative Economic Insect Report*. In 1976, the publication was expanded to include reports of diseases, nematodes, and weeds under the new title *Cooperative Plant Pest Report*. Both publications were given widespread distribution. These publications ended in 1980, although the American Phytopathological Society assumed responsibility for the continued publication of *Plant Disease Reporter* under the new title PLANT DISEASE. These surveys and publications provided useful information and, perhaps more importantly, valuable insight into characteristics needed to improve future surveys and reporting procedures.

With the experience gained from previous surveys, the USDA in cooperation with state departments of agriculture, land-grant universities, and relevant scientific societies, including the American Phytopathological Society, began a series of pilot projects and studies to determine appropriate procedures for a comprehensive national survey program designed to detect new pests and pathogens and to monitor important endemic pests and pathogens. A plant disease pilot project was initiated in 10 Upper Midwest states in 1977. Plant pathologists located in these states survey fields to detect possible new diseases of corn, soybeans, and small grains and to monitor the severity and distribution of important endemic diseases of these crops. Also in 1977 a pilot project was initiated to survey important food crops, forages, citrus trees, and home gardens for new plant pests, diseases, and noxious weeds around major U.S. ports of entry. In 1978 a pilot project was started in the northeastern United States designed to train survey specialists with broad skills in plant pathology and entomology who would coordinate survey activities for a wide range of plant pests and diseases. In 1980 the Intersociety Consortium for Plant Protection completed a study for the USDA of national and international plant pest information needs. In 1981 an ad hoc technical review group composed of representatives from state departments of agriculture and land-grant universities was established by the USDA to advise in the planning of a nationwide pest and disease survey program.

These pilot projects and planning activities, along with the experience gained from previous national survey efforts, helped identify characteristics needed in future national survey programs. Accordingly, any future national survey should be cooperative, with federal, state, and private personnel fully involved in the planning and the implementation. Also, it will need to be supported by automated data processing capabilities to handle rapidly large volumes of data and by standard survey procedures to make information comparable among states and regions. Personnel and equipment will be needed to accurately and quickly identify exotic as well as endemic insects, nematodes, pathogens, and weeds.

Although subject to change, plans in 1980 called for a cooperative federal/state plant pest and disease survey program to begin in a few states in 1982 and gradually be expanded into a national system over the next several years.

Once survey personnel detect a new plant pest or pathogen, organization capabilities should be available to provide rapid follow up action. This action should include assessments of the destructive potential of the new find and the possibility of implementing effective control measures and, when appropriate, deployment of emergency control measures, i.e., eradication or containment. For example, in recent years the USDA and affected states have had to deploy personnel and equipment on an emergency basis to investigate the first U.S. finds of a new race of the Scleroderris canker fungus in New York, Vermont, and New Hampshire; soybean rust in Puerto Rico; sugarcane smut in Florida; sugarcane rust in Louisiana; and the pinewood nematode in Missouri. The reestablishment of previously eradicated plant pests such as the Mediterranean fruit fly in California also required rapid cooperative federal or state action. The action may vary from 1) a massive federal/state cooperative effort, such as the program mounted in 1981 to eradicate the Mediterranean fruit fly in California, to 2) regulating the interstate movement of plants or plant parts from infested states to disease-free states, such as for sugarcane smut in Florida, to 3) minimal action of informing states of the presence of a new plant pest or pathogen, such as the pinewood nematode, when survey results indicated that it had already become so widespread that federal control efforts would not be effective.

Action Programs—Control of Established Plant Pests and Pathogens

As has been pointed out, plant quarantine efforts involving exclusion and detection are not always successful, and new plant pests and pathogens do become established in the United States. When this happens the federal government and affected state governments turn to the fourth line of defense to determine whether action programs are possible and warranted to eradicate, contain, or slow the spread of new plant pests and pathogens. Action programs also may involve cooperative federal/state efforts to control periodic outbreaks of important endemic pests and pathogens, particularly when large areas of federal land are involved. An example is the cooperative program involving federal/state and farmer participation in the control of grasshopper outbreaks on rangeland in western states.

Since the early part of the 1910 decade, cooperative federal/state action programs have been implemented to control white pine blister rust, flag smut of wheat, phony peach disease, Dutch elm disease, potato golden nematode, soybean cyst nematode, witchweed, gypsy moth, and Khapra beetle, to name a few. Two of the most massive control efforts involved plant diseases: citrus canker and wheat stem rust. Citrus canker was eradicated from seven Gulf states in a cooperative federal/state program that lasted from 1913 until the late 1940s and required the destruction of millions of citrus trees. The barberry eradication campaign, begun in 1918 to reduce the losses caused by stem rust to wheat and other small grains, was conducted in 19 major wheat-producing states. Although barberry has not been completely eliminated from the major U.S. grain-growing region, it has been reduced to low populations. This, combined with the widespread use of resistant cultivars, has contributed to the greatly decreased severity of stem rust. In fact, the United States has not experienced a major widespread stem rust epidemic for nearly 30 years.

Exclusion and detection programs are required continuously as part of comprehensive plant protection efforts. Conversely, individual action programs, particularly those directed at one organism, are of a finite nature, having logical points to begin and end based on program goals. The citrus canker action program was concluded when extensive surveys indicated that the disease was no longer present in the country and thus the program goal to eradicate the disease had been met. Occasionally it may be necessary to modify the goals of action programs. For example, the original goal of the federal/state golden nematode action program was eradication. However, as more knowledge was gained concerning the nematode and its control, eradication did not seem to be feasible. In 1981 the goal was modified so that the program now is designed to contain the nematode while reducing its population on crop land to nondetectable numbers, primarily through the use of resistant cultivars.

In the future, it will be necessary to start new action programs as new pests and pathogens occasionally slip through the first three lines of defense. Extensive planning has been under way recently to determine the organisms that are most likely to be introduced into the United States, the damage they might cause if they become established, and the best control approach. In addition, because of environmental concerns, pest and pathogen resistance to pesticides, and rapidly accelerating costs of pesticides and of energy to apply pesticides, future action programs will focus on biological control and other nonchemical control alternatives.

INTERNATIONAL PERSPECTIVE

Regional and International Recognition of Plant Quarantine Importance

Since 1951, the year that the International Plant Protection Convention was approved by the Sixth Conference of the Food and Agriculture Organization of the United Nations (FAO) (7,10), the importance of plant quarantines to agricultural production of developing countries as well as developed countries has been increasingly recognized nationally and internationally. Many countries also recognized that if increased agriculture production was to be achieved, compliance with provisions as set forth under the articles of the International Plant Protection Convention and the establishment or strengthening of existing regional plant protection organizations was essential.

Consequently, beginning in 1951, the following plant protection organizations came into being:

1. European and Mediterranean Plant Protection Organization (EPPO)—1951 (8)
2. Organismo Internacional Regional de Sanidad Agropecuaria (OIRSA)—1953 (3). (Central America, Mexico, and Panama)
3. Plant Protection Committee for the South East Asia and Pacific Region (PPC/SEAP)—1956 (11)
4. Inter-African Phytosanitary Council (IAPSC)—1956 (1)
5. Near East Plant Protection Commission (NEPPC)—1963
6. Comite Interamericano de Proteccion Agricola (CIPA)–1965 (Argentina, Bolivia, Brazil, Chile, Paraguay, and Uruguay)
7. Junta del Acuerdo de Cartagena (JUNAC)—1969 (Bolivia, Colombia, Ecuador, Peru, and Venezuela)
8. Caribbean Plant Protection Commission (CPPC)—1967 (3)
9. North American Plant Protection Organization (NAPPO)—1976

These organizations represent a worldwide network of regional plant protection bodies, with almost all countries being members of the organization established for their specific region.

Changing Attitudes Toward Plant Quarantine Enforcement

Before 1951, depending on the country and the degree of development of its plant quarantine service, attitudes toward the importation of plant products and germ plasm reflected a closed-door policy, an open-door policy, or a compromise between the two, frequently based more on needs for agricultural products and other commodities than on phytosanitary risks (5). During the past 30 years, however, technical assistance provided by FAO and financed by a number of sources including FAO, the United Nations Development Programme, other international or regional organizations, bilateral aid agencies, and individual donor countries has resulted in improved plant quarantine services among many developing countries. Developed countries also have improved their plant quarantine services. There also has been a greater tendency to adhere more closely to basic plant quarantine principles when announcing plant import regulations, although a better understanding of these principles frequently is still required. Most important, it is gradually being recognized by the plant quarantine leadership in many countries that an effective plant quarantine system should serve to filter out dangerous, exotic plant pests and pathogens and not to block movement of plant products and germ plasm (2).

Looking into the future, we should be aware that agricultural production must continue to expand, especially in developing countries where continued crop diversification is needed to reduce the trend toward monoculture cropping systems. Many of the world's most destructive pests and pathogens have not achieved worldwide distribution. Thus, it seems apparent that an efficient, truly international plant quarantine system is urgently needed to assist in the worldwide movement of agricultural commodities and germ plasm while at the same time reducing the pest and pathogen risk associated with their movement.

The Future Direction and Responsibilities of Plant Quarantine in International Plant Protection

There should be general agreement concerning the need for an effective plant quarantine system to protect a country's agricultural resources from the introduction and establishment of highly destructive, exotic plant pests and pathogens. Two major objectives of plant quarantine systems involve 1) the protection of agriculture from the introduction and establishment of pests and pathogens of quarantine importance and 2) the protection of germ plasm collections themselves from the destructive attacks of domestic and foreign pests and pathogens (2,6). To meet these objectives, plant quarantine services should be performed under basic legislation and plant import regulations conforming to internationally and biologically accepted principles. Quarantine services should have trained inspectors to recognize and assess pest and pathogen introductions and the necessary physical facilities, including postentry plant quarantine facilities, to safely and efficiently handle these organisms. Information and public relations programs also should be available to develop public awareness and cooperation. Unfortunately, many countries do not possess all of these capabilities.

The assistance provided by international organizations, regional plant protection bodies, and bilateral aid agencies has contributed much toward achieving basic plant quarantine objectives on an international basis. Nevertheless, much more must and can be done. Some countries have progressed rapidly while others have remained almost inactive in the development of an effective plant quarantine service. This can be corrected by utilizing the existing plant quarantine foundation of regional plant protection organizations established during the past 30 years.

Although regional plant protection organizations have made substantial contributions to improved plant health in the past 30 years, they should take on additional responsibilities in future years to protect agriculture from the spread of pests and diseases (4). These responsibilities could include: 1) maintaining open channels of communication, 2) establishing an international data base to provide all plant protection organizations access to timely information concerning occurrence and distribution of pests and pathogens in individual countries or regions, 3) providing member importing countries information concerning an exporting country's pest and pathogen situation, and 4) being in contact with international agricultural research centers located around the world concerning germ plasm being shipped to cooperating countries and any possible pest or pathogen risk associated with these shipments.

Establishment of a Network of Postentry and Intermediate Plant Quarantine Stations

In view of the worldwide need for new or improved cultivars, every effort should be made to eliminate existing barriers to the movement of genetic stock.

Certainly, the uncontrolled exchange of germ plasm should not occur, but at present there are so few postentry and intermediate plant quarantine stations worldwide that there exists no possibility of properly screening the large volume of plant propagating material moving in international research and commerce (2). Obviously the establishment of additional postentry and intermediate plant quarantine stations is needed soon to reduce this problem. Some stations could be located in developed countries to provide services for countries lacking economic and technical means to establish their own facilities (2). Since the establishment of these stations requires a substantial amount of time, countries with limited economic resources but qualified professional personnel should establish as a first step modest postentry quarantine facilities. Recently Venezuela, Colombia, and Mexico, and the Dominican Republic have established postentry quarantine facilities.

Financing the construction of postentry stations, regardless of whether the facilities may be reduced in size (as in Colombia) or extensive and capable of providing services on a regional scale (as in Kenya and Nigeria), is an obstacle frequently difficult to surmount. In the past a limited amount of economic and technical assistance has been provided by FAO, the U.N Development Programme, and the USDA. However, this assistance has not been sufficient to solve the problem. In the future it is hoped that bilateral aid agencies, donor countries to international organizations such as FAO, international agricultural research centers, regional plant protection organizations, and others will recognize the urgency of this problem and, to the extent possible, establish or support the establishment of these facilities.

LITERATURE CITED

1. Addoh, P. G. 1977. The International Plant Protection Convention: Africa. FAO Plant Prot. Bull. 25:164-166.
2. Berg, G. H. 1977. Post-entry and intermediate quarantine stations. Pages 315–326 in: Plant Health and Quarantine in International Transfer of Genetic Resources. W. B. Hewitt and L. Chiarappa, eds. CRC Press, Inc., Boca Raton, FL. 346 pp.
3. Berg, G. H. 1977. The International Plant Protection Convention: Central America and the Caribbean. FAO Plant Prot. Bull. 25:160-163.
4. Food and Agriculture Organization. 1981. Strengthening plant quarantines and related programmes—Central America, Mexico and the Caribbean area. AG:DP/RLA/74/050, terminal rep. United Nations, Rome, Italy. 73 pp.
5. Josling, T. 1981. World Food Production, Consumption and International Trade: Implications for Agriculture. In: Food and Agricultural Policy for the 1980s. D. G. Johnson, ed. Am. Enterprise Inst. for Public Policy Res., Washington. 229 pp.
6. Kahn, R. P. 1977. Plant quarantine: Principles, methodology and suggested approaches. Pages 289–307 in: Plant Health and Quarantine in International Transfer of Genetic Resources. W. B. Hewitt and L. Chiarappa, eds. CRC Press, Inc., Boca Raton, FL. 346 pp.
7. Karpati, J. F., 1981. History and current status of the International Plant Protection Convention (IPPC). Working paper. Sixth Session Caribbean Plant Protection Commission, Paramaribo, Suriname. 18 pp.
8. Mathys, G. 1977. European and Mediterranean Plant Protection Organization. FAO Plant Prot. Bul. 25:152-156.
9. McCubbin, W. A. 1954. The Plant Quarantine Problem. Ejnar Munksgaard, Copenhagen. 255 pp.
10. Mulders, J. M. 1977 The International Plant Protection Convention: 25 Years Old. FAO Plant Prot. Bull. 25:149-151.
11. Reddy, D. B. 1977. The International Plant Protection Convention: Plant Protection Committee for the South East Asia and Pacific Region. FAO Plant Prot. Bull. 25:157-159.
12. Rohwer, G. G. 1979. Plant quarantine philosophy of the United States. Pages 23–34 in: Plant Health—The Scientific Basis for Administrative Control of Plant Diseases and Pests. D. L. Ebbels and J. E. King, eds. Blackwell Scientific Publ.,

Oxford. 322 pp.
13. Shirakawa, H. S. 1981. Working paper. Sixth Session Caribbean Plant Protection Commission, Paramaribo, Suriname. 13 pp.
14. Spears, J. F. 1974. A Review of Federal Domestic Plant Quarantines. Animal and Plant Health Inspection Serv., U.S. Dep. Agric., Hyattsville, MD. 95 pp.

CHAPTER 46

GOVERNMENT PROGRAMS

J. F. FULKERSON

It is widely recognized that government programs per se have played a significant role in plant health maintenance efforts in this nation. This being so, it is probably a useful exercise to examine briefly the nature of this role thus far, in the fondest of hopes that it might thereby allow us to contribute more effectively to the outcome of things in the future.

Some disclaimers and definitions relating to the scope and scale of the subject matter to be treated in this brief chapter seem to be essential. The literature in this subject matter area in its broadest terms is staggering. Thus, for the purposes at hand, government programs and government actions are considered roughly equivalent, and our attention will be limited to only certain actions, notably those thought to be of major consequence. Similarly, plant health maintenance is assumed herein to be roughly equivalent to plant protection, with especial attention to plant pathology. Thus, a shorthand title of government actions and their impact on the practices and profession of plant pathology is appropriate for this purpose.

The major activities of plant pathologists, i.e., teaching, extension, and research, have all been influenced by an array of government actions. However, the interdependence of these activities has generally meant that the stage was set by the influences on the research component. Certainly there are exceptions, but only in recent years, with a heavy intrusion of social and nonagricultural interests and expertise in extension, have we seen government influence initiated irrespective of actions in the research community. Thus, in the main, the influence of government actions on plant pathology as a profession has been an integral function of its research component.

Plant pathology as a profession has been developed largely as a public enterprise in this country, and almost exclusively so in the rest of the world. The profession and its practice is to an overwhelming degree the result of governmental action. These actions have involved governments at all levels, i.e., local, state, and federal, with local actions being directed primarily to the extension activity of the profession.

The primary government actions responsible for this unique governmental enterprise all occurred in the last century. Who would have thought that an action of a federal government (Act of 15 May 1862) that contained such prosaic mandates as "...to procure, propagate, and distribute among the people new and valuable seeds and plants. . ." would ultimately provide a mandate for plant pathology as a science, and, for the most part, for the agricultural sciences broadly?

It is remarkable that in a year filled with great sorrow and strife for the nation that another action (Act of 2 July 1862) of importance to plant pathology, and in this instance science more broadly, was consumated. Here again the very simple and direct language of providing for ". . .the endowment, support, and maintenance of at least one college. . ." in each state with especial concerns for agriculture, set the stage for some unique developments. These developments proved to be unparalleled in other nations and even in other areas of science in this nation. In one stroke, this action joined the federal and state governments in an effort that set the stage for the development of a form of institutionalized science that has been recognized as unparalleled in its effectiveness. Not only were such essential elements as independence, pluralism, and diversity accommodated, but also support and maintenance of purpose over time were mandated.

The third step in this remarkable set of actions occurred in 1887 with the passage of the Hatch Act, which provided for the improvement of the endowment over time of this newly formed public enterprise in science. It provided an organizational and administrative structure on a national basis for the agricultural sciences, i.e., the agricultural experiment stations. It should not go unnoticed that this landmark legislation contained some very specific instructions important to plant pathology as a profession, i.e., that there would be an object and duty in the public interest to "...conduct original researches..." of plants and ". . .the diseases to which they are severally subject, with the remedies for the same." Government action in this instance not only provided a charter for the development of plant pathology efforts as a public interest, but it also reinforced the institutionalization of this science, provided a fabric within which it could develop, and gave it endowments over time.

It is noteworthy that the government actions cited this far provided a basis for the development of a part of the scientific community of this nation that was not only unique in this country but equally unique worldwide. The actions were woven jointly through federal and state governments, inclusive of the entire nation, and were structured in a fashion to provide for change, yet carefully institutionalized. The whole process was clearly a reflection of the Constitution itself in that only those things that could not or would not be done by others were being undertaken, and that authorities and administrations were broadly based, irrespective of size or economic importance of the various parties to the enterprise.

Thus the plant pathology efforts developed as part of an extensive network of scientific competence, addressing a vast array of scientific questions and agricultural problems for the nation and indeed for the world.

It again is instructive to note that the early government actions were inclusive of teaching, extension, and research. With time, legislation dealing directly with teaching and extension were devised. Additional actions over time improved the endowment, emphasized basic research, and further developed the system. There were also government actions at local, state, and federal levels directed to a wide range of disease problems. Among these were rust of wheat, Dutch elm disease, blueberry virus, soybean cyst nematode, acid rain, corn blight, and fireblight.

Government actions directed at purposes and goals other than the agricultural sciences and their practices have left very large impacts at times and indeed continue to be very important to plant pathology interests. Production and marketing credit systems that suddenly alter practices or dictate practices that generate disease problems are well recognized. Actions involving land use per se and such things as grain handling and marketing have left substantial marks on disease problems. Even some very minor government actions directed at rather specific crop plants, such as corn and soybeans, have resulted in the spread of major disease problems.

Major government actions in more recent times would seem to have had a massively undesirable impact, while on the other hand, they have left us with great challenges and more than a small glimmer of hope. It's curious, indeed, that during the times of great economic hardship of the late 1800s and the 1930s, agricultural research efforts were strengthened, whereas in what were apparently "good times" economically, they were vastly eroded. Thus, the 1960s and 1970s saw a substantial weakening in their relative status, especially as agricultural sciences were caught up in the general antiscience mood that gripped the nation. The fact that agricultural sciences' great benefits were so badly managed by society seemed to go unnoticed for a time.

Coupled with the other problems mentioned above was a government action in the fall of 1965 that had very special potentials for damage to the scientific community, and especially to agricultural sciences as public trusts, and these were largely realized. This action, a federal executive order designed to provide a systems approach to planning and budgeting throughout government (followed in most respects by similar actions in state governments), became translated into a mechanism for centralization of control. Its basic effect was to make budget determine science, and decision processes were moved far away from the sites of action and from the scientific community. This was especially so in the agricultural sciences. If it had been possible to take such steps against professions such as law, politics, or religion, there would have been a great public outcry. But then, agricultural sciences have never been too great at public outcrying.

But Mr. A. Lincoln, who as President saw the wisdom of this great public trust in agricultural research, also admonished us to realize that the public will not be fooled all of the time, and the corrective processes show some hopeful signs.

REFERENCE

Knoblauch, H. C., Law, E. M., and Meyer, W. P. 1962. State agricultural experiment stations: A history of research policy and procedure. U.S. Dep. of Agric. Misc. Publ. 904. U.S. Govt. Printing Office, Washington, DC. 262 pp.

CHAPTER 47

INTERNATIONAL COOPERATION IN AGRICULTURAL RESEARCH: THE BASIS FOR FEEDING MORE PEOPLE AND FEEDING PEOPLE MORE

GEORGE A. ZENTMYER, H. DAVID THURSTON, and JOHN S. NIEDERHAUSER

International cooperation in plant pathology is as old as the science itself. Rare is the research project, field demonstration, or teaching class that does not share some benefit from collaboration with scientists in other countries. Few phytopathological problems are unique to a single country, and even fewer solutions are reached on a strictly national basis. Pathogens ignore political boundaries.

International cooperation provides a striking demonstration of the scientist's responsibility toward the betterment of humankind. As Philip Handler, past president of the National Academy of Sciences, has eloquently stated (3):

> . . . science, which has revealed the most awesome and profound beauty we have yet beheld, is also the principal tool that our civilization has developed to mitigate the condition of man.

And to bring this challenge home to our own profession, William Snyder in 1971 (14) stated that "the obligation the plant pathologist has accepted is to society—world society—not to a crop or a culture."

Though we are not demographers, as world citizens we are all aware that any chance of success in the conquest of hunger and malnutrition in the world depends upon the stabilization of population growth, especially in tropical countries. Norman Borlaug emphasized this relationship in his Nobel Peace Prize address (5) in 1970:

> If fully implemented, the Green Revolution can provide sufficient food for sustenance during the next three decades; but the frightening power of human reproduction must also be curbed, otherwise the success of the Green Revolution will be ephemeral only . . . Malthus signaled the danger a century and a half ago . . . .

Predictions on world population during the next century have peaks ranging from 10.5 to 14.5 billion, with most of the increase occurring in the less developed regions of southern Asia, Africa, and Latin America.

During the past few decades, advances in world agricultural production have been dramatic. However, most of these successes have been in the more developed countries of temperate regions. In the modern agriculture of these regions, improved seeds are planted and cultivated by machines, irrigated and fertilized at optimum levels, protected from pests and diseases, and harvested mechanically for distribution in sophisticated marketing systems. These highly efficient and impressive agricultural production programs are the result of the combined research efforts of scientists in the biological, physical, and social sciences. International cooperation has provided a major contribution to the necessary coordination and stimulation for these remarkable advances.

It must suffice here merely to note these great advances in agricultural productivity in the countries of the developed world. Our concern in this paper is concentrated on those areas of the developing world where agricultural production and productivity remain low. Despite a few highly successful food production breakthroughs in these less developed areas, most of the agriculture of these countries remains comparatively unaffected by modern technology.

Within the past 30 years, many institutions and innovative strategies have been created to accelerate the development of agricultural production and productivity in the developing world; these efforts are of mutual benefit to all participants. Plant pathologists have participated in this worldwide campaign and have made some brilliant contributions. It is our purpose here to examine this record of international collaboration; to analyze some of the results; and to propose ways that plant pathologists can improve and expand their contributions to world food production.

EXISTING CHANNELS FOR INTERNATIONAL COOPERATION

There are several kinds of institutions whose activities include international cooperation in scientific and agricultural development: 1) international agricultural research centers, 2) governmental organizations, 3) inter-governmental organizations, 4) universities, 5) private institutions, 6) scientific societies, and 7) foundations. It is impossible here to identify all of these organizations. To give an idea of the scope of this subject, in the *Acronym List of International Organizations Related to Agriculture, Economic Development and Pest Management* (4), prepared by the Consortium for International Crop Protection, there are 543 organizations listed under four categories: national, state, and local (124); regional and international (125); bilateral and multilateral technical assistance and economic assistance (122); and education research and reference (172). The publication by Kriesberg (10), the University of Missouri Bulletin M 107 (17), and the University of California publication on a hungry world (1) are valuable references to this proliferating area.

International Agricultural Research Centers

The cooperative agricultural program in Mexico, which began in 1943 as a joint effort of the Rockefeller Foundation and the Mexican government, provided the role model and the impetus for the subsequent establishment of 13 international centers. The original program in Mexico, led by plant pathologist J. George Harrar, involved research on four basic food crops: wheat, maize, potatoes, and beans. This was a highly successful program, not only in improving the quantity and quality of Mexico's food production but in training several hundred Mexican agricultural scientists. This success led to the establishment of similar Rockefeller Foundation programs in Colombia, Chile, and India. Use of germ plasm developed in Mexico triggered the remarkable increase in wheat production on the Indo-Gangetic plain in the 1960s that was labeled the Green Revolution. This splendid achievement actually preceded the formation of most of the international centers.

The first international center was the International Rice Research Institute (IRRI), established in 1960 by the Ford and Rockefeller Foundations in collaboration with the government of the Philippines. Then, in 1966, the Centro Internacional de Mejoramiento de Maiz y Trigo (CIMMYT), also funded by the Ford and Rockefeller Foundations, was established in Mexico as an outgrowth of 20 years of cooperation in wheat and maize research between the Rockefeller Foundation and the Mexican Ministry of Agriculture. Subsequently, the Ford and Rockefeller Foundations developed the International Institute of Tropical Agriculture (IITA) in Nigeria in 1968 and the International Center of Tropical Agriculture (CIAT) in Colombia in 1969.

In 1971, the Consultative Group on International Agricultural Research (CGIAR), was established to coordinate funding for existing international centers and to establish additional centers. This consortium of donors is sponsored by the Food and Agriculture Organization of the United Nations (FAO), the World Bank, and the United Nations Development Program. Donor members include 35 countries, international and regional agencies, and private foundations.

During the past 10 years, nine additional international centers have been established by CGIAR (Fig. 1) (16), dedicated to improving the quantity and quality of food production and the standard of living of poor people in the developing countries, and (2)

> to arm the developing countries with superior varieties of essential crops and improved farming systems for the production of food plants and animals.

Other international centers, such as the Asian Vegetable Research and Development Center and the International Soybean Program are not supported by the CGIAR system but produce results important to developing countries.

Governmental Organizations

Many of the developed countries have organizations whose basic aim is to provide assistance in worldwide agricultural development, with particular reference to the developing countries.

The United States is involved in agricultural assistance projects in many countries, through the Agency for International Development (AID) in the State Department and the Office of International Cooperation and Development (OICD) in the U.S. Department of Agriculture (USDA). The major channel for U.S. assistance to developing countries is through AID. For the fiscal year 1982, a total of $8.1 billion was requested for all foreign and economic assistance; of this amount, nearly 25% ($1.9 billion) was requested for AID development assistance (12). AID also provides more than 25% of the funding for the CGIAR network and has made extensive use of contracts with U.S. universities and colleges and the USDA to carry out its mandate.

The OICD was created by the Secretary of Agriculture in 1978 by combining several units with international programs from separate USDA agencies. Its primary role is "to plan, coordinate, and evaluate USDA's policies and programs aimed at alleviating hunger and malnutrition throughout the world." To this end OICD works with other government agencies, U.S. universities, and international organizations, including the United Nations.

Canada's International Development Research Centre (IDRC) is an independent public corporation established in 1971 to aid scientific research in developing countries. The Canadian International Development Agency (CIDA) is similar to the U.S. AID. Under the German Agency for Technical Cooperation (Deutsche Gesellschaft für Technische Zusammenarbeit or GTZ), the Federal Republic of Germany operates technical cooperation projects in agriculture with partners in developing countries in Africa, Asia, and Latin America.

With primary, but not exclusive reference to the Commonwealth countries, England has for more than 60 years provided extensive international agricultural services, through the Commonwealth Agricultural Bureaux (CAB) and the Overseas Development Administration (ODA). In the CAB, the Commonwealth

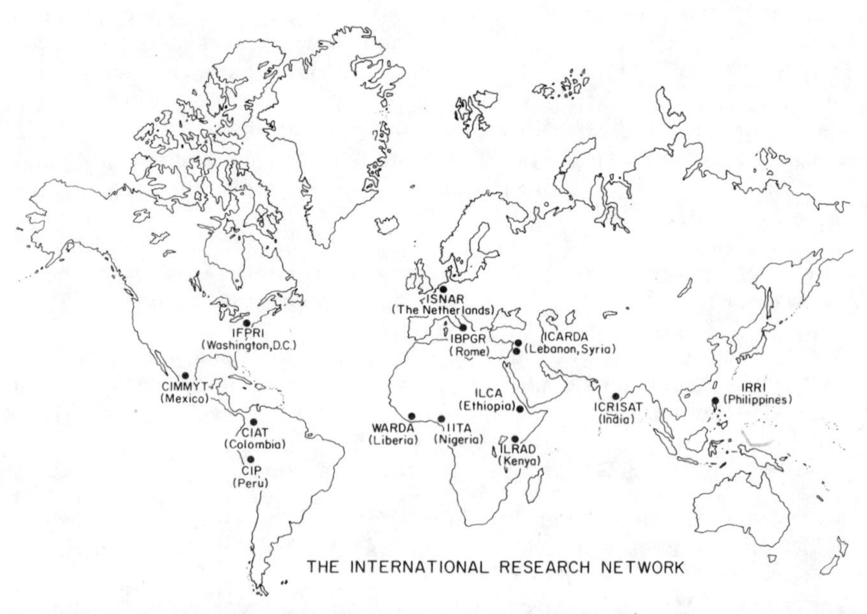

Fig. 1. Distribution of the international centers (adapted from reference 2). Identity of the centers is as follows: CIAT (Centro Internacional de Agricultura Tropical), CIP (Centro Internacional de la Papa), CIMMYT (Centro Internacional de Mejoramiento de Maiz y Trigo), IBPGR (International Board for Plant Genetics Resources), ICARDA (International Center for Agricultural Research in the Dry Areas), ICRISAT (International Crops Research Institute for the Semi-Arid Tropics), IFPRI (International Food Policy Research Institute), IITA (International Institute of Tropical Agriculture), ILRAD (International Laboratory for Research on Animal Diseases), ILCA (International Livestock Centre for Africa), IRRI (International Rice Research Institute), ISNAR (International Service for National Agricultural Research), WARDA (West Africa Rice Development Association).

Mycological Institute (CMI) is the primary agency relevant to plant pathology, with its information service (Review of Applied Plant Pathology), its identification service and investigations, including the culture collection, and its training programs.

France has had an active international agricultural program for many years, with eight institutes involving countries in Africa, the Indian Ocean, Latin America, Southwest Asia, and overseas French Territories. In 1979 the eight institutes were coordinated under GERDAT (Le Groupement d'Etudes et de Recherches pour le Developpement de l'Agronomie Tropicale). Also, ORSTOM (Office de le Recherche Scientifique et Technique Outre-Mer) covers overseas agricultural research.

Denmark has an excellent Institute of Seed Pathology for Developing Countries, with headquarters in Copenhagen. The Netherlands has a long history of involvement in the tropics. The Royal Tropical Institute, established in 1910, is justly famed for its contributions to the development and improvement of agricultural planning, production, and processing in tropical and subtropical countries, and it publishes *Tropical Abstracts.* Many other developed countries are involved in agricultural assistance programs, including Australia, Belgium, China, East Germany, Japan, Norway, Spain, Sweden, Switzerland, and the USSR.

Intergovernmental Organizations

The primary organization concerned with agriculture under the United Nations (UN) is the FAO, which has a membership of 144 countries. Some coordination and considerable funding for FAO's developmental activities are provided by the UN Development Program (UNDP) and the World Bank Group. The FAO was established in 1945, with the objective of raising the levels of nutrition and standards of living throughout the developing world. Chiarappa (6) has recently provided an excellent summary of the role of FAO in research and control of plant disease problems in many countries.

The Organization of American States has a longstanding concern and interest in agricultural development. The Inter-American Institute of Agricultural Cooperation (IICA) was established in Costa Rica in 1942; it now has members from 25 countries in the Western Hemisphere. The research and training center at Turrialba became the Centro Agronomico Tropical de Investigacion y Enseñanza (CATIE) in 1973, with cooperation between IICA and the government of Costa Rica, and support from many agencies. Emphasis is on improving the standard of living of small-scale farmers in the American tropics.

One of the earliest international organizations was the South Pacific Commission (SPC), founded in February 1947 by an agreement signed by the governments of Australia, France, the Netherlands, New Zealand, the United Kingdom, and the United States. Although The Netherlands withdrew in 1962, several independent Pacific states have since been admitted to membership. The SPC's purpose is to promote the economic and social welfare and advancement of the 20 island countries and territories of the peoples of the South Pacific region (7); a regional plant protection service is one of its activities.

The International Plant Protection Convention is an organization specializing in international cooperation in plant health; it was approved under FAO in 1951, and 79 countries are now contracting parties to the Convention. The primary aim is to strengthen international efforts to combat important pests affecting plants and plant products and to prevent their spread across international boundaries. Examples of the regional organizations include the European and Mediterranean Plant Protection Organization (EPPO) and the North American Plant Protection Organization (NAPPO).

University Programs

As early as 1876, Massachusetts State College (now the University of Massachusetts) helped to found and develop Sapporo Agricultural School in Japan. A prototype of U.S. cooperation with foreign universities was formed in 1924 between the N.Y. State College of Agriculture of Cornell University and the University of Nanking in China. It would be impossible to catalog in this chapter all of the overseas university projects even in the past three decades.

Formal U.S. technical assistance programs began in January 1949 when President Truman introduced his Point IV program. Since then technical assistance has been important in U.S. foreign policy, and many U.S. universities, with major funding by AID, have been deeply involved in cooperative projects with colleges and universities in other countries, especially in Latin America and Asia, with emphasis on institution building. A noteworthy example of the role of plant pathologists in an overseas program is the North Carolina State University program in Peru with the National Agricultural University and the Peruvian Ministry of Agriculture. French and Apple (8) provide a perceptive analysis of the typical problems encountered.

Since 1971, AID has had a general technical services contract with the University of California with the aim of improving pest management in developing countries. This comprehensive project includes pesticide management workshops, quality control programs, development of pest management and environmental protection systems, training programs, development of coordinated pest management research projects, and informational services. Other cooperating universities are: Cornell, Florida, Hawaii, Miami, Minnesota, North Carolina State, Oregon State, Puerto Rico, and Texas A&M. These universities have become incorporated as the Consortium for International Crop Protection (CICP) with headquarters in Berkeley, CA.

There have also been many international research projects initiated by individual plant pathologists in U.S. universities, often sponsored by these institutions directly. A few of the more prominent of those that have made a significant impact internationally are those dealing with: *Pseudomonas* (Kelman, Sequeira, et al, Universities of Wisconsin and North Carolina), *Fusarium* (Snyder, Hansen, et al, University of California, Berkeley), *Puccinia* (Stakman et al, University of Minnesota), *Phytophthora* (Fawcett, Klotz, Zentmyer, et al, University of California, Riverside).

Private Institutes and Private Industry[1]

An outstanding example of international cooperation sponsored by the private sector is the American Cocoa Research Institute (ACRI), supported by the Chocolate Manufacturers Association of America. Although none of the

[1]We are indebted to E. P. Imle for the material in this section.

U.S. chocolate manufacturers own cacao plantations, they have nevertheless, through ACRI, supported research, training, and dissemination of information and of germ plasm that has greatly aided the producers of cacao beans, most of whom are small-scale farmers. Similar groups in the United Kingdom and in Europe have established their own programs and have collaborated with ACRI.

Another example of private industry's contribution to crop production in the developing countries is in banana and plantain production, where the primary concern has been disease control. Formerly the banana fruit companies owned their own plantations and supported large-scale research efforts. Even though banana production is now almost entirely in the hands of independent producers in the tropics, production research by the larger fruit companies has continued and results are available to independent producers. Some of these fruit companies have made considerable contributions to development of new crops to replace or alternate with bananas. An example was the United Fruit Company's introduction, testing, and large-scale planting of the African Oil Palm in Central and South America; this has become one of the major cash crops of the American tropics.

Rubber manufacturers have also made substantial contributions to rubber production in the tropics, through research by the larger companies on their own plantations. Research results have been transferred to small growers.

International Scientific Societies

Another significant source of international cooperation is through international scientific societies, with their international congresses and symposia, working groups, and committees. These provide excellent avenues for exchange of scientific information, discussion, and many aspects of international cooperation. A prime example is the International Society for Plant Pathology (ISPP), which was organized and held its first International Congress in England in 1968. Forty phytopathological societies, with more than 12,000 members now belong to the ISPP.

CRITIQUE OF CONVENTIONAL PLANT DISEASE CONTROLS IN INTERNATIONAL COOPERATION

After many disillusioning experiences, it has been generally recognized that temperate zone agricultural technology or "modern" agriculture, including pest control methods, cannot be transferred directly to the "traditional" agricultural systems of the tropics in the effort to increase food production in those regions. We must learn not only what practices work in our "modern" system of agriculture, but also must attempt to analyze and understand the crop protection components of "traditional" agricultural systems, often with mixed and multiple cropping.

Disease Prevention and Quarantines

Never before have people moved crop plants from country to country and continent to continent at the rate that is taking place today. Improved communication and increased scientific interchange provide opportunities for scientists anywhere in the world to obtain cultivars of crop plants with ease.

Hundreds of tons of seed of improved crops move internationally, as for example, seed of the improved wheat and rice cultivars that resulted in the Green Revolution. The vast majority of this movement and exchange is highly beneficial, but potential disease hazards exist when new germ plasm is moved around the world.

The threat of introducing new pests is probably greatest in tropical areas because of the tremendous diversity of crop plants grown there and the general lack of knowledge of the potential of various pathogens in new areas and of their means of spread. Coffee rust is a dramatic example of a destructive disease that was restricted to Africa and Asia until its discovery in South America in 1970; it is now causing serious losses and economic disruption of the industry throughout the Americas.

Examples of pathogens that have been confined to certain continents or regions include the virulent South American leaf blight fungus on rubber, *Microcyclus ulei*, which has not spread from the Americas to the Asian rubber-producing areas; the cacao swollen shoot virus that occurs only in West Africa; and the witches' broom fungus on cacao that occurs only in some parts of South America. Some of this restricted movement is undoubtedly related to the great care taken in disseminating pathogen-free stock in the extensive worldwide interchange of rubber and cacao germ plasm.

Regulations, including quarantines, have been developed, and regional plant protection organizations are operating under the International Plant Protection Convention approved under FAO in 1951. These provide mechanisms for attempting to eradicate and to prevent the spread of pests within and between countries. A key question is: how useful and effective are quarantines and eradication programs?

By their nature, quarantines demand international cooperation. Their general appeal is based on the understandable desire of a country to exclude potentially dangerous pathogens and pests. Unfortunately, in their application quarantines often have not been successful in satisfying their declared objectives, and occasionally they have been created or enforced for reasons other than plant protection. Sometimes the strict application of quarantines prevents the entry of small samples of plant materials for research purposes, while tons of similar plant materials are imported through commercial channels with little or no attention or interference from quarantine officials.

How can the quarantine and plant protection situation be improved? There are several possible approaches:

1. A more realistic appraisal should be made of those pathogens that are clearly localized and might be needlessly exported by failure to use other control measures, certification, or clean stock programs, thus providing a realistic and biologically sound basis for the quarantines.

2. If it is determined that pathogens are already present in the countries concerned, they should be removed from quarantine regulations, unless clearcut evidence exists of potentially dangerous races; certain common potato viruses (such as PVX, PVY, and PVLR) are examples of viruses that have already been distributed worldwide by seed-exporting countries.

3. Plant pathologists should support efforts to establish international standards for the movement of seed materials. Certification measures are already being used to some extent to ensure the production of pathogen-free planting material (germ plasm), including heat treatment and use of plant tissue culture.

There are institutions and laboratories equipped to clean up and multiply seed and propagating materials free from pathogens. An example is the international movement of disease-free strawberry stolons from California. This aspect needs expansion and international regulation so that the materials can have an internationally recognized label certifying that established standards are met. Unless such an international system for distribution of pathogen-free seed and propagating material is created, the current quarantine programs will continue, even though costly and often ineffective.

Resistance

The use of disease resistance has been emphasized in many crops in the tropics and with good reason, as a durable, effective resistance provides the ultimate in disease control with usually little cost to the farmer. The high-yielding, disease-resistant wheat and rice cultivars that have been so successful in many developing countries provide outstanding examples. Also the international potato late blight resistance trials in the Toluca Valley, Mexico, have provided a valuable screening mechanism for field resistance so highly desired by plant breeders all over the world.

Chemical Control

The international centers have not emphasized the chemical control of diseases, primarily because of the very limited resources of the small farmer in developing countries. With simplified application techniques and new, highly effective chemicals becoming available, additional research emphasis on chemical control in the developing countries would seem warranted.

Problems with chemical control include the tendency of many of the modern fungicides to quickly lose their effectiveness because of resistance in the pathogen, and the strikingly unequal approaches to regulation of chemicals in different countries—ranging from overregulation to underregulation. Attempts should be made for more uniform regulations, so that, for example, chemicals that are definitely proved to be injurious to humans and banned in one country are not used with impunity in other areas.

Other Methods of Control

A variety of other control methods and integrated systems of disease control are worthy of close attention as we search for solutions to critical food production problems. We can learn much from traditional systems that have been developed by small farmers over centuries of experience. Agricultural systems in the developing countries that are oriented toward the small farmer should be encouraged and publicized, with more attention given to breeding lines based, for example, on the needs of the subsistence farmers. The high-producing, high-fertilizer-requiring cultivars have greatly increased production in many areas, but their practical value to small farmers may be limited by other factors.

Perhaps one of the most promising opportunities to increase food supplies is in the reduction of postharvest losses. It is appalling to observe the hard-won harvest of grains, tubers, and fruits subjected to pests and pathogens that substantially reduce or spoil the already scarce food supply in many tropical countries. Efficient grain and food storage procedures are known to reduce these losses and could make a major contribution to alleviating the world hunger problem. This area needs much further study and emphasis if we are to apply and adapt the basic principles of drying and storage developed in the temperate zones to the special problems of the tropics. Is it not possible that the application of a practical, effective technology for the preservation of harvested crops would increase available food supplies even more than the conventional production techniques in agronomy, entomology, genetics, and plant pathology that currently receive most of our attention?

ASSESSMENT AND FUTURE OF INTERNATIONAL COOPERATION

Impressive gains have been made in developing countries in increasing national production, but the basic goal of benefiting the poorest people has not yet been met. According to Robert McNamara (11), former president of the World Bank, nearly 800 million people (40% of the total two billion in developing countries) survive on incomes estimated at 30 cents per day (in U.S. purchasing power), relegating them to conditions of malnutrition, illiteracy, and squalor.

There is a growing awareness in the world today that any solution to the hunger problem is not only linked to a solution for income inequality and poverty, but that hunger is a *result* of poverty. Obviously any stable progress in the struggle against hunger would be dependent upon a much broader, multidisciplinary approach that so far has eluded us. Though agricultural research has a great influence on agricultural production, there is frequent criticism by sociologists and economists that secondary problems have been created by the "Green Revolution," problems such as labor displacement, inequitable income distribution, and the favoring of large farmers.

These problems are real and detract from the tangible success of production programs. However, the solutions to these newly created, subsequent problems are basically in the province of the same social scientists who so correctly point them out. Fortunately, in recent years the need for closer collaboration between the economist, social scientist, and the production-oriented agricultural scientist has been generally recognized. But even today, as Peter Jennings has commented (9),

> . . . economists have developed few guidelines to help the production scientists establish research strategies to reduce the adverse consequences of increased production. Until these strategies are formulated, we have no choice but to pursue the simplistic goal of increased food production. To do otherwise is to risk the disaster of world famine.

Is it not preferable to face the problems of unequal distribution of the benefits of increased production than to confront the hopeless dilemma of lack of food?

Successes

There are several impressive examples of successes in international cooperation in plant pathology resulting from the international research center programs.

International Disease Nurseries

CIMMYT has established a comprehensive network of international nurseries for testing the disease resistance of cultivars and lines in the breeding programs, following the extensive international system of wheat rust resistance nurseries initiated by E. C. Stakman and colleagues at the University of Minnesota and the USDA, the longest continuous international cooperation project in plant pathology. Priority has been given to trials for resistance to stem rust, leaf rust, and stripe rust of wheat. These international and regional nurseries provide a dynamic connection in the gene flow between breeding in Mexico and the regional or national programs, which contribute local cultivars to enrich the gene pool. Multilineal cultivars have been developed that provide a mosaic of different genetic sources of resistance, as well as being high-yielding and adapted in many countries.

Downy Mildew of Maize

CIMMYT has also made substantial contributions to control of downy mildew of maize, through its breeding program based on the world's most complete collection of maize germ plasm. This disease, a limiting factor in production of maize in Africa, S. E. Asia, and the Americas, has been brought under control in many tropical areas by the use of good agronomic cultivars with high resistance. Through the cooperative efforts of CIMMYT and the Rockefeller Foundation, leadership for this research has been concentrated in Thailand and has been extended to several other areas including the Philippines, Indonesia, Nepal, and more recently countries in Latin America. Better control of this disease has been a major factor in the sharp increase of maize yields in S. E. Asia.

Rice Diseases

By 1966, IRRI had released its first high-yielding cultivar, IR8, to launch the Green Revolution in rice production, and today IRRI has the largest rice germ plasm bank in the world, with more than 60,000 cultivars. Drawing on this bank, scientists at IRRI have produced high-yielding multiple-disease-resistant breeding lines with useful levels of tolerance or resistance to six of the most important pests and diseases of rice, including some resistance to blast, bacterial blight, tungro, and grassy stunt, By 1980, about 30% of all of the rice land in the tropics, and most of the irrigated land, had been planted to IR8 or one of the cultivars descended from this early selection, especially IR36. IRRI also has an extensive worldwide network of cooperative disease nurseries in 76 cooperating countries.

PROBLEMS YET TO BE SOLVED

The international centers constitute a remarkable international network for study of a wide range of agricultural problems. What has really been the total impact of the center-generated technology on food production in the developing world? Is the basic purpose of the centers met by the generation of technology of potential application in national food production programs? Or is the purpose met only when this technology is applied and results in increased food production or when the challenging and difficult goal is met of feeding more people and feeding people more?

These questions are raised for a much-needed clarification and not as a criticism of the programs or achievements of the centers. It is generally recognized that production impact must be accomplished through the activities of the national programs. However, national research capabilities have not generally been strengthened to keep pace with the international research of the centers. The "delivery system" of the generated technology is deficient or lacking. Strong and active national extension systems are needed, to bring the results from the centers and from the national research institutions to the small farmer. Wortman and Cummings (18) point this up very cogently:

> The gap between what we have learned and what is applied in the countryside remains enormous. . . . We place an especially high premium on the role of technology and the importance of research, but only if it is organized to reach and benefit great numbers of rural people.

The International Service for National Agricultural Research (ISNAR) was recently established with the objective of strengthening the national programs, in collaboration with the international centers.

During the past few decades, international scientific cooperation has strengthened national agricultural research programs in several developing countries. Stronger institutions, highly qualified scientists, and internationally minded administrators make these countries capable of expanded collaboration as equal partners and as leaders in specific international cooperative research projects of widespread importance. These national programs have not only the scientists and facilities to provide this leadership, but in certain places are better qualified by special characteristics of the environment or materials to make a vital contribution to further progress.

For example, Mexico is a unique location for international cooperative research on potato late blight: first, because the wild potato species indigenous to central Mexico have provided the major source of germ plasm used in breeding for resistance; and second, because central Mexico is the original home of this notorious pathogen attacking potatoes all over the world, and the sexual stage of the fungus pathogen (*Phytophthora infestans*) exists in nature only in Central Mexico, where a wide range of pathogenic races occur. The recent initiative taken by the Mexican government to establish an international cooperative potato late blight project in Mexico, enlisting the collaboration of interested scientists all over the world, is a praiseworthy effort that augurs well for future progress and success in the control of this important disease. Similar opportunities for such expanded international cooperation to solve other specific disease problems undoubtedly exist in other national programs in tropical countries.

A serious problem with many overseas university programs has been the lack of continuity. Foreign assignments have usually been for periods of 2 years or less, and about the time a scientist became adjusted and productive at an overseas post, it was time to come home. Universities need long-term commitments of funds for continuity of highly qualified personnel on overseas assignments, and these have seldom been available.

International scientific cooperation is fostered and even dependent upon the education and training of students from other countries. U.S. universities have been very active and effective in this vital phase of international cooperation, but additional thought and careful planning is needed. Selection of students from developing countries is difficult for U.S. universities and should be studied by the APS Teaching Committee. Logically, the most highly qualified and motivated students, who plan to return to the tropics and work in programs to increase food production, should be given priority.

A broad training is needed for foreign students, with less emphasis on specialization and basic research. More courses in tropical plant pathology and exposure to the pertinent literature in the field should be available. Students from developing countries should be encouraged to do some of their thesis research in their home country; this is possible in all too few U.S. universities. For U.S. students, doctoral recipients should have at least a reading, and preferably also a speaking, knowledge of at least one foreign language; this is more important now from the standpoint of communication, not primarily for translation of scientific articles. The virtual abandonment of language requirements is a step backward in the development of international cooperation and understanding.

The outstanding success of U.S. agriculture is due in large part to the effective combination of research, education, and extension in our land grant system of colleges and universities. Unfortunately, this combination is seldom found in developing nations, and too often there is little cooperation among educational, research, and extension agencies. Thus, although education was greatly strengthened through "institution building," research and extension development often lagged behind.

THE ROLE OF THE AMERICAN PHYTOPATHOLOGICAL SOCIETY IN INTERNATIONAL COOPERATION IN PLANT HEALTH PROGRAMS

Many of the architects of the most spectacular example of agricultural development in recent decades, the Green Revolution in wheat and rice, were trained as plant pathologists (e.g., J. G. Harrar, Norman Borlaug, and Peter Jennings). There is general agreement that the APS should continue this tradition of leadership in international cooperation in agricultural research, even though the definition of its role and the extent of involvement sometimes provoke spirited debate. Following are some suggestions for a future role of the APS as an international scientific society seeking solutions for a food and population problem that is shared by the entire world.

1. Encourage APS to occasionally hold meetings in tropical countries of the Americas. We are the *American* Phytopathological Society. The meeting in Mexico in 1972 was a great success. Puerto Rico, Costa Rica, Venezuela, or Colombia might be possible sites for the annual meeting. There is no better way to encourage interest and to foster understanding in the tropics and in the problems of developing countries than firsthand experience.
2. Strengthen the ties of the APS with the Caribbean Division. Is it not time to consider establishing a new division of APS in South America?
3. Expand and continue sponsoring national and international symposia, workshops, etc., on themes and issues relating to international cooperation.
4. Continue to work with and support ISPP in its many activities in international cooperation.
5. Strengthen and encourage the activities of the International Cooperation Committee and the Tropical Plant Pathology Committee of APS. Consider the formation of an advisory committee to the APS on keeping society members

up-to-date on the latest developments and problems in world food production.

6. Facilitate communication among plant pathologists. The *World Directory of Plant Pathologists*, edited by Fran E. Fisher, grew out of a project of the APS International Cooperation Committee. How can APS help to update and improve this directory to better facilitate international cooperation? Computerization of the list, indexing in various ways, etc., would improve the usefulness of this document.

7. Develop a mechanism whereby APS can interact with and give advice and recommendations on priorities to AID and BIFAD (Board for International Food and Agricultural Development) since these agencies provide most of the funds available in the United States for work in developing countries.

8. Consider the formation of an APS committee to study how our society can help students from developing countries receive more appropriate training related to their national agricultural problems and how they might be helped to find the best place to receive such training.

9. Study whether "group membership" in APS can be made more useful to scientists in developing countries.

10. Keep members aware of the potential for early large-scale tragedy in this population/resource imbalance: a) Remind members of the importance of serving on projects related to research for developing countries. b) Raise questions about the advisability of direct food aid and food subsidies for developing countries, which only relieve immediate hunger and provide no long-range solutions. Such aid may even undermine local agricultural production incentives and aggravate the effects of the population explosion. c) Foster contacts with social scientists in the complex arena of food production, distribution, and marketing. Two past presidents of APS, L. H. Purdy (13) and J. G. Tammen (15), have signaled the need for increased awareness of cultural, economic, and social constraints if a significant impact is to be made on the world food-poverty-population problem. Without a coordinated effort between the biological and social sciences, the war on world hunger cannot be won.

As noted above, the growing food crisis is related directly to and aggravated by the increase in population. Therefore, it is pertinent here to mention the recommendation made recently that the APS refrain from any international cooperation with countries with an annual population growth rate above 2% (see Paddock, Chapter 4, this volume).

An international scientific society such as the APS should avoid such a negative policy. Such threats are unrealistic and do not inspire a country to change policy or launch a program to comply with the conditions of the threat. We believe it is morally wrong to deny a country the technology available to produce more food, just as it would be inhumane to deny medical technology to certain countries because their rate of population growth is judged to be too great.

We believe in a more positive approach and one that is consistent with the view that the APS is an international scientific society designed to serve the world community as a whole. While we do recognize the urgent need to reduce the rate of growth of the world's population, we also believe that we must concentrate and improve our efforts to feed this future population. We believe that it is within the capacity of the physical and human resources of the world to meet this monumental challenge. This international cooperative effort is worthy of the finest scientific talent that can be mustered. Both as a society and as individuals, plant pathologists should join other citizens of the world in the conquest of hunger.

ACKNOWLEDGMENTS

We appreciate suggestions and information provided by the following: J. G. Bald, L. Chiarappa, P. Crill, I. D. Firman, S. Fuentes, E. P. Imle, A. Johnston, L. J. Littlefield, D. Bap Reddy, B. L. Renfro, R. B. Stevens, and K. Verhoeff.

LITERATURE CITED

1. Anonymous. 1974. A Hungry World: The Challenge to Agriculture. Univ. Calif. Div. Agric. Sci., Berkeley. 327 pp.
2. Anonymous. 1980. Consultative Group on International Agricultural Research. CGIAR Secretariat, Washington, DC. 50 pp.
3. Anonymous. 1981. ALS Lifelines, Newsletter of the Assembly of Life Sciences. Vol. 7. Nat. Acad. Sci., Washington, DC.
4. Anonymous. 1981. Acronym List of International Organizations Related to Agriculture, Economic Development and Pest Management. Pest Management and Related Environmental Protection Project. Univ. Calif., Berkeley. 124 pp.
5. Borlaug, N. E. 1972. The green revolution, peace and humanity. (Nobel Peace Prize speech, 1970). Pages 12–39 in: CIMMYT Reprint and Translation Series No. 3. CIMMYT, El Bataan, Mexico. 39 pp.
6. Chiarappa, L. 1980. The role of FAO in research and control of plant diseases. Plant Dis. 64:362-367.
7. Firman, I. D. 1978. Plant pathology in the region served by the South Pacific Commission. Rev. Plant Pathol. 57:85-90.
8. French, E. R., and Apple, J. L. 1974. The commodity in-depth projects of North Carolina State University in Peru: A model bilateral aid programme. FAO Plant Prot. Bull. 22:42-47.
9. Jennings, P. R. 1974. Plant breeding, the green revolution, and food production in the developing countries. Donald F. Jones Mem. Lecture. Conn. Agric. Exp. Stn., New Haven. 16 pp.
10. Kriesberg, M. 1977. International organizations and agricultural development. Foreign Agric. Econ. Rep. 131. U.S. Dep. Agric., Washington, DC. 135 pp.
11. McNamara, R. S. 1973. Poverty in the developing world. From: Address to the Board of Governors, World Bank, Nairobi, Kenya.
12. McPherson, M. P. 1981. America: Continuing a proud tradition of helping others. Agenda 4(4):25-28.
13. Purdy, L. F. 1980. Plant pathology, change and the future. Plant Dis. 64:982-983.
14. Snyder, W. C. 1971. Plant pathology today. Annu. Rev. Phytopathol. 9:1-6.
15. Tammen, J. F. 1981. Food, society, and plant pathology. Plant Dis. 65:7.
16. Thurston, H. D. 1977. International crop development centers: A pathologist's perspective. Annu. Rev. Phytopathol. 15:223-247.
17. Warnken, P. F., Ragsdale, R. M., and Flaim, C. M. 1978. International agriculture: Challenging opportunities. Bull. M107. Univ. Mo., Columbia. 166 pp.
18. Wortman, S., and Cummings, R. W., Jr. 1978. To Feed This World; The Challenge and the Strategy. Johns Hopkins Univ. Press, Baltimore. 440 pp.

CHAPTER 48

EDUCATION IN PLANT PATHOLOGY: PROBLEMS AND CHALLENGES AHEAD

WILLIAM MERRILL

One of the three basic ploys in science fiction writing is, "*If this continues* . . ."; one extrapolates a trend to a point of seemingly utter absurdity and bases the story on the final result. The same technique also is quite appropriate for crystal-ball gazing. *If* trends over the past two decades continue for the remaining years of the twentieth century, what problems and challenges lie ahead for educators in plant pathology?

Some particularly obvious trends relate to technological changes outside of agriculture, the "knowledge explosion," changes in student body size and composition, curriculum demands, and some in-house problems.

NONAGRICULTURAL TECHNOLOGY

The biggest trend involves solid-state miniaturization and micro-miniaturization of electronics, coupled with decreasing cost per unit so that sophisticated instrumentation is within the economic reach of a large segment of the population. In the classroom, the most visible harbingers are hand-held calculators. In 1975, about 10% of my students had hand calculators. In 1981, about 80% had them. Within a decade, most students will have access to private computers nearly as sophisticated as many of today's research computers.

These computers will interface with large research libraries, video units, and word processors. No longer will a student have to pay thousands of dollars to sit in a cramped, sweaty, noisome classroom and listen (amid the snoring of his peers) to the confused ramblings of some second rate (or worse) professor expounding on Chaucer, statistics, or basic genetics. Students will enroll in a course taught by one of the two or three best teachers of the subject, sit at home in a recliner and watch tapes as often and as long as desired on a video player. The student will interface the computer through the telephone lines with the computer of the University Center offering the course and will take the examinations by writing out the answers on the student's own word processor. Many of the regurgitation examinations (90% of all college exams are nothing more) will be corrected by the computer. In a programmed mastery learning set-up, the student will study and retake the examination until mastery of the subject matter is achieved. Students will proceed at their own rates, free from the distractions of typical college classrooms. They may take Chaucer from a professor at Columbia, chemistry from a professor at Harvard, and statistics from a professor at Iowa. The degree may be from the Cooperative University of North America.

Futuristic? Perhaps. But the technology to do this exists today; it merely must be applied.

When the student has to do a term paper (say, in chemistry), instead of spending long hours in the library poring through the chemistry abstracts, the student will punch key words into a home computer and dial the research library handling chemistry. The library computer will sift through the key words and its computerized abstracts (more of this later) and identify key papers. The citations and abstracts of these papers will be displayed on the student's video screen. Selected papers to be read will be identified, and the library computer will feed these papers from microfiches across the screen of the student's video unit.

Abstracting and scientific journals as we know them will be obsolete. A chemist will submit a manuscript to a chemistry journal. The manuscript will be reviewed and edited. But instead of being printed, it will be entered onto microfiche at the International Chemistry Library (ICL). The ICL also will abstract the paper and cross-reference it in its computer. The author will pay a fee, similar to page charges. Users will pay a fee for abstract searches, title print-out, and microfiche transmission. Way-out? Not really. Today's technology can already do most of what I envision. Today's abstracting journals represent an *extremely* archaic and cumbersome technology!

Other areas also will feel the impact of electronics. Major crop production systems will be "wired for sound" and data fed to small grower-owned home computers. Instruments in the grower's fields or greenhouses will monitor environmental conditions and feed them periodically into the computer. The computer will feed the data through a series of predictors, and ring a warning bell. The printout machine will say, "Conditions favorable for development of downy mildew in tobacco fields one and three. Spray immediately with . . ." The grower then will follow directions for the control program. Visionary?? Not really. The hardware is available and within economic reach of some growers. The problem lies in the fact that we pathologists have not developed and refined sufficiently reliable software packages!

Picture the following scenario: John and Mary are apple growers and there is something wrong with the foliage on their trees. They sit down at their computer, which is interfaced with a large university computer, and tell it they have a problem with apples. The big computer asks what kind of a problem. They answer that leaves are affected. The computer asks what the leaves look like. The growers answer that leaves have spots. After a series of questions and answers about the nature of the spots, the computer switches through a series of video segments that appear in sequence on the screen in the owners' home, until they see one that looks similar. The computer then shows them a complete video cassette on the problem, including variations in symptom development on different cultivars, etc. The growers agree that the problem has been diagnosed. The computer then flashes the control recommendations on the screen, or they are printed out on the word processor at the owners' home. Again, the technology to do most of this exists today; it merely must be applied.

The typical extension specialist represents another archaic technology. The bulk of the specialist's time is wasted traveling frantically and continuously over the state, primarily to "hold the hand" of the county extension staff. If examined critically, I doubt if these activities are cost-effective. It would be far less expensive to link the main extension offices with county or regional centers via live closed-circuit television. The specialists could hold periodic discussions with clientele in several counties simultaneously. (I envision a setup similar to that of the MacNeil-Lehrer Report on U.S. National Public Television.) Much of the traditional extension work will be taken over by private practitioners.

No doubt many will scoff at these futuristic visions. Yet, primary-grade children today play with hand-held calculators, whereas people in my age group never even *saw* a calculator until they were in college! Beginning graduate students do not know how to use the Monroe calculator—in my laboratory that was the newest thing on the market only 15 years ago! Third-and fourth-grade students in some schools already are playing with computers! What will it be like in the next decade when primary-school children have access to word processors, video units, and computers that interface with research computers at universities, research centers, and the Library of Congress?

Is the college and university system as we know it an outmoded technology? In many respects, yes! We could do far better with a system such as I have described, at least for the first 2 years of a student's career. (I suspect that for most activities, primary and secondary schools also are an outmoded technology and could best be replaced. They would exist for one major function: to teach social interactions. Unfortunately, such schools appear to be doing worse in this than in traditional academic subjects!)

One thing will prevent universities from becoming totally obsolete: the need for hands-on experiences. There is no realistic way to replace hands-on laboratory experiences, whether in chemistry, engineering, or plant pathology, with an electronic system. Thus, for the final 2 years, students will continue to come to a university center for laboratory courses and practicum-type work experiences.

What will happen to the over-developed physical plants of many universities and colleges? Will they crumble to dust? Become museums to the past? Be converted to retirement homes?

THE KNOWLEDGE EXPLOSION

Few plant pathologists seem to have contemplated the knowledge explosion that has occurred over the past two to three decades. Concomitantly, there has been a tremendous increase in the degree of sophistication in doing research. Many techniques and types of equipment used in physiology, epidemiology, virology, and other subdisciplines only 10 years ago are obsolete today. For example, the gas chromatograph I purchased in 1967, the most sophisticated model available at that time, was obsolete by 1972 and was scrapped in 1975. The data acquisition system acquired by my former colleague, F. A. Wood, and his students in 1965–66 actually was obsolete even before the unit was fully on-line! This great increase both in amount of factual knowledge and in the degree of sophistication needed to do modern research has led to specialization in, and fragmentation of, our profession. The more astute among us predicted this years ago by observing what had happened to the medical profession in the previous half century. This specialization and fragmentation will increase even more in years ahead. This is inevitable in the evolution of a complex science.

Invariably accompanying this fragmentation is the increasing provincialism that heralds the onset of isolation and fossilization. To pick on my own kind for an example, few forest pathologists have any awareness of advances made in other crop areas. The pervasive feeling is that if it has not been done with a tree species, then it just hasn't been done! This leads to shallow and inept research and to poor science. Specialists talk only with fellow specialists and develop their own esoteric jargon, often seemingly as a camouflage to fend off nonspecialists.

Many members of the American Phytopathological Society (APS) rail against PHYTOPATHOLOGY, saying they can no longer understand any of the papers. Again, this is *inevitable* when a journal covers a broad subject area that is fragmented into many subdisciplines. It also is an obvious symptom of professional obsolescence caused by the knowledge explosion and increasing sophistication of research techniques.

What does this mean for education in plant pathology? We need to spend less time on "the facts." "Facts" change. "Biological truth" is only an approximation. The more precise and sophisticated our techniques become, the closer we will approach "truth." But "truth" lies at the center of a spiral; because a spiral extends to infinity, we will never reach "truth." We should be emphasizing theories, concepts, the thought processes by which these evolved and were tested, what they mean if right (or if wrong). We also must *train the student in self-education.* And we must teach the student to have a skepticism bordering on the unhealthy for published "facts."

In graduate education far too much emphasis has been placed on teaching techniques that rapidly become obsolete. Most departments primarily turn out technicians, rather than scholars capable of continuing self-education. Indeed, rather than selecting for or encouraging the development of an inquiring mind, the hallmark of a scholar, our graduate programs deliberately select against them or turn them off; any student who dares voice serious doubts about our sacred dogmas will fail candidacy or comprehensive examinations. Any paper questioning the orthodox or proposing any radical idea will be rejected by our "peers."

Our profession has not faced the problem of the knowledge explosion. It is going to require longer times for the students to acquire this knowledge, unless they become very narrow specialists. Unfortunately, we have been *training* increasingly specialized technicians for at least the past two decades, as anyone who has served on search committees during that time can affirm.

The fragmentation of plant pathology into many subdisciplines also means that no department can adequately cover the entire subject. Some departments will excel in certain subdisciplines; others will excel elsewhere. Departments should trade students, rather than attempt to handle them in an area in which the department has minimal or no expertise. Few are ready to accept this view! In 1980, D. R. MacKenzie and I polled approximately 10% of the APS members in teaching institutions as to how they ranked graduate departments in terms of 28 subdisciplines. Numerous replies stated, "Our department can train a Ph.D. in *any* area of plant pathology." Train a virologist in a department lacking a virologist? Or a quantitative epidemiologist in a department lacking one? A few replies were similar to one from a senior member of a large midwestern department: "Our department ranks number one in all categories, and no one else even comes close to being number two!" Hardly examples of stark realism!

With the increase in specialization has come a proliferation of plant pathology courses. Every specialist must have his or her own course to teach. Whether we admit it even to ourselves, we all are "prima donnas." Too often we seek to put our own imprint onto courses and students. The closest thing to true cloning of humans exists in graduate education, to the great detriment of the students—and of society. The struggle for identity also has meant that there are few (very rare) examples of truly interdisciplinary courses. Students become locked into programs by our own demands for identity and student teaching hours rather than by what is best for the student. This is why a truly integrated curriculum in plant health has never been developed in the United States and probably will not be for years to come. More of this later.

The "bottom line" of the knowledge explosion and increasing sophistication in research techniques is professional obsolescence. That is the stark reality facing all professionals. A new Ph.D. peaks 3 to 5 years after receiving the degree and becomes increasingly obsolete thereafter, except in one (increasingly narrow) niche. The medical profession has faced this reality through accreditation and certification and the requirement for periodic retraining to maintain certification. *The continuing education of ourselves is a major challenge.* The teach-ins organized by the Teaching Committee of APS are a preliminary step in the right direction but have involved only a very small percentage of the membership. Those who most need the retraining are those who rarely attend such meetings!

The knowledge explosion is another reason for developing educational systems outside of the classroom. We need these programs not only for ourselves, but also for county staff, industry representatives, and, as they become increasingly prevalent, private practitioners.

STUDENT BODY

There have been pronounced changes in our undergraduate and graduate clientele. Thirty years ago most agriculture students had rural backgrounds and some type of agricultural experience. Today our students come primarily from urban-suburban environments and lack any exposure to production agriculture, as opposed to subsistence or subsubsistence survival on "family farms." As agriculture becomes increasingly intensive and consolidated into large production units, increasingly fewer students will be exposed to production agriculture. How do we somehow expose these students to agriculture, with all of its problems and challenges?

Undergraduate enrollment in agriculture has peaked and either leveled off or already started to decline. In the early 1970s we rode the "save-the-environment" boom that brought some top students into agriculture, students who previously would have gone into medicine or engineering. The current generation of students is far more pragmatic—witness the burgeoning enrollments in electrical engineering, computer science, coal mining technology, and petroleum engineering. Thus fewer top-notch scholars will enter agriculture, and with smaller agriculture enrollments, there will be fewer good students from among whom to recruit graduate students. Even today, the best qualified graduate students seldom originate from agricultural campuses but come from biology programs in small, private liberal arts colleges. Students seeking to enter graduate work in plant pathology from most curricula in agronomy, forestry, or horticulture currently are deficient in about 1 year of undergraduate coursework. They usually have insufficient credits in mathematics, chemistry, biochemistry, and physics; many lack courses in statistics and computer science; and most have no or insufficient knowledge in plant physiology, genetics, and plant anatomy.

As the average age of the population increases, we will be faced with older, part-time students seeking to widen and improve their own professional backgrounds or perhaps seeking new degrees. Modern electronic facilities will enable us to reach this clientele in their own homes.

For example, video cassettes will take these students on "field trips" and teach not only diagnosis, but even prognosis. But hands-on laboratory work still will be needed in many instances. Can we solve this need with very intensive short courses of 40–60 hours per week?

I recognize that some credit and noncredit college courses are being offered via television (e.g., see *Time*, 5 Oct 1981, p. 46). These are barely a beginning, and to my knowledge no College of Agriculture has really made a *serious* commitment to pursue this type of educational programming.

CURRICULUM DEMANDS

The knowledge explosion, plus the chaotic conditions in our primary and secondary schools, have resulted in the diminution of professional education in the agricultural colleges. Further, agricultural colleges are not very demanding of students and will become less demanding as enrollments drop! Today students who cannot even write a grammatically correct sentence of more than two words graduate with honors! Agriculture students frequently have been viewed as societal square pegs in round holes. There are increasing demands to require our students to take more "general education" credits in arts, humanities, and social sciences. Simultaneously, there are increasing amounts of professional and technological knowledge to which these students must be exposed. For years I have been telling everyone who would listen (and many who would not) that all college students should be required to take courses in statistics and computer science. One cannot intelligently interpret any news report without an appreciation of the misuse and abuse of statistics. Few facets of our lives remain unaffected by computers; an educated person should know how these things function—and misfunction. I am pleased to note that one of the Ivy League schools now requires all students to take computer science. Agricultural colleges, who should have blazed the way, will catch up in about another 20 years. Liberal arts departments in land grant universities never will catch up.

I agree that agriculture students, indeed, all college students, need more education in arts, humanities, and the social sciences. They also need much better communication skills; they need more basic science, such as mathematics, chemistry, biochemistry, and physics; they need statistics and computer science. But agriculture students also need the professional courses, plus a grounding in breadth and depth in modern production agriculture. How can we do all of this in a 4-year program? Obviously, we can't! I will return to this in a bit.

IN-HOUSE PROBLEMS

A major problem of our university system is the departmental structure. Courses are labeled by department, professional curricula require xxx credits of these departmental courses, and departments receive funding based on student credit hours generated. Thus departments encourage students to take as much course work within the department as possible, to generate more teaching credit hours, to obtain more funding, etc., in a vicious circle. Students seldom are encouraged to cross departmental lines and often are actively discouraged from crossing college lines. The brick walls housing the departments, although pierced by windows and doors, have become almost as impenetrable as the Berlin Wall. They work to the detriment of the education of our students and all too frequently hinder research. In spite of all pretenses to the contrary, there is not today a single program in the United States that is *really* educating and training plant health specialists, the M.D.s of the plant world. Such educational programs must of necessity be truly interdisciplinary. Interdisciplinary programs seldom work and rarely work well when run by committees, the members of which owe their primary allegiances elsewhere, i.e., to departments.

ONE SOLUTION

Any fool can ask more questions in 5 minutes than 10 wise men can answer in their lifetimes. Thus it behooves me to offer at least one possible solution to a few of these problems.

Let us convert our professional curricula in agriculture to 5-year programs, abolish all departmental majors at the undergraduate level, and create teaching cadres responsible directly to the Dean for Resident Instruction.

The first 2 years (for all college students) would include an intensive integrated core sequence on the evolution and development of western civilization, including history, economics, philosophy, arts, humanities, and religion, among others. They would have a *demanding* sequence of courses in spoken and written communications, as well as in statistics and computer science. They would also receive their basic mathematics, chemistry, biochemistry, physics, and biology. Much of this could be done via the electronic media that I have suggested, but it could also be done by traditional methods. *Only* after completing this sequence would a student be allowed to declare a college of interest.

The third year for all agricultural students would include intensive coverage of modern agricultural technology, production systems, and economics and of international agriculture. The students would have required hands-on work programs during the academic year, in the barns, fields, and greenhouses. *Only* after completing this crucial third year would a student be allowed to opt for animal science, plant science, agricultural engineering, or agricultural related fields (rural sociology, agricultural economics, or agricultural education).

This program would change the entire face of education in colleges of agriculture and drastically upgrade the products of the system.

In the area of plant health, students would then proceed to a 3-year program leading to the Doctor of Plant Medicine, *an interdisciplinary program with its own faculty* drawn from plant pathology, entomology, weed science, nematology, plant nutrition, genetics, pesticide chemistry, toxicology, etc., and including the extension specialists in all of these fields. After all, the extension specialists should be the best prepared people to handle such a teaching program. I recognize the universally preached orthodox and sacrosanct canon that "extension specialists do not have time to do their extension work and teach." By linking extension specialists to regional centers around the state via closed circuit TV as suggested, specialists would not have to spend at least 50% of their working hours in travel and hence would have sufficient time. A sufficiently well-trained county or regional staff and private practitioners could handle the bulk of the actual field contact with growers.

About four universities would be involved in such a program. The program would require an internship in another area of the country before certification. Periodic retraining and recertification would be necessary. The graduates would be employed in county and regional extension centers, in agricultural industries, and as private practitioners.

Students who were research oriented would then proceed to a Ph.D. program administered by a graduate faculty in the area of specialization, such as plant pathology, or entomology. Here the student would specialize in some broad aspect of plant pathology and, upon completion of the degree, use a postdoctoral position to become highly specialized in some subdiscipline.

This program would provide education and training for the generalist, specialist, and superspecialist as part of a well-defined sequence with cut-outs and employment opportunities at each level. Will it ever happen? Eventually, for this appears to be the pattern of evolution in other fields, such as human medicine. But given the ultraconservative nature of land grant institutions in general and of agricultural college bureaucracy specifically, it will not occur soon, no matter how desirable the outcome might be for society.

THE CHALLENGES AHEAD

Thomas Edison once said that he could not predict exactly what was going to happen but that he knew without doubt electricity would have a great impact on society. He would be amazed at what has happened in the past two decades—and even more amazed at what is going to happen in the next two. The faces of agriculture and education are going to be altered profoundly. At least one company already is producing video cassettes for kindergarten and primary education. Others will soon follow. The college and university system, locked to its archaic traditions, will be the last to adopt these; this will occur about 20 years hence. Why 20 years? That is when the current kindergarten and primary-age generation, now using these techniques, will become young professors in the system. Witness: colleges and universities are just beginning a concerted effort to develop TV courses; this occurred *only* after a generation of students (the TV generation) had grown up and become professors. So have no doubt—it will occur.

I throw the following challenges to you: What plant pathologist will be the first to use video cassettes to reach clientele studying independently at home? Who will pioneer in the use of live closed-circuit TV for statewide extension meetings? Who will develop the first really functional computerized diagnostic service for plant diseases and pests? Who will develop the first comprehensive computerized abstracting and microfiche journal services? Which three or four departments will cooperate and develop a comprehensive curriculum for a Doctor of Plant Medicine? (This almost happened, until the heads of the two departments involved warped the program into an alternative for the Ph.D. and lost support of the faculties and administrations!) When will plant pathologists collectively realize that the general practitioner requires preparation vastly different from that of a research scientist? And that the same is true of the extension specialist? Or that fragmentation of our profession into subdisciplines is inevitable? And that all Doctors of Philosophy in plant pathology should have a broad background and be generally conversant with the entire field? And hence that superspecialization should occur at the postdoctoral level rather than at the M.S. or Ph.D. level? Which college of agriculture will be the first to face the reality that professional degrees in agriculture require a 5-year program? (Hasn't it ever seemed strange to *anyone* that baccalaureate degrees in civil engineering, agronomy, and English literature all require the same number of credit hours? Are not these degrees awarded on the basis of time spent in class, rather than upon what a student in these fields should and must know?)

Finally, I urge the reader to remember that the above thoughts are not Divine Revelation, but merely one iconoclast's interpretation of trends. I also urge you to examine what has happened to human medicine—plant pathology today stands where human medicine stood only about 75 years ago. I see us traveling down the same path, for better or for worse.

INDEX